Advanced Structured Materials

Volume 46

Series editors

Andreas Öchsner, Southport Queensland, Australia
Lucas F.M. da Silva, Porto, Portugal
Holm Altenbach, Magdeburg, Germany

More information about this series at http://www.springer.com/series/8611

Holm Altenbach · Robert V. Goldstein
Evgenii Murashkin
Editors

Mechanics for Materials and Technologies

Springer

Editors
Holm Altenbach
Institut für Mechanik
Otto-von-Guericke-Universität
Magdeburg
Germany

Robert V. Goldstein
Mechanics of Strength and Fracture of Materials and Structures
Ishlinsky Institute for Problems in Mechanics of the Russian Academy of Sciences
Moscow
Russia

Evgenii Murashkin
Modelling in Solid Mechanics
Ishlinsky Institute for Problems in Mechanics of the Russian Academy of Sciences
Moscow
Russia

ISSN 1869-8433 ISSN 1869-8441 (electronic)
Advanced Structured Materials
ISBN 978-3-319-56049-6 ISBN 978-3-319-56050-2 (eBook)
DOI 10.1007/978-3-319-56050-2

Library of Congress Control Number: 2017936025

Printed on acid-free paper

This Springer imprint is published by Springer Nature
The registered company is Springer International Publishing AG
The registered company address is: Gewerbestrasse 11, 6330 Cham, Switzerland

Preface

Professor Alexander V. Manzhirov is one of the leading Russian researchers in solid mechanics and applied mathematics, a worldwide authority who has exerted a major influence on research in the areas of *Mechanics of Growing Solids*, *Mechanical Analysis of Additive Manufacturing*, *Contact Mechanics and Tribology*, *Viscoelasticity*, *Creep* and *Integral Equations and their Applications*. This volume of the Advanced Structured Materials Series is dedicated to his sixtieth birthday and contains a selection of research papers written by friends and colleagues from various countries, including China, Germany, India, Russia, South Africa, the UK, and the USA. Some of these papers deal with research fields close to Prof. Manzhirov's scientific activities, while the other ones show new trends in solid mechanics which could be combined with his scientific interests in the future.

Professor Manzhirov was born on 24 May 1957 in Rostov-on-Don, Russia. His scientific carrier is related to different places:

- In 1979 he obtained his diploma (equivalent to M.Sc. degree) in Mechanics (Diploma with Honors) at the Rostov State University, Department of Mechanics and Mathematics, where he studied from 1974 - 1979
- In 1982 he was employed at the Institute for Problems in Mechanics of the Academy of Sciences of the USSR (at present the Ishlinsky Institute for Problems in Mechanics of the Russian Academy of Sciences)
- In 1983 he defends his doctoral thesis (candidate of physico-mathematical sciences) at the Moscow Institute of Electronic Engineering Industry (he was a PhD student at the Kuibyshev Moscow Institute of Civil Engineering from 1979 to 1982)

- In 1993 he defends his second doctoral thesis (doctor of physico-mathematical sciences, equivalent to the DSc degree or doctor habilitatus) at the Institute for Problems in Mechanics of the Russian Academy of Sciences) which was accepted by the Higher Attestation Commission of the Russian Federation
- Since 2014 he is Foreign Member of the National Academy of Sciences of the Republic of Armenia

In 1998 he obtained the certificate of professor in mathematics by the Higher Certification Committee of the Russian Federation.
At the moment he serves as Deputy Director and Head of Department for Modeling in Solid Mechanics at the Ishlinsky Institute for Problems in Mechanics of the Russian Academy of Sciences, as Head of Branch Department of Applied Mathematics at the Bauman Moscow State Technical University, as Professor of Department of Higher Mathematics at the National Research Nuclear University (Moscow Engineering Physics Institute) and as Professor of Department of Higher Mathematics at the Moscow Technological University.
At the same time he is Vice Chairman of the Solid Mechanics Scientific Council of the Russian Academy of Sciences, a member of the Mathematics and Mechanics Expert Council of the Higher Attestation Commission of the Russian Federation, and a member of the Mathematics, Mechanics and Computer Science Expert Council of the Russian Foundation for Basic Research. In addition, he is a member of:

- Russian National Committee on Theoretical and Applied Mechanics (RNCTAM)
- European Mechanics Society (EUROMECH)
- American Society of Mechanical Engineers (ASME)
- American Mathematical Society (AMS)
- Gesellschaft für Angewandte Mathematik und Mechanik (GAMM)
- International Association of Engineers (IAENG)
- Research Board of Advisors of the American Biographical Institute (ABI)
- Editorial Boards of about a dozen scientific journals, including Mechanics of Solids (MTT), Proceedings of National Academy of Sciences of Armenia - Mechanics, Computational Mechanics of Continua and of the scientific website The World of Mathematical Equations (EqWorld)

His main awards

- Russian State Scientific Scholarship for Outstanding Scientists 1997, 2003
- First Competition of the Science Support Foundation, 2001
- Certificate of Honor of the Russian Foundation for Basic Research, 2016
- Certificate of Honor of the Higher Certification Committee of the Ministry of Education and Science of the Russian Federation, 2016

He is an author or co-author of 19 books and textbooks in English, German and Russian including some best-selling handbooks. In addition, he published more than 200 research papers and obtained two inventor's certificates. Selected publications can be find at the end of the preface.
All of these publications, scientific achievements and other contributions have given us great respect for him, and we regard him as a perfect scientist. At the same time,

he is a man of humor. He is a husband, father of three beautiful girls, loves travelling and likes to get new and unexpected impressions. We are grateful to Prof. Manzhirov for his scientific contributions to solid mechanics and applied mathematics, and believe that he will continue to have an impact in research.
We have to thank all contributors for their perfect job. Last but not least, we gratefully acknowledge Dr. Christoph Baumann (Springer Publisher) for support of the book project.

Magdeburg, Moscow,
May 2017

Holm Altenbach
Robert Goldstein
Evgenii Murashkin

60 Selected Publications of Prof. A. V. Manzhirov

Selected Monographs and Textbooks

- N. Kh. Arutyunyan, A. V. Manzhirov, and V. E. Naumov, Contact Problems in Mechanics of Growing Solids [in Russian], Nauka Publishers, Moscow, 1991
- A. D. Polyanin and A. V. Manzhirov, Handbook of Integral Equations: Exact Solutions [in Russian], Factorial Press, Moscow, 1998
- A. D. Polyanin and A. V. Manzhirov, Handbook of Integral Equations, CRC Press, Boca Raton, 1998
- A. V. Manzhirov and A. D. Polyanin, Methods for the Solution of Integral Equations: A Handbook [in Russian], Factorial Press, Moscow, 1999
- N. Kh. Arutyunyan and A. V. Manzhirov, Contact Problems in Creep Theory of Creep [in Russian], National Academy of Sciences of Armenia, Erevan, 1999
- A. D. Polyanin and A. V. Manzhirov, Handbuch der Integralgleichungen: Exakte Lösungen, Spektrum Akademischer Verlag, Heidelberg, Berlin, 1999
- A. V. Manzhirov and A. D. Polyanin, Handbook of Integral Equations: Solution Methods [in Russian], Factorial Press, Moscow, 2000
- A. D. Polyanin and A. V. Manzhirov, Handbook of Integral Equations. Second Edition, Chapman & Hall/CRC Press, Boca Raton, 2007
- N. K. Gupta and A. V. Manzhirov, Eds., Topical Problems in Solid Mechanics, Elite Publishing, New Delhi, 2008
- A. D. Polyanin and A. V. Manzhirov, Handbook of Mathematics for Engineers and Scientists, Chapman & Hall/CRC Press, Boca Raton–London, 2008
- A. I. Chernoutsan, A. V. Egorov, A. V. Manzhirov, etc. A Concise Handbook of Mathematics, Physics, and Engineering Sciences. Ed. by A. D. Polyanin and A. I. Chernoutsan. Boca Raton-London: Chapman & Hall/CRC Press, 2010
- A. V. Manzhirov, N. K. Gupta, and D. A. Indeitsev, Eds., Topical Problems in Solid and Fluid Mechanics, Elite Publishing, New Delhi, 2011
- N. K. Gupta, A. V. Manzhirov, and R. Velmurugan, Eds., Topical Problems in Theoretical and Applied Mechanics, Elite Publishing, New Delhi, 2013

- A. D. Polyanin and A. V. Manzhirov, Integral Equations. In Two Volumes. Handbook for High School Students. Second Edition, Revised and Expanded [in Russian]. Urait Publishing House, Moscow, 2017

Selected Papers

- Manzhirov, A.V., Kovalenko, E.V.: Contact problem for a two-layer aging viscoelastic foundation. Journal of Applied Mathematics and Mechanics. 46(4) (1982)
- Manzhirov, A.V.: Axisymmetric contact problems for non-uniformly aging layered viscoelastic foundations. Journal of Applied Mathematics and Mechanics. 47(4) (1983)
- Manzhirov, A.V.: Plane and axisymmetric problems of the action of loads on a thin nonuniform viscoelastic layer. Journal of Applied Mechanics and Technical Physics. 24(5) (1983)
- Manzhirov, A.V.: On a method of solving two-dimensional integral equations of axisymmetric contact problems for bodies with complex rheology. Journal of Applied Mathematics and Mechanics. 49(6) (1985)
- Grishin, S.A., Manzhirov, A.V.: Contact problems for a thin layer in nonlinear steady creep. Mechanics of Solids. 21(6) (1986)
- Manzhirov, A.V.: Some formulations and solutions of contact problems of creep for arbitrary systems of dies. Mechanics of Solids. 22(3) (1987)
- Manzhirov A.V.: Contact problems of the interaction between viscoelastic foundations subjected to ageing and systems of stamps not applied simultaneously. Journal of Applied Mathematics and Mechanics. 51(4) (1987)
- Aleksandrov, V.M., Manzhirov, A.V.: Two-dimensional integral equations in applied mechanics of deformable Solids. Journal of Applied Mechanics and Technical Physics. 28(5) (1988)
- Arutyunyan, N.Kh., Manzhirov, A.V.: Contact problems of the mechanics of bodies with accretion. Journal of Applied Mathematics and Mechanics. 53(1) (1989)
- Manzhirov, A.V.: The torsion of a growing cylinder by a rigid stamp. Journal of Applied Mathematics and Mechanics. 54(5) (1990)
- Manzhirov, A.V., Chernysh, V.A.: Contact problem for a layered inhomogeneous aging cylinder reinforced by a rigid ring. Journal of Applied Mechanics and Technical Physics. 31(6) (1990)
- Manzhirov, A.V., Chernysh, V.A.: The contact problem of the discrete fitting of an inhomogeneous viscoelastic ageing cylinder with a system of rigid collars. Journal of Applied Mathematics and Mechanics. 55(6) (1991)
- Manzhirov A. V., Chernysh V.A.: Contact problem for a layered inhomogeneous aging cylinder reinforced by a rigid ring. Journal of Applied Mechanics and Technical Physics. 31(6) (1991)
- Manzhirov, A.V.: The general non-inertial initial-boundary value problem for a viscoelastic ageing solid with piecewise-continuous accretion. Journal of Applied Mathematics and Mechanics. 59(5) (1995)

- Polyanin, A.D.; Manzhirov, A.V.: The model solution method in the theory of linear integral equations. Doklady Mathematics. 55(3) (1997)
- Kazakov, K.E., Manzhirov, A.V.: Conformal contact between layered foundations and punches. Mechanics of Solids. 43(3) (2008)
- Manzhirov, A.V., Lychev, S.A.: Mathematical modeling of growth processes in nature and engineering: A variational approach. Journal of Physics: Conference Series. 181(1) (2009)
- Lychev, S.A., Manzhirov, A.V.: Differential operators associated with the equations of motion and nondissipative heat conduction in the Green-Naghdi theory of thermoelasticity. Journal of Physics: Conference Series. 181(1) (2009)
- Manzhirov, A.V.: The wear contact problem for an elastic foundation with an inhomogeneous coating. Proceedings of World Tribology Congress 2009. 6-11, September, 2009. Kyoto, Japan. Kyoto: Japanese Society of Tribologist. (2009)
- Lychev, S.A., Manzhirov, A.V., Joubert, S.V.: Closed solutions of boundary-value problems of coupled thermoelasticity. Mechanics of Solids. 45(4) (2010)
- Lychev, S.A., Lycheva, T.N., Manzhirov, A.V.: Unsteady vibration of a growing circular plate. Mechanics of Solids. 46(2) (2011)
- Kuznetsov, S.I., Manzhirov, A.V., Fedotov, I.: Heat conduction problem for a growing ball. Mechanics of Solids. 46(6) (2011)
- Manzhirov, A.V., Lychev, S.A.: The mathematical theory of growing solids: Finite deformations. Doklady Physics. 57(4) (2012)
- Levitin, A.L., Lychev, S.A., Manzhirov, A.V. et al.: Nonstationary vibrations of a discretely accreted thermoelastic parallelepiped. Mechanics of Solids. 47(6) (2012)
- Manzhirov, A.V., Shatalov, M., Fedotov, I.: On the resonant behaviour of longitudinally vibrating accreting rods. 8th South African Conference on Computational and Applied Mechanics (SACAM2012), Johannesburg, South Africa, 3–5 September 2012. (2012)
- Lychev, S.A., Manzhirov, A.V.: The mathematical theory of growing bodies. Finite deformations. Journal of Applied Mathematics and Mechanics. 77(4) (2013)
- Lychev, S.A., Manzhirov, A.V.: Reference configurations of growing bodies. Mechanics of Solids. 48(5) (2013)
- Manzhirov, A.V., Lychev, S.A., Gupta, N.K.: Nonlinear Models of Growing Solids. Proceedings of the Indian National Science Academy. 79(4) (2013)
- Manzhirov, A.V.: Mechanics of growing solids and phase transitions. Key Engineering Materials. 535-536 (2013)
- Joubert, S.V.; Shatalov, M.Y.; Manzhirov, A.V.: Bryan's effect and isotropic nonlinear damping. Journal of Sound and Vibration. 332(23) (2013)
- Manzhirov, A.V.: Multi-body contact problem for a nonhomogeneous elastic coated foundation with wear. Conference: 5th World Tribology Congress, WTC 2013, 3 (2014)
- Manzhirov, A.V., Lychev, S.A.: Mathematical modeling of additive manufacturing technologies Lecture Notes in Engineering and Computer Science. 2212(1) (2014)

- Lychev, S.A., Manzhirov, A.V.: Discrete and continuous growth of hollow cylinder. Finite deformations. Lecture Notes in Engineering and Computer Science. 2212(1) (2014)
- Manzhirov, Alexander V.: Mechanics of growing solids: new track in mechanical engineering. Proceedings of the ASME International Mechanical Engineering Congress and Exposition, 9 (2014)
- Manzhirov, A. V., Parshin, D. A.: Influence of the Erection Regime on the Stress State of a Viscoelastic Arched Structure Erected by an Additive Technology under the Force of Gravity. Mechanics of Solids. 50(6) (2015)
- Manzhirov, A. V., Parshin, D. A.: Arch Structure Erection by an Additive Manufacturing Technology under the Action of the Gravity Force. Mechanics of Solids. 50(5) (2015)
- Manzhirov, A.V.: A Method for Mechanical Design of AM Fabricated Viscoelastic Parts. Transactions on Engineering Technologies. Springer:Singapore. (2016)
- Manzhirov, A.V.: A mixed integral equation of mechanics and a generalized projection method of its solution. Doklady Physics. 61(10) (2016)
- Manzhirov, A.V., Parshin, D.A.: Application of prestressed structural elements in the erection of heavy viscoelastic arched structures with the use of an additive technology. Mechanics of Solids. 51(6) (2016)
- Manzhirov, A.V: Fundamentals of Mechanical Design and Analysis for AM Fabricated Parts. Procedia Manufacturing. 7 (2016)
- Manzhirov, A.V: Mechanical Design of AM Fabricated Prismatic Rods under Torsion. MATEC Web of Conferences. 95 (2017)
- Manzhirov, A.V: Advances in the Theory of Surface Growth with Applications to Additive Manufacturing Technologies. Procedia Engineering. 173 (2017)
- Kazakov, K.E., Kurdina, S.P., Manzhirov, A.V.: Contact interaction between surface nonuniform bases and regular systems of rigid punches. Procedia IUTAM (in press). (2017)
- Lychev, S.A., Manzhirov, A.V., Shatalov, M., Fedotov, I.: Transient Temperature Fields in Growing Bodies Subject to Discrete and Continuous Growth Regimes. Procedia IUTAM (in press). (2017)
- Manzhirov, A.V.,Gupta, N.K.: Fundamentals of Continuous Growth Processes in Technology and Nature. Procedia IUTAM (in press). (2017)
- Manzhirov, A.V., Kazakov, K.E.: Contact problem with wear for a foundation with a surface nonuniform coating. Doklady Physics (in press). (2017).

Contents

List of Contributors

Leonid D. Akulenko
Ishlinsky Institute for Problems in Mechanics RAS, Vernadskiy Ave, 101 bldg 1, 119526 Moscow, Russia, e-mail: kumak@ipmnet.ru

Sergei Alexandrov
Ishlinsky Institute for Problems in Mechanics RAS, Vernadskiy Ave, 101 bldg 1, 119526 Moscow, Russia, e-mail: sergei_alexandrov@yahoo.com

Holm Altenbach
Institut für Mechanik, Otto-von-Guericke-Universität Magdeburg, Universitätsplatz 2, 39106 Magdeburg, Germany, e-mail: holm.altenbach@ovgu.de

Boris D. Annin
Lavrentyev Institute of Hydrodynamics of Siberian Branch of the RAS, Lavrentyev Ave. 15 630090 Novosibirsk, Russia, e-mail: annin@hydro.nsc.ru

Marcus Aßmus
Institut für Mechanik, Otto-von-Guericke-Universität Magdeburg, Universitätsplatz 2, 39106 Magdeburg, Germany, e-mail: marcus.assmus@ovgu.de

Olga M. Babeshko
Kuban State University, Stavropolckia st. 149, 350040 Krasnodar, Russia, e-mail: babeshko49@mail.ru

Vladimir A. Babeshko
Southern Scientific Center RAS, Chehov st. 41, 344006 Rostov-on-Don & Kuban State University, Stavropolckia st. 149, 350040 Krasnodar, Russia, e-mail: babeshko41@mail.ru

Alexandra S. Begun
Institute of automation and control processes of FEB RAS, 690041 Vladivostok, Russia, e-mail: asustinova@dvo.ru

Stefan Bergmann
Institut für Mechanik, Otto-von-Guericke-Universität Magdeburg, Universitätsplatz 2, 39106 Magdeburg, Germany, e-mail: stefan.bergmann@st.ovgu.de

Judith Nomantande Maureen Bidie
Department of Mathematics and Statistics, Tshwane University of Technology, South Africa, e-mail: bidiejnm@tut.ac.za

Alexander A. Burenin
Institute of Machine Science and metallurgy of FEB RAS, 681005 Komsomol'sk-on-Amur, Russia, e-mail: burenin@dvo.ru

Johanna Eisenträger
Institut für Mechanik, Otto-von-Guericke- Universität Magdeburg, Universitätsplatz 2, 39106 Magdeburg, Germany, e-mail: johanna.eisentraeger@ovgu.de

Victor A. Eremeyev
Southern Federal University, Institute of Mathematics, Mechanics and Computer Science, Milchakova Street 8a, Rostov-on-Don, 344090, Russia and Rzeszów University of Technology, Faculty of Mechanical Engineering and Avionics, al. Powstańców Warszawy 8, 35959 Rzeszów, Poland, e-mail: eremeyev.victor@gmail.com; veremeyev@prz.edu.pl

Olga V. Evdokimova
Southern Scientific Center RAS, Chehov st. 41, 344006 Rostov-on-Don, Russia, e-mail: evdokimova.olga@mail.ru

Igor A. Fedotov
Department of Mathematics and Statistics, Tshwane University of Technology, South Africa, e-mail: FedotovI@tut.ac.za

Gennady V. Garkushin
Joint Institute for High Temperatures of the Russian Academy of Sciences, Izhorskaya st. 13 bldg 2, 125412 Moscow, Russia

Dimitri V. Georgievskii
Lomonosov Moscow State University, Leninskie Gory 1, 119991 Moscow, Russia, e-mail: georgiev@mech.math.msu.su

Robert V. Goldstein
Ishlinsky Institute for Problems in Mechanics RAS, Vernadskiy Ave, 101 bldg 1, 119526 Moscow, Russia, e-mail: goldst@ipmnet.ru

Kondagunta Veeraraghava Nagendra Gopal
Department of Aerospace Engineering, Indian Institute of Technology, Madras, India, e-mail: gopal@ae.iitm.ac.in

Valentin A. Gorodtsov
Ishlinsky Institute for Problems in Mechanics RAS, Vernadskiy Ave, 101 bldg 1, 119526 Moscow, Russia, e-mail: gorod@ipmnet.ru

Irina G. Goryacheva
Ishlinsky Institute for Problems in Mechanics RAS, Vernadskiy Ave, 101 bldg 1, 119526 Moscow, Russia, e-mail: goryache@ipmnet.ru

Narinder Kumar Gupta
Department of Applied Mechanics, Indian Institute of Technology, Delhi, India, e-mail: narinder_gupta@yahoo.com

Dmitry A. Indeitsev
Institute for Problems in Mechanical Engineering RAS, V.O., Bolshoy pr., 61, 199178 St. Petersburg, Russia, e-mail: dmitry.indeitsev@gmail.com

Elena Ivanova
Peter the Great Saint-Petersburg Polytechnic University, Politekhnicheskaja 29, 195251 St.-Petersburg, Russia, e-mail: elenaivanova239@gmail.com

Stephan V. Joubert
Department of Mathematics and Statistics, Tshwane Univesity of Technology, South Africa, e-mail: JoubertSV@tut.ac.za

Gennady I. Kanel
Joint Institute for High Temperatures of the Russian Academy of Sciences, Izhorskaya st. 13 bldg 2, 125412 Moscow, Russia, e-mail: kanel@ficp.ac.ru

Julius Kaplunov
Keele University, Great Britain, e-mail: j.kaplunov@keele.ac.uk

Alexander V. Kaptsov
Ishlinsky Institute for Problems in Mechanics RAS, Vernadskiy Ave, 101 bldg 1, 119526 Moscow, Russia, e-mail: kaptsov@ipmnet.ru

Vladimir I. Karev
Ishlinsky Institute for Problems in Mechanics RAS, Vernadskiy Ave, 101 bldg 1, 119526 Moscow, Russia, e-mail: wikarev@ipmnet.ru

Kirill E. Kazakov
Ishlinsky Institute for Problems in Mechanics RAS, Vernadskiy Ave, 101 bldg 1, 119526 Moscow & Bauman Moscow State Technical University, Baumanskaya st. 5 bldg 1, 105005 Moscow, Russia, e-mail: kazakov-ke@yandex.ru

Dmitry M. Klimov
Ishlinsky Institute for Problems in Mechanics RAS, Vernadskiy Ave, 101 bldg 1, 119526 Moscow, Russia

Alexander G. Kolpakov
Siberian State University of Telecommunications and Informatics, Kirov st. 86, 630009 Novosibirsk, Russia, e-mail: algk@ngs.ru

Yury F. Kovalenko
Ishlinsky Institute for Problems in Mechanics RAS, Vernadskiy Ave, 101 bldg 1, 119526 Moscow, Russia, e-mail: perfolinkgeo@yandex.ru

Larisa V. Kovtanyuk
Institute of automation and control processes of FEB RAS, Vladivostok Russia, 690041, e-mail: lk@dvo.ru

Sergey A. Kumakshev
Ishlinsky Institute for Problems in Mechanics RAS, Vernadskiy Ave, 101 bldg 1, 119526 Moscow, Russia, e-mail: kumak@ipmnet.ru

Ilya I. Kudish
Department of Mathematics, Kettering University, Flint, MI, 48504, USA, e-mail: ikudish@kettering.edu

Svetlana P. Kurdina
Bauman Moscow State Technical University, Baumanskaya st. 5 bldg 1, 105005 Moscow, Russia, e-mail: svetlana-ka@yandex.ru

Masodi Lesiba Glen Lekalakala
Department of Mathematics, Tshwane University of Technology, South Africa, e-mail: glen@vut.ac.za

Sergei A. Lychev
Ishlinsky Institute for Problems in Mechanics RAS, Vernadskiy Ave, 101 bldg 1, 119526 Moscow, Russia, e-mail: lychevsa@mail.ru

Yulia A. Mochalova
Institute for Problems in Mechanical Engineering RAS, V.O., Bolshoy pr., 61, 199178 St. Petersburg, Russia, e-mail: yumochalova@yandex.ru

Nikita F. Morozov
St. Petersburg State University, Universitetskaya nab. 7/9, 199034 St.Petersburg, Russia, e-mail: morozov@nm1016.spb.edu

Wolfgang H. Müller
Institute of Mechanics, Chair of Continuum Mechanics and Materials Theory, Technical University of Berlin, Einsteinufer 5, 10587 Berlin, Germany e-mail: whmueller1000@gmail.com

Evegnii V. Murashkin
Ishlinsky Institute for Problems in Mechanics RAS, Vernadskiy Ave, 101 bldg 1, 119526 Moscow, Bauman Moscow State Technical University, Baumanskaya st. 5 bldg 1, 105005 Moscow, National Research Nuclear University "MEPhI" (Moscow Engineering Physics Institute), Kashirskoye shosse 31, 115409 Moscow, & Moscow Technological University, Vernadsky Ave 78, 119454 Moscow, Russia, e-mail: murashkin@ipmnet.ru, evmurashkin@gmail.com

Konstantin Naumenko
Institut für Mechanik, Otto-von-Guericke-Universität Magdeburg, Universitätsplatz 2, 39106 Magdeburg, Germany, e-mail: konstantin.naumenko@ovgu.de

Galina L. Panchenko,
Institute of Machine Science and metallurgy of FEB RAS, 681005 Komsomolsk-on-Amur, Russia

Dmitry A. Parshin
Ishlinsky Institute for Problems in Mechanics RAS, Vernadskiy Ave, 101 bldg 1, 119526 Moscow & Bauman Moscow State Technical University, Baumanskaya st. 5 bldg 1, 105005 Moscow, Russia, e-mail: parshin@ipmnet.ru

Alexander V. Pavlenko
Joint Institute for High Temperatures of the Russian Academy of Sciences, Izhorskaya st. 13 bldg 2, 125412 Moscow, Russia

Danila A. Prikazchikov
School of Computing and Mathematics, Keele University, Keele, Staffordshire, ST5 5BG, UK, e-mail: d.prikazchikov@keele.ac.uk

Alexander N. Prokudin
Institute of Machine Science and metallurgy of FEB RAS, 681005 Komsomolsk-on-Amur, Russia, e-mail: sunbeam_85@mail.ru

Yuri N. Radayev
Ishlinsky Institute for Problems in Mechanics RAS, Vernadskiy Ave, 101 bldg 1, 119526 Moscow, Russia, e-mail: radayev@ipmnet.ru, y.radayev@gmail.com

Sergei I. Rakin
Siberian State Transport University, Dusi Koval'chuk st. 191, 630049 Novosibirsk, Russia,
e-mail: rakinsi@ngs.ru

Abishera Ravichandra Rajkumar
Department of Aerospace Engineering, Indian Institute of Technology, Madras, India

Sergey V. Razorenov
Joint Institute for High Temperatures of the Russian Academy of Sciences, Izhorskaya st. 13 bldg 2, 125412 Moscow, Russia

Andrey S. Savinykh
Joint Institute for High Temperatures of the Russian Academy of Sciences, Izhorskaya st. 13 bldg 2, 125412 Moscow, Russia

Olga Sergushova
School of Computing and Mathematics, Keele University, Keele, Staffordshire, ST5 5BG, UK, e-mail: o.sergushova@keele.ac.uk

Michael Y. Shatalov
Department of Mathematics and Statistics, Tshwane Univesity of Technology, South Africa, e-mail: ShatalovM@tut.ac.za

Efim I. Shifrin
Ishlinsky Institute for Problems in Mechanics RAS, Vernadskiy Ave, 101 bldg 1, 119526 Moscow, Russia, e-mail: shifrin@ipmnet.ru

Hiltje Spoelstra
Department of Mathematics and Statistics, Tshwane Univesity of Technology, South Africa, e-mail: SpoelstraH@tut.ac.za

Feodor I. Stepanov
Ishlinsky Institute for Problems in Mechanics RAS, Vernadskiy Ave, 101 bldg 1, 119526 Moscow, Russia, e-mail: stepanov_ipm@gmail.ru

Elena V. Torskaya
Ishlinsky Institute for Problems in Mechanics RAS, Vernadskiy Ave, 101 bldg 1, 119526 Moscow, Russia, e-mail: torskaya@ipmnet.ru

Petr E. Tovstik
St. Petersburg State University, Universitetskaya nab. 7/9, 199034 St.Petersburg, Russia, e-mail: peter.tovstik@mail.ru

Tatyana P. Tovstik
Institute of Problems of Mechanical Engineering RAS, Bol'shoi pr. 61, St. Petersburg, 199178, Russia, e-mail: tovstik_t@mail.ru

Konstantin B. Ustinov
Ishlinsky Institute for Problems in Mechanics RAS, Vernadskiy Ave, 101 bldg 1, 119526 Moscow, Russia, e-mail: ustinov@ipmnet.ru

Ramachandran Velmurugan
Department of Aerospace Engineering, Indian Institute of Technology, Madras, India, e-mail: ramanv@iitm.ac.in

Elena Vilchevskaya
Institute for Problems in Mechanical Engineering of the Russian Academy of Sciences, Bol'shoy pr. 61, V.O., 199178 St. Petersburg, Russia & Peter the Great Saint-Petersburg Polytechnic University, Politekhnicheskaja 29, 195251 St.-Petersburg, Russia, e-mail: vilchevska@gmail.com

Si-Yu Wang
Shanghai Institute of Applied Mathematics and Mechanics, Shanghai University, Yanchang Road 149, 200072 Shanghai, China, e-mail: 343396676@qq.com

Heng Xiao
Shanghai Institute of Applied Mathematics and Mechanics, Shanghai University, Yanchang Road 149, 200072 Shanghai, China, e-mail: xiaoheng@shu.edu.cn

Zheng-Nan Yin
Shanghai Institute of Applied Mathematics and Mechanics, Shanghai University, Yanchang Road 149, 200072 Shanghai, China, e-mail: znyin@shu.edu.cn

Lin Zhan
Shanghai Institute of Applied Mathematics and Mechanics, Shanghai University, Yanchang Road 149, 200072 Shanghai, China, e-mail: 302974155@qq.com

Chapter 1
Multi-Mode Symmetric and Asymmetric Solutions in the Jeffery-Hamel Problem for a Convergent Channel

Leonid D. Akulenko, Dimitri V. Georgievskii & Sergey A. Kumakshev

Abstract A complete solution of the boundary-value problem for a convergent channel is constructed and investigated over the entire range of admissible values of the channel angle for small, moderately large, and large Reynolds numbers. Both analytical and numerical methods are applied to construct the solution. Special attention is given to solutions that can be regularly continued for small *Re*. To construct such solutions, a high-accuracy numerical-analytical technique has been developed. This technique is based on a rapidly convergent iteration method (accelerated-convergence method) combined with the continuation with respect to a parameter. The existence of three-mode solutions for sufficiently large channel angles is established. These solutions are symmetric with respect to the bisector of the channel angle and have source and sink domains. The evolution of the velocity profile as *Re* increases without limit is studied. The solutions for the critical value of the channel angle, in which case the regular continuation with respect to a parameter is impossible, are investigated. Multi-mode symmetric and asymmetric solutions, singular with respect to *Re* for asymptotically small *Re*, are constructed and analyzed. A number of new hydromechanical phenomena are established and accounted for.

Key words: Multi-Mode · Jeffery-Hamel problem · Convergent channel · Reynolds number · Steady flow · Viscoplastic material

---

Leonid D. Akulenko
Ishlinsky Institute for Problems in Mechanics RAS, Vernadskiy Ave, 101 bldg 1, 119526 Moscow, Russia,
e-mail: `kumak@ipmnet.ru`

Dimitri V. Georgievskii
Lomonosov Moscow State University, Leninskie Gory 1, 119991 Moscow, Russia,
e-mail: `georgiev@mech.math.msu.su`

Sergey A. Kumakshev
Ishlinsky Institute for Problems in Mechanics RAS, Vernadskiy Ave, 101 bldg 1, 119526 Moscow, Russia,
e-mail: `kumak@ipmnet.ru`

H. Altenbach et al. (eds.), *Mechanics for Materials and Technologies*,
Advanced Structured Materials 46, DOI 10.1007/978-3-319-56050-2_1

1.1 Introduction and Statement of the Problem

The boundary-value problem for steady radial flows of a viscous incompressible fluid in a plane convergent channel (see, e.g., Jeffery, 1915; Hamel, 1917; Rosenhead, 1940; Kochin et al, 1965; Batchelor, 1970; Landau and Lifschitz, 1987; Loitsyanskii, 1966) is one of a few of non-one-dimensional problems of hydrodynamics of Newtonian fluids that have an exact self-similar solution. Therefore, a complete solution of this problem over the entire range (including the limiting values) of the dimensionless parameters - the convergent channel angle 2β and the Reynolds number Re - are of great interest from the academic and applied points of view.
The history of the investigation of this problem can be conventionally divided into several stages. The first stage is associated with the names of George Barker Jeffery and Georg Karl Wilhelm Hamel and their articles Jeffery (1915); Hamel (1917).
Jeffery was the first to formulate the boundary-value problem for the steady-state flows of a viscous incompressible fluid between two intersecting plane walls and has given an implicit solution of this problem in terms of Jacobi elliptic functions and elliptic integrals (Jeffery, 1915). The adhesion condition was assumed on the walls and, in addition, a constant flow rate was prescribed. This system was called the divergent channel if the flow rate was positive and the convergent channel if the flow rate was negative.
Independently of Jeffery's studies and somewhat later, Hamel established the possibility of radial plane flows of a viscous fluid. He also derived the ordinary differential equation governing such a flow and found the solution of this equation in terms of Weierstrass elliptic functions. The possibilities and ways for the physical implementation of such a flow have not been investigated. Neither Hamel's results involve the statement of the boundary-value problem. Apparently, the term *Jeffery–Hamel flows* applied to radial plane flows of fluid is historically justified. This is just the term that has been mostly utilized in the scientific literature.
In our opinion, the statement of the classical boundary-value problem for radial flows of a viscous incompressible fluid in a plane divergent or convergent channel (between two intersecting plane walls) with fixed flow rate and the implicit analytical solution of this problem are due to Jeffery (1915). Therefore, it is appropriate to call the examined boundary-value problem with the additional integral condition by his name (*Jeffery problem*).
Subsequently, a great number of studies appeared in which the integral condition of constancy of the flow rate was replaced by various local conditions, for example, by that of prescribed velocity on the channel axis. (Such studies are continuing to appear nowadays.) These reformulations substantially simplify the boundary-value problem and, which is the most important, lead to solutions with quite different quantitative features. In what follows, by the Jeffery problem we will understand the original statement of Jeffery (1915) with the constant flow rate condition and will not consider other problems.
The second stage can be associated with the theoretical studies by Rosenhead (1940) and Fraenkel (1962, 1963), which are being broadly cited. In these, as well as in some other studies in the middle of the 20th century, a number of profiles of the

flow inside a divergent or convergent channel have been constructed analytically or numerically. Among these profiles there are multi-mode and asymmetric ones. In the domain of the parameters "channel half-angle–Reynolds number", regions with qualitatively different behavior of the flow have been constructed. The qualitative difference takes into account such properties as the number of modes of the flow, the presence of singular points, the existence of a critical angle, stability, and some other aspects. The studies of Rosenhead (1940) and Fraenkel (1962, 1963) have had an important value. They have shown that the boundary-value problem for the Jeffery–Hamel flows is highly difficult to solve and have indicated a great number of new aspects of the solution of mathematical problems subjected to an additional integral condition. In all these publications, the complexity of the problem and incompleteness of the investigation were indicated.

The current stage we associate with the development of an appropriate numerical-analytical method and a high-accuracy computational algorithm for the construction of solutions of the boundary-value problem for a wide range of the parameters. The approach to the solution of the basic problem involves direct integration of a sequence of initial value (Cauchy) problems with high degree of accuracy. Then the accelerated convergence Newton-type method is utilized to determine the constants of integration on the basis of the boundary conditions. These constants have a clear interpretation in terms of the problem.

To investigate the evolution of the flow behavior, we utilize the continuation with respect to a parameter. This continuation is performed with respect to the Reynolds number for fixed channel angle or vice versa. The aforementioned constants completely determine the solution of the boundary-value problem on the basis of the solution of the initial value problem.

This approach provides an effective technique for constructing the velocity and pressure profiles, as well as other kinematic and dynamic characteristics of the flow of a viscous fluid, with a relative error of 10^{-9} to 10^{-8} for any channel angle and $Re \sim 10^3$. The proposed algorithm does not utilize tabulated elliptic functions and elliptic integrals. The tables of these functions are not complete and precise enough for the solution of the problem under consideration. It should be noted that the estimation of the actual accuracy on the basis of an implicit analytical approach is complicated and has not been given in the cited publications.

Consider a plane steady flow of a viscous incompressible fluid in an unbounded domain $\Omega = \{(r,\theta) : r > 0, -\beta < \theta < \beta\}$ (Fig. 1.1) with a source-sink singularity of power $-Q$ at the point O. The domain Ω is referred to as a plane divergent channel for $Q < 0$ and a plane convergent channel for $Q > 0$. The fluid is characterized by the density ρ and the dynamic viscosity μ.

There are no other characteristic parameters in the problem, apart from the four quantities β, Q, ρ, and μ. Since there are no three dimensionally independent quantities among these parameters (since $[Q] = [\mu]/[\rho]$), a complete nondimensionalization of the governing equations in this problem is impossible. The solution will depend on two dimensionless parameters, the channel angle 2β $(0 < \beta \leq \pi)$ and $Re = \rho Q/\mu$, $|Re| < \infty$.

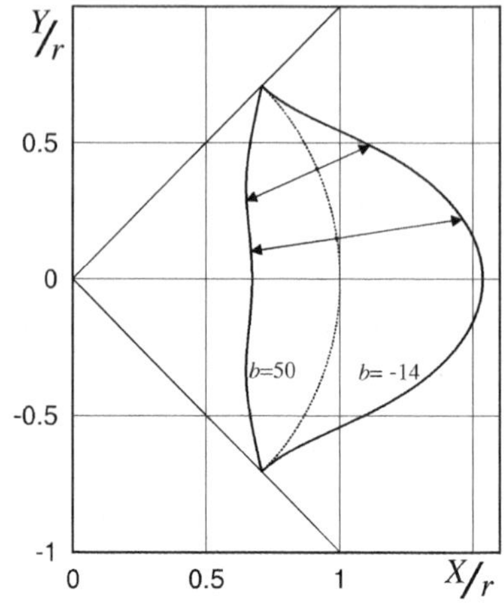

Fig. 1.1 Characteristic radial single-mode velocity profiles for viscous fluid flow in divergent ($b = -14$) and convergent ($b = 50$) channels for the angle $2\beta = 90°$, $b = 2\beta Re$.

It is known that for the plane case, one can seek a solution, which describes the motion of the fluid, in a self-similar form. The velocity field $\mathbf{v} = (v_r, v_\theta)$ is radial, with

$$v_r = -\frac{Q}{r}V(\theta), \quad v_\theta \equiv 0. \tag{1.1}$$

The velocity field of (1.1) satisfies the incompressibility condition.

Two Navier–Stokes equations (the radial and angular projections of the vector Navier–Stokes equation) lead to the ordinary differential equation for the dimensionless scalar function $V(\theta)$ and the expression for the pressure p

$$V'' + 4V - ReV^2 = C, \quad p = \frac{\rho Q^2}{2r^2 Re}(C - 4V). \tag{1.2}$$

The adhesion condition for the fluid of the boundary of Ω makes clear the physical meaning of the unknown constant C in (1.2),

$$V(\pm\beta) = 0, \quad C = V''(\pm\beta). \tag{1.3}$$

In addition, we have the integral relation

$$\int_{-\theta}^{\theta} V(\theta)\,d\theta = 1. \tag{1.4}$$

This relation implies that the flow rate through any surface surrounding the source (or sink) is the same. Using the boundary conditions of (1.3) and the integral condition of (1.4), one can determine the integration constants and the unknown C.

The analytical study of the problem (1.2)–(1.4) has been dealt with in a large number of publications on hydrodynamics, including monographs and textbooks (see, e.g., Rosenhead, 1940; Millsaps and Pohlhausen, 1953; Kochin et al, 1965; Batchelor, 1970; Landau and Lifschitz, 1987; Loitsyanskii, 1966). It should be noted that sometimes the term Jeffery–Hamel flows is applied to problems different from that of (1.2)–(1.4). For example in Millsaps and Pohlhausen (1953), the integral condition of (1.4) has been replaced by the "local" condition $V(0) = V_0 = 1$, i.e., the flow velocity on the axis of the divergent-convergent channel, rather than the flow rate, is prescribed. This leads to substantial changes in the numerical-analytical methods for solving the problem and in the solution properties.

Note that the relations of (1.2)–(1.4) involve only dimensionless known and unknown functions and constant parameters. This is due to the self-similarity (Loitsyanskii, 1966), rather than nondimensionalization[1].

Equation (1.2) admits a reduction in order by multiplying by V' and separating total derivatives. This equation has the first integral

$$\frac{V'^2}{2} + 2V^2 - Re\frac{V^3}{3} - CV = \frac{V'^2(\mp\beta)}{2}. \tag{1.5}$$

The classical results are associated with the integration of Eq. (1.5) in terms of elliptic functions. The resulting elliptic integrals depend on several parameters, as well as on the roots of a system of three transcendental equations for three constants of integration. Analysis of these integrals is complicated. It is hardly possible to construct a closed-form solution of the boundary-value problem with the constant-flow-rate condition for arbitrary β and Re.

The solution of the Jeffery problem for the divergent channel ($Q < 0$) has been considered to be more complex and diverse than the solution for the convergent channel ($Q > 0$). For the former problem, multi-mode solutions with symmetric and asymmetric velocity profiles relative to the axis $\theta = 0$ have been known. Here and in what follows, by multi-mode solutions (multi-mode flows), the flows in which the source ($V(\theta) > 0$) and sink ($V(\theta) < 0$) domains alternate are understood. The existence of multi-mode and asymmetric solutions of the nonlinear boundary-value problem with the constant-flow-rate condition for a convergent channel indicates the nonuniqueness of solution, since this problem always (for $0 < \theta \leq \pi/2$, see Sects. 1.2–1.7) has a single-mode symmetric solution. A number of theorems on the uniqueness of the classical single-mode profile for asymptotically small Re have been proved by Rivkind and Solonnikov (2000). For sufficiently large Re, the steady-state solution for a divergent channel becomes unstable (Eagles, 1966), which leads to the occurrence of developed turbulence.

Various aspects of analytical, numerical or experimental investigation of the flows of a viscous fluid in plane convergent or divergent channels and their modern modifications have been presented in numerous publications in the last 15–20 years. We will enumerate some of them.

Instability and various bifurcations of the Jeffery–Hamel flows have been studied by Sobey and Drazin (1986). It is shown that the unique symmetric flow, existing for small Re, becomes unstable for large Re. A bifurcation leads to the appearance of

[1] As has already been said, a complete nondimensionalization of the problem on the basis of its characteristic parameters is impossible.

two stable asymmetric solutions. For large *Re*, eight asymmetric stable solution have been found and the existence of seven symmetric unstable solutions has been proved. For very large *Re*, time-periodic solutions of this problem have been identified and the possibility of the Hopf bifurcation has been discussed.

Small 2D perturbations of the classical single-mode symmetric Jeffery–Hamel flows have been studied by Banks et al (1988); Eagles (1988). It has been noticed that the development of the perturbations depends on whether the channel angle is less or greater than a certain critical angle, the value of which depends on the Reynolds number. Proceeding from the studies by Rosenhead (1940) and Fraenkel (1962), they performed a formal classification of the flows on the basis of the analysis of roots of a cubic polynomial. On the parameter plane αRe, where α is the channel half-angle and *Re* is the Reynolds number, they construct (for small *Re*) the curves separating the regions with different flow types. Then they consider the influence of slight curvature of the walls on the flow parameters and the Fraenkel approximation associated with this influence have been discussed (Fraenkel, 1962, 1963).

The motion in a channel with a piecewise-straight boundary has been analyzed by Tutty (1996). A critical value of the angle between the walls has been found. If this critical value is exceeded, the Jeffery–Hamel flows does not exist. The solution of the problem in this case has a shape of a wave with large amplitude, with vortices appearing now at one, then at the other wall of the channel. If the angle between the walls is less than the critical value, the existence of such a wave depends on the geometry of the walls near the sink point. In Kerswell et al (2004), the results by Tutty (1996) have been developed on the basis of numerical methods and the stability of steady nonlinear waves has been studied.

A thorough numerical analysis of the Jeffery–Hamel flows have been performed by Dennis et al (1997) in order to investigate the existence and uniqueness of symmetric solutions and identify the beginning of an instability.

For the converging flow, neither multi-mode behavior nor an asymmetry of the velocity profile have been established in the classical publications. The issues of the existence of such profiles (which, in fact, is reduced to the uniqueness analysis of the converging flow) and the properties of such flows had remained a challenge.

Using a modification of the analytical-numerical accelerated-convergence method (Akulenko et al, 2002c), we have constructed and completely investigated single-mode and multi-mode convergent flows for a broad range of parameters β and *Re*. This investigation is based on the previous particular results (Akulenko et al, 2002b,a, 2004, 2000, 2003b).

In the boundary-value problem of (1.2)–(1.4), we can proceed from the function V of the angle θ to the new dimensionless function y of the normalized variable x

$$y = 2\beta V, \quad x = \frac{1}{2}\left(\frac{\theta}{\beta} + 1\right), \quad 0 \le x \le 1. \tag{1.6}$$

This change of variables reduces the problem of (1.2)–(1.4) with the integral condition to the form

$$y'' + a^2 y - b y^2 = \lambda, \quad a = 4\beta, \quad b = 2\beta Re,$$
$$y(0) = y(1) = 0, \quad \int_0^1 y(x)\,\mathrm{d}x = 1. \tag{1.7}$$

The constants λ and γ defined by

$$\lambda = 8\beta^3 C = y''(0) = y''(1), \quad \gamma = y'(0)$$

are unknown and have to be determined for prescribed a and b. The parameter λ can be eliminated by differentiation, but this leads to a nonlinear third-order equation. Introduce the function $z(x)$ to replace the integral condition in (1.7) by the differential equation with boundary conditions

$$z' = y - 1, \quad z(0) = z(1) = 0. \tag{1.8}$$

The boundary-value problem of (1.7) and (1.8) can be represented in the form of standard fourth-order boundary-value problem (without the integral condition). From (1.2) we find that the pressure p and the velocity profile $y(x)$ are related by

$$p(r,\theta) = \frac{2\rho Q^2}{r^2}\frac{\lambda - a^2 y(x)}{a^2 b}. \tag{1.9}$$

It is apparent from (1.9) that the pressure increases without limit as $b \to 0$. Nevertheless, the solution of the problem of (1.7) and(1.8) can be regular. For the radial flow to be implemented in practice, the pressure at any distance r must be a function of x (or the angle θ). This function is unknown in advance.

1.2 Analytical Expressions, Asymptotic Expansions, and Integral Estimates of the Solutions

We will present analytical expressions for $y(x)$ for various values of the parameters a and b (β and Re).

1.2.1 Perturbation Method for Small Re

The relations $0 < Re \ll 1$ correspond to slow flow of a very viscous fluid. This is the case for numerous technological processes (e.g., drawing of metal sheets or extrusion of plastic materials) and natural phenomena (e.g., flow of rocks or glaciers). We can seek the solution $y(x)$ for arbitrary angles β in the form of a series in powers of b for any finite a,

$$y(x) = \sum_{n=0}^{\infty} b^n y_n(x), \quad \lambda = \sum_{n=0}^{\infty} b^n \lambda_n, \quad \gamma = \sum_{n=0}^{\infty} b^n \gamma_n \tag{1.10}$$

with first several terms being preserved in this series. For example, for $n = 0$, we have (Kochin et al, 1965)

$$\begin{gathered} y_0(x) = \frac{2\beta}{D}\left[\cos(2\beta(2x-1)) - \cos 2\beta\right], \\ D = \sin 2\beta - 2\beta\cos 2\beta, \quad \lambda_0 = -\frac{32\beta^3}{D}\cos 2\beta, \quad \gamma_0 = \frac{8\beta^2}{D}\sin 2\beta. \end{gathered} \tag{1.11}$$

The zero approximation of (1.11) does not make sense for $D = 0$, i.e., for the values of β that are roots of the equation $\tan 2\beta = 2\beta$. The least positive root of this equation is $\beta = \beta^* \approx 2.247$, which corresponds to a convergent channel with an aperture angle exceeding that of 180°. Such convergent channels are realistic from a physical point of view.

For the subsequent coefficients $y_n(x)$, $n \geq 1$, we have linear boundary value problems

$$\begin{gathered} y_n'' + a^2 y_n = \lambda_n + F_n(y_0, y_1, \ldots, y_{n-1}), \\ y_n(0) = y_n(1) = 0, \quad \int_0^1 y_n(x)\,\mathrm{d}x = 0 \end{gathered} \tag{1.12}$$

with unknown constants $\lambda_n = y_n''(0) = y_n''(1)$ and known function F_n; for example, $F_1(y_0) = y_0^2$. For $\beta \neq \beta^*$, the problems of (1.12) can be readily solved, with the solution being uniquely defined. For $n \geq 1$, the calculations become cumbersome (Akulenko et al, 2000) and for $\beta \sim 1$, unpractical because of low accuracy. For $\beta = \beta^*$, the problem of (1.12) has no solutions and, therefore, another approach is required (Sect. 1.6).

1.2.2 Perturbation Method for Small Aperture Angles

The relations $0 < \beta \ll 1$ and $0 < Re \sim 1$ or $0 < a \ll 1$ and $0 < b \ll 1$ correspond to the case where the convergent channel walls are almost parallel. In the problem of (1.7), the regular asymptotic expansions in powers of β

$$y(x) = \sum_{n=0}^{\infty} \beta^n y_n(x), \quad \lambda = \sum_{n=0}^{\infty} \beta^n \lambda_n, \quad \gamma = \sum_{n=0}^{\infty} \beta^n \gamma_n \tag{1.13}$$

lead to the approximation of the solution by polynomials with respect to x, say for $n = 0, 1$

$$y_0(x) = 6x(1-x), \; y_1(x) = \frac{6Re}{35}x(14x^5 - 42x^4 + 35x^3 - 9x + 2)$$
$$\lambda_0 = -12, \quad \lambda_1 = -\frac{108}{35}Re. \tag{1.14}$$

The constants γ_n (1.13) can be obtained by differentiating $y_n(x)$ at $x = 0$. The parabolic profile $y_0(x)$ in (1.14) corresponds to the Poiseuille flow between two parallel walls. The degree of the polynomials $y_n(x)$ increases rather rapidly and is equal to $4n + 2$.

1.2.3 Asymptotic Behavior of the Solution for Large Re

The case of $Re \gg 1$ and $\beta \sim 1$ formally corresponds to the limiting case of an ideal fluid. For $Re = \infty$, the solution is discontinuous at the boundary points and has the form

$$y_\infty(x) \equiv 1, \; 0 < x < 1, \quad y_\infty(0) = y_\infty(1) = 0.$$

The occurrence of the small coefficient $1/b$ of the highest derivative in (1.7) requires singular asymptotic methods to be involved (van Dyke, 1964). In the first approximation in terms of $1/b$, the solution can be represented by Kochin's formula (Georgievskii, 1998)

$$y(x) = 1 - \frac{6}{1 + \coth\left[\coth^{-1} 5 + \sqrt{b/2}(1 - |2x - 1|)\right]},$$
$$\lambda(b) = -b, \quad \gamma(b) = \sqrt{\frac{4b}{3}}. \tag{1.15}$$

This formula provides acceptable accuracy even for $b \sim 10^3 \div 10^4$.

Let $0 < \beta \ll 1$ but $b \sim 1$, which corresponds to the flow with large Reynolds numbers between almost parallel walls. In the zero approximation, the left-hand side of (1.7) does not contain the term $a^2 y$. However, the integration of the resulting boundary-value problem, as is the case in general, leads to elliptic functions. The only minor simplification is that the solution depends of a single parameter b.

1.2.4 Integral Estimates

The boundary value problem of (1.7) is equivalent to the isoperimetric variational problem

$$\frac{1}{2}\int_0^1 (y'^2 - a^2y^2 + \frac{2}{3}by^3)\,\mathrm{d}x \to \text{extr},$$
$$y(0) = y(1) = 0, \qquad \int_0^1 y(x)\,\mathrm{d}x = 1. \tag{1.16}$$

On the basis of (1.16), one can construct integral estimates for $y(x)$, λ, and γ. Any continuously differentiable function $f(x)$ satisfies the relation

$$\int_{x_1}^{x_2} (y'' + a^2y - by^2)f'\,\mathrm{d}x = \lambda\left(f(x_2) - f(x_1)\right). \tag{1.17}$$

where $0 \le x_1 < x_2 \le 1$. Let x_1 an x_2 be arbitrary different roots of the function $y(x)$, i.e.,

$$y(x_1) = y(x_2) = 0. \tag{1.18}$$

Let $f(x) = y(x)$. In this case, the integration by parts in (1.17) with reference to (1.18) leads to the important relations

$$y'(x_2) = \pm y'(x_1). \tag{1.19}$$

If $x_1 = 0$ and $x_2 = 1$, then the lower sign in (1.19) corresponds to the classical single-mode solution and the upper sign corresponds to a multi-mode solutions with odd number of modes. The condition $y'(1) = y'(0)$ implies the multi-mode character of the flow.

Let $f' = y$ and $f(x_1) = 0$. In this case, with reference to the Friedrichs inequality (Rektorys, 1980)

$$\int_{x_1}^{x_2} y'^2\,\mathrm{d}x \ge \frac{\pi^2}{(x_2 - x_1)^2}\int_{x_1}^{x_2} y^2\,\mathrm{d}x \tag{1.20}$$

and the relation of (1.17), we obtain the estimate for the parameter λ

$$\lambda\int_{x_1}^{x_2} y\,\mathrm{d}x \le \left(a^2 - \frac{\pi^2}{(x_2 - x_1)^2}\right)\int_{x_1}^{x_2} y^2\,dx - b\int_{x_1}^{x_2} y^3\,\mathrm{d}x. \tag{1.21}$$

The estimate of (1.21) turns out to be fairly precise for the case of the single-mode solution ($x_1 = 0$, $x_2 = 1$), provided that b is relatively small.

Let $f = y'$. In this case, integrating (1.17) by parts yields

$$n\lambda y'(x_1) = \int_{x_1}^{x_2} (y''^2 - a^2y^2 + 2byy')\,\mathrm{d}x \tag{1.22}$$

where $n = 0$ ($n = -2$) for the functions $y(x)$ for which one should choose the upper (lower) sign in (1.19). The functions $y(x)$ with the boundary conditions (for these functions, $n = 0$)

$$y^{(m)}(x_1) = y^{(m)}(x_2), \quad m = 0, 1, 2 \tag{1.23}$$

satisfy one more Friedrichs inequality

$$\int_{x_1}^{x_2} y''^2 \, dx \geq \frac{4\pi^2}{(x_2 - x_1)^2} \int_{x_1}^{x_2} y'^2 \, dx. \tag{1.24}$$

The relations of (1.22) and (1.24) imply the estimate

$$\int_{x_1}^{x_2} \left[\frac{4\pi^2}{(x_2 - x_1)^2} - a^2 + 2by \right] y'^2 \, dx \leq 0. \tag{1.25}$$

For some x_1, x_2, a, and b, the relation of (1.24) holds with the equality sign. This is the case, for example, for $x_1 = 0, x_2 = 1, a = 2\pi$, and $b = 0$. Of course, the conditions of (1.23) hold in this case.

1.3 Numerical-Analytical Accelerated Convergence Method and Continuation with Respect to a Parameter

The ranges of the angle $\beta \sim 0.1 - 1$ and Reynolds number $Re \sim 0.1 - 10^3$ ($b \sim 10^{-2} - 10^3$) are rather important for applications. However, for these ranges, the relations of (1.10)–(1.15) lead to unacceptably large errors. These relations can serve as rough estimates or initial approximations for high-accuracy numerical methods. Therefore, it is necessary to develop efficient numerical-analytical approaches to the solution of the problem of (1.7) and (1.8).

We propose a modified Newton-type accelerated convergence method (Akulenko et al, 2000, 2003b, 2002c) combined with the continuation with respect to a parameter. This method enables one to calculate the desired quantities with high degree of accuracy in one or two iterations with moderate computational effort. The relative error of the solution ranges between 10^{-8} and 10^{-7}. In many cases, such a solution can be considered to be practically exact. If necessary, the error can be reduced. The algorithm of this method involves the calculation of the unknown values $\gamma = y'(0)$ and $\lambda = y''(0)$ which determine the function $y(x)$ as the solution of the initial value (Cauchy) problem of (1.7) and (1.8) for fixed a and b (and, hence, for fixed β and *Re*).

To start the algorithm, one should have known initial guesses γ_0 and λ_0 for the desired quantities γ and λ. These initial guesses should be fairly precise and can be found on the basis of the variational approach with the functionals of the type of (1.16), provided that the choice of the trial function $y_0(x)$ has been successful

(Akulenko et al, 2000, 2003b, 2002c). Calculations for the single-mode flow show that the function $y(x,a,b)$ has a fairly simple form. A rather convenient procedure is that of continuation with respect to the parameters a and b (or β and Re). For example, for a fixed $a = 4\beta > 0$, one can take γ_0 and λ_0 as the initial approximations $\gamma_0(b_1)$ and $\lambda_0(b_1)$ for a sufficiently small $b_1 > 0$. Using the accelerated convergence recurrent algorithm, one can calculate $\gamma_1 = \gamma(b_1)$ and $\lambda_1 = \lambda(b_1)$ with the desired degree of accuracy. These values are taken as the initial approximations $\gamma_0(b_2)$ and $\lambda_0(b_2)$, and so on. The availability of the sets $\{\gamma_k\}$ and $\{\lambda_k\}$ of fairly precise values of γ_k and λ_k enables one to increase the interval $\delta b_{k+1} = b_{k+1} - b_k$ by means of a polynomial extrapolation of the initial approximations $\gamma_0(b_{k+1})$ and $\lambda_0(b_{k+1})$.

The iterative algorithm for refining the desired values $\gamma(b)$ and $\lambda(b)$ at the first and subsequent steps involves the solution of two initial value problems for sixth-order systems or one initial value problem for a ninth-order system. These problems involve common equations of (1.7) and (1.8) subjected to the conditions

$$y(0) = z(0) = 0,\ y'(0) = \gamma_n(b),\ \lambda = \lambda_n(b),\ n = 0,1,\dots,\ b \in \{b_k\} \tag{1.26}$$

and the equations for the sensitivity functions v, w and u, s (defined as the derivatives of the solution $y(x)$, $z(x)$ with respect to γ and λ, respectively)

$$\begin{aligned} v'' + a^2 v - 2byv = 0, \quad w' = v, \quad v(0) = w(0) = 0, \quad v'(0) = 1, \\ u'' + a^2 u - 2byu = 1, \quad s' = u, \quad u(0) = u'(0) = s(0) = 0. \end{aligned} \tag{1.27}$$

The initial value problems of (1.7), (1.8), (1.26), and ((1.27) can be solved either simultaneously or separately. As a result, we obtain the functions $y_n(x)$ and $z_n(x)$. To refine the quantities $\gamma_n(b)$ and $\lambda_n(b)$ it is necessary to know the functions y_n, z_n, v_n, w_n, u_n, and s_n at the end point $x = 1$ for $b \in \{b_k\}$. The standard procedure of Newton's method has the form

$$\begin{aligned} \gamma_{n+1}(b) = \gamma_n(b) + \delta\gamma_n(b), \quad \lambda_{n+1}(b) = \lambda_n(b) + \delta\lambda_n(b), \\ \delta\gamma_n = -[y_n(1)s_n(1) - z_n(1)u_n(1)]\Delta_n^{-1}(1), \\ \delta\lambda_n = [y_n(1)w_n(1) - z_n(1)v_n(1)]\Delta_n^{-1}(1), \quad n = 0,1,\dots, \\ \Delta_n(x) = v_n(x)s_n(x) - u_n(x)w_n(x), \quad \Delta_n(1) \neq 0. \end{aligned} \tag{1.28}$$

The recurrent process of (1.26)–(1.28) with respect to n continues until the required degree of accuracy has been attained, i.e., until the residuals $y_n(1)$ and $z_n(1)$ become sufficiently small. The convergence of the algorithm requires a thorough verification at each step and is associated with the values of the aforementioned residuals and the determinant $\Delta_n(1)$ of the sensitivity coefficient matrix of (1.28). A specific feature of the proposed modification of the accelerated convergence method is the additional check of the convergence of the process with respect to the abscissas ξ_n and η_n, i.e., with respect to $|\varepsilon_n|$ and $|\mu_n|$,

$$\varepsilon_n = 1 - \xi_n, \quad \mu_n = 1 - \eta_n; \qquad \xi_n = \arg y_n(x), \quad \eta_n = \arg z_n(x). \tag{1.29}$$

The values ξ_n and η_n are the roots (zeros) of the functions $y_n(x)$ and $z_n(x)$ nearest to $x = 1$. The conditions $|\varepsilon_0| \ll 1$, $|\mu_0| \ll 1$, $|\varepsilon_n| \to 0$, and $|\mu_n| \to 0$ indicate the existence of the desired solution of the boundary-value problem of (1.7) and (1.8) and the convergence of the algorithm of (1.26)–(1.28). These conditions are easy to verify. The convergence has an accelerated (quadratic) character, with

$$\begin{aligned} y_n(1) &= -\gamma_n \varepsilon_n, \quad |\varepsilon_{n+1}| \sim \varepsilon_n^2, \\ z_n(1) &= -\mu_n, \quad |\mu_{n+1}| \sim \mu_n^2, \ n = 0, 1, \ldots. \end{aligned} \tag{1.30}$$

The convergence is determined by the smallness of the absolute values of the parameters $\varepsilon = \varepsilon_0$ and $\mu = \mu_0$, that is, by the nearness of the abscissas ξ_0 and η_0 to $x = 1$. This nearness can be provided by a successful choice of the initial approximation $\gamma_0(b)$ and $\lambda_0(b)$, $b \in \{b_k\}$. To ensure the convergence from this initial approximation the step δb_k of the continuation with respect to b should be sufficiently small. To increase the accuracy, one can use an extrapolation from previous steps. In addition, the condition $|\Delta_n(1)| \geq C > 0$ is required to hold; see (1.28). A noticeable violation of the character of the estimates of (1.29) and (1.30) needs an additional analysis. This violation, as a rule, is associated with the degeneration of the standard algorithm in the case of $\Delta_n(1) \approx 0$ (for example, as $b \to \infty$). The simultaneous calculation of the residuals $y_n(1)$, $z_n(1)$ and ε_n, μ_n enables one to perform efficient high-accuracy calculations of the desired parameters $\gamma(b)$ and $\lambda(b)$ for any fixed β from the interval $0 < \beta \leq \pi$. The continuation with respect to the parameter β, $\beta \in \{\beta_j\}$, is realized in a similar way. If necessary, one can construct the functions $\gamma(\beta, Re)$ and $\lambda(\beta, Re)$ by means of the continuation with respect to β and *Re*.
It has been established during the calculations, that the solution (y, z) of the problem of (1.7) and (1.8) is determined with a relative error of 10^{-7} to 10^{-8} in several (usually two or three) iterations without the extrapolation with respect to b_k. It should be noted that $\Delta_n(1) \approx 0.1$ at the beginning of the calculations (for $b \ll 1$). This requires the initial steps in b to be relatively small, $\delta b_k \approx 0.01$. As b increases, the value of $\Delta_n(1)$ increases, which enables one to increase the step to $\delta b_k \approx 0.1 - 1$ already at early iterations (for $b \sim 0.5$). The computational practice shows that the smallness of the parameters ε_0 and μ_0 is the most essential condition for the rapid convergence of the iterative algorithms. Usually, it is sufficient to have $\varepsilon_0 \sim 0.1$ and $\mu_0 \sim 0.1$.
We will present a high-accuracy solution of the boundary-value problem constructed by means of the accelerated (quadratic) convergence method described above.

1.4 Solutions Regularly Depending on the Reynolds Number

The accelerated convergence algorithm outlined in Sect. 1.3 provides a relatively simple technique for constructing the solutions of the boundary-value problem that admit regular indefinite continuation with respect to b ($0 < b < \infty$), i.e., for $Re > 0$.

As the initial point, we take $b = 0$, for which the solution has the form of (1.11) and depends also on the parameter a, i.e., on the angle β. The function $y_0(x)$ is smooth in x ($0 \le x \le 1$) for all β from the interval $0 < \beta \le \pi$, apart from $\beta = \beta^*$ defined by

$$\beta^* = \arg D(\beta) \approx 2.2467047 \approx (128.7)^\circ. \tag{1.31}$$

For $\beta = \beta^*$, the denominator D in (1.11) vanishes. Note that β^* is independent of the argument x.

The singular solution in the case of $b \to 0$ and $a \to a^* = 4\beta^*$ needs a separate numerical analysis (Sect. 1.6), different from the analysis in the regular case occurring for all $b \ge 0$, provided that $a \ne a^*$. In what follows, we analyze the flows for the full range of the channel angle $0 < 2\beta \le 2\pi$. This analysis is of significant interest from the viewpoint of both mathematics and hydrodynamics. In particular, we study the flows for "exotic" cases of $\beta = \pi/2$ and $\beta = \pi$. The case of $\beta = \pi/2$ corresponds to the flow in a half-space with the slot sink and $\beta = \pi$ corresponds to the flow in the entire space with two indefinitely close half-planes between which the sink of the fluid occurs.

Most of the previous studies cover the range of small angles $\beta \sim 5^\circ - 10^\circ$, which is important for technological applications. The study of the flow for larger β is of interest for natural science. It should be noted that the increase in the angle between the convergent channel walls leads to an essential restructuring in the velocity profile $y(x)$ and new qualitative phenomena.

The proposed approach to the study of the flow involves the determination of the lacking values of the parameters $\gamma(\beta) = y'(0)$ and $\lambda(\beta) = y''(0)$. These parameters determine the solution of the nonlinear boundary-value problem of (1.7) and (1.8) as a solution of an initial value problem. These parameters depend on a and, hence, on the angle β. For $b = 0$, we have $\gamma(0) = \gamma_0(\beta)$ and $\lambda(0) = \lambda_0(\beta)$, where $\gamma_0(\beta)$ and $\lambda_0(\beta)$ are defined in (1.11). To facilitate the comprehension of the further construction, we plotted the functions $\gamma_0(\beta)$ and $\lambda_0(\beta)$ in Fig. 1.2. At $\beta = \beta^*$ of (1.11), these functions undergo a discontinuity. They do not have a limit as $\beta \to \beta^*$ and, hence, the expressions for these functions make no sense at this point. In the case of $b = 0$ and $a = a^*$, the linear boundary-value problem does not have a solution, and the corresponding solution for $b > 0$ cannot be constructed by means of regular perturbations or continuation with respect to a parameter.

Although the critical value $\beta = \beta^*$ corresponds to a realistic convergent channel (with the angle β^* greater than straight angle, $3\pi/2 > 2\beta^* \approx 4.5 > \pi$), the absence of a solution and, hence, the impossibility of the steady flow can be accounted for by the limiting case of $Re \to 0$. For $Re > 0$, this singularity disappears. The fact that this singularity does not manifest itself in the case where $Re = 0$ but $\beta \ne \beta^*$ should be regarded as a mathematical result. Note that in accordance with (1.2) and (1.9), the pressure p increases without limit as $Re \to 0$. The singular behavior of the flow profile $y(x)$ and its derivatives $\gamma(b)$ and $\lambda(b)$ at $x = 0$ for $\beta = \beta^*$ and $b \to 0$ is a hydrodynamic effect. This effect should be validated theoretically and experimentally.

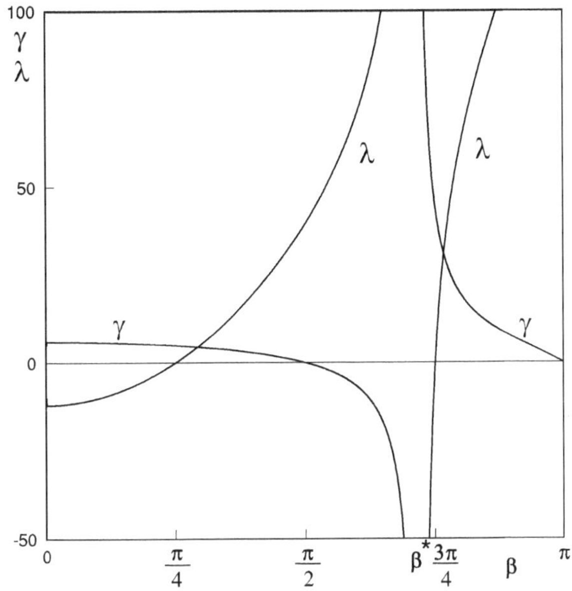

Fig. 1.2 The missing parameters $\gamma_0(\beta) = y'(0)$ and $\lambda_0(\beta) = y''(0)$ as functions of the convergent channel half-angle β for $b = 0$. These parameters determine the solutions obtained by continuation with respect to Re.

Based on the functions $\gamma_0(\beta)$ and $\lambda_0(\beta)$ plotted in Fig. 1.2, one can divide the interval $0 < \beta \leq \pi$ into five qualitatively different intervals, corresponding to the changes of these functions in sign,

$$1)\ 0 < \beta \leq \frac{\pi}{4},\quad 2)\ \frac{\pi}{4} < \beta \leq \frac{\pi}{2},\quad 3)\ \frac{\pi}{2} < \beta < \beta^*,$$
$$4)\ \beta^* < \beta \leq \frac{3\pi}{4},\quad 5)\ \frac{3\pi}{4} < \beta \leq \pi. \tag{1.32}$$

Figures 1.3 and 1.4 plot the lacking values of γ_i (solid curves) and λ_i (dashed curves), $i = 1,\ldots,8$, as a function of b for characteristic increasing values of β_i from the intervals of (1.32) and moderately large values of the parameter b, $0 \leq b \leq 300$. To fix our ideas, we consider the following values of β_i:

$$\beta_1 = \frac{\pi}{180} = 1^\circ,\ \beta_2 = \frac{\pi}{4} = 45^\circ,\ \beta_3 = \frac{\pi}{2} = 90^\circ,$$
$$\beta_4 = \beta^- = 2.2 \approx (126.2)^\circ,\ \beta_5 = \beta^+ = 2.3 \approx (131.6)^\circ, \tag{1.33}$$
$$\beta_6 = \frac{3\pi}{4} = 135^\circ,\ \beta_7 = 2.88 \approx 165^\circ,\ \beta_8 = \pi = 180^\circ.$$

Such a choice corresponds to the partition of (1.32) and is commented in what follows when analyzing the modes of the solutions. The case of $\beta = 10^\circ$ is investigated separately in Sect. 1.7.

The construction of the functions $\gamma_i(b)$ and $\lambda_i(b)$ for β_i of (1.33) on the basis of the algorithm of Sect. 1.3 is one of the basic computational results that facilitate the

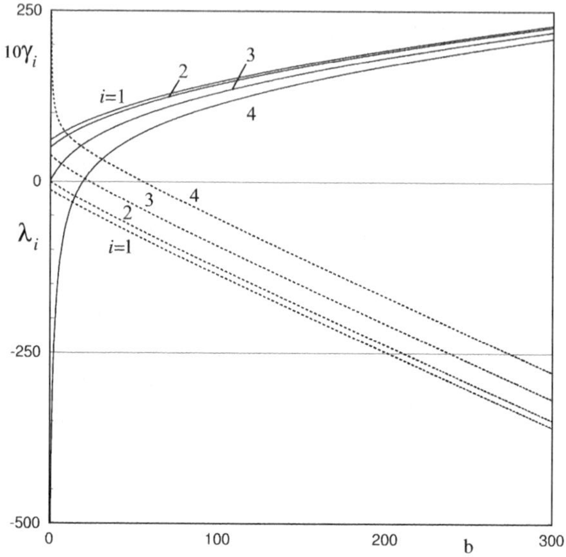

Fig. 1.3 The missing parameters $\gamma_i(b) = y_i'(0)$ (the solid curves) and $\lambda_i(b) = y_i''(0)$ (the dashed curves) as functions of the parameter $b = 2\beta Re$ for fixed values of the angle β_i: $\beta_1 = 1^\circ$, $\beta_2 = 45^\circ$, $\beta_3 = 90^\circ$, and $\beta_4 = \beta^- = 2.2\text{rad} \approx 126.2^\circ$. The variables γ_i are presented after multiplying by 10 ($10\gamma_i$).

determination of all essential characteristics of the flows and enable one to compare various flow modes. A common property of the families of curves $\gamma_i(b)$ and $\lambda_i(b)$ is that all these curves follow Kochin's asymptotic behavior of (1.15) for $b \gg 1$. Significant deviations from the asymptotic curve occur for $b \sim 1 - 100$.

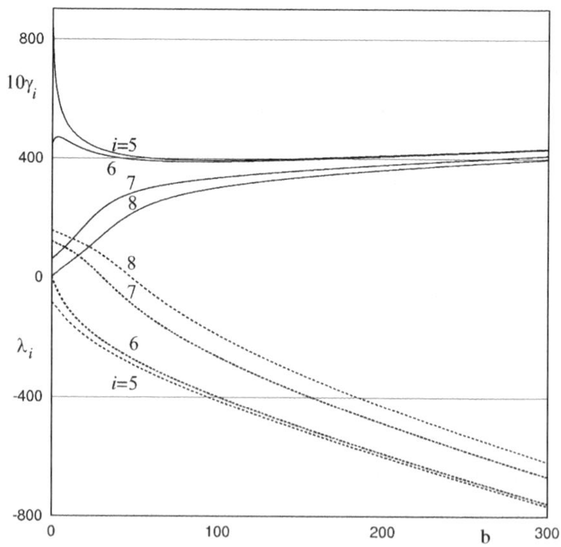

Fig. 1.4 The missing parameters $\gamma_i(b) = y_i'(0)$ (the solid curves) and $\lambda_i(b) = y_i''(0)$ (the dashed curves) as functions of the parameter $b = 2\beta Re$ for fixed values of the angle β_i: $\beta_5 = \beta^+ = 2.3\text{rad} \approx 131.6^\circ$, $\beta_6 = 135^\circ$, $\beta_7 = 165^\circ$, and $\beta_8 = 180^\circ$. The variables γ_i are presented after multiplying by 10 ($10\gamma_i$).

The functions $\gamma_i(b)$ monotonically increase and $\lambda_i(b)$ monotonically decrease as $b \to \infty$ for $i = 1,2,3,4$. The behavior of the functions $\gamma_i(b)$ and $\lambda_i(b)$ for $i = 5,6,7,8$ substantially changes, which reflects the changes in the flow modes.
The functions $\gamma_i(b)$ and $\lambda_i(b)$ have fairly simple shapes. However, the construction of these functions requires rather sophisticated high-precision calculations, with a relative error of 10^{-8} to 10^{-7}. Beginning with $b \sim 100$, typical boundary-layer effects occur. Attempts to construct the desired solutions using the tabulated elliptic integrals fail to provide the required accuracy even for $b \sim 10$.
The further investigation of the motion of a viscous fluid in a convergent channel is associated with the analysis of the velocity profiles $y_i(x)$ for the values β_i of (1.33) and various $b \geq 0$ to follow the evolution and bifurcations of the flow modes depending on $Re > 0$.

1.5 Construction of the Velocity Profiles and Analysis of the Fluid Flow Modes

The analysis of the kinematic and dynamic properties of the flows on the basis of the functions $y(x)$ and $V(\theta)$ related by (1.6) enables one to identify a number of interesting hydrodynamic effects. Using the functions $y(x)$ and $V(\theta)$, one can determine the pressure $p(r,\theta)$, according to (1.9), the components of the strain rate and stress tensors, and other dynamic characteristics of the flow of a viscous fluid in a convergent channel. The expressions for the strain rate and stress tensor components are given in Sect. 1.8, see (1.36) and (1.37). To construct the numerical-analytical solution of the nonlinear boundary-value problem, it is convenient to utilize the normalized variables x and y with the parameters a and b and unknown γ and λ and make use of the accelerated convergence method.
This solution results in the functions $\gamma_i(b)$ and $\lambda_i(b)$ constructed with high degree of accuracy for β_i of (1.33). These functions are plotted in Figs. 1.3 and 1.4. The corresponding families of the velocity profiles $y_i(x)$ are presented in Figs. 1.5 and 1.6 for a number of typical values of b. The basic common property of these functions is that they are symmetric with respect to $x = 1/2$ and, hence, the flow velocity profiles are symmetric with respect to the channel axis $\theta = 0$.
The family of curves $y_1(x)$ for $\beta = \beta_1$ (see (1.33)) and $b = 0,10,300$ (Fig. 5) illustrates the evolution of the flow mode corresponding to high-viscosity fluid ($Re \approx 0$) to the mode corresponding to the low-viscosity fluid ($Re \sim 10^4$). This case of small angle β between the channel walls characterizes the transition from the Poiseuille flow to the Jeffery–Hamel flows. The increase in this angle ($0 < \beta < \pi/4$) does not lead to new qualitative features in the flow mode. For the point β_2, Fig. 1.5 presents the family of curves $y_2(x)$ for $b = 0,10,300$. The flow mode in this case is close to the mode for $\beta = 1^\circ$ considered previously. The curves are presented to illustrate this fact. The major difference is that the curvature of the curve $y_2(x)$ vanishes at the points $x = 0$ and $x = 1$ for $b = 0$, i.e., $\lambda_2 = 0$. As b increases ($b > 0$), this feature disappears and the curve acquires the typical shape (convex upward).

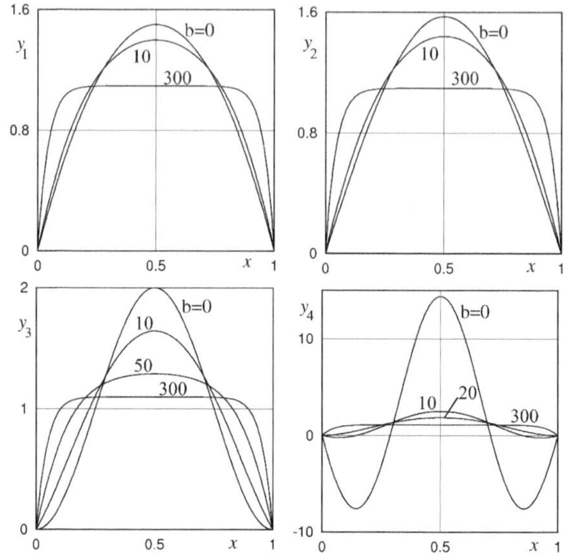

Fig. 1.5 The evolution of the velocity profiles $y_i(x)$ with respect to the parameter b ($0 \leq b \leq 300$) for the angles β_1, β_2, β_3, and β_4. The values of b are given under the curves.

The further increase in the angle β ($\beta_2 < \beta < \beta_3 = \pi/2$) leads to the change of λ in sign and a sharp increase in the local curvature of $y(x)$ at the points $x = 0$ and $x = 1$. For relatively small β, the inequality $\lambda(b) > 0$ holds. However, the parameter $\gamma_0(\beta)$ decreases as β increases, and at the limiting point $\beta_3 = \pi/2$, corresponding to a plane with the sink slot, the quantity γ_3 vanishes. In the middle portion of the interval $0 < x < 1$, for example, at $x = 1/2$, the curve $y(x)$ is convex upward, as

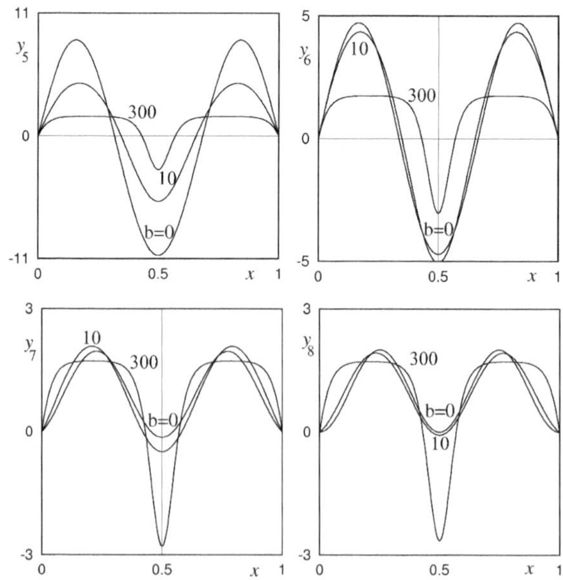

Fig. 1.6 The evolution of the velocity profiles $y_i(x)$ with respect to the parameter b ($0 \leq b \leq 300$) for the angles β_5, β_6, β_7, and β_8. The values of b are given under the curves.

was the case previously (the first interval of (1.32). The increase in the parameter b decreases λ and increases γ (see curves γ_3 and λ_3 in Fig. 1.3); the curves $y(x)$ become convex upward for sufficiently large b. The curves $y(x)$ corresponding to $\beta_3 = \pi/2$ are shown in Fig. 1.5 for $b = 0, 10, 50, 300$.

Unexpected qualitative features in the solution behavior begin to manifest themselves as the angle β continues to increase and becomes greater than the straight angle. For the third interval of (1.32), $\pi/2 < \beta < \beta^*$, we have constructed two families of the curves $y(x)$, corresponding to $\beta_3 = \pi/2$ (this family has been described previously) and β_4. As is apparent from (1.33) and Fig. 1.5, $\beta_4 < \beta^*$. The increase in β leads to a decrease in γ_0 ($\gamma_0 = \gamma(0) < 0$) and the further increase in λ_0 ($\lambda_0 = \lambda(0) > 0$). There appear backward flow regions (sources) at the ends of the interval $0 < x < 1$. These regions correspond to $y(x) < 0$ and are arranged symmetrically with respect to the middle point $x = 1/2$. At the middle portion of the interval, we have a sink ($y(x) > 0$). The width of these regions and velocity amplitudes monotonically increase without limit as $\beta \to \beta^*$. For a fixed β (we have taken β_4 close to β^*), as *Re* or b increase, the profile first takes on the shape corresponding to the second interval of (1.32) (for example, for $b = 20$) and then, for sufficiently large b (for example, $b = 300$), the shape corresponding to the first interval.

As β increases beyond β^* ($\beta^* < \beta < \pi$), the behavior of the solutions changes cardinally (Fig. 1.2). The desired lacking parameters $\gamma_i(b)$ and $\lambda_i(b)$ for the characteristic values β_i of (1.33) for $i = 5, 6, 7, 8$ are presented in Fig. 1.4, while the families of the velocity profiles $y_i(x)$ for similar values of b are shown in Fig. 1.6. These curves are substantially different from the curves considered previously for $i = 1, 2, 3, 4$, i.e., for $0 < \beta < \beta^*$. All these solutions are three-mode ones and this property does not disappear as b or the Reynolds number increase. In addition, there are a sink near the convergent channel walls and a source (as is the case for a divergent channel) in a neighborhood of the axis $x = 1/2$ ($\theta = 0$). For both the source and the sink, the velocity amplitudes increase without limit as $\beta \to \beta^*$ and $b \to 0$. The total flow rate of the source and the sink, of course, is equal to unity (with an error of 10^{-9} to 10^{-7}). Note that the sink velocity amplitude monotonically decreases for b large enough and tends to unity as $b \to \infty$, in accordance with Kochin's asymptotic relation of 1.15. The source velocity amplitude depends on b nonmonotonically and can substantially increase as *Re* increases. The degree of nonmonotonicity of the sink is substantially lower. On the other hand, for large $b \gg 1$, in accordance with (1.15), the velocity amplitude tends to 2 and the width of the source region tends to zero, as is observed for $b = 300$.

We will illustrate the aforementioned specific features of the solutions for a convergent channel with large angle ($\beta > \beta^*$) by examples (Fig. 1.6). For the fourth interval of (1.32) for β, this figure shows two families of curves, $y_5(x)$ and $y_6(x)$, corresponding to $\beta = \beta_5$ and $\beta = \beta_6$ in (1.33). The value $\beta_5 = \beta^+$ slightly exceeds the critical value and leads to substantial amplitudes of the velocity $y_5(x)$ of the source and the sink for $b \ll 1$(the curve $y_5(x)$ for $b = 0$ is presented.) The increase in the parameter $b \sim 1$ leads to a fairly rapid decrease in these amplitudes, since $\gamma_5(b)$ rapidly decreases as b increases (the curve $y_5(x)$ for $b = 10$ is presented). As b increases further, the behavior of the curves $y_5(x)$ changes in accordance with the asymptotic relations of (1.15) (the curve for $b = 300$ is presented). The behavior of the curves $\gamma_5(b)$ and $\lambda_5(b)$ also corresponds to the asymptotic relations.

Of considerable interest from the viewpoint of hydrodynamics is the family of velocity profiles y_6 at the limiting point of the fourth interval of (1.32), $\beta = \beta_6 = 3\pi/4$. As is apparent from Fig. 1.4, we have $\lambda_6 = 0$ at this point and, in addition, the curve $\gamma_6(b)$ is fairly smooth, which indicates that the evolution of the family $y_6(x)$ for $b \sim 10$ is relatively slow. Compare, for example, the change in the amplitudes for $b = 0$ and $b = 10$ at $\beta = \beta^+$ and $\beta = \beta_6$. The sink amplitude monotonically decreases as b increases, whereas the dependence of the source velocity amplitude on this parameter is nonmonotonic.

The fifth interval of values of the angle β in (1.32) corresponds to limiting large flow regions. The major specific feature of the flows in this case, for example, for $\beta = \beta_7$ and $\beta = \beta_8$ (Fig. 1.6) is that the velocity profile for $b \sim 10$ evolves relatively slowly, which is similar to the case of $\beta = \beta_6$ considered previously. As b increases ($b = 300$), the velocity amplitude of the backward (source) flow noticeably increases, whereas the sink velocity amplitudes at the edges (walls) of the convergent channel decrease. In the limit as $b \to \infty$, the asymptotic relation of (1.15) is valid. It is of interest to note the clearly pronounced maxima of the sink velocity that occur at $x \approx 1/4$ and $x \approx 3/4$ (i.e., at $\theta = \pm 90°$) for the values of b up to 100.

Using the obtained solutions $y(x)$ and $V(\theta)$, one can construct geometrical easy-to-perceive (non-normalized) pictures for the profiles or the so-called physical pictures. For small channel angles $\beta \leq \pi/4$, one can confine oneself to the profiles $y_i(x)$ ($i = 1,2$) presented in Fig. 1.5. For large angles $\beta > \pi/2$, a substantial deformation of the curves $y_i(x)$, $i \geq 3$, occurs due to substantial distortion of the zero profile $y \equiv 0$.

We will represent the desired physical profile by the relations

$$X = (r + v_r\,\delta t)\cos\theta, \quad Y = (r + v_r\,\delta t)\sin\theta \tag{1.34}$$

where X and Y are fixed Cartesian coordinates, δt is a time interval, the component $v_r(r,\theta)$ is defined by (1.1) ($v_\theta = 0$), and the value of r is fixed. Substitute the expression of (1.1) for v_r into (1.34) and then normalize the resulting relations by r to obtain the parametric equations for the desired curve

$$\begin{gathered}\frac{X}{r} = \xi(\theta) = \left(1 - \frac{\varepsilon y(x)}{2\beta}\right)\cos\theta, \quad \frac{Y}{r} = \eta(\theta) = \left(1 - \frac{\varepsilon y(x)}{2\beta}\right)\sin\theta, \\ x = \frac{1}{2}\left(1 + \frac{\theta}{\beta}\right), \quad \varepsilon = \frac{Q}{r^2}\delta t = \mathrm{const} > 0.\end{gathered} \tag{1.35}$$

The parameter ε is allowed to be varied and can be chosen so as to make the picture easier to perceive.

Figures 1.7 and 1.8 present the profiles of (1.35) for the cases of $\beta_3 = \pi/2$ and $\beta_4 = \pi$, respectively, with $\varepsilon = 1$. To be specific, in order not to overload the picture, we have presented the curves for the limiting values of b ($b = 0$ and $b = 300$). (These

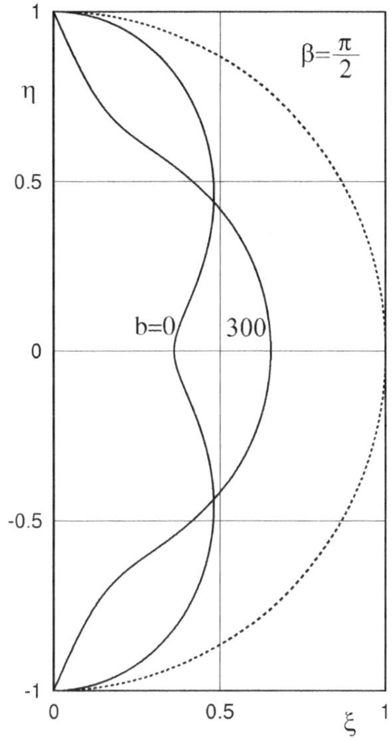

Fig. 1.7 Physical profiles of the flow velocity in the Cartesian coordinates $\xi\eta$ for an "exotic" case where the walls of the convergent channel form a flat angle ($\beta = \beta_3 = \pi/2$). This case corresponds to a plane with the sink slot at the origin. (The outflowing fluid is on the right of the η-axis.) The half-circumference of unit radius corresponds to zero reference level. The profiles are presented for $b = 0$ and $b = 300$.

values label the respective profiles.) These curves demonstrate unusual behavior and indicate interesting hydrodynamic phenomena.

1.6 Numerical-Analytical Solution of the Problem for the Critical Value of the Channel Angle

As has been established previously, the limiting solution of the nonlinear boundary-value problem of (1.11) as $b \to 0$, $y_0(x)$, does not exist for the critical angle β^* of (1.31). For $b > 0$, the desired solution exists but cannot be constructed by means of the standard procedure of continuation with respect to a parameter presented in Sect. 1.3. For $\beta = \beta^*$ and a fixed $b = b^* > 0$, one can apply the accelerated convergence method to the generating solutions corresponding to $\beta^- < \beta^*$ and $\beta^+ > \beta^*$ sufficiently close to β^*; see (1.33). To be specific, we took $b^* = 10$. Then the continuation with respect to the parameter b is applied for $0 < b < b^*$ and $b^* < b < \infty$. Using this approach, we constructed the desired solution — the functions $\gamma^*(b)$, $\lambda^*(b)$, and $y(x)$ — for the critical angle β^* of (1.31) and b ranging in the interval $0.1 \le b \le 300$. This solution was constructed with high degree of accuracy, the relative error ranging between 10^{-9} and 10^{-7}.

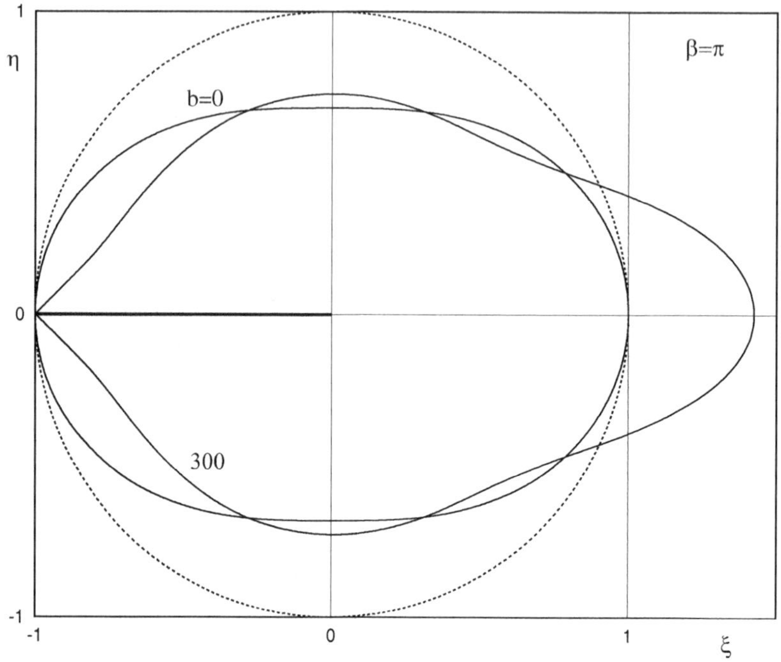

Fig. 1.8: Physical profiles of the flow velocity in the Cartesian coordinates $\xi\eta$ for an "exotic" case where the convergent channel walls form a round angle ($\beta = \beta_8 = \pi$). This case corresponds to two infinitely close half-planes, shown by the negative ξ-half-axis (thick line), with a sink between. The fluid occupies the entire space. The half-circumference of unit radius corresponds to zero reference level. The profiles are presented for $b = 0$ and $b = 300$. For $b = 300$, a substantial source is observed in the region adjacent to the channel axis.

The calculations indicate that to the angle β^* there correspond two solutions — $\gamma^* = \gamma^{\mp}(b)$, $\lambda^* = \lambda^{\mp}(b)$ — shown in Fig. 1.9. These solutions are continuations of the solutions $\gamma_{4,5}(b)$ and $\lambda_{4,5}(b)$, constructed previously for $\beta_{4,5} = \beta^{\pm}$ (Figs. 1.3 and 1.4), with respect to the parameter β. It is interesting that $\gamma^{\mp}(b) \to \mp\infty$ and $\lambda^{\mp}(b) \to \pm\infty$ as $b \to 0$ and, hence, the steady flow degenerates for sufficiently small *Re*. In this case, the blocking-type phenomenon is observed, with the pressure and the total force increasing without limit; for more detail, see Sect. 1.8. This phenomenon is of interest from the viewpoint of hydrodynamics. Note that the function $\gamma^{-}(b)$ has a rather shallow minimum at $b \sim 1000$ ($\gamma^{\mp} \sim \sqrt{4b/3} \to \infty$ as $b \to \infty$), whereas the function $\lambda^{+}(b)$ has a sharp maximum at $b \sim 3$ ($\lambda^{\mp} \sim -b \to -\infty$ as $b \to \infty$).

The families of the velocity profiles $y^* = y^{\mp}(x)$ corresponding to different solutions $\gamma^{\mp}$ and $\lambda^{\mp}$ are shown in Fig. 1.10 for $b = 0.1, 0.2, 0.3$. The analysis of these solutions for $b \sim 1$ and $b \ll 1$ leads to the conclusions similar to those for the profiles $y_4(x)$ and $y_5(x)$. However, for $b \ll 1$ (we took $b = 0.1$), a large swing in the source

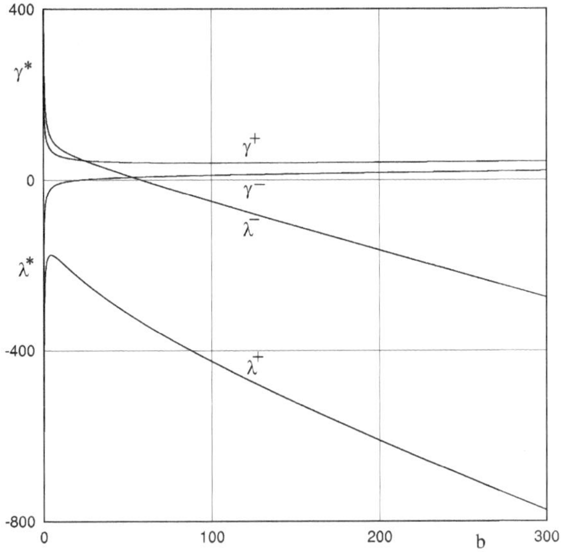

Fig. 1.9 The missing parameters $\gamma^* = \gamma^{\pm}(b)$ and $\lambda^* = \lambda^{\pm}(b)$ corresponding to the critical value $\beta = \beta^*$ as functions of the parameter $b = 2\beta^* Re$.

and sink velocities is observed, this swing increasing without limit as $b \to 0$. The calculations for $b < 0.1$ are complicated, since the quantities $\gamma^{\mp}$ and $\lambda^{\mp}$ sharply increase.

Thus, we have established a new property of steady flows in a plane convergent channel. This property needs further theoretical and experimental investigations. The occurrence of two flow modes for $\beta = \beta^*$ and $b > 0$ makes it necessary to investigate the stability of these solutions and their physical feasibility.

The physical profiles $\xi^{\mp}(\theta)$ and $\eta^{\mp}(\theta)$, corresponding to two solutions $y^{\mp}(x)$, have been constructed on the basis of the relations of (1.35). The respective curves are shown in Figs. 1.11 and 1.12 for the limiting values of b, $b = 0.1$ and $b = 300$. These values are indicated in the figures. Since the variation of the functions $y^{\mp}(x)$ in the case of $b = 0.1$ is an order of magnitude greater than that in the case of $b \ll 1$, we have taken $\varepsilon = 0.1$ to make the curves easier to perceive. For $b = 300$, we have taken $\varepsilon = 1$, as was the case previously (Figs. 1.7 and 1.8).

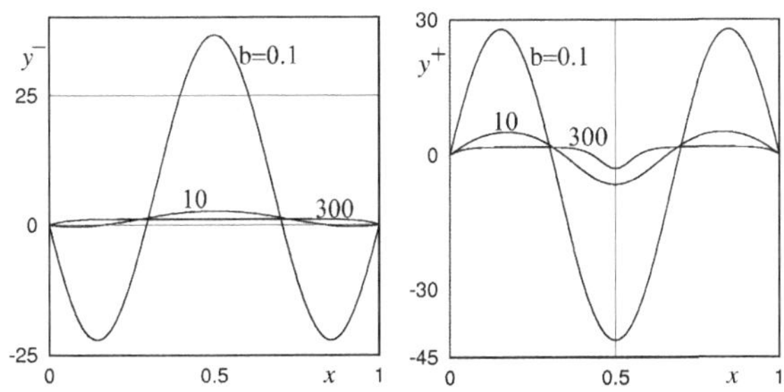

Fig. 1.10 The evolution of the velocity profiles $y^{\pm}(x)$ for the angle $\beta = \beta^*$ depending on the parameter $b = 2\beta^* Re$. The profile $y^-(x)$ becomes single-mode as b increases (at $b \approx 20$), while the profile $y^+(x)$ is three-mode for all $b > 0$.

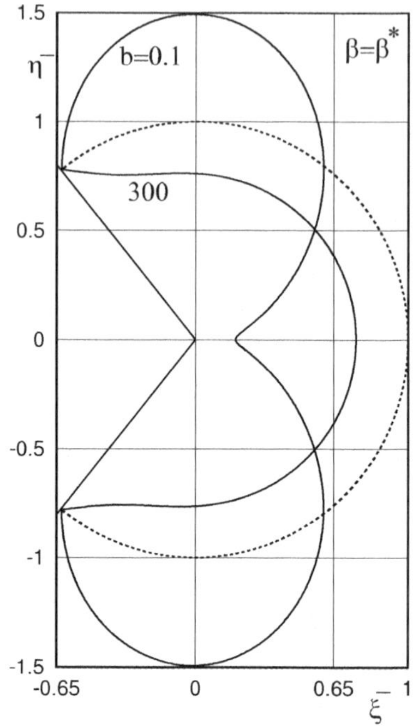

Fig. 1.11 Physical profiles of the flow velocity components $\xi^-(\theta)$ and $\eta^-(\theta)$, corresponding to $\gamma^-(b)$, $\lambda^-(b)$, and $y^-(x)$, in the Cartesian coordinates $\xi\eta$ for the critical value of $\beta = \beta^*$. The portion of the circumference of unit radius corresponds to zero reference level. The profiles are presented for $b = 0.1$ and $b = 300$. For $b = 0.1$, noticeable backward flows (source) occur. For $b = 300$, a sink characteristic of low-viscosity fluid flow is observed.

Note that three-mode solutions demonstrate rather exotic behavior of the profile and, undoubtedly, are of considerable interest for mechanics of viscous fluids. The source effect for the profile of ξ^+ and η^+ turns out to be significant for all b that have been considered, including $b \sim 10^2 - 10^3$.

1.7 New Multi-Mode Asymmetric Solutions that Cannot be Regularly Continued with Respect to *Re*

To fix our ideas, we will consider the channel angle $\beta = 10°$, i.e., $a \approx 0.7$, which is frequently the case in applications. Numerous calculations with high degree of accuracy were performed also for other values of a, with b being varied in broad range, $0 < b \leq 200$, which corresponds to $0 < Re \leq 600$. The high-accuracy solution (with a relative error of 10^{-8} to 10^{-7} or an absolute error of an order of 10^{-5}) of the problem of (1.7) and (1.8) can be constructed by means of the modified accelerated convergence Newton-type algorithm combined with the continuation with respect to a parameter. This algorithm has been successfully applied to the classical problem of single-mode symmetric flow, for which the exact limiting (as $b \to 0$) solution is available in closed form for any a.

Substantial difficulties encountered when constructing multi-mode velocity profiles are accounted for by the degeneration of the problem as $b \to 0$. In this case, the de-

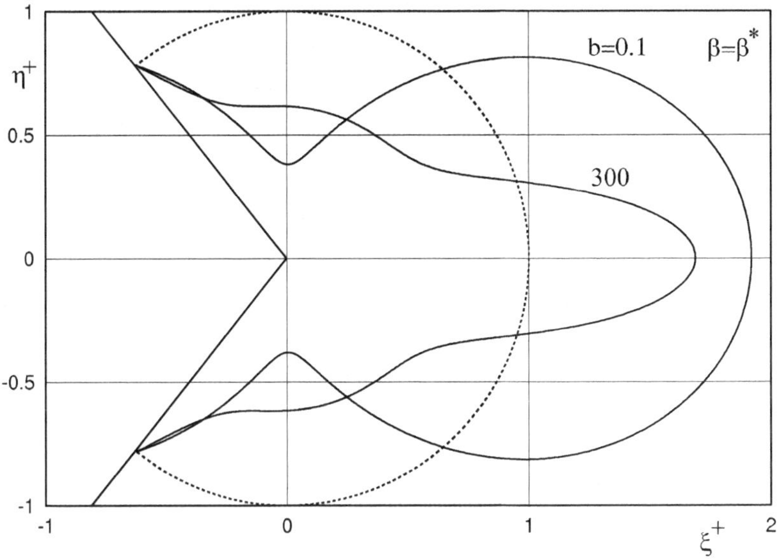

Fig. 1.12: Physical profiles of the flow velocity components $\xi^+(\theta)$ and $\eta^+(\theta)$, corresponding to $\gamma^+(b)$, $\lambda^+(b)$, and $y^+(x)$, in the Cartesian coordinates $\xi\eta$ for the critical value of $\beta=\beta^*$. The portion of the circumference of unit radius corresponds to zero reference level. The profiles are presented for $b=0.1$ and $b=300$. For $b=0.1$, a sink of the fluid is observed near the walls of the convergent channel, whereas noticeable backward flows (source) occur near the channel axis. These backward flows are significant also for $b=300$.

sired parameters $\gamma(b)$ and $\lambda(b)$, necessary for solving the corresponding initial value problem, increase without limit. For some fixed value of $b=b_0$ (usually $b_0\approx 10$) and the selected mode number ($n=2,3,4,\ldots$), we search for the unknown parameters γ and λ and calculate these parameters with high degree of accuracy to satisfy the boundary conditions with a prescribed error. Then we apply the continuation with respect to the parameter b to construct the universal curves $\gamma(b)$ and $\lambda(b)$ for $0<b<b_0$ and $b_0<b<\infty$. The computational difficulties are aggravated by the fact that γ_n and λ_n are several orders of magnitude greater that γ_1 and λ_1, respectively, and become very large for $b=1$; for example, $\gamma_5\sim 10^4$ and $\lambda_5\sim 10^5$. Possibly, it is for this reason that trustworthy results related to the identification and analysis of multi-mode solutions of the Jeffery problem have been absent so far.

In what follows, we give and comment a graphical representation of the results of the numerical-analytical investigation of multi-mode solutions for $n=2,3,4,5$. The curves corresponding to the multi-mode solutions are compared with the respective curves corresponding to the classical solution ($n=1$) for various values of a. It has been established that for $n\geq 2$, the solutions corresponding to the modes with odd n are symmetric (with respect to $x=1/2$, which corresponds to $\theta=0$), whereas the solutions corresponding to even n are asymmetric.

Analysis indicates certain structural properties of the flows. Specifically, the positive maxima ($n \geq 3$) and the negative minima ($n \geq 4$) of the functions $y_n(x,b)$ have the same absolute values. In addition, at all points x_i, $i = 1,\ldots,n+1$, at which the function $y_n(x,b)$ vanishes (for fixed n and b), the derivatives y_n' are equal in absolute values and, hence, $\gamma_n(b) = \pm y_n'(x_i,b)$. Thus, the multi-mode solution can be regarded as a combination (superposition) of a single-mode and two-mode solutions, which, apparently, is common to radial flows.

Figures 1.13 and 1.14 present the graphs of the functions $\gamma_n(b)$ and $\lambda_n(b)$ on different scales for $1 \leq b \leq 10$ and $10 \leq b \leq 200$. These curves characterize the solution of the boundary-value problem of (1.7) and (1.8) obtained by integrating the initial value problems for fixed a (i.e., for fixed β). These curves group in pairs for $n = 2,3$ and $n = 4,5$. They demonstrate an interesting behavior near $b = 0$ and have vertical asymptotes, specifically, $\gamma_n \to +\infty$ and $\lambda_n \to -\infty$ as $b \to 0$. Recall that in the classical single-mode solution, the quantities γ_1 and λ_1 have finite limits as $b \to 0$. Note also that γ_n and λ_n have large absolute values for $b \sim 1$ and that the differences

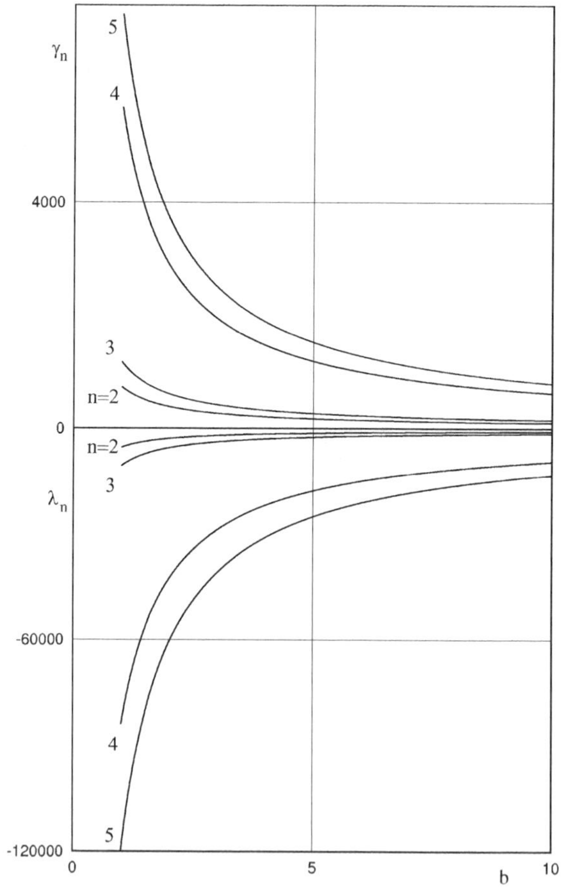

Fig. 1.13 The missing parameters $\gamma_n(b) = y_n'(0)$ and $\lambda_n(b) = y_n''(0)$ as functions of the parameter $b = 2\beta Re$ for n-mode flows ($n = 2,3,4,5$) in the convergent channel with the opening half-angle $\beta = 10°$. The curves $\gamma_n(b)$ ($\gamma_n > 0$) and $\lambda_n(b)$ ($\lambda_n < 0$) are presented on different scales for the range $1 \leq b \leq 10$. These flow modes cannot be continued with respect to the Reynolds number as $Re \to 0$.

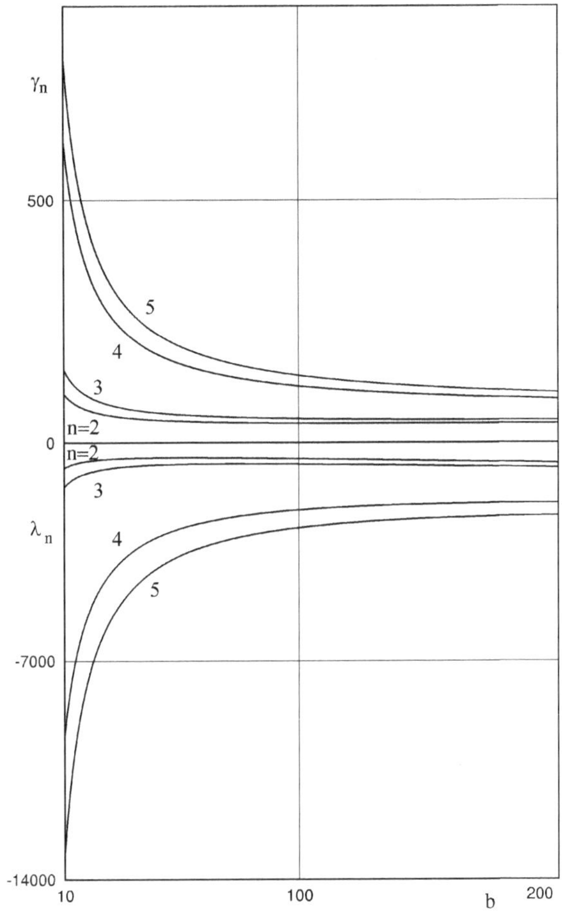

Fig. 1.14 The missing parameters $\gamma_n(b) = y_n'(0)$ and $\lambda_n(b) = y_n''(0)$ as functions of the parameter $b = 2\beta Re$ for n-mode flows ($n = 2,3,4,5$) in the convergent channel with the opening half-angle $\beta = 10°$. The curves $\gamma_n(b)$ ($\gamma_n > 0$) and $\lambda_n(b)$ ($\lambda_n < 0$) are presented in different scales for the range $10 \leq b \leq 200$. The starting points of the curves in this figure correspond to the end points of the respective curves in Fig. 1.13. The scales along the b-axis are substantially different for Figs. 1.13 and 1.14.

$\gamma_n - \gamma_{n-1}$ and $\lambda_n - \lambda_{n-1}$ increase without limit as $b \to 0$ or $n \to \infty$. Each curve γ_n attains a minimum at some fairly large b ($b = b_n^\gamma \sim 10^2 - 10^3$) and then rather slowly converges from above to the asymptote $\gamma_\infty = \sqrt{4b/3}$. In a similar way, the curves λ_n reach a maximum at some $b = b_n^\lambda \sim 10^2$ and then rather slowly converge from below to the asymptote $\lambda_\infty = -b$. This implies that all steady flow modes, including the basic mode ($n = 1$), converge (in an appropriate metric) to an ideal fluid flow for $0 < x < 1$ as $b \to \infty$ ($Re \to \infty$). This observation is important from the standpoint of hydromechanics.

In fact, the curves $\gamma_n(b)$ and $\lambda_n(b)$ are the major result of the investigations, on the basis of which, by solving the initial value problem for Eq. (1.7) subject to the initial condition of (1.8), one can obtain various characteristics of steady flows in a convergent channel, such as the velocity profile of (1.1), the pressure of (1.2), and the components of the strain rate and stress tensors; see Sect. 1.8. The shape of the curves $\gamma_n(b)$ and $\lambda_n(b)$ is rather simple, but the construction of these curves

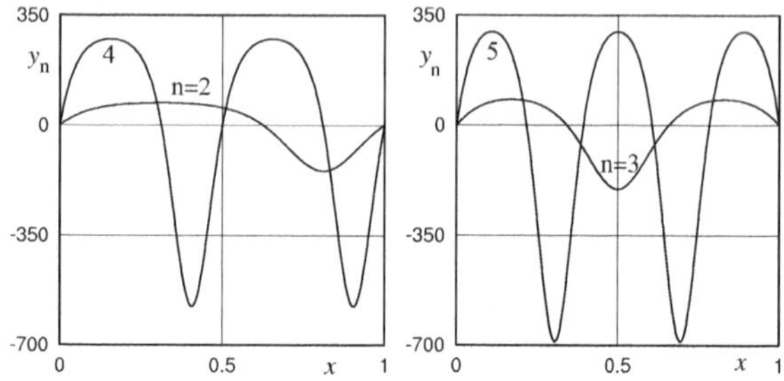

Fig. 1.15 The velocity profiles $y_n(x)$ of multi-mode flows for relatively small b ($b = 1$). Substantial source regions and large ranges of the functions y_n are characteristic of this case; $y_4(1/2) = 0$.

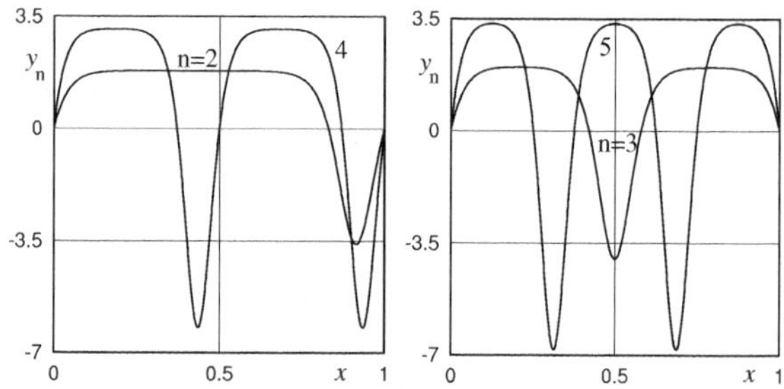

Fig. 1.16 The velocity profiles $y_n(x)$ of multi-mode flows for relatively large b ($b = 200$). The source regions are becoming narrower and large ranges of the functions are substantially reduced (approximately, by a factor of 100), as compared with Fig. 1.15; $y_4(1/2) = 0$.

needs rather complicated calculations with high degree of accuracy, the complexity of these calculations increasing as $b \to 0$ or $b \to \infty$. The solution of the problem under consideration demonstrates clearly expressed boundary-layer and internal-layer phenomena. The computational algorithms based on the familiar methods of the functional analysis (e.g., the Bubnov–Galerkin method), finite element method, or finite-difference methods fail to provide acceptable results.

The character of the multi-mode flows is illustrated in Figs. 1.15 and 1.16. In these figures, the velocity profiles $y_n(x)$ for $n = 2,3,4,5$ are plotted for two values of b, for relatively small $b = 1$ (Fig. 1.15) and for relatively large $b = 200$ (Fig. 1.16). For small b (or Re), the amplitude of oscillation of the velocity functions $y_n(x)$ between positive and negative values is large. The positive ($y_n > 0$) and negative ($y_n < 0$) values of the velocity define the sink and source regions, respectively. As the parameter b increases, the swing of the functions $y_n(x)$ is reduced and the source amplitude decreases. For large b ($b \sim 10^2$ — 10^3), the modes corresponding to the convergent flow of a low-viscosity fluid are pronounced. The deviation of the velocity profile from the rectangular profile corresponding to the flow of a zero-viscosity (ideal) fluid tends to zero (in a certain metric) for $0 < x < 1$, as b increases, with $\max y \to 1$, and $\min y \to -2$. Typical boundary-layer phenomena are observed near the points $x = 0$ ($\theta = -\beta$) and $x = 1$ ($\theta = \beta$).

1.8 Kinematic and Force Characteristics of Steady Flows

Having determined the function $y(x)$, one can calculate the pressure p of (1.9) and the velocity component $v_r(r,\theta)$ of (1.1), using the relation of (1.6) between y and V. Then one can determine the kinematic and dynamic characteristics. For example, the strain rate tensor components have the form

$$v_{rr} = -v_{\theta\theta} = \frac{Q}{r^2}V(\theta), \quad v_{r\theta} = -\frac{Q}{2r^2}V'(\theta). \tag{1.36}$$

By substituting (1.36) into the constitutive relations for viscous fluid one can find the stress tensor components

$$\sigma_{rr;\theta\theta} = -p \pm \frac{2\rho Q^2}{r^2 Re}V(\theta), \quad \sigma_{zz} = -p, \quad \sigma_{r\theta} = -\frac{\rho Q^2}{r^2 Re}V'(\theta). \tag{1.37}$$

As a result, one can find the force characteristics of the fluid. In accordance with (1.37), the components P_r and P_θ of the stress vector $\mathbf{P}$ at any point of the circular arc $r = \text{const}$, $|\theta| < \beta$ have the form

$$\begin{aligned} P_r(r,\theta) = \sigma_{rr}(r,\theta) = \frac{\rho Q^2}{r^2 Re}\left(4V(\theta) - \frac{C}{2}\right), \\ P_\theta(r,\theta) = \sigma_{r\theta}(r,\theta) = -\frac{\rho Q^2}{r^2 Re}V'(\theta). \end{aligned} \tag{1.38}$$

At the boundary points of this arc ($\theta = \pm\beta$), in accordance with (1.3) and (1.38), we have

$$P_r(r,\pm\beta) = -\frac{\rho Q^2 C}{2r^2 Re}, \quad P_\theta(r,\pm\beta) = -\frac{\rho Q^2}{r^2 Re}V'(\pm\beta). \tag{1.39}$$

It is apparent from these expressions that the values of P_r are the same at both ends of the arc, whereas the values of P_θ are different for $n = 1,3,5,\ldots$.

In accordance with (1.38), the components F_r and F_θ of the resultant force $\mathbf{F}$ (more precisely, the density of this force along the z-axis) at the distance r away from the vertex are expressed in terms of the constant C and other parameters by

$$F_r(r) = \int_{-\beta}^{\beta} rP_r\,\mathrm{d}\theta = \frac{\rho Q^2}{rRe}(4-\beta C), \quad F_\theta(r) = \int_{-\beta}^{\beta} rP_\theta\,\mathrm{d}\theta = 0. \tag{1.40}$$

By analogy with (1.38), we calculate the density N of the power of the forces due to the stresses $\mathbf{P}$

$$N = \int_{-\beta}^{\beta} r\sigma_{rr}v_r\,\mathrm{d}\theta = \frac{\rho Q^3}{r^2 Re}\left(\frac{C}{2} - \frac{4}{Re}\left(V'(\beta) - V'(-\beta) + 4 - 2\beta C\right)\right). \tag{1.41}$$

The relations of (1.36)–(1.41) determine the force characteristics and the power required for sustaining the steady flow mode.

1.9 Conclusions

An interest in the Jeffery problem, its numerous modern modifications, and other classical problems of continuum mechanics has increased in recent years. This can be explained, on the one hand, by increased potentials of computers and the development of relevant software and, on the other hand, by practical requirements of having solved a wide class of problems associated, for example, with unsteady motions, non-isothermal deformation, and extrusion of non-Newtonian and viscoplastic materials from plane slots and convergent channels. The solution of the Jeffery problem is a reference (zero) approximation for such motions of a medium. Therefore, it is extremely important to know the characteristic features of this solution in various ranges of the parameters.

The numerical and analytical solutions of the boundary-value problem found in the present paper can serve as reference (zero) approximations for more complicated problems related to heat and mass transfer (Millsaps and Pohlhausen, 1953), essentially unsteady flows (Shapeev, 2004), and extrusion of an incompressible viscoplastic material with low yield stress from a plane convergent channel (Akulenko et al, 2003a).

Acknowledgements We thank Professor N.N. Bolotnik for his useful comments on the text of this paper.

This research was partly supported by grants of the Russian Foundation for Basic Research (Nos. 17-01-00538, 16-01-00412 and 15-01-00848).

References

Akulenko LD, Georgievskii DV, Kumakshev SA, Nesterov SV (2000) Numerical and analytical investigation of steady-state flow of viscous fluid in a plane confuser. Doklady Physics 45:467–471

Akulenko LD, Georgievskii DV, Kumakshev SA, Nesterov SV (2002a) The flow of a viscous fluid inside a confuser with large opening angle. Doklady Physics 47:698–701

Akulenko LD, Georgievskii DV, Kumakshev SA, Nesterov SV (2002b) New nonsymmetric and multimode solutions of the problem of the flow of a viscous fluid in a plane confuser. Doklady Physics 47:219–223

Akulenko LD, Kumakshev SA, Nesterov SV (2002c) Effective numerical-analytical solution of isoperimetric variational problems of mechanics by an accelerated convergence method. J Appl Maths Mechs 66:693–708

Akulenko LD, Georgievskii DV, Klimov DM, Kumakshev SA, Nesterov SV (2003a) Stationary flow of a viscoplastic medium with small yield stress in a plane confusor. Russ J Math Phys 10:381–397

Akulenko LD, Georgievskii DV, Kumakshev SA, Nesterov SV (2003b) Nonsymmetric and multi-mode converging nozzle flows in the Jeffery–Hamel problem. Moscow Univ Mechanics Bull 58(2):1–3

Akulenko LD, Georgievskii DV, Kumakshev SA, Nesterov SV (2004) Solutions of the Jeffery–Hamel problem regularly extendable in the Reynolds number. Fluid Dynamics 39:12–28

Banks WHH, Drazin PG, Zaturska MB (1988) On perturbation of Jeffery–Hamel flow. J Fluid Mech 186:559–581

Batchelor GK (1970) An Introduction to Fluid Dynamics. Cambridge Univ. Press, Cambridge

Dennis SCR, Banks WHH, Drazin PG, Zaturska MB (1997) Flow along a diverging channel. J Fluid Mech 336:183–202

van Dyke M (1964) Perturbation Methods in Fluid Mechanics. Acad. Press

Eagles PM (1966) The stability of a family of Jeffery–Hamel solutions for divergent channel flow. J Fluid Mech 24:199–207

Eagles PM (1988) Jeffery–Hamel boundary-layer flows over curved beds. J Fluid Mech 186:583–597

Fraenkel LE (1962) Laminar flow in symmetric channels with slightly curved walls. Part I. On the Jeffery–Hamel solutions for flow between plane walls. Proc Royal Soc London A267:119–138

Fraenkel LE (1963) Laminar flow in symmetric channels with slightly curved walls. Part II. An asymptotic series for the stream function. Proc Royal Soc London A272:406–428

Georgievskii DV (1998) Stability of Viscoplastic Solids Deformation Processes (in Russ.). URSS Publ., Moscow

Hamel G (1917) Spiralförmige Bewegungen zähen Flüssigkeiten. Jahr-Ber Deutsch Math Ver 25:34–60

Jeffery GB (1915) The two-dimentional steady motion of a viscous fluid. Phil Mag Ser 6 29(172):455–465

Kerswell RR, Tutty OR, Drazin PG (2004) Steady nonlinear waves in diverging channel flow. J Fluid Mech 501:231–250

Kochin NE, Kibel IA, Roze NV (1965) Theoretical Hydrodynamics. John Wiley, New York

Landau LD, Lifschitz EM (1987) Fluid Mechanics, 2nd edn. Pergamon Press, Oxford

Loitsyanskii LG (1966) Mechanics of Liquids and Gases. Pergamon Press, Oxford

Millsaps K, Pohlhausen K (1953) Thermal distributions in Jeffery–Hamel flows between nonparallel plane walls. J Aeronaut Sci 20:187–196

Rektorys K (1980) Variational Methods in Mathematics, Science and Engineering, 2nd edn. Springer Netherlands

Rivkind L, Solonnikov VA (2000) Jeffery–Hamel asymptotics for steady state Navier–Stokes flow in domains with sector-like outlets in infinity. J Math Fluid Dynamics 2:324–352

Rosenhead L (1940) The steady two-dimensional radial flow of viscous fluid between two inclined plane walls. Proc Royal Soc London A175:436–467

Shapeev AV (2004) Unsteady self-similar flow of a viscous incompressible fluid in a plane divergent channel. Fluid Dynamics 39:36–41

Sobey IJ, Drazin PG (1986) Bifurcation of two-dimensional channel flows. J Fluid Mech 171:263–287

Tutty OR (1996) Nonlinear development of flow in channels with non-parallel walls. J Fluid Mech 326:265–284

Chapter 2
Riemann's Method in Plasticity: a Review

Sergei Alexandrov

Abstract This paper deals with models of pressure-independent and pressure dependent plasticity under plane strain conditions and provides a review of quantities satisfying the equation of telegraphy. This equation can be solved by the method of Riemann. In particular, the Green's function for the equation of telegraphy is the Bessel function of zero order. An advantage of using the method of Riemann for solving boundary value problems is a high accuracy of solutions. Therefore, solutions found by this method can be used for verifying the accuracy of other methods. Some results presented in this paper are restricted to rigid plastic solids whereas others are independent of whether elastic strains are included. The last section of the paper concerns with the theory of ideal flows.

Key words: Telegraph equation · Riemann method · Plane strain · Pressure-Independent plasticity · Pressure-Dependent plasticity · Ideal flows

2.1 Preliminary Remarks

The present paper focuses on three systems of constitutive equations of plastic solids under plane strain conditions. One of these systems is the classical theory of perfectly plastic solids. A great account on this model has been given in Hill (1950). The results for this model are valid for any yield criterion in the case of rigid plasticity. However, they are independent of whether elastic strains are included for the Tresca yield criterion. The other systems are based on pressure-dependent yield criteria. One of these criteria is the Coulomb-Mohr yield criterion. This criterion together with the stress equilibrium equations under plane strain conditions forms

Sergei Alexandrov
Ishlinsky Institute for Problems in Mechanics RAS, Vernadskiy Ave, 101 bldg 1, 119526 Moscow, Russia,
e-mail: sergei_alexandrov@yahoo.com

H. Altenbach et al. (eds.), *Mechanics for Materials and Technologies*,
Advanced Structured Materials 46, DOI 10.1007/978-3-319-56050-2_2

a statically determinate system. Therefore, the results presented for this system are restricted to stress analysis but are independent of any flow rule that may be chosen to calculate the deformation and also independent of whether elastic strains are included. Reviews of models of pressure-dependent plasticity based on the Coulomb-Mohr yield criterion have been provided in Cox et al (2008) and Goddard (2014). The other pressure-dependent yield criterion adopted in the present paper has been proposed in Druyanov (1993). This criterion is appropriate for powder and porous metals. The corresponding plane strain criterion is derived by means of the associated flow rule. The results for this model are restricted to rigid plasticity. The classical theory of rigid perfectly plastic solids based on the Tresca yield criterion permits a class of solutions named ideal flows. In particular, ideal flows are solenoidal smooth plastic flows in which an eigenvector field associated everywhere with the greatest principal strain rate is fixed in the material. A review of the ideal flow theory has been provided in Chung and Alexandrov (2007). Under plane strain conditions the yield criterion is immaterial. Therefore, the results given in the present paper for ideal flows are valid for any yield criterion.

All the systems of equations considered in the present article are hyperbolic. In what follows, characteristic-based coordinate systems will be denoted by (α,β).

There are three basic boundary value problems. Those are: (i) two intersecting characteristic curves are given, (ii) the stresses are given along a certain curve (this curve should not coincide with a characteristic curve), and (iii) one characteristic curve together with a curve along with the orientation of the $\alpha-$line is known is given. Many quantities involved in the aforementioned theories satisfy the equation of telegraphy:

$$\frac{\partial^2 f}{\partial\alpha\partial\beta}+f=0. \tag{2.1}$$

This equation can be solved by the method of Riemann. In particular, the Green's function for the equation of telegraphy is (Hill, 1950)

$$G(a,b,\alpha,\beta)\equiv J_0\left[2\sqrt{(a-\alpha)(b-\beta)}\right] \tag{2.2}$$

where $J_0(z)$ is the Bessel function of zero order. A property of the Green's function is

$$\oint\left[\left(G\frac{\partial f}{\partial\alpha}-f\frac{\partial G}{\partial\alpha}\right)\mathrm{d}\alpha+\left(f\frac{\partial G}{\partial\beta}-G\frac{\partial f}{\partial\beta}\right)\right]\mathrm{d}\beta=0. \tag{2.3}$$

Here the integral is taken round any closed contour. Choosing an appropriate contour the aforementioned boundary value problems can be solved by means of equation (2.3).

The intention given in the present paper has been to define each new symbol where it first appears in the text. There are, however, certain symbols that re-appear consistently throughout the text. These symbols are defined here. In particular, σ_1 is the algebraically greatest principal stress, ψ is the angle between the axis corresponding to the principal stress σ_1 and the x of Cartesian coordinates (x,y) measured from the x-axis anticlockwise, ϕ is the angle between the α-direction of the characteristic based coordinate system (α, β) and the x-axis measured from the x-axis anticlockwise, R is the radius of curvature of the α-characteristic lines, and S is the radius of curvature of the β-characteristic lines. Note that R and S are algebraic quantities whose signs depend on the sense of space derivatives.

2.2 Pressure-Independent Plasticity

A great account on the theory of pressure-independent rigid perfectly plastic solids has been given in Hill (1950). Plane strain deformation of such solids is described by a hyperbolic system of equations. Any plane strain yield criterion can be written in the form

$$(\sigma_{xx} - \sigma_{yy})^2 + 4\sigma_{xy}^2 = 4k^2 \tag{2.4}$$

where σ_{xx}, σ_{yy} and σ_{xy} are the components of the stress tensor in Cartesian coordinates (x, y) and k is the shear yield stress, a material constant. The results presented below are also valid for elastic/plastic deformation if the Tresca yield criterion is adopted. The flow rule associated with the yield criterion (2.4) is

$$\xi_{xx} = \lambda(\sigma_{xx} - \sigma_{yy}), \quad \xi_{yy} = -\lambda(\sigma_{xx} - \sigma_{yy}), \quad \xi_{xy} = 2\lambda\sigma_{xy}, \tag{2.5}$$

where ξ_{xx}, ξ_{yy} and ξ_{xy} are the components of the strain rate tensor in the Cartesian coordinates and λ is a non-negative multiplier. Eliminating λ in (2.5) and expressing the strain rate components in terms of the velocity components relative the Cartesian coordinates, u_x and u_y, yield

$$\frac{\partial u_x}{\partial x} + \frac{\partial u_y}{\partial y} = 0, \quad 2\sigma_{xy}\left[\frac{\partial u_x}{\partial x} - \frac{\partial u_y}{\partial y}\right] = (\sigma_{xx} - \sigma_{xy})\left[\frac{\partial u_x}{\partial y} + \frac{\partial u_y}{\partial x}\right]. \tag{2.6}$$

The equilibrium equations are

$$\partial\sigma_{xx}/\partial x + \partial\sigma_{xy}/\partial y = 0, \quad \partial\sigma_{xy}/\partial x + \partial\sigma_{yy}/\partial y = 0. \tag{2.7}$$

The system of equations consisting of (2.4), (2.6) and (2.7) is hyperbolic. The characteristics of the stresses and the velocities coincide. Therefore, there are only two distinct characteristic directions at a point. The characteristic curves are determined by the equations:

$$\mathrm{d}y/\mathrm{d}x = \tan(\psi - \pi/4), \mathrm{d}y/\mathrm{d}x = \tan(\psi + \pi/4). \tag{2.8}$$

The orientation of the algebraically greatest principal stress and characteristic curves relative to the Cartesian coordinate system is illustrated in Fig. 2.1.

In what follows it is assumed that the two families of characteristics are curved. In this case the equilibrium equations (2.7) are equivalent to Hill (1950)

$$p - p_0 + 2k(\psi - \psi_0) = 4k\beta, \quad p - p_0 - 2k(\psi - \psi_0) = -4k\alpha. \tag{2.9}$$

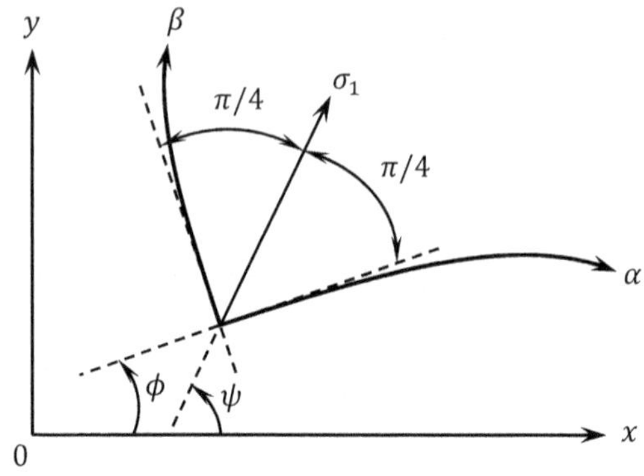

Fig. 2.1 Orientation of the algebraically greatest principal stress σ_1 and characteristic based coordinates (α,β) relative to Cartesian coordinates (x,y) for yield criterion (2.4)

Here $p = -(\sigma_{xx}+\sigma_{yy})/2$, p_0 and ψ_0 are constant. Solving (2.9) for $p-p_0$ and $\psi-\psi_0$ yields

$$p-p_0 = 2k(\beta-\alpha), \quad \psi-\psi_0 = \beta+\alpha. \tag{2.10}$$

Let ϕ be the anticlockwise angular rotation of the α-line from the x-axis (Fig. 2.1). It is seen from (2.8) that the characteristic curves are orthogonal. Therefore, the anticlockwise angular rotation of the β-line from the x-axis is $\phi+\pi/2$. The radii of curvature R and S of the α- and β-lines are defined by the equations:

$$R^{-1} = \partial\phi/\partial s_\alpha, \quad S^{-1} = -\partial(\phi+\pi/2)/\partial s_\beta = -\partial\phi/\partial s_\beta \tag{2.11}$$

where $\partial/\partial s_\alpha$ and $\partial/\partial s_\beta$ are space derivatives taken along the α- and β-lines respectively. It is seen from (2.8) that

$$\psi = \phi + \pi/4. \tag{2.12}$$

Therefore, Eqs. (2.11) are equivalent to

$$R^{-1} = \partial\psi/\partial s_\alpha, \quad S^{-1} = -\partial\psi/\partial s_\beta. \tag{2.13}$$

It follows from the geometry of Fig. 2.1 that

$$\begin{aligned}\partial x/\partial s_\alpha &= \cos(\psi-\pi/4),\ \partial x/\partial s_\beta = -\sin(\psi-\pi/4),\\ \partial y/\partial s_\alpha &= \sin(\psi-\pi/4),\ \partial y/\partial s_\beta = \cos(\psi-\pi/4),\end{aligned}$$

Using (2.10) and (2.13) these equations are transformed to

$$\begin{aligned}\partial x/\partial\alpha &= R\cos(\psi-\pi/4),\ \partial x/\partial\beta = S\sin(\psi-\pi/4),\\ \partial y/\partial\alpha &= R\sin(\psi-\pi/4),\ \partial y/\partial\beta = -S\cos(\psi-\pi/4).\end{aligned} \tag{2.14}$$

The compatibility equations are

$$\frac{\partial^2 x}{\partial\alpha\partial\beta} = \frac{\partial^2 x}{\partial\beta\partial\alpha}, \quad \frac{\partial^2 y}{\partial\alpha\partial\beta} = \frac{\partial^2 y}{\partial\beta\partial\alpha} \tag{2.15}$$

Substituting (2.14) into these equations and taking into account (2.10) result in

$$\begin{aligned}\frac{\partial R}{\partial\beta}\cos\left(\psi-\frac{\pi}{4}\right) - R\sin\left(\psi-\frac{\pi}{4}\right) &= \frac{\partial S}{\partial\alpha}\sin\left(\psi-\frac{\pi}{4}\right) + S\cos\left(\psi-\frac{\pi}{4}\right),\\ \frac{\partial R}{\partial\beta}\sin\left(\psi-\frac{\pi}{4}\right) + R\cos\left(\psi-\frac{\pi}{4}\right) &= -\frac{\partial S}{\partial\alpha}\cos\left(\psi-\frac{\pi}{4}\right) + S\sin\left(\psi-\frac{\pi}{4}\right)\end{aligned}$$

Solving these equations for $\partial R/\partial\beta$ and $\partial S/\partial\alpha$ gives

$$\partial R/\partial\beta = S, \quad \partial S/\partial\alpha = -R. \tag{2.16}$$

It is evident from these equations that R and S separately satisfy (2.1). Once R and S have been found, the dependence of x and y on α and β is determined from (2.14) where ψ should be eliminated by means of (2.10). The dependence of the stresses on α and β is found from

$$\sigma_{xx} = -p + \cos 2\psi, \quad \sigma_{yy} = -p - \cos 2\psi, \quad \sigma_{xy} = k\sin 2\psi, \tag{2.17}$$

where ψ and p should be eliminated by means of (2.10). Thus the dependence of the stresses on x and y is given in parametric form with α and β being the parameters.

The Mikhlin coordinates $\bar{x}$ and $\bar{y}$ are defined by the equations

$$\bar{x} = x\cos\phi + y\sin\phi, \quad \bar{y} = -x\sin\phi + y\cos\phi. \tag{2.18}$$

These quantities are the coordinates of the point P under consideration referred to axes passing through the origin O and parallel to the characteristic directions at P (Fig. 2.2). Differentiating the first equation in (2.18) with respect to β and the second with respect to α leads to

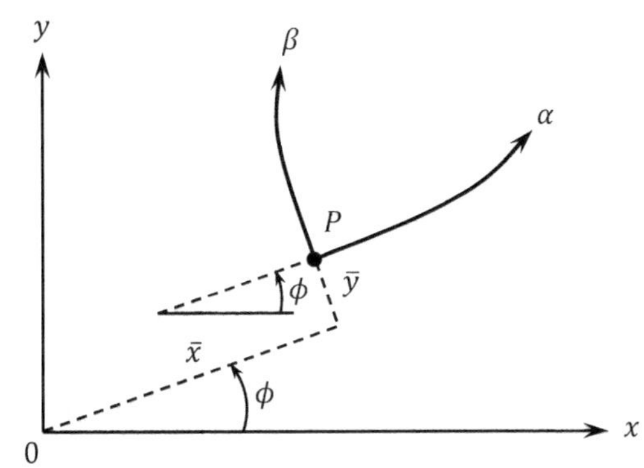

Fig. 2.2 Illustration of Mikhlin coordinates

$$\frac{\partial \bar{x}}{\partial \beta} = \frac{\partial x}{\partial \beta}\cos\phi - x\sin\phi + \frac{\partial y}{\partial \beta}\sin\phi + y\cos\phi,$$
$$\frac{\partial \bar{y}}{\partial \alpha} = -\frac{\partial x}{\partial \alpha}\sin\phi - x\cos\phi + \frac{\partial y}{\partial \alpha}\cos\phi - y\sin\phi. \quad (2.19)$$

It has been taken into account here that $\partial\phi/\partial\alpha = \partial\phi/\partial\beta = 1$ due to (2.10) and (2.12). Using (2.12) equations (2.8) can be rewritten as

$$\frac{\partial y}{\partial \alpha}\cos\phi - \frac{\partial x}{\partial \alpha}\sin\phi = 0, \quad \frac{\partial y}{\partial \beta}\sin\phi + \frac{\partial x}{\partial \beta}\cos\phi = 0. \quad (2.20)$$

Substituting (2.18) and (2.20) into (2.19) gives

$$\partial\bar{x}/\partial\alpha = -\bar{x}, \quad \partial\bar{x}/\partial\beta = \bar{y}. \quad (2.21)$$

It is evident from these equations that $\bar{x}$ and $\bar{y}$ separately satisfy (2.1). Once $\bar{x}$ and $\bar{y}$ have been found, the dependence of x and y on α and β is determined from (2.18) where ϕ should be eliminated by means of (2.12) and then by means of (2.10). Thus using (2.10), (2.12) and (2.17) the dependence of the stresses on x and y is given in parametric form with α and β being the parameters.

Let us introduce a principal lines based coordinate system (ξ, η) where ξ- and η-lines are trajectories of the principal stresses σ_ξ and σ_η, respectively. Let h be the Lame coefficient for the ξ-lines. Then, it is always possible to choose the parameterization of the η-lines such that the Lame coefficient for these lines is $1/h$ (Sadowsky, 1941). Assume that $\sigma_\xi \equiv \sigma_1$. Then, it follows from the geometry of Fig. 2.1 that

$$\frac{\partial x}{\partial \xi} = h\cos\psi, \ \frac{\partial x}{\partial \eta} = -\frac{1}{h}\sin\psi, \qquad \frac{\partial y}{\partial \xi} = h\sin\psi, \ \frac{\partial y}{\partial \eta} = \frac{1}{h}\cos\psi. \quad (2.22)$$

The compatibility equations are

$$\frac{\partial^2 x}{\partial \xi \partial \eta} = \frac{\partial^2 x}{\partial \eta \partial \xi}, \quad \frac{\partial^2 y}{\partial \xi \partial \eta} = \frac{\partial^2 y}{\partial \eta \partial \xi}$$

Substituting (2.22) into these equations and rotating the (x, y) coordinate system such that $\psi = 0$ result in

$$\frac{\partial h}{\partial \eta} + \frac{1}{h}\frac{\partial \psi}{\partial \xi} = 0, \quad \frac{1}{h^3}\frac{\partial h}{\partial \xi} + \frac{\partial \psi}{\partial \eta} = 0. \quad (2.23)$$

Using a standard procedure it is possible to show that this system of equations is hyperbolic. Its characteristic lines are given by

$$\mathrm{d}\eta/\mathrm{d}\xi = \mp h^2 \quad (2.24)$$

and the relations along these lines by

$$\mathrm{d}h \mp h\mathrm{d}\psi = 0. \quad (2.25)$$

The upper sign in (2.24) and (2.25) corresponds to the α- characteristic lines and the lower sign to the β- characteristic lines (Fig. 2.1). Equation (2.25) can be immediately integrated to give

$$\ln h - (\psi - \psi_0) = s(\beta), \quad \ln h + (\psi - \psi_0) = r(\alpha). \tag{2.26}$$

Here $r(\alpha)$ is an arbitrary function of α, $s(\beta)$ is an arbitrary function of β, and ψ_0 is constant. Assume that the two families of characteristic lines are curved. In this case, different choices of the functions $r(\alpha)$ and $s(\beta)$ involved in (2.26) merely change the scale of the α- and β-characteristic lines, respectively. Therefore, without loss of generality it is possible to select $r(\alpha) = 2\alpha$ and $s(\beta) = -2\beta$. Then, (2.26) becomes

$$\ln h - (\psi - \psi_0) = -2\beta, \quad \ln h + (\psi - \psi_0) = 2\alpha.$$

Solving these equations for h and $\psi - \psi_0$ yields

$$h = \exp(\alpha - \beta), \quad \psi - \psi_0 = \alpha + \beta. \tag{2.27}$$

Equation (2.24) is equivalent to

$$\partial\eta/\partial\alpha = -h^2\partial\xi/\partial\alpha, \quad \partial\eta/\partial\beta = h^2\partial\xi/\partial\beta. \tag{2.28}$$

Let us introduce the new variables μ and v defined by

$$\xi = \mu\exp(\beta - \alpha), \quad \eta = \omega\exp(\alpha - \beta). \tag{2.29}$$

Substituting (2.29) into (2.28) and eliminating h by means of (2.27) yields

$$\partial(\mu + \omega)/\partial\alpha = \mu - \omega, \quad \partial(\omega - \mu)/\partial\beta = \mu + \omega. \tag{2.30}$$

Differentiating the first of these equations with respect to β and the second with respect to α and then eliminating the first derivatives with respect to α and β by means of (2.30) lead to

$$\frac{\partial^2\mu}{\partial\alpha\partial\beta} + \frac{\partial^2\omega}{\partial\alpha\partial\beta} = -\mu - \omega, \quad \frac{\partial^2\omega}{\partial\alpha\partial\beta} - \frac{\partial^2\mu}{\partial\alpha\partial\beta} = \mu - \omega$$

Solving these equations for $\partial^2\mu/\partial\alpha\partial\beta$ and $\partial^2\omega/\partial\alpha\partial\beta$ gives

$$\frac{\partial^2\omega}{\partial\alpha\partial\beta} + \omega = 0, \quad \frac{\partial^2\mu}{\partial\alpha\partial\beta} + \mu = 0. \tag{2.31}$$

Thus ω and μ separately satisfy (2.1). It follows from the geometry of Fig. 2.3 and (2.13) that

$$R = \sqrt{2}h\frac{\partial\xi}{\partial\alpha}, \quad S = -\sqrt{2}h\frac{\partial\xi}{\partial\beta}$$

Substituting (2.27) and (2.29) into these equations and using (2.30) yield

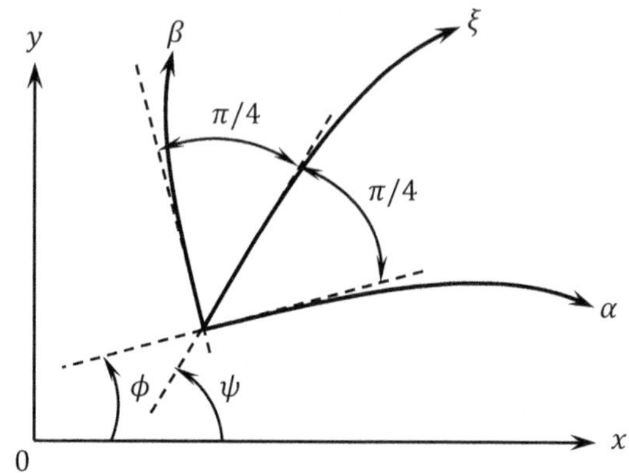

Fig. 2.3 Orientation of ξ-coordinate curves relative to (α,β) and (x,y) coordinates

$$
\begin{aligned}
R &= \sqrt{2}\left(\frac{\partial\mu}{\partial\alpha}-\mu\right)=-\sqrt{2}\left(\frac{\partial\omega}{\partial\alpha}+\omega\right),\\
S &= -\sqrt{2}\left(\frac{\partial\mu}{\partial\beta}+\mu\right)=-\sqrt{2}\left(\frac{\partial\omega}{\partial\beta}-\omega\right).
\end{aligned}
\tag{2.32}
$$

Using (2.27), (2.32) and the solution of (2.31) the right hand sides of the equations in (2.14) can be expressed as functions of α and β. Therefore, the dependence of x and y on α and β can be found from (2.14) by integration along any path. Then, using (2.29) and the solution of (2.31) it is possible to determine ξ and η as functions of x and y in parametric form with α and β being the parameters.

Let u and v be the velocity components referred to the α- and β- characteristic lines. The Geiringer equations are equivalent to Hill (1950)

$$\partial u/\partial\alpha - v = 0, \quad \partial v/\partial\beta + u = 0. \tag{2.33}$$

It is evident from this equation that u and v separately satisfy (2.1).

2.3 Pressure-Dependent Plasticity

The Mohr-Coulomb yield criterion is widely used in the mechanics of granular and other materials (Cox et al, 2008; Goddard, 2014). The system of equations comprising this criterion and the stress equilibrium equations under plane strain conditions forms a statically determinate system. This system is hyperbolic (Spencer, 1964). The Mohr-Coulomb yield criterion under plane strain conditions reads

$$q - p\sin\varphi = c\cos\varphi \tag{2.34}$$

where c is the cohesion, φ is the angle of internal friction and

$$2p = -(\sigma_{xx} + \sigma_{yy}), \quad 2q = \sqrt{(\sigma_{xx} - \sigma_{yy})^2 + 4\sigma_{xy}^2}.$$

Both c and φ are constant. The characteristic lines are given by Spencer (1964)

$$\mathrm{d}y/\mathrm{d}x = \tan(\psi - \pi/4 - \varphi/2), \quad \mathrm{d}y/\mathrm{d}x = \tan(\psi + \pi/4 + \varphi/2). \tag{2.35}$$

The first equation determines the α-lines and the second the β-lines. The characteristic relations along the α- and β-lines are

$$\cot\varphi \mathrm{d}q + 2q\mathrm{d}\psi = 0, \text{ and } \cot\varphi \mathrm{d}q - 2q\mathrm{d}\psi = 0, \tag{2.36}$$

respectively.

In what follows it is assumed that the two families of characteristics are curved. In this case equation (2.36) is equivalent to

$$\cot\varphi \ln(q/q_0) + 2(\psi - \psi_0) = 4\beta\cos\varphi, \quad \cot\varphi \ln(q/q_0) - 2(\psi - \psi_0) = -4\alpha\cos\varphi,$$

where q_0 and ψ_0 are constant. It immediately follows from these equations that

$$\ln(q/q_0) = 2(\beta - \alpha)\sin\varphi, \quad \psi - \psi_0 = (\beta + \alpha)\cos\varphi. \tag{2.37}$$

It is seen from (2.35) that the angle between the direction of the algebraically greatest principal stress and each of the characteristic directions is constant and is equal to $\pi/4 + \varphi/2$ (Fig. 2.4). Therefore, Eqs. (2.11) and (2.13) for the radii of curvature of the characteristic lines are valid. It follows from the geometry of Fig. 2.4 that

$$\frac{\partial x}{\partial s_\alpha} = \cos\left(\psi - \frac{\pi}{4} - \frac{\varphi}{2}\right), \quad \frac{\partial x}{\partial s_\beta} = -\sin\left(\psi - \frac{\pi}{4} + \frac{\varphi}{2}\right),$$
$$\frac{\partial y}{\partial s_\alpha} = \sin\left(\psi - \frac{\pi}{4} - \frac{\varphi}{2}\right), \quad \frac{\partial y}{\partial s_\beta} = \cos\left(\psi - \frac{\pi}{4} + \frac{\varphi}{2}\right).$$

Using (2.13) and (2.37) these equations are transformed to

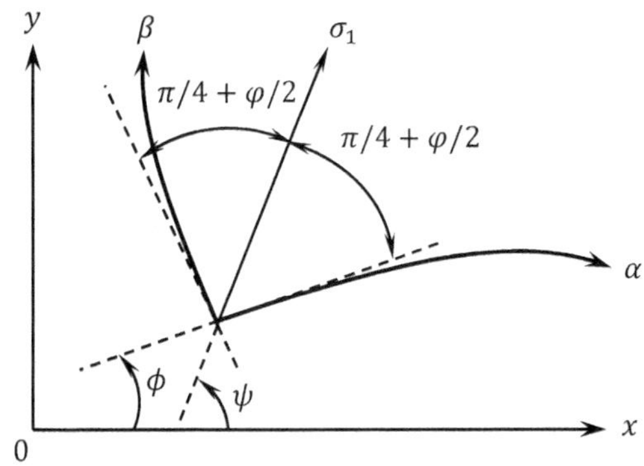

Fig. 2.4 Orientation of the algebraically greatest principal stress σ_1 and characteristic based coordinates (α, β) relative to Cartesian coordinates (x, y) for yield criterion (2.34)

$$\begin{aligned}\frac{\partial x}{\partial \alpha} &= R\cos\varphi\cos\left(\psi-\frac{\pi}{4}-\frac{\varphi}{2}\right), \quad \frac{\partial x}{\partial \beta} = S\cos\varphi\sin\left(\psi-\frac{\pi}{4}+\frac{\varphi}{2}\right),\\ \frac{\partial y}{\partial \alpha} &= R\cos\varphi\sin\left(\psi-\frac{\pi}{4}-\frac{\varphi}{2}\right), \quad \frac{\partial y}{\partial \beta} = -S\cos\varphi\cos\left(\psi-\frac{\pi}{4}+\frac{\varphi}{2}\right).\end{aligned} \tag{2.38}$$

Substituting these equations into the compatibility equations (2.15) and using (2.37) yield

$$\begin{aligned}&\frac{\partial R}{\partial \beta}\cos\left(\psi-\frac{\pi}{4}-\frac{\varphi}{2}\right) - R\sin\left(\psi-\frac{\pi}{4}-\frac{\varphi}{2}\right)\cos\varphi = \\ &\quad \frac{\partial S}{\partial \alpha}\sin\left(\psi-\frac{\pi}{4}+\frac{\varphi}{2}\right) + S\cos\left(\psi-\frac{\pi}{4}+\frac{\varphi}{2}\right)\cos\varphi,\\ &\frac{\partial R}{\partial \beta}\sin\left(\psi-\frac{\pi}{4}-\frac{\varphi}{2}\right) + R\cos\left(\psi-\frac{\pi}{4}-\frac{\varphi}{2}\right)\cos\varphi = \\ &\quad -\frac{\partial S}{\partial \alpha}\cos\left(\psi-\frac{\pi}{4}+\frac{\varphi}{2}\right) + S\sin\left(\psi-\frac{\pi}{4}+\frac{\varphi}{2}\right)\cos\varphi.\end{aligned}$$

Solving these equations for $\partial R/\partial \beta$ and $\partial S/\partial \alpha$ results in

$$\partial R/\partial \beta + R\sin\varphi = S, \quad \partial S/\partial \alpha - S\sin\varphi = -R. \tag{2.39}$$

It is evident that these equations reduce to (2.16) at $\varphi = 0$. The further analysis is facilitated by the use of the following transformation equations

$$R = R_0\exp\left[(\alpha-\beta)\sin\varphi\right], \quad S = S_0\exp\left[(\alpha-\beta)\sin\varphi\right]. \tag{2.40}$$

where R_0 and S_0 are new functions of α and β. Substituting (2.40) into (2.39) gives

$$\partial R_0/\partial \beta = S_0, \quad \partial S_0/\partial \alpha = -R_0. \tag{2.41}$$

It is evident from these equations that R_0 and S_0 separately satisfy (2.1). This result has been obtained in Alexandrov (2015). Once R_0 and S_0 have been found, the dependence of x and y on α and β is determined from (2.38) where R and S should be eliminated by means of (2.40) and then ψ by means of (2.37). The dependence of p, q and ψ on α and β immediately follows from (2.34) and (2.37). Then, the stresses are found from

$$\sigma_{xx} = -p + q\cos 2\psi, \quad \sigma_{yy} = -p - q\cos 2\psi, \quad \sigma_{xy} = q\sin 2\psi.$$

These relations and the solution for R_0 and S_0 combine to give the dependence of the stresses on x and y in parametric form with α and β being the parameters.

The yield criterion

$$\frac{|\sigma_i - \sigma_j|}{2\tau_s} + \frac{|\sigma|}{p_s} = 1, i, j = 1, 2, 3, i \neq j \tag{2.42}$$

is used for porous and powder materials Druyanov (1993). In equation (2.42), σ_1, σ_2 and σ_3 are the principal stresses, and τ_s and p_s are prescribed functions of the porosity. In what follows, it is assumed that the porosity is uniformly distributed and therefore τ_s and p_s are constant. Assuming that the axis corresponding to the stress σ_3 is perpendicular to planes of flow, $\sigma_1 > \sigma_2$ and $\sigma_1 + \sigma_2 + \sigma_3 < 0$ the yield criterion (2.42) under plane strain conditions reduces to Druyanov (1993)

$$\sigma_1(3p_s - 4\tau_s) - \sigma_2(3p_s + 2\tau_s) = 6\tau_s p_s. \tag{2.43}$$

Using the transformation equations for stress components it is possible to find that (Fig. 2.5 a)

$$\begin{aligned} \sigma_{xx} &= \frac{\sigma_1+\sigma_2}{2} + \frac{\sigma_1-\sigma_2}{2}\cos 2\psi, \ \sigma_{yy} = \frac{\sigma_1+\sigma_2}{2} - \frac{\sigma_1-\sigma_2}{2}\cos 2\psi, \\ \sigma_{xy} &= \frac{\sigma_1-\sigma_2}{2}\sin 2\psi. \end{aligned} \tag{2.44}$$

Substituting these equations into the equilibrium equations (2.7) yields to

$$\begin{aligned} &\frac{\partial(\sigma_1+\sigma_2)}{\partial x} + \frac{\partial(\sigma_1-\sigma_2)}{\partial x}\cos 2\psi - 2(\sigma_1-\sigma_2)\sin 2\psi\frac{\partial \psi}{\partial x} + \\ &\qquad \frac{\partial(\sigma_1-\sigma_2)}{\partial y}\sin 2\psi + 2(\sigma_1-\sigma_2)\cos 2\psi\frac{\partial \psi}{\partial y} = 0, \\ &\frac{\partial(\sigma_1+\sigma_2)}{\partial y} - \frac{\partial(\sigma_1-\sigma_2)}{\partial y}\cos 2\psi - 2(\sigma_1-\sigma_2)\sin 2\psi\frac{\partial \psi}{\partial y} + \\ &\qquad \frac{\partial(\sigma_1-\sigma_2)}{\partial x}\sin 2\psi + 2(\sigma_1-\sigma_2)\cos 2\psi\frac{\partial \psi}{\partial x} = 0, \end{aligned} \tag{2.45}$$

One can always rotate the (x,y) coordinate system so that its axes coincide with the principal stress directions at a given point (Fig. 2.5 b). Then $\psi = 0$ and equation (2.45) becomes

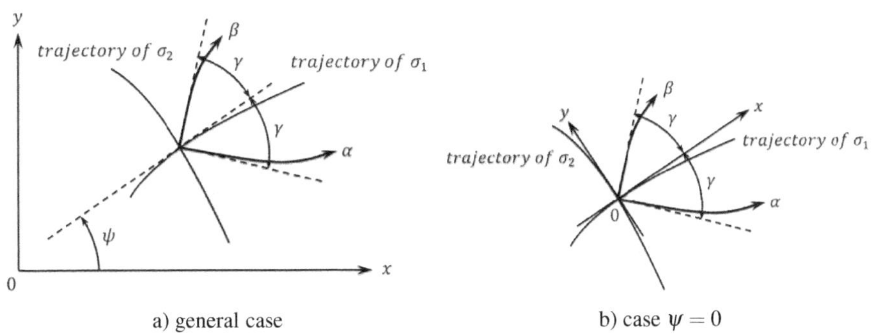

Fig. 2.5: Orientation of the algebraically greatest principal stress σ_1 and characteristic based coordinates (α, β) relative to Cartesian coordinates (x, y) for yield criterion (2.43)

$$\frac{\partial\sigma_1}{\partial x}+(\sigma_1-\sigma_2)\frac{\partial\psi}{\partial y}=0,\quad \frac{\partial\sigma_2}{\partial y}+(\sigma_1-\sigma_2)\frac{\partial\psi}{\partial x}=0.$$

Using (2.43) to eliminate σ_2 in these equations gives

$$\frac{\partial\sigma_1}{\partial x}+\frac{6\tau_s(\sigma_1+p_s)}{3p_s+2\tau_s}\frac{\partial\psi}{\partial y}=0,\quad (3p_s-4\tau_s)\frac{\partial\sigma_1}{\partial y}+6\tau_s(\sigma_1+p_s)\frac{\partial\psi}{\partial x}=0. \tag{2.46}$$

Appling a standard procedure it is possible to find that the angle γ between the principal stress axis corresponding to the stress σ_1 and each of the characteristic lines is (Fig. 2.5 b).

$$\gamma=\arctan\sqrt{\frac{3p_s-4\tau_s}{3p_s+2\tau_s}}. \tag{2.47}$$

Thus the system (2.46) is hyperbolic if $3p_s>4\tau_s$. This condition is usually satisfied for porous and powder materials Druyanov (1993). Returning to the (x,y) coordinate system shown in Fig. 2.5 a) the equations for characteristic lines can be written as

$$dy/dx=\tan(\psi\mp\gamma). \tag{2.48}$$

Here and in what follows the upper sign corresponds to the α-lines and the lower sign to the β-lines (Fig. 2.5). The characteristic relations are found from (2.46) and (2.47) as

$$\sqrt{\frac{(3p_s-4\tau_s)(3p_s+2\tau_s)}{6\tau_s(\sigma_1+p_s)}}d\sigma_1\mp d\psi=0. \tag{2.49}$$

In what follows it is assumed that the two families of characteristics are curved. In this case integration in (2.49) yields

$$\begin{aligned}
-\frac{\sqrt{(3p_s-4\tau_s)(3p_s+2\tau_s)}}{6\tau_s}\ln\left[\frac{\sigma_1+p_s}{\sigma_0}\right]+(\psi-\psi_0)&=2\beta\sin2\gamma,\\
-\frac{\sqrt{(3p_s-4\tau_s)(3p_s+2\tau_s)}}{6\tau_s}\ln\left[\frac{\sigma_1+p_s}{\sigma_0}\right]-(\psi-\psi_0)&=-2\alpha\sin2\gamma,
\end{aligned}$$

where σ_0 and γ are constant. It immediately follows from these equations that

$$\begin{aligned}
\psi-\psi_0&=(\alpha+\beta)\sin2\gamma,\\
-\frac{\sqrt{(3p_s-4\tau_s)(3p_s+2\tau_s)}}{6\tau_s}\ln\left[\frac{\sigma_1+p_s}{\sigma_0}\right]&=(\beta-\alpha)\sin2\gamma.
\end{aligned} \tag{2.50}$$

It is seen from (2.47) that the angle between the direction of the algebraically greatest principal stress and each of the characteristic directions is constant. Therefore, equations (2.11) and (2.13) for the radii of curvature of the characteristic lines are valid. It follows from the geometry of Fig. 2.5 a) that

$$\frac{\partial x}{\partial s_\alpha}=\cos(\psi-\gamma),\ \frac{\partial x}{\partial s_\beta}=\cos(\psi+\gamma),\ \frac{\partial y}{\partial s_\alpha}=\sin(\psi-\gamma),\ \frac{\partial y}{\partial s_\beta}=\sin(\psi+\gamma).$$

Using (2.13) and (2.50) these equations are transformed to

$$\partial x/\partial \alpha = R\sin 2\gamma\cos(\psi-\gamma),\ \partial x/\partial \beta = -S\sin 2\gamma\cos(\psi+\gamma),$$
$$\partial y/\partial \alpha = R\sin 2\gamma\sin(\psi-\gamma),\ \partial y/\partial \beta = -S\sin 2\gamma\sin(\psi+\gamma), \tag{2.51}$$

Substituting these equations into (2.15) and using (2.50) yield

$$\frac{\partial R}{\partial \beta}\cos(\psi-\gamma) - R\sin 2\gamma\sin(\psi-\gamma) = -\frac{\partial S}{\partial \alpha}\cos(\psi+\gamma) + S\sin 2\gamma\sin(\psi+\gamma),$$
$$\frac{\partial R}{\partial \beta}\sin(\psi-\gamma) + R\sin 2\gamma\cos(\psi-\gamma) = -\frac{\partial S}{\partial \alpha}\sin(\psi+\gamma) - S\sin 2\gamma\cos(\psi+\gamma).$$

Solving these equations for $\partial R/\partial \beta$ and $\partial S/\partial \alpha$ results in

$$\partial R/\partial \beta - R\cos 2\gamma = S, \quad \partial S/\partial \alpha + S\cos 2\gamma = -R. \tag{2.52}$$

It is evident that these equations reduce to (2.16) at $\gamma = \pi/4$. The further analysis is facilitated by the use of the following transformation equations

$$R = R_0 \exp\left[(\beta-\alpha)\cos 2\gamma\right], \quad S = S_0 \exp\left[(\beta-\alpha)\cos 2\gamma\right], \tag{2.53}$$

where R_0 and S_0 are new functions of α and β. Substituting (2.53) into (2.52) gives

$$\partial R_0/\partial \beta = S_0, \quad \partial S_0/\partial \alpha = -R_0. \tag{2.54}$$

It is evident from these equations that R_0 and S_0 separately satisfy (2.1). A similar result has been obtained in Alexandrov and Lyamina (2015). Once R_0 and S_0 have been found, the dependence of x and y on α and β is determined from (2.51) where R and S should be eliminated by means of (2.53) and ψ by means of (2.50). The dependence of σ_1 and σ_2 on α and β follows from (2.43) and (2.50). Then, the stresses in the Cartesian coordinates are found from (2.44) and (2.50). These relations and the solution for R_0 and S_0 combine to give the dependence of σ_{xx}, σ_{yy} and σ_{xy} on x and y in parametric form with α and β being the parameters.

2.4 Planar Ideal Flows

Ideal plastic flows are those for which all material elements undergo minimum work paths (Chung and Richmond, 1994). The theory of bulk ideal flow has been developed for rigid perfectly plastic solids satisfying Tresca's yield condition and its associated flow rule. However, in the case of plane strain deformation any criterion can be written in the form of Eq. (2.4). The existence of steady three-dimensional ideal flows has been demonstrated in Hill (1967). This result has been extended to non-steady flows in Richmond and Alexandrov (2002). A comprehensive overview on the ideal flow theory and its applications has been provided in Chung and Alexandrov (2007).

In the case of steady flows the ideal flow condition is that the steamlines are everywhere coincident with principal strain rate (and stress) directions. The general theory for this type of ideal flow has been developed in Aleksandrov (2000). The magnitude of the velocity vector is proportional to the scale factor h involved in (2.22). Therefore, the solution for ξ and η given by (2.29) and (2.31) determines the streamlines and the velocity field.

In the case of non-steady flow the ideal flow condition is that the principal lines coordinate system (ξ, η) may be taken to be fixed in the material so that it is Lagrangian. Therefore, the solution for ξ and η given by (2.29) and (2.31) together with the solution of (2.14) where R and S should be eliminated by means of (2.32) and ψ by means of (2.27) supplies the mapping between the Largangian coordinates and the Eulerian coordinates (x, y) in parametric form with α and β being the parameters. The general theory for this type of ideal flow has been developed in (Richmond and Alexandrov, 2000).

2.5 Conclusions

It has been demonstrated that a number of quantities in pressure-independent and pressure-dependent plasticity satisfy the telegraph equation (2.1). These quantities are shown in (2.16), (2.21), (2.31), (2.33), (2.41), and (2.54). In the case of ideal flows ξ and η that are found by means of (2.29) and (2.31) have an additional physical meaning explained in the previous section. The solution of (2.1) can be found by means of (2.3) with a high accuracy. In particular, solutions found by this method can be used to verify the accuracy of solutions found by other methods (Hill et al, 1951).

Acknowledgements The author acknowledges financial support of this research through the grant RSCF-16-49-02026.

References

Aleksandrov SE (2000) Steady-state plane ideal flows in the theory of plasticity. Mechanics of Solids 35(2):114–119

Alexandrov S (2015) Geometry of plane strain characteristic fields in pressure-dependent plasticity. Zeitschrift für Angewandte Mathematik und Mechanik 95(11):1296–1301

Alexandrov S, Lyamina E (2015) Riemann method for the plane strain of a homogeneous porous plastic material. Mechanics of Solids 50(2):171–175

Chung K, Alexandrov S (2007) Ideal flow in plasticity. In: Applied Mechanics Reviews, vol 60 (6), American Society of Mechanical Engineers, pp 316–335

Chung K, Richmond O (1994) The mechanics of ideal forming. Trans ASME J Appl Mech 61:176–181

Cox GM, Thamwattana N, McCue SW, Hill JM (2008) Coulomb–mohr granular materials: quasi-static flows and the highly frictional limit. In: Applied Mechanics Reviews, vol 61 (6), American Society of Mechanical Engineers, p 060802
Druyanov BA (1993) Technological Mechanics of Porous Bodies. Oxford University Press, USA
Goddard JD (2014) Continuum modeling of granular media. In: Applied Mechanics Reviews, vol 66 (5), American Society of Mechanical Engineers, p 050801
Hill R (1950) The Mathematical Theory of Plasticity. Claredon Press
Hill R (1967) Ideal forming operations for perfectly plastic solids. Journal of the Mechanics and Physics of Solids 15(3):223–227
Hill R, Lee EH, Tupper SJ (1951) A method of numerical analysis of plastic flow in plane strain and its application to the compression of a ductile material between rough plates. Trans ASME J Appl Mech 18(1):46–52
Richmond O, Alexandrov S (2000) Nonsteady planar ideal plastic flow: general and special analytical solutions. Journal of the Mechanics and Physics of Solids 48(8):1735 – 1759
Richmond O, Alexandrov S (2002) The theory of general and ideal plastic deformations of treasca solids. Acta mechanica 158(1):33–42
Sadowsky MA (1941) Equiareal pattern of stress trajectories in plane plastic strain. J Appl Mech 8:A74–A76
Spencer AJM (1964) A theory of the kinematics of ideal soils under plane strain conditions. Journal of the Mechanics and Physics of Solids 12(5):337–351

Chapter 3
Homogenization of Corrugated Plates Based on the Dimension Reduction for the Periodicity Cell Problem

Boris D. Annin, Alexander G. Kolpakov & Sergei I. Rakin

Abstract It is demonstrated that for corrugated plates, more generally, for cylindrical plates three-dimensional periodicity cell problem of the homogenization theory can be reduced to two-dimensional problem on the cross section of the periodicity cell. The transition to two-dimensional problem significantly simplifies the numerical analysis of corrugated plates (other non-numericalmethods are not effective if plate is thick). Significant simplification of the problem is achivied in the case of the equality of Poisson's ratios of the plate components, in particular, for a plate made of single homogeneous material. We present results of numerical analysis for a plate with a sinusoidal corrugation. Both thin and thick plates are studied. For thin plates our results demonstrate good agreement with the results presented in the recent paper (Ye et al, 2014).

Key words: Corrugated plate · Elasticity theory · Homogenization · Periodicity cell · Dimension reduction · Effective stiffnesses · Local stress/strain state · Universal relations

---

Boris D. Annin
Lavrentyev Institute of Hydrodynamics of Siberian Branch of the RAS, Labrentyev Ave. 15 630090 Novosibirsk, Russia,
e-mail: `annin@hydro.nsc.ru`

Alexander G. Kolpakov
Siberian State University of Telecommunications and Informatics, Kirov st. 86, 630009 Novosibirsk, Russia,
e-mail: `algk@ngs.ru`

Sergei I. Rakin
Siberian State Transport University, Dusi Koval'chuk st. 191, 630049 Novosibirsk, Russia,
e-mail: `rakinsi@ngs.ru`

H. Altenbach et al. (eds.), *Mechanics for Materials and Technologies*,
Advanced Structured Materials 46, DOI 10.1007/978-3-319-56050-2_3

3.1 Introduction

The problem of the computation of effective stiffness of corrugated plate attracts attention of many researchers. The history of the problem is presented in Ye et al (2014), where the beginning of the study is referred to 1923, noted an existence of extensive literature on the subject, and actuality of the problem until now.

The actuality of the problem is explained by the following. On the one hand, corrugated plates are widely used in the modern engineering structures. On the other hand, although geometry of corrugated plate is simple, it is impossible to obtain solution of the three-dimensional elasticity problem for corrugated plate in the explicit form. The use of approximate approaches leads to discrepancies in the results obtained in the frameworks of various theories (the comparison of formulas for approximate computation of effective stiffness of corrugated plate may be found in Ye et al (2014); Xia et al (2012). The use in modern engineering structures of plates with corrugated core (Buannic et al, 2003; Talbi et al, 2009; Bartolozzi et al, 2013) gives additional stimulus for the development of methods for accurate computation of corrugated plates.

The most progress in the computational analysis of corrugated plates was made on the basis of two-dimensional shell theory (Ye et al, 2014; Xia et al, 2012) and the homogenization theory (Andrianov et al, 1998; Arkhangelskii and Gorbachev, 2007; Lee and Yu, 2011). The first approach works if the plate thickness is substantially less both the length and the radius of curvature of corrugation. If this condition is not met, the homogenization theory based on the three-dimensional elasticity theory model should be applied (Caillerie, 1984; Kohn and Vogelius, 1984; Kalamkarov and Kolpakov, 1997).

The starting point of our analysis is the observation that for corrugated plate, which shape locally is similar to hollow cylinder, the three-dimensional periodicity cell problems of the homogenization theory can be reduced to two-dimensional problems on the cross-section of the periodicity cell. This makes it possible to perform numerical analysis for corrugated plates of arbitrary thickness, including computation of the effective stiffness and the local strain/stress concentration tensors, using modern engineering software (ANSYS, etc.) with high accuracy.

The most simplification of the original problem takes place in the case of the equality of Poisson ratios for the plate components, in particular for plates made of a homogeneous material.

An addition simplification of the original problem takes place if corrugation is symmetric. For this case, we obtain the universal relationships between the averaging stiffness.

Previously, a transition to a problem on the cross-section of plate was done in Lewinski (1985) for plates with piecewise flat cross-section. The analysis presented in Lewinski (1985) was based on the assumption that solution to the periodicity cell problem has the form $\Theta(\mathbf{y}) = T(y_1, y_2)$ in the tension mode and $\Theta(\mathbf{y}) = T(y_1, y_2) + y_3 Z(y_1, y_2)$ in the bending mode (see Eqs. (16) in Lewinski, 1985) that is typical for flat-plate theory. Our numerical experiments demonstrate that this assumption may be accepted for plate with piecewise flat cross-section if the thickness of the plate remains constant at the distance 10-20 times more than the characteristic thickness of the plate. The assumption from Lewinski (1985) cannot be applied to curvilinear plates, in particular, to corrugated plates. Results of numerical computations presented in this paper illustrate this thesis.

Our method is based on the reduction of dimension of problem as a consequence of the invariance of the problem in spatial variables. An approach based on the similar idea was previously used in the homogenization procedure for helical-shaped beams in Frikha et al (2013). The mentioned paper is the most close to our approach in the sense of general idea – the use of invariance in the homogenization procedure for thin-walled structures (plates, beams, etc.) when it exists. For solid composites the mentioned idea was widely used, for example, for laminated composites. In the technical sense, the problem considered in Frikha et al (2013), as well as the problems related to solid composites, strongly differ from the problem under consideration. The difference results from the difference of periodicity cell problems for solid composites, plates and beams (Caillerie, 1984).

In Lee and Yu (2011), it was note that the problem of computation of stiffness of corrugated plate, which is uniform along one of the in-plane directions, may be associated with two-dimension problem. This conclusion was done in the frameworks of the variational asymptotic method (Lee and Yu, 2011), which allows omit one variable (this is variable x_1 in our notations, see Fig. 3.1) from the very beginning of the analysis of the problem. It is not the case, which one meets by applying the homogenization theory to the analysis of corrugated plates. In the homogenization theory, the unit cell problems depend on all three spatial variables (see, e.g., exact solutions to the unit cell problems for uniform plate in Kolpakov, 2010), while the derivative of solution to unit cell problem depend on two variables. The account of dependence of solution to unit cell problem on all variables is important for correct reduction of dimension in the homogenization theory. More exactly, for some unit cell problems we may simply omit x_1 variable because it makes no contribution to the unit cell problems. But for several unit cell problems, x_1 makes contribution to the corresponding two-dimensional problems. The last cases correspond to the bending of the plate in x_1 direction and torsion of the plate. All the mentioned problems are discussed in detail below.

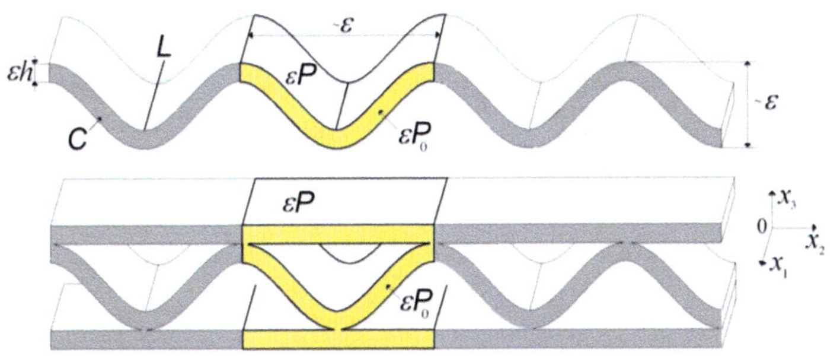

Fig. 3.1: Corrugated plate (top) and plate with corrugated core (bottom)

3.2 Statement of the Problem

Consider a plate of periodic structure, which occupies three-dimensional cylindrical region P formed by moving straight line L, called generatrix, perpendicularly two-dimensional domain S of periodic structure – the cross section of the plate, see. Fig. 3.1. Denote the periodicity cell of the plate by εP (drawn in bold lines in Fig. 3.1), where ε means the characteristic size of the periodicity cell (ε is also the characteristic thickness of the corrugated plate "in whole", do not confuse ε with the thickness of the plate εh, see Fig. 3.1).

Without loss of generality, we can assume that the generatrix L is parallel to the axis Oy_1 and the periodicity cell may be represented in the form $\varepsilon P = [0,1] \times \varepsilon P_0$, where εP_0 is the periodicity cell of the cross-section C. If the characteristic dimension ε is small in comparison with the in-plane size of plate, the original plate may be substituted by a flat plate possessing so-called effective stiffness, see for details (Caillerie, 1984; Kohn and Vogelius, 1984; Kalamkarov and Kolpakov, 1997).

To compute the effective stiffness of corrugated plate, we use the homogenization method as applied to plates, developed originally in Caillerie (1984); Kohn and Vogelius (1984). In accordance with the homogenization method as applied to plates, we introduce the "fast" variables $\mathbf{y} = \mathbf{x}/\varepsilon$ and it is solved the so-called periodicity cell problems on the periodicity cell $P = [0,1] \times P_0$ (see Fig. 3.2), that have the following form: find function $\mathbf{N}^{\mathrm{AB}\nu}(\mathbf{y})$ as solution of the boundary-value problem

$$\begin{cases} (a_{ijkl}(\mathbf{y}) N^{\mathrm{AB}\nu}_{k,l}(\mathbf{y}) + (-1)^{\nu} a_{ij\mathrm{AB}}(\mathbf{y}) y_3^{\nu})_{,j} = 0 \quad \text{in} \quad P, \\ (a_{ijkl}(\mathbf{y}) N^{\mathrm{AB}\nu}_{k,l}(\mathbf{y}) + (-1)^{\nu} a_{ij\mathrm{AB}}(\mathbf{y}) y_3^{\nu}) n_j = 0 \quad \text{on} \quad \Gamma, \\ \mathbf{N}^{\mathrm{AB}\nu}(\mathbf{y}) \quad \text{periodic in} \quad y_1, y_2 \in S. \end{cases} \tag{3.1}$$

Here Γ is the lateral (free) surface of the periodicity cell; S is the projection of periodicity cell onto the plane Oy_1y_2 (see Fig. 3.2.), a_{ijkl} are elastic constants; $\mathbf{n}$ is outer vector-normal on surface Γ. Hereafter Latin indices take the values 1, 2, 3. In (3.1) A and B take the values 1 and 2; μ and ν take the values 0 and 1.

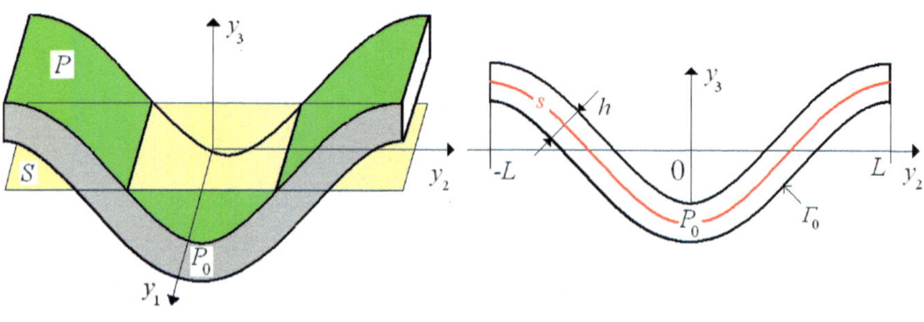

Fig. 3.2: Periodicity cell (left) and its cross-section (right) in the "fast" variables $\mathbf{y} = \mathbf{x}/\varepsilon$

Our consideration is concentrated on the plates made of isotropic material(s). Nevertheless it is convenient to write the elastic constants in the general form, as a_{ijkl}, keeping in mind that

$$a_{ijkl} = \frac{E\nu_0}{(1+\nu_0)(1-2\nu_0)}\delta_{ij}\delta_{kl} + \frac{E}{2(1+\nu_0)}(\delta_{ik}\delta_{jl} + \delta_{il}\delta_{jk}),$$

where E and ν_0 are Young's modulus and Poisson ratio, respectively (we use notation ν_0 for the Poisson ratio in order to distinguish this from the index ν, which is traditionally used in the periodicity cell problem Caillerie, 1984; Kohn and Vogelius, 1984; Kalamkarov and Kolpakov, 1997).

Having solved problem (3.1), the effective stiffness $D^{\nu+\mu}_{AB\Gamma\Delta}$ are computed by the formula (Caillerie, 1984; Kohn and Vogelius, 1984; Kalamkarov and Kolpakov, 1997) $(A, B, \Gamma, \Delta = 1,2; \mu, \nu = 0, 1)$

$$D^{\nu+\mu}_{AB\Gamma\Delta} = \frac{1}{|S|}\int_P (a_{ABkl}(\mathbf{y})N^{\Gamma\Delta\nu}_{k,l}(\mathbf{y}) + (-1)^\nu a_{AB\Gamma\Delta}(\mathbf{y})y_3^\nu)(-1)^\mu y_3^\mu d\mathbf{y} \tag{3.2}$$

where S is projection of the periodicity cell P to the Oy_1y_2-plane, see Fig. 3.2 (left); $|S|$ means the area of S. Values of the indices ν and μ such that $\nu+\mu=0$ corresponds to in-plane stiffness $D^0_{AB\Gamma\Delta}$, $\nu+\mu=1$ corresponds to coupled (nonsymmetrical) stiffness $D^1_{AB\Gamma\Delta}$, and $\nu+\mu=2$ corresponds to bending stiffness $D^2_{AB\Gamma\Delta}$. The computation of stiffness in the way described above guarantees the closeness of displacements and deflections of the original corrugated plate and flat plate possessing the stiffness $D^{\nu+\mu}_{AB\Gamma\Delta}$ (3.2), see for details Caillerie (1984); Kohn and Vogelius (1984); Kalamkarov and Kolpakov (1997).

3.3 Dimension Reduction for the Periodicity Sell Problem

We assume that the elastic constants do not depend on the variable y_1, i.e., they have the form $a_{ijkl}(y_2,y_3)$. The coefficients of such kind describes the most practically important cases: plates made of homogeneous materials ($a_{ijkl} = const$ in this case) and plates made of several homogeneous plates ($a_{ijkl}(y_2,y_3)$ is piecewise-constant function in this case). We demonstrate now, that problem (3.1) with the coefficients $a_{ijkl}(y_2,y_3)$ in the cylindrical periodicity cell $P=[0,1]\times P_0$ can be reduced to two-dimensional problems on the cross section P_0 of the periodicity cell. Denote $\mathbf{M} = \frac{\partial \mathbf{N}^{AB\nu}}{\partial y_1}$ the derivative of $\mathbf{N}^{AB\nu}(\mathbf{y})$ with respect to y_1 $(A, B = 1,2; \nu = 0, 1)$. Differentiating equations in (3.1) with respect to y_1 and taking into account that the derivatives of y_3 and $a_{ijkl}(y_2,y_3)$ in y_1 are zero and the normal n_j does not depend on y_1, we arrive at the problem

$$\begin{cases} (a_{ijkl}(y_2,y_3)M_{k,l}(\mathbf{y}))_{,j} = 0 & \text{in } P_0, \\ (a_{ijkl}(y_2,y_3)M_{k,l}(\mathbf{y}))n_j = 0 & \text{on } \Gamma_0, \\ \mathbf{M}(\mathbf{y}) \ \text{periodic in} & y_2 \in [-L,L]. \end{cases} \tag{3.3}$$

L is displayed in Fig. 3.2 ($2L$ is the period of the cross-section of the plate).

It is known (Caillerie, 1984; Kohn and Vogelius, 1984; Kalamkarov and Kolpakov, 1997) that solution to problem (3.3) is an arbitrary constant. This constant does not affect the value of effective stiffness (3.2) because constant vanish under the differentiation in formula (3.2). Take the constant equal to zero. Then $\mathbf{M}(\mathbf{y}) = 0$. It means that $\mathbf{N}^{\mathrm{AB}\nu}(\mathbf{y})$ does not depend on y_1, being a function of two variables y_2 and y_3: $\mathbf{N}^{\mathrm{AB}\nu}(\mathbf{y}) = \mathbf{N}^{\mathrm{AB}\nu}(y_2, y_3)$. Then the periodicity cell problem (3.1) takes the form (in (3.4) $\delta, \kappa = 2, 3$; i, k =1,2,3; $\mathrm{A}, \mathrm{B} = 1, 2$; ν =0, 1)

$$\begin{cases} (a_{i\delta k\kappa}(y_2,y_3)N^{\mathrm{AB}\nu}_{k,\kappa} + (-1)^\nu a_{i\delta \mathrm{AB}}(y_2,y_3)y_3^\nu)_{,\delta} = 0 & \text{in } P_0, \\ (a_{i\delta k\kappa}(y_2,y_3)N^{\mathrm{AB}\nu}_{k,\kappa} + (-1)^\nu a_{i\delta \mathrm{AB}}(y_2,y_3)y_3^\nu)n_\delta = 0 & \text{on } \Gamma_0, \\ \mathbf{N}^{\mathrm{AB}\nu}(y_2,y_3) \text{ is periodic in } y_2 \in [-L,L]. \end{cases} \tag{3.4}$$

Periodicity cell problem for AB $= \mathbf{11}, 22$.

Consider problem (3.4) for $i = 1$. For isotropic material $a_{1\delta \mathrm{AB}} = 0$ $(\mathrm{A}, \mathrm{B} = 1, 2)$, and only $a_{1\delta 1\delta} \neq 0$ among $a_{1\delta k\kappa}$ in (3.4) (in (3.4) Latin $k = 1, 2, 3$ and Greek $\delta, \kappa = 2, 3$; ν =0, 1). Then, for $i = 1$, we obtain from (3.4) the following problem with respect to function $N_1^{\mathrm{AB}\nu}$ $(\mathrm{A}, \mathrm{B} = 1, 2$; $\mathrm{A}, \mathrm{B} = 1, 2$; ν =0, 1; $\delta = 2, 3)$:

$$\begin{cases} (a_{1\delta 1\delta}(y_2,y_3)N^{\mathrm{AB}\nu}_{1,\delta})_{,\delta} = 0 & \text{in } P_0, \\ (a_{1\delta 1\delta}(y_2,y_3)N^{\mathrm{AB}\nu}_{1,\delta})n_\delta = 0 & \text{on } \Gamma_0, \\ N_1^{\mathrm{AB}\nu}(y_2,y_3) \text{ is periodic in } y_2 \in [-L,L]. \end{cases} \tag{3.5}$$

Solution to (3.5) is a constant, which may be taken equal to zero: $N_1^{\mathrm{AB}\nu}(y_2, y_3) = 0$. The remaining equations in (3.4) form the following two-dimensional elasticity theory problem (in (3.6) $\gamma, \delta, \iota, \kappa$ take values 2 and 3; $\mathrm{A}, \mathrm{B} = 1, 2$; ν =0, 1):

$$\begin{cases} (a_{\gamma\delta\iota\kappa}(y_2,y_3)N^{\mathrm{AB}\nu}_{\iota,\kappa} + (-1)^\nu a_{\gamma\delta \mathrm{AB}}(y_2,y_3)y_3^\nu)_{,\delta} = 0 & \text{in } P_0, \\ (a_{\gamma\delta\iota\kappa}(y_2,y_3)N^{\mathrm{AB}\nu}_{\iota,\kappa} + (-1)^\nu a_{\gamma\delta \mathrm{AB}}(y_2,y_3)y_3^\nu)n_\delta = 0 & \text{on } \Gamma_0, \\ \mathbf{N}^{\mathrm{AB}\nu}(y_2,y_3) \text{ is periodic in } y_2 \in [-L,L]. \end{cases} \tag{3.6}$$

Demonstrate that for AB $= \mathbf{11}, 22$, expressions $a_{\gamma\delta \mathrm{AB}}(y_2, y_3)y_3^\nu$ may be represented in the form $a_{\gamma\delta\alpha\beta}(y_2, y_3)u_{\alpha,\beta}$ (remind that indices $\alpha, \beta, \gamma, \delta$ take values 2 and 3). For this aim the system of algebraic equations

$$a_{\gamma\delta\alpha\beta}(y_2,y_3)e_{\alpha\beta} = a_{\gamma\delta \mathrm{AB}}(y_2,y_3)y_3^\nu \tag{3.7}$$

must be solved. In (3.7), $e_{\alpha\beta}$ means $u_{\alpha,\beta}$ (by virtue of the symmetry of the elastic constants, they are equal to the infinitezimal strains).

Construction of displacements $\mathbf{u}$ *for* AB $= \mathbf{11}$. For this case, (3.7) takes the form (ν =0, 1)

$$\begin{cases} a_{2222}(y_2,y_3)e_{22} + a_{2233}(y_2,y_3)e_{33} = a_{2211}(y_2,y_3)y_3^\nu, \\ a_{2323}(y_2,y_3)e_{23} = 0, \\ a_{3322}(y_2,y_3)e_{22} + a_{3333}(y_2,y_3)e_{33} = a_{3311}(y_2,y_3)y_3^\nu. \end{cases} \tag{3.8}$$

Since it is assumed the material is isotropic, we obtain the following solution to (3.8)

$$e_{22} = e_{33} = \frac{a_{2211}(y_2, y_3)}{a_{2222}(y_2, y_3) + a_{2233}(y_2, y_3)} y_3^{\nu}, \quad e_{23} = 0 \tag{3.9}$$

For isotropic materials

$$\frac{a_{2211}(y_2, y_3)}{a_{2222}(y_2, y_3) + a_{2233}(y_2, y_3)} = \nu_0(y_2, y_3),$$

where $\nu_0(y_2, y_3)$ is the Poisson's ratio.

Next we consider plate made of a material with constant Poisson's ratio. It may be a plate made of a single homogeneous material like in Fig. 3.1 (top) or a plate assembled of several structural elements made of materials with the same Poisson's ratios like in Fig. 3.1 (bottom). For this case fraction in (3.9) takes constant value equal to ν_0 and Eq. (3.9) takes the form $e_{22} = e_{33} = \nu_0 y_3^{\nu}$, $e_{23} = 0$. Strains, which are linear functions of coordinates, satisfy the compatibility equation (Kolpakov, 2010), thus there exist displacements, which generate the strains above. Compute these displacements. For $\nu = 0$, Eq. (3.9) take the form

$$u_{2,2} = \nu_0, \quad u_{3,3} = \nu_0, \quad u_{2,3} + u_{3,2} = 0 \tag{3.10}$$

Solution to (3.10) is evident

$$u_2 = \nu_0 y_2, \quad u_3 = \nu_0 y_3 \tag{3.11}$$

The case $\nu = 1$ needs some more computations. For this case (3.9) has the form

$$u_{2,2} = \nu_0 y_3^{\nu}, \quad u_{3,3} = \nu_0 y_3^{\nu}, \quad u_{2,3} + u_{3,2} = 0 \tag{3.12}$$

From the first and second equations in (3.12), we derive the following representations for u_2 and :

$$u_2 = \nu_0 y_2 y_3 + f(y_3), \quad u_3 = \frac{\nu_0}{2} y_3^2 + g(y_2). \tag{3.13}$$

Substituting (3.13) into the third equation in (3.12), we obtain the following equation: $\nu_0 y_2 + f'(y_3) + g'(y_2) = 0$. This equation is satisfied by the functions $g(y_2) = -\nu_0 y_2^2/2$ and $f(y_3) = 0$. Finally, we obtain solution to (3.9) for $\nu = 1$:

$$u_2 = \nu_0 y_2 y_3, \quad u_3 = \frac{\nu_0}{2} y_3^2 - \frac{\nu_0}{2} y_2^2 \tag{3.14}$$

Construction of displacements **u** *for* AB $= 22$**.** This case corresponds to the standard homogenization procedure as applied to plates and it was considered in Kolpakov (2010). In this connection, we present the results in condensed form. The system of equations (3.7) has in this case the form (ν =0, 1)

$$\begin{cases} a_{2222}(y_2,y_3)e_{22}+a_{2233}(y_2,y_3)e_{33}=a_{2222}(y_2,y_3)y_3^{\nu}, \\ a_{2323}(y_2,y_3)e_{23}=0, \\ a_{3322}(y_2,y_3)e_{22}+a_{3333}(y_2,y_3)e_{33}=a_{3322}(y_2,y_3)y_3^{\nu}. \end{cases} \tag{3.15}$$

Solution to (3.15) is

$$e_{22}=y_3^{\nu}, \quad e_{33}=0, \; e_{23}=0 \tag{3.16}$$

For $\nu=0$, displacements corresponding to (3.16) are

$$u_2=y_2, \qquad u_3=0 \tag{3.17}$$

For $\nu=1$, (3.16) takes the form

$$u_{2,2}=y_3, \qquad u_{3,3}=0, \quad u_{2,3}+\; u_{3,2}=0 \tag{3.18}$$

From the first and the second equations in (3.18), we obtain

$$u_2=y_2y_3+f(y_3), \qquad u_3=g(y_2) \tag{3.19}$$

Substituting (3.19) into Eq. $(3.18)_3$, we obtain $y_2+f'(y_3)+g'(y_2) = 0$. One can take functions $g(y_2)=-y_2^2/2$ and $f(y_3)=0$ as solution to this equation. Finally, we obtain

$$u_2=y_2y_3, \qquad u_3=-\frac{1}{2}y_2^2 \tag{3.20}$$

Next, we will mark the solutions (3.11), (3.14), (3.17), and (3.20) by the corresponding indices AB and ν, and write $\mathbf{u}(y_2,y_3)$ in the form $\mathbf{u}^{\mathrm{AB}\nu}(y_2,y_3)$.

For the function $\mathbf{M}^{\mathrm{AB}\nu}(y_2,y_3)=\mathbf{N}^{\mathrm{AB}\nu}(y_2,y_3)+(-1)^{\nu}\mathbf{u}^{\mathrm{AB}\nu}(y_2,y_3)$, problem (3.6) takes the form

$$\begin{cases} (a_{\gamma\delta\iota\kappa}(y_2,y_3)M^{\mathrm{AB}\nu}_{\iota,\kappa})_{,\delta}=0 \quad \text{in} \quad P_0, \\ a_{\gamma\delta\iota\kappa}(y_2,y_3)M^{\mathrm{AB}\nu}_{\iota,\kappa}(y_2,y_3)n_{\delta}=0 \quad \text{on} \quad \Gamma_0, \\ \mathbf{M}^{\mathrm{AB}\nu}(y_2,y_3)-(-1)^{\nu}\mathbf{u}^{\mathrm{AB}\nu}(y_2,y_3) \quad \text{periodic in} \quad y_2\in[-L,L]. \end{cases} \tag{3.21}$$

Problem (3.21) is more convenient, especially for numerical analysis, than the original periodicity cell problem (3.6) because it does not involve the "mass" and "surface" forces, which present in the original periodicity cell problem (3.6) in the form not usual for engineering software.

Periodicity cell problem for AB $=12$

In the case under consideration, for $i=1$, we obtain from (3.5) the following problem with respect to one function $N_1^{12\nu}(y_2,y_3)(\delta=2,3;\nu=0,\,1)$:

$$\begin{cases} (a_{1\delta1\delta}N^{12\nu}_{1,\delta}+(-1)^{\nu}a_{1\delta12}y_3^{\nu})_{,\delta}=0 \quad \text{in} \quad P_0, \\ (a_{1\delta1\delta}N^{12\nu}_{1,\delta}(y_2,y_3)+(-1)^{\nu}a_{1\delta12}y_3^{\nu})n_{\delta}=0 \quad \text{on} \quad \Gamma_0, \\ N_1^{12\nu}(y_2,y_3) \quad \text{periodic in} \quad y_2\in[-L,L]. \end{cases} \tag{3.22}$$

The remaining $(i=\gamma=2,3)$ equations for the following two-dimensional elasticity theory problem $(\gamma,\delta,\iota,\kappa=2,3;\;\nu=0,\,1)$:

$$\begin{cases} (a_{\gamma\delta\iota\kappa}N^{12\nu}_{\iota,\kappa}+(-1)^{\nu}a_{\gamma\delta 12}y_3^{\nu})_{,\delta}=0 \quad \text{in} \quad P_0, \\ (a_{\gamma\delta\iota\kappa}N^{12\nu}_{\iota,\kappa}(y_2,y_3)+(-1)^{\nu}a_{\gamma\delta 12}y_3^{\nu})n_{\delta}=0 \quad \text{on} \quad \Gamma_0, \\ \mathbf{N}^{12\nu}(y_2,y_3) \quad \text{periodic in} \quad y_2\in[-L,L]. \end{cases} \tag{3.23}$$

Since for isotropic material $a_{\gamma\delta 12}=0$ (γ,δ =2,3), problem (3.23) takes the form ($\gamma,\delta,\iota,\kappa=2,3;\nu$ =0, 1)

$$\begin{cases} (a_{\gamma\delta\iota\kappa}N^{12\nu}_{\iota,\kappa})_{,\delta}=0 \quad \text{in} \quad P_0, \\ a_{\gamma\delta\iota\kappa}N^{12\nu}_{\iota,\kappa}(y_2,y_3)n_{\delta}=0 \quad \text{on} \quad \Gamma_0, \\ \mathbf{N}^{12\nu}(y_2,y_3) \quad \text{periodic in} \quad y_2\in[-L,L]. \end{cases}$$

Solution to this problem is vector-constant (Caillerie, 1984; Kohn and Vogelius, 1984; Kalamkarov and Kolpakov, 1997), which may be taken equal to zero. Thus, the periodicity cell problem for AB = 12 is reduced to the problem (3.22) for one differential equation with respect to scalar function $N_1^{12\nu}(y_2,y_3)$.

Problem (3.22) describes anti-plane deformation (Barber, 2011). Also, it may be treated as thermo-conductivity problem. We shall check if it is possible to represent the free term in (3.22) in the form $(-1)^{\nu}a_{1\delta 1\delta}w_{,\delta}$ ($\delta=2,3$). Write the equation ($\nu=0$, 1)

$$a_{1\delta 1\delta}w_{,\delta}=(-1)^{\nu}a_{1\delta 12}y_3^{\nu} \tag{3.24}$$

For $\delta=2$ and $\delta=3$, we obtain from (3.24) that $a_{1212}w_{,2}=(-1)^{\nu}a_{1212}y_3^{\nu}$ and $a_{1313}w_{,3}=0$. From these equalities, we obtain the following system of differential equations

$$w_{,2}=(-1)^{\nu}y_3^{\nu},\ w_{,3}=0 \tag{3.25}$$

For $\nu=0$, the system (3.25) takes the form $w_{,2}=1$, $w_{,3}=0$. This system is integrable and its solution is $w(y_2,y_3)=y_2$. Introducing function $M^{120}(y_2,y_3)=N_1^{120}(y_2,y_3)+y_2$, we can rewrite (3.22) in the form of boundary-value problem without "mass" and "surface" forces:

$$\begin{cases} \Delta M^{120}=0 \quad \text{in} \quad P_0, \\ \dfrac{\partial M^{120}}{\partial \mathbf{n}}=0 \quad \text{on} \quad \Gamma_0, \\ M^{120}(\pm L,y_3)-y_2 \quad \text{periodic in} \quad y_2\in[-L,L]. \end{cases} \tag{3.26}$$

For $\nu=1$, the system (3.25) takes form $w_{,2}=-y_3$, $w_{,3}=0$. This system is not integrable. Really, the necessary integrability condition is not satisfied for this system because $w_{,23}=-y_{3,3}=1=w_{,32}=0$. For $\nu=1$, we can rewrite problem (3.22) in the form

$$\begin{cases} \Delta N_1^{121}=0 \quad \text{in} \quad P_0, \\ \dfrac{\partial N_1^{121}}{\partial \mathbf{n}}(y_2,y_3)-y_3n_2=0 \quad \text{on} \quad \Gamma_0, \\ N_1^{121}(y_2,y_3) \quad \text{periodic in} \quad y_2\in[-L,L] \end{cases} \tag{3.27}$$

suitable for computations below.

Multi-connected cross-section

In multi-element structure, for example, plate with corrugate core, the cross-section of structure is multi-connected domain, i.e. contains internal openings (holes), see Fig. 3.1 (bottom). The normal stress on the boundaries Γ_l of the internal openings is zero. It leads to the arising of additional boundary condition, which sounds as ($\gamma,\delta,\iota,\kappa = 2,3$;A,B $= 1,2$; ν =0, 1)

$$(a_{\gamma\delta\iota\kappa}(y_2,y_3)N^{\mathrm{AB}\nu}_{\iota,\kappa}(y_2,y_3) + (-1)^{\nu}a_{\gamma\delta\mathrm{AB}}(y_2,y_3)y_3^{\nu})n_\delta = 0 \quad \text{on } \Gamma_l$$

By using the functions $\mathbf{u}^{\mathrm{AB}\nu}$, we can rewrite this boundary condition as

$$a_{\gamma\delta\iota\kappa}(y_2,y_3)M^{\mathrm{AB}\nu}_{\iota,\kappa}(y_2,y_3)n_\delta = 0 \quad \text{on } \Gamma_l$$

3.4 Symmetric Corrugation

Many corrugated plate used in industry have symmetric structure - the cross-section P_0 of the periodicity cell is symmetric with respect to Oy_3-axis, see Figs. 3.2 and 3.3. As a consequence, solutions to the problems (3.6) and (3.22) have the same symmetry. Consider problem (3.6), which is two-dimensional elasticity problem. By virtue of the symmetry, the displacements $N_2^{\mathrm{AB}\nu}$ (A,B $= 1,2;\nu$ =0, 1) and the normal stresses vectors $\sigma_{\mathbf{n}}$on the boundaries of a single periodicity cell should look as it is shown in Fig. 3.3 on the left. By virtue of the periodicity, the displacement $N_2^{\mathrm{AB}\nu}$ and the normal stresses vector $\sigma_{\mathbf{n}}$ on the boundaries of adjacent periodicity cell must look as it is shown in Fig. 3.3 on the right.

Continuous conjugation of the displacement and normal stress on the boundaries of the adjacent cells shown in Fig. 3.3 is possible only if $N_2^{\mathrm{AB}\nu}$and σ_{23} are zero on the boundaries of the adjacent cells, i.e.,

$$N_2^{\mathrm{AB}\nu}(\pm L, y_3) = 0, \; \sigma_{23}(\pm L, y_3) = 0 \tag{3.28}$$

Conditions (3.28) replace the periodicity conditions in (3.21).

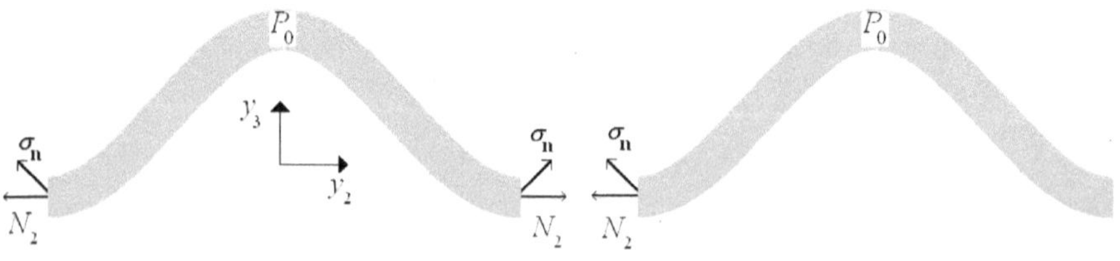

Fig. 3.3: Interface frequency adjacent cells which are symmetrical about the axis Oy_3

Rewrite Eqs. (3.28) in the terms of functions $\mathbf{M}^{AB\nu} = \mathbf{N}^{AB\nu} + (-1)^\nu \mathbf{u}^{AB\nu}(y_2, y_3)$ (A, B $= 1, 2$; $\nu = 0, 1$). The stresses σ_{23}, corresponding to displacement $\mathbf{u}^{AB\nu}(y_2, y_3)$, by construction are equal to $a_{23AB}(y_2, y_3)y_3^\nu$, see (3.7). For isotropic materials $a_{23AB} = 0$ for indices AB $=11$, 12, 21, 22,. Then the second equality in (3.28) holds both for $\mathbf{N}^{AB\nu}(y_2, y_3)$ and $\mathbf{N}^{AB\nu}(y_2, y_3)$. As far as first equality in (3.28), there exist two different cases: AB $=$ 11 and AB $=$ 22. For the case AB $=$ 11, taking into account definition of the function $\mathbf{u}^{11\nu}(y_2, y_3)$ (3.11) and (3.14), we obtain the following equations on the boundaries $y_2 = \pm L(\nu = 0, 1)$:

$$M_2^{11\nu}(\pm L, y_3) = \pm \nu_0 L y_3^\nu$$

For AB $=$ 22, taking into account definition of the function (3.17) and (3.20), we obtain the following equations on the boundaries $y_2 = \pm L$:

$$M_2^{22\nu}(\pm L, y_3) = \pm L y_3^\nu$$

Finally, we arrive at the problem

$$\begin{cases} (a_{\gamma\delta\iota\kappa}(y_2, y_3)M_{\iota,\kappa}^{AB\nu})_{,\delta} = 0 \quad \text{in } P_0, \\ a_{\gamma\delta\iota\kappa}(y_2, y_3)M_{\iota,\kappa}^{AB\nu}(y_2, y_3)n_\delta = 0 \quad \text{on } \Gamma_0, \\ \sigma_{23} = a_{23\iota\kappa}(\pm L, y_3)M_{\iota,\kappa}^{AB\nu}(\pm L, y_3) = 0, \\ \mathbf{M}^{AB\nu}(\pm L_2, y_3) = \pm c L y_3^\nu, \end{cases} \tag{3.29}$$

where $c = \nu_0$ for AB $=$ 11 and $c = \nu_0$ for AB $=$ 22 ($\nu = 0, 1$).

Problem (3.22) for $\nu = 0$ or equivalent problems (3.26), may be transformed into

$$\begin{cases} \Delta M^{120} = 0 \quad \text{in} \quad P_0, \\ \dfrac{\partial M^{120}}{\partial \mathbf{n}} = 0 \quad \text{on } \Gamma_0, \\ M^{120}(\pm L, y_3) = \pm L. \end{cases}$$

Elementary transformations do not allow transform problem (3.22) for $\nu = 1$ (or equivalent problem (3.27)) into a problem without "mass" and "surface" forces. Problem (3.27) may be transformed into suitable form by introduction of conjugate function (see below).

Stiffness $D_{2222}^{\nu+\mu}$, $D_{1111}^{\nu+\mu}$, $D_{1122}^{\nu+\mu}$, $D_{2211}^{\nu+\mu}$

Computing the stiffness indicated above, we use solutions to periodicity cell with indices AB $= 11, 22$. Write formula for the computation of the effective stiffness in the terms of solution to the reduced problems. Taking into account that in the case under consideration $\mathbf{N}^{\alpha\beta\nu}(\mathbf{y}) = (0, N_2^{\alpha\beta\nu}(y_2, y_3), N_3^{\alpha\beta\nu}(y_2, y_3))$, we obtain from (3.2) (A, B, $\Gamma, \Delta = 1, 2$; $\nu, \mu = 0, 1$)

$$D_{AB\Gamma\Delta}^{\nu+\mu} = \frac{1}{|S|} \int_P (a_{AB\iota\kappa}(\mathbf{y}) N_{\iota,\kappa}^{\Gamma\Delta\nu}(\mathbf{y}) + (-1)^\nu a_{AB\Gamma\Delta}(\mathbf{y}) y_3^\nu)(-1)^\mu y_3^\mu d\mathbf{y} \tag{3.30}$$

In order to compute stiffness $D_{1111}^{\nu+\mu}$, problem (3.29) is solved for AB = 11. After that the stiffness $D_{1111}^{\nu+\mu}$ are computed in accordance with (3.30) as

$$D_{1111}^{\nu+\mu} = \frac{1}{|S_0|}\int_P (a_{11\iota\kappa}(y_2,y_3)N_{\iota,\kappa}^{11\nu}(y_2,y_3) + (-1)^\nu a_{1111}(y_2,y_3)y_3^\nu)(-1)^\mu y_3^\mu dy_2dy_3.$$

Substituting $\mathbf{N}^{11\nu}(y_2,y_3) = \mathbf{M}^{11\nu}(y_2,y_3) - (-1)^\nu \mathbf{u}^{11}(y_2,y_3)$, we obtain

$$\begin{aligned} D_{1111}^{\nu+\mu} = \frac{1}{|S|}\int_P (a_{11\iota\kappa}(y_2,y_3)M_{\iota,\kappa}^{11\nu}(y_2,y_3) - (-1)^\nu a_{11\iota\kappa}(y_2,y_3)u_{\iota,\kappa}^{11\nu}(y_2,y_3)+ \\ +(-1)^\nu a_{1111}(y_2,y_3)y_3^\nu)(-1)^\mu y_3^\mu dy_2dy_3 \end{aligned} \tag{3.31}$$

Taking into account (3.9), we obtain

$$\begin{aligned} &-a_{11\iota\kappa}u_{\iota,\kappa}^{11\nu} + a_{1111}y_3^\nu \\ &\quad = -\frac{(a_{1122}(y_2,y_3) + a_{1133}(y_2,y_3))a_{2211}(y_2,y_3)}{a_{2222}(y_2,y_3) + a_{2233}(y_2,y_3)}y_3^\nu + a_{1111}(y_2,y_3)y_3^\nu. \end{aligned}$$

For isotropic material, this expression is equal to $E(y_2,y_3)y_3^\nu$ (E is Young's modulus) and (3.31) takes the form

$$D_{1111}^{\nu+\mu} = \frac{1}{|S|}\int_P (a_{11\iota\kappa}(y_2,y_3)M_{\iota,\kappa}^{11\nu}(y_2,y_3) + (-1)^\nu E(y_2,y_3)y_3^\nu)(-1)^\mu y_3^\mu dy_2dy_3 \tag{3.32}$$

In order to compute stiffness $D_{2222}^{\nu+\mu}$, problem (3.29) is solved for AB = 22. After that the stiffness are computed in accordance to the formula (3.30) as

$$D_{2222}^{\nu+\mu} = \frac{1}{|S|}\int_P a_{22\iota\kappa}(y_2,y_3)M_{\iota,\kappa}^{22\nu}(y_2,y_3)(-1)^\mu y_3^\mu dy_2dy_3. \tag{3.33}$$

Bearing into mind that $a_{\alpha\beta\iota\kappa}(y_2,y_3)M_{\iota,\kappa}^{AB\nu}(y_2,y_3)$ is stress $\sigma_{\alpha\beta}(y_2,y_3)$, we can rewrite formula (3.33) as

$$D_{2222}^{\nu+\mu} = \frac{1}{|S|}\int_P \sigma_{22}^{22\nu}(y_2,y_3)(-1)^\mu y_3^\mu dy_2dy_3 \tag{3.34}$$

Stiffness $D_{1122}^{\nu+\mu}, D_{2211}^{\nu+\mu}$

It is known, see, e.g., Caillerie (1984); Kohn and Vogelius (1984); Kalamkarov and Kolpakov (1997) that $D_{1122}^{\nu+\mu} = D_{2211}^{\nu+\mu}$ (ν,μ =0, 1). Then, we may computer one of these values. Consider

$$D_{2211}^{\nu+\mu} = \frac{1}{|S|}\int_P (a_{22\iota\kappa}(y_2,y_3)M_{\iota,\kappa}^{11\nu}(y_2,y_3) + (-1)^\nu a_{2211}(y_2,y_3)y_3^\nu)(-1)^\mu y_3^\mu dy_2dy_3.$$

Taking into account that $\mathbf{N}^{11\nu}(y_2,y_3) = \mathbf{M}^{11\nu}(y_2,y_3) - (-1)^{\nu}\mathbf{u}^{11\nu}(y_2,y_3)$, we obtain with regard to (7, AB = 11)

$$D_{2211}^{\nu+\mu} = \frac{1}{|S|}\int_P a_{22\iota\kappa}(y_2,y_3)M_{\iota,\kappa}^{11\nu}(y_2,y_3)(-1)^{\mu}y_3^{\mu}dy_2dy_3. \tag{3.35}$$

In the similar way, we obtain with regard to (7, AB = 22)

$$D_{1122}^{\nu+\mu} = \frac{1}{|S|}\int_P (a_{11\iota\kappa}(y_2,y_3)M_{\iota,\kappa}^{22\nu}(y_2,y_3)(-1)^{\mu+\nu}y_3^{\mu+\nu}dy_2dy_3. \tag{3.36}$$

3.5 Numerical Example 1 - Computation of Effective Stiffness of thin Corrugated Shells

The aim of or paper is analysis of thick (at least not very thin) plates. The theoretical analysis presented above (as well in the last section "Universal relations between the effective stiffness of corrugated plate") is universal, i.e., valid for plates of arbitrary thickness. We meet restriction for our method when we solve the unit cell problems numerically. This restriction is related not with the homogenization method, but with the fact that general finite element procedures are not adapted to thin domains. For thin plates, methods similar to one presented in Ye et al (2014) looks like the best. For not extremely thin plates out method gives stable computational results (except one case discussed below). It makes possible compare results of our computations with the results of other investigators.
Available results on the computation of the effective stiffness of corrugated plates are limited to thin plates. One of the latest papers on the subject is Ye et al (2014), which presents new formulas for computation of effective stiffness of thin corrugated plates as well as other formulas, which the authors (Ye et al, 2014) characterized as generally accepted.
We applied our method to compute effective stiffness of thin-walled corrugated plate considered in Ye et al (2014). The characteristics of corrugation were taken like in Ye et al (2014): it is sinusoidal corrugation with the length (period) $2L$ =0.64m and height 0.11m; thickness $h = 0.005$m, see Fig. 3.2; plate material characteristics: Young's modulus 30GPa, Poisson ratio 0.2. At the indicated size corrugation periodicity cell is a thin shell – the thin-wallness parameter $h/2L$ is 0.005/0.64=0.0078 (remind that a shell is accepted as thin-walled if the thin-walled parameter $h/2L$ is less than 1/20 Timoshenko, 1961).
Table 3.1 presents the effective stiffness computed in accordance to formula (3.33) and (3.31), and the effective stiffness computed in Ye et al (2014) by using the formulas derived in [1] (concerning D_{1212}^2, see below). All coupling stiffness $D_{\alpha\beta\alpha\beta}^1 = 0$ (α,β =1, 2) due to the symmetry of the plate under consideration with respect to the plane $y_3 = 0$. As demonstrated below (see formula (3.45) in the section "Universal relations between the effective stiffness of corrugated plate", for plate made of

Table 3.1: Effective stiffness

Formula	D^0_{2222} N/m	D^2_{2222} N·m	D^0_{1111} N/m	D^2_{1111} N·m	D^0_{1212} N/m	D^2_{1212} N·m	D^0_{1122} N/m	D^2_{1122} N·m
(3.33), (3.35), (3.43)	46458	251	187155658	1026941	50172461	161	92916	50
[1]	47613	261	187080000	1025540	50113000	162	9523	52

homogeneous material $D^{\nu+\mu}_{1122} = \nu_0 D^{\nu+\mu}_{2222}$. Similar formula was obtained early in Ye et al (2014). As a result, D^0_{1122} and D^2_{1122} coincide with the values listed in Ye et al (2014) with the accuracy proportional to the difference between D^0_{2222} and D^2_{2222} in Table 3.1 (i.e. with error less than 5%) and $D^1_{1122} = 0$.
In order to estimate the accuracy of numerical computations, stiffness D^0_{2222} and D^2_{2222} were computed in two different ways - by using formula (3.33) and formula (3.34). The computed values are presented in Table 3.2 From the mathematical analysis of the problem it follows that values computed by using formulas (3.33) and (3.34) must be equal. Relatively differences between the computed values are 0.33% and 0.375%, which is usually accepted as very good accuracy in the engineering computations.
In the numerical computations it was used about 25,000 two-dimensional triangular finite elements (about 10 elements through the thickness of the plate). Solution of the problem (3.29) took about 1 minute. Computations of the integrals (3.33) and (3.35) took negligible time.

3.6 Computation of the Effective Stiffnesses D^2_{1212}, D^2_{2121} for Thin Plates

The standard method of computation of the effective stiffness is based on the solution of boundary-value problem (3.27), and than applying Eq. (3.2), which may be written in the form

$$D^2_{1212} = -\frac{1}{|S|}\int\limits_P a_{121\kappa}(\mathbf{y})N^{121}_{1,\kappa}(\mathbf{y})y_3 d\mathbf{y} + \frac{1}{|S|}\int\limits_P a_{1212}(\mathbf{y})y_3^2 d\mathbf{y} \tag{3.37}$$

We found that standard method is not effective for thin shell. It is the results of appearance of contrast values in this computation scheme if plate is thin. In the exam-

Table 3.2: Effective stiffness

Formula	D^0_{2222}N/m	N·m
(3.33)	46458	250.9
(3.34)	46304	250
Relatively difference	0.33%	0.375%

ple considered above, numerical values of integrals in (3.37) are of order of 10^6 (for example, integral $|S|\int_P y_3^2 dy_2 dy_3/E$ =420629), while the effective stiffness D^2_{1212}, D^2_{2121} are values of the order of 10^2 (the torsion stiffness of flat plate of the thickness $h = 0.005$ may be used as estimate, it is $Eh^3/[24(1+\nu)] \approx 130$). In other words, the result value in (3.37) has of order of 0.01% (one hundred percent) of the subtracted values. Although the problem (3.29) is quite simple, ANSYS does not guarantee its solutions with the accuracy 0.01%. As a result, the computational procedure based on (3.29), (3.37) is unstable: small relative error in the computation of integrals (3.37) leads to a large relative error in computation of their difference. When using the approach based on (3.29), (3.37), the authors observed fluctuations of result in a range from 50 to 200 due to variations of parameters of ANSYS project, such as shape of finite element, methods for defining boundary conditions, etc., which slightly influence the final result in a stable computational procedure.

Note that the problem of contrast values presents in all problems formulated in the form (3.1), (3.2). The problem of contrast values is removed by the introduction of auxiliary functions $\mathbf{u}^{AB\nu}$ and w, whose addition to the solution, vanish "mass" and "surface" forces. Such the auxiliary functions exist for all the periodicity cell problems except only the problem under consideration. Really, it is possible to remove forces $(-1)^\nu a_{1\delta 12} y_3^\nu$ if and only if the system (3.25) is integrable. But it is *not* integrable for $\nu = 1$.

We propose a method of computation of effective stiffness D^2_{1212}, D^2_{2121}, which contains no contrast values for thin shells. We introduce the function ψ by the following conditions:

$$\psi_{,3} = N^{121}_{1,2} - y_3, \quad \psi_{,2} = -N^{121}_{1,3}. \tag{3.38}$$

Equations (3.38) are similar to the definition of conjugate function (Kolpakov, 2010). Equality $(N^{121}_{1,2} - y_3)_{,2} = -N^{121}_{1,33}$, which follows from the equation $\Delta N^{121}_1 = 0$ in (3.27), guarantees existence of the functions ψ possessing property (3.38).

Differentiating (3.38), we obtain

$$\psi_{,33} = N^{121}_{1,23} - 1, \quad \psi_{,22} = -N^{121}_{1,32}. \tag{3.39}$$

By adding the equations in (3.39), we obtain equation

$$\Delta\psi = -1 \text{ in } P_0 \tag{3.40}$$

Rewrite the boundary conditions on the top and the bottom boundaries Γ_0 in (3.29) in the form $(N^{121}_{1,2} - y_3)n_2 + N^{121}_{1,3} n_3 = 0$. This equation may be written as

$$\frac{\partial\psi}{\partial s} = \psi_{,3} n_2 - \psi_{,2} n_3 = 0 \tag{3.41}$$

where $\frac{\partial}{\partial s}$ means the derivative along the boundary (line) Γ_0. By virtue of (3.41), the function ψ is constant on the top and the bottom boundaries of the periodicity cell. Without loss of generality, we can assume that $\psi = 0$ on the bottom boundary.

Function $\partial N^{121}_1/\partial\mathbf{n}$ is periodic in $y_2 \in [-L, L]$. Since

$$\frac{\partial N_1^{121}}{\partial \mathbf{n}} = \psi_{,3}n_2 + y_3 n_2 - \psi_{,2}n_3 = \frac{\partial \psi}{\partial s} + \quad + y_3 n_2,$$

then $\partial \psi / \partial s$ is periodic $y_2 \in [-L, L]$. Then

$$\psi(y_2, y_3) = \int\limits_{H-h/2}^{y_2} \frac{\partial \psi}{\partial s} ds$$

take the same values for $y_2 = \pm L$. The quantity

$$\psi(L, H+h/2) = \int\limits_{H-h/2}^{H+h/2} \frac{\partial \psi}{\partial s} ds = \int\limits_{H-h/2}^{H+h/2} \frac{\partial M_1^{121}}{\partial \mathbf{n}} ds$$

is equal to the non-symmetric (coupling) stiffness D^1_{1212}, which is zero. This follows from the symmetry of the solution to the periodicity cell problem (3.26) (remind that D^1_{1212} may be computed through solution to problem (3.26) as well as solution to problem (3.27), see Eq. (3.2) and [9] for details). Then $\psi = 0$ on the top boundary.

As a result, we obtain the boundary conditions for equation (3.40)

$$\begin{cases} \psi(y_2, y_3) = 0 \quad \text{on } \Gamma_0, \\ \psi(y_2, y_3) \quad \text{is periodic in} \quad y_2 \in [-L, L]. \end{cases} \tag{3.42}$$

From (3.37) and (3.38), we obtain

$$D^2_{1212} = -\frac{1}{|S|} \int\limits_P a_{1212} \psi_{,3}(y_2, y_3) y_3 dy_2 dy_3 \tag{3.43}$$

There are no contrast values in the problem (3.40), (3.42) and formula (3.43). The stiffness D^2_{1212} presented in Table 3.1 was computed by using the method presented in this section.

3.7 Numerical Example 2 - Corrugated Plates of Arbitrary Thickness

We present results of our computation of effective stiffness D^0_{2222} for corrugated plates in wide diapason of thin-walled parameter $h/2L$. We consider corrugations which may be accepted as thin-walled and which cannot be accepted as thin-walled structures.

We analyze corrugated plate described in Example 1 (sinusoidal, length $2L$ =0.64m, height 0.11m; Young's modulus 30GPa, Poisson ratio 0.2). Now, the thickness of plate changes from 0.005 to 0.08. Remind that the critical value of the thin-walled parameter $h/2L$, which separates thin-walled shells from thick shells is accepted as 1/20 (Huber, 1923).

In Tables 3.3 and 3.4, we present values of effective stiffness D^0_{2222} computed by using our method and formula (3.12) from Ye et al (2014). Table 3.3 contains data for plates which may be accepted as thin-walled (the thickness of plate changes from 0.005 to 0.03, $h/2L$<1/20). Table 3.4 contains data for plates, which cannot be accepted as thin-walled (the thickness of plate takes values 0.06 and to 0.08, $h/2L$>0.09). We observe that the effective stiffness computed by using our method (developed with no restrictions on the plate thickness) and formulas from Ye et al (2014) (derived under condition that the plate is thin-walled structure) are close. For thin plates D^0_{2222} computed by using formula (3.12) from Ye et al (2014) is greater than the homogenized stiffness D^0_{2222} (3.33) $\nu+\mu=0$. If apply formula (3.12) from [1] for thick plates, one finds that D^0_{2222} approaches the homogenized stiffness D^0_{2222} (3.33) $\nu+\mu=0$ for $h/2L\approx 0.1$ and that became less than the homogenized stiffness. The "good agreement" between stiffness D^0_{2222} predicted by formula (3.12) from Ye et al (2014) and the homogenization theory is an example of surprises, which one can meet when apply approximate formulas.

Note that theoretically stiffness computed by using formula (3.12) from Ye et al (2014) and computed with the exact homogenization method must coincide asymptotically as $h\to 0$. But here we meet the situation when the approximated method from Ye et al (2014), which is asymptotic in nature, works better than the homogenization method coupled with the finite element computations. It is not an insufficient of the homogenization theory, but insufficient of finite element method, which usually loss accuracy in very thin domains. An alternative to the application of the finite element computations is the development of method based on the approximation of the unit cell problem of the homogenization theory by shell theory problem. This paper is devoted to thick plates rather than thin ones and we do not discuss approximation of the unit cell problem by shell theory problem here.

Present simple approximate formulas taken from Ye et al (2014) (the number of the formulas in Ye et al (2014) are (3.6) and (3.9))

Table 3.3: Stiffness D^0_{2222} for plates which may be accepted as thin-walled

h	0.005	0.007	0.01	0.015	0.02	0.03
$h/2L$	0.008	0.011	0.016	0.023	0.031	0.047
D^0_{2222} (3.33) $\nu+\mu=0$	46458	125963	368443	1228437	2919025	9760250
D^0_{2222} [1]	45215	124030	361400	1218000	2881300	9668700
Discrepancy %	3%	2%	2%	1%	1%	1%

Table 3.4: Stiffness D^0_{2222} for plats which cannot be accepted as thin-walled

h	0.06	0.08
$h/2L$	0.09	0.125
D^0_{2222} (3.33) $\nu+\mu=0$	75324063	172466250

$$D^0_{1111} = sEh, \quad D^0_{2222} = \frac{Eh^3}{6(1-\nu^2)T^2}, \quad D^0_{1122} = \nu D^0_{1111}, \quad D^0_{1212} = \frac{1}{s}\frac{Eh}{2(1+\nu)},$$

$$D^2_{1111} = EI, \quad D^2_{2222} = \frac{1}{s}\frac{Eh^3}{12(1-\nu^2)}, \quad D^2_{1122} = \nu D^2_{1111}, \quad D^2_{1212} = \frac{1}{s}\frac{Eh^3}{24(1+\nu)} \tag{3.44}$$

(s is the length of the corrugation divided by length of period, see Fig. 3.2, I is the moment of inertia along the corrugation direction) are presented.

Table 3.5 presents the effective stiffness D^2_{2222} computed using our method and formula from (3.44) for various values of the thickness h.

In Table 3.6 non-zero stiffness of sinusoidal corrugated plate for thickness $h = 0.06$ (this is the thick plate) are presented. Coupling stiffness $D^1_{\alpha\beta\alpha\beta} = 0$ due to the symmetry of the plate.

Note that formulas (3.44) for $h = 0.06$ and $h = 0.08$ provide us with the reference values, only, because these formulas were derived for thin plates. It is the reason why we select the most simple formulas from Ye et al (2014), although Ye et al (2014) presents numerous more accurate formulas. For this reason, we save in the reference values only the first two digits, to underline those approximate nature.

The significant difference between results of our computations and predictions of the theory of thin-walled shells (Kirchhoff-Love hypothesis (Huber, 1923), Hencky-type constraints (Lewinski, 1985)) are observed for the local stress/strain state of thick corrugations is analyzed. In thin-walled corrugations the stress/strain state follows to Kirchhoff-Love hypothesis, they are linear in the coordinate perpendicular to middle surface of the corrugation. This type of stress/strain state was observed in all our computations for thin walled corrugations.

Figure 3.4 displays local stress intensity S corresponding to solution to the cellular problem (3.29) for thickness $h = 0.08$. Solution corresponds to the overall deformation mode “tension of corrugated plate in the direction perpendicular to corrugations”, see Fig. 3.1. The corresponding thin-walled parameter is $h/2L$ =0.125>1/20.

Table 3.5: Stiffness D^2_{2222} for various value of the plate thickness

h	D^2_{2222} (3.44)	(3.33) $\nu+\mu=2$
0.005	266	259
0.06	450000	421292
0.08	1000000	1020016

Table 3.6: Stiffness of corrugated plate for $h = 0.06$

	D^0_{1111}	D^2_{1111}	D^0_{2222}	D^2_{2222}	D^0_{1212}	D^2_{1212}	D^0_{1122}	D^2_{1122}
	22425966437	13150570	75324063	421292	606535156	279887	15064813	84258
(3.44)	22000000000	12000000	92000000	450000	600000000	280000	18000000	90000

Values of the stress intensity S in the point marked in Fig. 3.4 are presented in Table 3.7. The maximum value of the stress intensity was observed in the top of corrugation on the inner side of the corrugation, marked by T in Fig. 3.4.

On the top of corrugation the ratio of the stress intensities on the inner and outer sides (marked by T and B, correspondingly) is $\frac{S_T}{S_B} = 1.48$. For many isotropic materials the strength criterion may be taken in the form $S = \sigma^*$ [21], where σ^* means the strength limit. The point marked in Fig. 3.4 by$T$ is the weakest point of the corrugation made of such materials. Computations carried out for other deformations modes (bending, shift) indicate the same point T as the point of maximum value of the stress intensity.

Figure 3.5 displays local stress intensity S corresponding to solution $\mathbf{N}^{221}(\mathbf{y})$ to the cellular problem (3.29) for thickness $h = 0.12$. The corresponding thin-walled parameter is $h/2L$ =0.125>1/20. Solution $\mathbf{N}^{221}(y_2, y_3)$ corresponds to overall deformation mode "bending of corrugated plate in the direction perpendicular to corrugations". Values of the stress intensity S in the point T and B in Fig. 3.5 are S_T =0.802GPa (it is the maximum value over the corrugation) and S_B =0.309GPa, respectively. The ratio $\frac{S_T}{S_B} = 2.6$.

Although $\mathbf{N}^{221}(\mathbf{y})$ corresponds to "pure bending" mode in whole, we observe in Fig. 3.4 no points with the stress intensity $S = 0$ (the minimum stress intensity is more than 7000000 Pa), which would be observed on the neutral surface. Position of the "overall" neutral surface is displayed in Fig. 3.5 as black horizontal line.

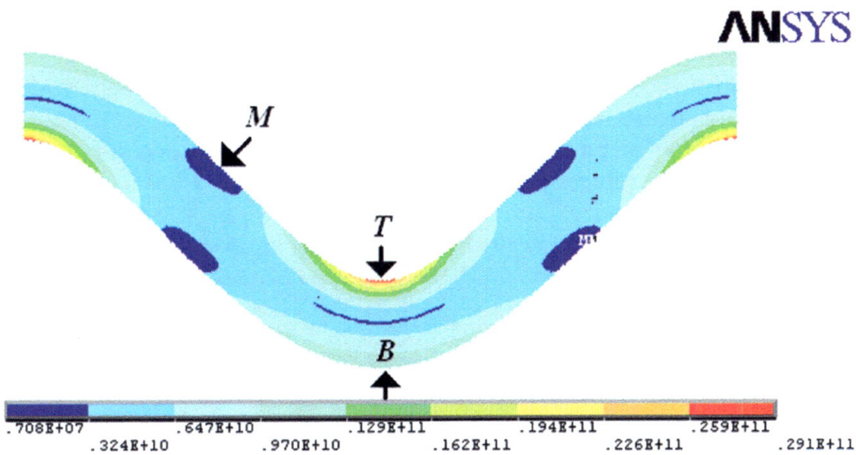

Fig. 3.4: Stress intensity corresponding to solution to the cellular problem for $h = 0.08$

Table 3.7: Values of the local stress intensity (Pa)

Reference point	B	T	M
S	1880000	2790000	241229

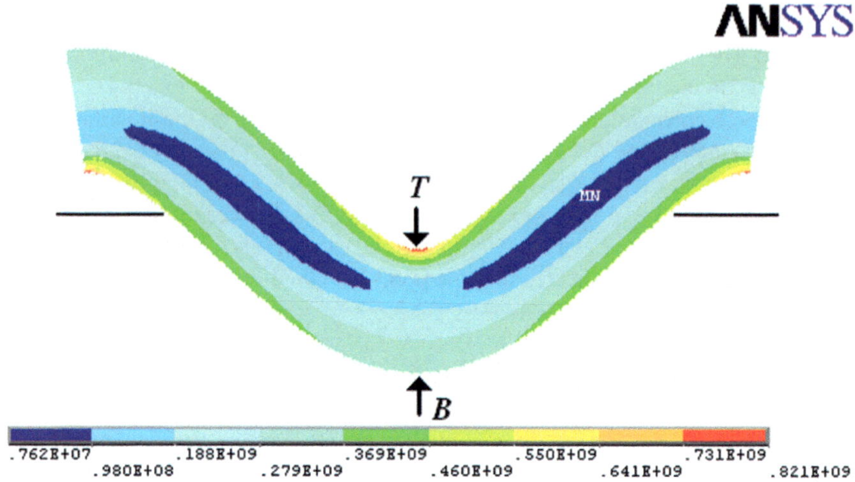

Fig. 3.5: Stress intensity corresponding to solution $\mathbf{N}^{221}(y_2, y_3)$ to the cellular problem for $h = 0.12$

For comparison, we present local stress intensity S corresponding to solution $\mathbf{N}^{221}(\mathbf{y})$ (the bending mode) to the cellular problem (3.29) for thickness $h = 0.005$, see Fig. 3.6. The corresponding thin-walled parameter is $h/2L$ =0.008<<1/20. In Fig. 3.6, the zoomed fragment of the bottom of the corrugation is displayed. The stress intensity has the similar distribution along the corrugation.

In the tensile mode (in the direction perpendicular to corrugations, along Ox_2-axis) distribution of stresses along the corrugation is non-uniform even for thin-walled corrugation. In Fig. 3.7, it is displayed the local stress intensity Scorresponding to solution $\mathbf{N}^{220}(\mathbf{y})$ to cellular problem (3.29) for thickness $h = 0.005$. In Fig. 3.7, numerical values of the local stress intensity are presented for selected points. The maximal value if the stress intensity is observed in the top of the corrugation.

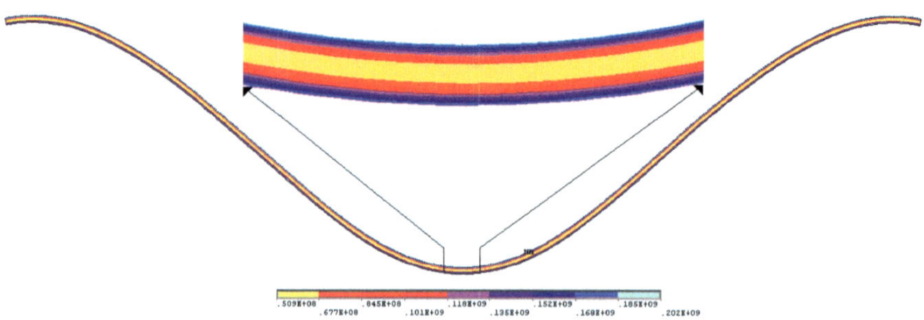

Fig. 3.6: Stress intensity corresponding to solution $\mathbf{N}^{221}(y_2, y_3)$ to the cellular problem for $h = 0.005$

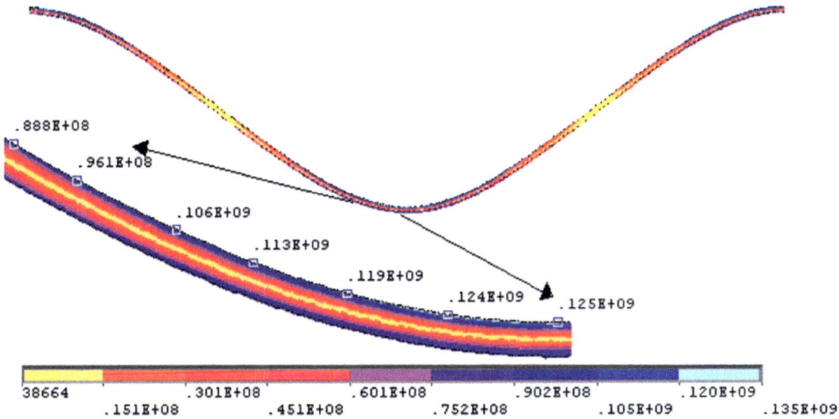

Fig. 3.7: Stress intensity corresponding to solution $\mathbf{N}^{220}(y_2, y_3)$ to the cellular problem for $h = 0.005$

Present examples of failure corrugated structures, which give arguments in the favor of our prediction that the top of corrugation is the weakest place of corrugated plates. Figure 3.8 displays failure roofing slate. The cracks have shapes close to right-lines and positions in the top of corrugation. An example of similar crack in metal corrugated tube can be found in the internet[1].

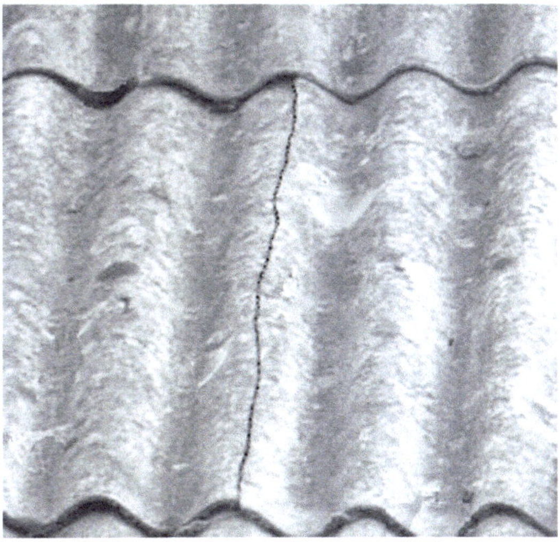

Fig. 3.8 Cracks in roofing slate

[1] http://www.flowcontrolnetwork.com/articles/q-a-getting-a-handle-on-metal-hose

3.8 Universal Relations Between the Effective Stiffness of Corrugated Plates made of Materials with the same Poisson's Ratio

Relations between the homogenized characteristics of composite material, which do not depend on the solutions to the periodicity cell problems are called universal connections (Dvorak, 2013) or universal relations (Bravo-Castillero et al, 2012). The universal relations may depend on the known or easily computed characteristics of composite, such as elastic constants, volume fractions of components, etc. In this section, we derive universal relations for the effective stiffness of corrugated plate made of materials with the same Poisson's ration (in particular, corrugated plates made of homogeneous material).

Consider problem (3.29). Taking into account definition of the parameter c in (3.29), we obtain the following relationship between solutions to the problem (3.29) for the indices AB $= 11$ and AB $= 22$ $(\nu,\mu = 0,1)$:

$$\mathbf{M}^{11\nu}(y_2,y_3) = \nu_0\mathbf{M}^{22\nu}(y_2,y_3) \tag{3.45}$$

By using (3.45) and (3.33), we transform formula (3.35) as follows:

$$\begin{aligned} D_{2211}^{\nu+\mu} &= \frac{1}{|S_0|}\int\limits_P a_{22\iota\kappa}(y_2,y_3)M_{\iota,\kappa}^{11\nu}(y_2,y_3)(-1)^{\mu}y_3^{\mu}dy_2dy_3 = \\ &= \frac{1}{|S_0|}\nu_0\int\limits_P a_{22\iota\kappa}(y_2,y_3)M_{\iota,\kappa}^{22\nu}(y_2,y_3)(-1)^{\mu}y_3^{\mu}dy_2dy_3 = \nu_0 D_{2222}^{\nu+\mu} \end{aligned}$$

and obtain universal relationship between effective stiffness $D_{2211}^{\nu+\mu}$ and $D_{2222}^{\nu+\mu}$:

$$D_{2211}^{\nu+\mu} = \nu_0 D_{2222}^{\nu+\mu} \tag{3.46}$$

Taking into account (3.46) and (3.32), we transform formula (3.36) as follows:

$$\begin{aligned} D_{1122}^{\nu+\mu} &= \frac{1}{|S_0|}\int\limits_P (a_{11\iota\kappa}(y_2,y_3)M_{\iota,\kappa}^{22\nu}(y_2,y_3)(-1)^{\mu+\nu}y_3^{\mu+\nu}dy_2dy_3 = \\ &= \frac{1}{|S_0|}\frac{1}{\nu_0}\int\limits_P (a_{11\iota\kappa}(y_2,y_3)M_{\iota,\kappa}^{11\nu}(y_2,y_3)(-1)^{\mu+\nu}y_3^{\mu+\nu}dy_2dy_3 = \\ &= \frac{1}{\nu_0}(D_{1111}^{\nu+\mu} - \frac{(-1)^{\mu+\nu}}{|S_0|}\int\limits_P E(y_2,y_3)y_3^{\mu+\nu}dy_2dy_3). \end{aligned} \tag{3.47}$$

Equality (3.47) provides us with the following relationship between effective stiffness $D_{1122}^{\nu+\mu}$ and $D_{1111}^{\nu+\mu}$:

$$D_{1122}^{\nu+\mu} = \frac{1}{\nu_0}(D_{1111}^{\nu+\mu} - \frac{(-1)^{\nu+\mu}}{|S_0|}\int\limits_P E(y_2,y_3)y_3^{\nu+\mu}dy_2dy_3). \tag{3.48}$$

Since$D_{1122}^{\nu+\mu} = D_{2211}^{\nu+\mu}$ [9], equations (3.46) and (48) lead to the following relationship between effective stiffness $D_{2222}^{\nu+\mu}$ and $D_{1111}^{\nu+\mu}$:

$$\nu_0 D_{2222}^{\nu+\mu} = \frac{1}{\nu_0}(D_{1111}^{\nu+\mu} - \frac{(-1)^{\nu+\mu}}{|S_0|}\int_P E(y_2,y_3)y_3^{\nu+\mu}dy_2dy_3). \tag{3.49}$$

Equations (3.46), (48) and (49) are universal – they hold for any corrugated plate, regardless of its specific cross-section shape, under unique condition of symmetry of the cross-section. These equations involve, except the effective stiffness, Poisson ratio and the moments

$$\frac{1}{|S_0|}\int_P E(y_2,y_3)y_3^{\mu+\nu}dy_2dy_3$$

of the Young's modulus.

3.9 Conclusions

For corrugated plate, more generally, for plate which occupies a cylindrical domain, three-dimensional periodicity cell problem of the homogenization theory can be reduced to a boundary-value problems with respect to function of two spatial variables determined on the cross-section of the periodicity cell. Depending on the index of the periodicity cell problem, we arrive at problem for anti-plane deformation or two-dimensional elasticity theory problem. Modern engineering software makes it possible to solve all 2D problems fast and accurate.

The reduced problems take very simple form for plates made of a homogeneous material or assembled of elements with equal Poisson ratios. If the periodicity cell is symmetric, additional simplifications of the reduced problems may be done.

It is found that the conventional procedure of the homogenization theory in application to the computation of the torsion effective stiffness of thin plates is not stable because it involves different values. It is proposed a modification of the procedure, which is stable for thin plates.

Our numerical computations demonstrate closeness of effective stiffness computed by our method and approximate methods given in [1, 2, 18, 19, 20]. Numerical computations also demonstrates linear distribution of local stress/strain over the thickness for thin plates.

The universal relations between the effective stiffness of corrugated plate for plates made of homogeneous material are found.

Acknowledgements This work was supported by the Russian Foundation for Basic Research (Project RFBR-East No. 15-41-0508, Project No 16-01-00679).

References

Andrianov IV, Diskovsky AA, Kholod EG (1998) Homogenization method in the theory of corrugated plates. Technische Mechanik 18:123–133

Arkhangelskii AF, Gorbachev VI (2007) Effective characteristics of corrugated plates. Mechanics of Solids 42(3):137–155

Barber JR (2011) Intermediate Mechanics of Materials. Springer, Berlin

Bartolozzi G, Pierini M, Orrenius U, Baldanzini N (2013) An equivalent material formulation for sinusoidal corrugated cores of structural sandwich panels. Composite Structures 100:173–185

Bravo-Castillero J, Rodríguez-Ramos R, Guinovart-Díaz R, Mechkour H, Brenner R, Camacho-Montes H, Sabina FJ (2012) Universal relations and effective coefficients of magneto-electro-elastic perforated structures. The Quarterly Journal of Mechanics and Applied Mathematics 65(1):61–85

Buannic N, Cartraud P, Quesnel T (2003) Homogenization of corrugated core sandwich panels. Composite Structures 59:299–312

Caillerie D (1984) Thin elastic and periodic plates. Mathematical Methods in Applied Sciences 6:159–191

Dvorak G (2013) Micromechanics of Composite Materials. Springer, Berlin

Frikha A, Cartraud P, Tryssede F (2013) Mechanical modeling of helical structures accounting for translation invariance. Part 1: Static behavior. Int J Solids Struct 50(9):1373–1382

Huber MT (1923) Die Theorie des kreuzweise bewehrten Eisenbetonplatten. Der Bauingenieur 4:354–360

Kalamkarov AL, Kolpakov AG (1997) Analysis, Design and Optimization of Composite Structures. John Wiley&Sons, Chichester

Kohn RV, Vogelius M (1984) A new model for thin plates with rapidly varying thickness. Int J Solids Struct 20:333–350

Kolpakov AG (2010) Stressed composite structures: Homogenized Models for Thin-walled Nonhomogeneous Structures with Initial Stresses. Springer, Berlin

Lee CY, Yu W (2011) Homogenization and dimensional reduction of composite plates with in-plane heterogeneity. Int J Solids Struct 48:1474–1484

Lewinski T (1985) Effective stiffness of transversally non-homogeneous plates with unidirectional periodic structure. Int J Solids Struct 32(19):3261–3287

Talbi N, Batti A, Ayad R, Guo YQ (2009) An analytical homogenization model for finite element modelling of corrugated cardboard. Composite Structures 88:280–289

Timoshenko S (1961) Theory of Elastic Stability. McGraw-Hill, New York

Xia Y, Friswell MI, Flores EIS (2012) Equivalent models of corrugated panels. Int J Solids Struct 49(13):1453–1462

Ye Z, Berdichevsky VL, Yu W (2014) An equivalent classical plate model of corrugated structures. Int J Solids Struct 51(11-12):2073–2083

Chapter 4
Consideration of Non-Uniform and Non-Orthogonal Mechanical Loads for Structural Analysis of Photovoltaic Composite Structures

Marcus Aßmus, Stefan Bergmann, Johanna Eisenträger, Konstantin Naumenko & Holm Altenbach

Abstract At natural weathering, terrestrial photovoltaic modules are exposed to mechanical loads, inter alia. These mechanical loads are primarily caused by wind and snow. They can have an essential influence on the durability of photovoltaic modules, what should be taken into account during the design process. Up to now, mechanical loads are considered as uniform and orthogonal in the majority of experimental and theoretical investigations, resulting in over- or underestimation of stresses in the components of photovoltaic modules. Regarding the mathematical-mechanical modelling, it is of particular interest to represent the applied loads in an adequate manner. Therefore, this work provides an approach to the mathematical description of location and orientation-dependent loads to take non-uniform and non-orthogonal mechanical impact into account. In addition, the integration in an efficient and effective computational solution strategy towards global structural analysis is presented, which is based on a newly developed direct approach for multiple layers. Thereby, we reflect on the governing equations and the implementation of non-uniform and non-orthogonal loading. Characteristic case studies complete the present treatise.

Key words: Photovoltaic module · Loading · Non-uniform · Non-orthogonal · Surface continuum · Polar medium · Multilayered surface structure · Finite element analysis

---

Marcus Aßmus · Stefan Bergmann · Johanna Eisenträger · Konstantin Naumenko · Holm Altenbach
Institut für Mechanik, Otto-von-Guericke- Universität Magdeburg, Universitätsplatz 2, 39106 Magdeburg, Germany,
e-mail: `marcus.assmus@ovgu.de`, `stefan.bergmann@st.ovgu.de` `johanna.eisentraeger@ovgu.de`, `konstantin.naumenko@ovgu.de`, `holm.altenbach@ovgu.de`

H. Altenbach et al. (eds.), *Mechanics for Materials and Technologies*,
Advanced Structured Materials 46, DOI 10.1007/978-3-319-56050-2_4

4.1 Introduction

4.1.1 Motivation

During the last decades, renewable energy technologies are on the rise such that photovoltaic modules gain more and more importance, too. Typically, photovoltaic modules comprise several layers, as depicted in Fig. 4.1 on the left-hand side. The skin layers protect sensitive internal components, i.e. the solar cells and electrical conductors, from external loads resulting from climatic conditions like precipitations or wind flow. Due to this, transparent and stiff materials like glass are usually used for the skin layers. In addition, there is a core layer, which includes an encapsulation, the solar cells, and electrical wiring. The encapsulation is made of soft polymers and is relatively thin compared to the skin layers.

In order to perform structural mechanics analyses of photovoltaic modules, various possibilities exist. Basically, it is possible to distinguish three-dimensional (3D) and two-dimensional (2D) approaches. Most common to three-dimensional approaches is the use of the finite element method (FEM) based on a classical (three-dimensional CAUCHY) continuum. The use of brick elements applied to this purpose is computationally expensive. However, the approaches for thin-walled structures are more efficient since the underlying body is slender ($L_1 \approx L_2 \gg H$, with overall thickness H). Within the framework of these approaches, an model structure is assumed, depicted in Fig. 4.1, centred. Considering aforementioned dimensions and properties, the present structure can be considered as an anti-sandwich ($h^c < h^s$, $G^c < G^s$, indices s and c for skin and core layers, while $2h^s + h^c = H$ holds, and G is the shear modulus). Potential candidates to treat problems at such structures are the first order shear deformation theory (e.g. Reissner, 1945; Mindlin, 1951), zig-zag theories, and layerwise theories (Carrera, 2003a). With regard to their applicability to the present problem, the different approaches are discussed in Naumenko and Eremeyev (2014) and Eisenträger et al (2015a). Since closed form solutions of thin-walled structural theories are available for a small set of problems only, numerical solution strategies using FEM, like presented in Eisenträger et al (2015b), are used. Typically, such approaches are based on 2D finite elements (Carrera, 2002), as illustrated in Fig. 4.1 on the right-hand side. However, it emerges that the eXtended LayerWise Theory (XLWT) presented in Naumenko and Eremeyev (2014) has the largest field of application and the least computational effort (Aßmus et al, 2016), thus forming the appropriate framework for global structural analysis at photovoltaic modules.

Mechanical loading on photovoltaic modules seems to be a major topic, as is apparent studying the relevant literature, e.g. Köntges et al (2014) or Ferrara and Philipp (2012). Caused by mechanical loading, damage in photovoltaic module components can occur, which influences the electrical power yield. If one is restricted to quasi-static loading, the two essential loads wind and snow can be identified. Close to the ground, wind is acting horizontally, while snow loads act in direction of the gravitational force, i.e. vertically. Since photovoltaic modules are tilted in dependence of

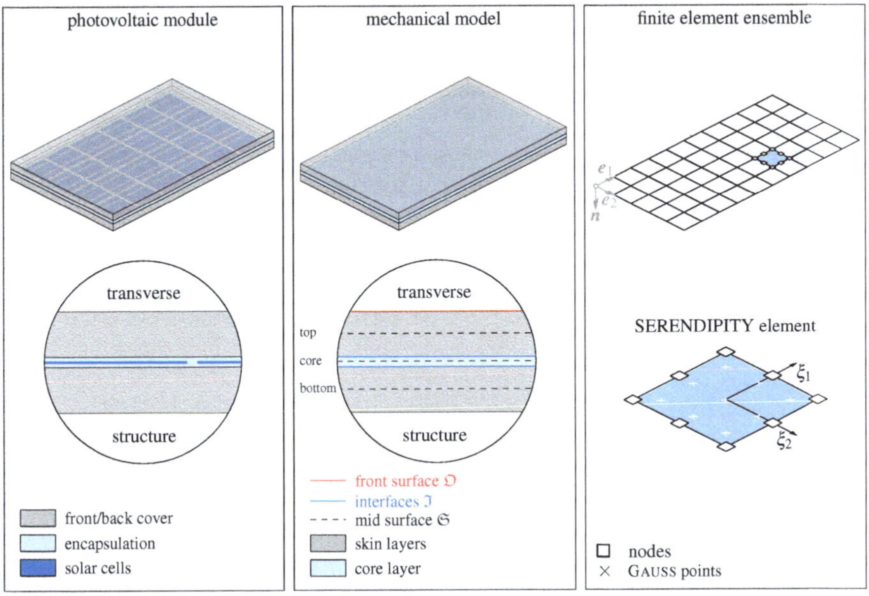

Fig. 4.1: Physical structure, mechanical model, and computational imitation to solve global boundary value problems at photovoltaic modules

latitude to generate best electrical yields, both loads will not act orthogonal to the plane mapped by the photovoltaic module front surface. Considering experimental, e.g. Moore (1977); Aßmus et al (2011); Kajari-Schröder et al (2011), and theoretical investigations, e.g. Sander et al (2013); Aßmus et al (2016), on mechanical loading at photovoltaic modules, it is clear that drastic simplifications are made with regard to the load assumptions. In most cases, the relevant standards IEC 62125[1] or IEC 61646[2] are referenced, which form the normative background for market approval of photovoltaic modules. Therein, it is demanded to apply a homogeneous load, orthogonal to the photovoltaic module mid plane with defined intensity and duration of loading in order to prove the structural integrity. Based on the assumptions for orthogonality and homogeneity of applied loads, the tests proposed therein are considered as unrealistic. General remarks concerning the determination of more realistic mechanical loads are given in EN 1991-1-3[3] for snow and EN 1991-1-4[4] for wind, for example. Although these standards are based on more realistic load as-

[1] Standard IEC 61215: Crystalline silicon terrestrial photovoltaic (PV) modules – Design qualification and type approval, International Electrotechnical Commission, Genf, 2nd edition (2005)

[2] Standard IEC 61646: Thin-film terrestrial photovoltaic (PV) modules – Design qualification and type approval, International Electrotechnical Commission, Genf, 2nd edition (2008)

[3] Standard EN 1991-1-3 (Eurocode 1): Eurocode 1 - Actions on structures - Part 1-3: General actions - Snow loads, European Committee for Standardisation (CEN), Brussels, 2010

[4] Standard EN 1991-1-4 (Eurocode 1): Eurocode 1 - Actions on structures - Part 1-4: General actions - Wind loads, European Committee for Standardisation (CEN), Brussels, 2010

sumptions concerning intensity, resulting loads proposed therein are merely rough estimations of actual loading conditions, especially concerning load distributions. While field data from outdoor exposure is an almost critical point in the area of photovoltaic research, different studies are concerned with the experimental identification of natural loading scenarios caused by precipitation and wind flow, e.g. Aßmus and Köhl (2012). However, this aspect does not seem to be finished completely since the number of different build-up possibilities and locations for photovoltaic modules is almost unlimited and difficult to generalise.

Since there is a lack of specific information on realistic loads until today, the next step is the mathematical modelling for an adequate representation of exemplaric loads on photovoltaic modules. This is not only of interest in present theoretical investigations because the problem involves experimental studies on mechanical strength as well. However, the aim of the present work is a mathematical approach to account for non-uniform and non-orthogonal actions on photovoltaic modules. This is of particular interest since engineers would like to know already during the planning phase how their design will withstand loading. Therefore, the application of approaches for adequate load representation will be shown in the context of the XLWT (Naumenko and Eremeyev, 2014), which provides an efficient framework to solve boundary value problems at the global scale of photovoltaic modules (Aßmus et al, 2016). Since closed form solutions of this theory can only be found for a small class of problems, a computational approach with the newly developed finite element proposed in Eisenträger et al (2015b) is used. All around, the present work provides an efficient tool, which is required by every engineer, working on the design of photovoltaic modules.

4.1.2 Objective and Structure

The present contribution deals with an approach to the description of non-uniform and non-orthogonal mechanical loads at photovoltaic modules. In addition, the effects of these load assumptions differing from previous, usual investigations are examined. For this purpose, a structural mechanics theory for thin-walled structures is used and extended to a three layered composite structure. The solution of the resulting boundary value problem is performed by means of the FEM. Thereby, a specially developed SERENDIPITY element is used.

The outline of the paper is as follows. In Sect. 4.2, we present a literature review for quasi-static mechanical loads resulting from natural weathering conditions. As mentioned, the mechanical loads resulting from snow and wind are handled. In the sequel, a mathematical concept for the treatment of natural mechanical loading is introduced, while we distinguish between wind and snow loading due to different directions of action.

In Sect. 4.3, we present general statements of the linear XLWT with a layerwise superposed membrane, bending, and shear state. Since solar cells embedded in the core layer do not contribute to the overall stiffness according to Naumenko and

Eremeyev (2014), it is possible to abstract the present structure to a three layered composite structure with homogenous layers. Based on a single layer which is mechanically represented by a midlayer, we enlarge our framework to a three layered system, where a coordinate-free description is used. In contrast to Naumenko and Eremeyev (2014), MINDLIN kinematics are considered for every layer in accordance with Eisenträger et al (2015b), even if transverse shear effects in the skin layers are negligible. The governing equations are disclosed, whereby the loads applied find particular attention.
Section 4.4 is concerned with an approach to efficient numerical implementation towards a finite element solution where we follow the track of Eisenträger et al (2015b). Special emphasis is placed on the derivation of the equations which are fundamental to the computational solution of the boundary value problem posed. For reasons of computational tractability, the governing equations are given in vector-matrix notation as usual in finite element representations. Here, only one element in transverse direction is used, considering the degrees of freedom (DOF's) of several layers at every node.
Finally, in Sect. 4.5 a representative case study is accomplished, where we issue the resulting DOF's, kinetics, and kinematics additionally. For the sake of simplicity, the test structure used in this case study is restricted to symmetry in transverse direction. The work concludes with a summary of this comprehensive treatment and an outlook on further investigations.

4.1.3 Preliminaries and Notation

Throughout the whole text, a direct tensor notation is preferred. Only if it is conducive to the clarity of the representation and to avoid additional formal definitions, we drop this convention and use index notation instead. Zero rank tensors are symbolised by italic letters (e.g. a), first rank tensors by italic lowercase bold letters (e.g. $\boldsymbol{a} = a_i \boldsymbol{e}_i$ or $\boldsymbol{b} = b_j \boldsymbol{e}_j$), second rank tensors by italic uppercase bold letters (e.g. $\boldsymbol{A} = A_{lm} \boldsymbol{e}_l \otimes \boldsymbol{e}_m$ or $\boldsymbol{B} = B_{no} \boldsymbol{e}_n \otimes \boldsymbol{e}_o$), and fourth rank tensors by italic uppercase bold calligraphic letters (e.g. $\boldsymbol{\mathcal{A}} = A_{pqrs} \boldsymbol{e}_p \otimes \boldsymbol{e}_q \otimes \boldsymbol{e}_r \otimes \boldsymbol{e}_s$), where EINSTEIN summation convention is applied. Letters in blackboard bold (e.g. $\mathbb{R}$ or $\mathbb{Z}$) are reserved for sets and spaces. Letters in broken script (Fraktur), e.g. $\mathfrak{D}$, $\mathfrak{I}$, or $\mathfrak{S}$, are used for material body manifolds. Considering a CARTESian coordinate system (COOS) and orthonormal bases, e.g. $\{\boldsymbol{e}_i\}$, basic operations for tensors used in this paper are the scalar product

$$\boldsymbol{a} \cdot \boldsymbol{b} = a_i b_j \boldsymbol{e}_i \cdot \boldsymbol{e}_j = a_i b_i = \alpha \qquad \alpha \in \mathbb{R}, \tag{4.1}$$

the dyadic product

$$\boldsymbol{a} \otimes \boldsymbol{b} = a_i b_j \boldsymbol{e}_i \otimes \boldsymbol{e}_j = \boldsymbol{C}, \tag{4.2}$$

the composition of a second and a first rank tensor

$$\boldsymbol{A}\cdot\boldsymbol{a}=A_{lm}a_i\boldsymbol{e}_l\otimes\boldsymbol{e}_m\cdot\boldsymbol{e}_i=A_{li}a_i\boldsymbol{e}_l=\boldsymbol{d}\,, \tag{4.3}$$

the composition of two second rank tensors

$$\boldsymbol{A}\cdot\boldsymbol{B}=A_{lm}B_{no}\boldsymbol{e}_l\otimes\boldsymbol{e}_m\cdot\boldsymbol{e}_n\otimes\boldsymbol{e}_o=A_{lm}B_{mo}\boldsymbol{e}_l\otimes\boldsymbol{e}_o=\boldsymbol{D}\,, \tag{4.4}$$

the double scalar product between a fourth and a second rank tensor

$$\begin{aligned}\mathcal{A}:\boldsymbol{B}&=A_{pqrs}B_{no}\boldsymbol{e}_p\otimes\boldsymbol{e}_q\otimes\boldsymbol{e}_r\otimes\boldsymbol{e}_s:\boldsymbol{e}_n\otimes\boldsymbol{e}_o\\&=A_{pqrs}B_{sr}\boldsymbol{e}_p\otimes\boldsymbol{e}_q=\boldsymbol{F}\,,\end{aligned} \tag{4.5}$$

the cross product between two first rank tensors

$$\boldsymbol{a}\times\boldsymbol{b}=a_ib_j\boldsymbol{e}_i\times\boldsymbol{e}_j=a_ib_j\varepsilon_{ijk}\boldsymbol{e}_k=\boldsymbol{c}\,, \tag{4.6}$$

and the cross product between a second and a first rank tensor

$$\boldsymbol{A}\times\boldsymbol{b}=A_{lm}b_j\boldsymbol{e}_l\otimes\boldsymbol{e}_m\times\boldsymbol{e}_j=A_{lm}b_j\varepsilon_{mjk}\boldsymbol{e}_l\otimes\boldsymbol{e}_k=\boldsymbol{G}\,, \tag{4.7}$$

where ε_{ijk} is the permutation symbol with

$$\varepsilon_{ijk}=\begin{cases}+1 & \text{if } (i,j,k) \text{ is an even permutation of } (1,2,3)\,,\\-1 & \text{if } (i,j,k) \text{ is an odd permutation of } (1,2,3)\,,\\0 & \text{if } (i,j,k) \text{ is not a permutation of } (1,2,3)\,.\end{cases} \tag{4.8}$$

The vectorial invariant of a second rank tensor $\boldsymbol{A}$ is defined by the box product $\boxtimes$ which represents a cross-scalar product in present treatise.

$$\begin{aligned}\boldsymbol{A}\boxtimes\boldsymbol{1}&=\boldsymbol{A}\times\cdot\boldsymbol{1}\\&=(A_{kl}\boldsymbol{e}_k\otimes\boldsymbol{e}_l)\times\cdot(\boldsymbol{e}_i\otimes\boldsymbol{e}_i)\\&=A_{kl}\,\delta_{li}\boldsymbol{e}_k\times\boldsymbol{e}_i\\&=A_{ki}\boldsymbol{e}_k\times\boldsymbol{e}_i\,.\end{aligned} \tag{4.9}$$

For symmetric tensors (coincidence of $\boldsymbol{A}$ and $\boldsymbol{A}^\top$, while superscript $^\top$ marks the transposed tensor such that $\boldsymbol{a}\cdot\boldsymbol{A}^\top\cdot\boldsymbol{b}=\boldsymbol{b}\cdot\boldsymbol{A}\cdot\boldsymbol{a}$ holds), the box product is the first rank zero tensor $\boldsymbol{o}$. Furthermore, the box product in the context of the connection with the vectorial invariant possesses the following properties

$$\boldsymbol{A}\boxtimes\boldsymbol{1}=\boldsymbol{A}\times\cdot\boldsymbol{1}=\boldsymbol{1}\cdot\times\boldsymbol{A}=-\boldsymbol{1}\cdot\times\boldsymbol{A}^\top=-\boldsymbol{A}^\top\times\cdot\boldsymbol{1}=-\boldsymbol{A}^\top\boxtimes\boldsymbol{1} \tag{4.10}$$

In Eq. (4.9), the KRONECKER delta δ_{ij} is used to represent the second rank unit tensor

$$\boldsymbol{1}=\delta_{ij}\boldsymbol{e}_i\otimes\boldsymbol{e}_j=\boldsymbol{e}_i\otimes\boldsymbol{e}_i \qquad \text{with} \quad \delta_{ij}=\begin{cases}1 & \text{if } i=j\,,\\0 & \text{if } i\neq j\,.\end{cases} \tag{4.11}$$

Using the tensor notation, Latin indices (e.g. i, j, k, l) run through the values 1, 2, and 3, while Greek indices (e.g. $\alpha, \beta, \gamma, \delta$) run through the values 1 and 2. The vector valued nabla operator is defined as $\nabla = \boldsymbol{e}_\alpha \, {}^{\partial}/_{\partial x_\alpha}$ for the planar considerations and $\nabla = \boldsymbol{e}_i \, {}^{\partial}/_{\partial x_i}$ at three dimensions. $\nabla \cdot \Box$ is the divergence, and $\nabla\Box$ is the gradient of a tensor. $\nabla^{\mathrm{sym}}\Box = {}^1/_2[\nabla\Box + \nabla^\top\Box]$ is the symmetric part of the associated gradient, where $\Box$ holds true for every differentiable tensor field. The transposed gradient is defined as $\nabla^\top\Box = [\nabla\Box]^\top$ where $\Box$ holds for all first rank tensors. An extended overview of tensor algebra and analysis is given in basic textbooks on continuum mechanics featuring mathematical propaedeutics, e.g. in Altenbach (2015); Lai et al (2009); Bertram (2012), or even in a more general manner in Lebedev et al (2010).

In vector-matrix notation, vectors are denoted as upright lowercase sans serif bold letters (e.g. displacement vector $\mathbf{u} = [u_1 \; u_2 \; u_3]^\top$) and matrices as upright uppercase sans serif bold letters (e.g. stiffness matrix $\mathbf{K}$ or matrix of shape functions $\mathbf{N}$).

4.2 Mechanical Loads at Photovoltaic Modules

4.2.1 Loading at Natural Weathering

The lifetime of photovoltaic modules is limited, particularly due to the impact of different mechanical loads in natural weathering. Such loads lead to damages, which induce power and subsequent yield losses. Some examples for damages are cracks at solar cells or plastic deformations at cell connectors (Köntges et al, 2014). Wind and snow loads represent the most important external mechanical actions on photovoltaic modules. For the sake of simplicity, the considerations in the present treatise are restricted to time-independent loadings. Influences caused by rain or hail, or even wind-induced vibrations are thus excluded.

In order to analyse the effects of non-uniform and non-orthogonal mechanical loading at photovoltaic modules, the considerations are restricted to loads on the outer surface of the front layer such that loads acting on the back layer are excluded. Nevertheless, loads acting at the outer surface may include positive and negative pressure. Every load will be characterised by its direction, amplitude, and spatial distribution. However, an approximation of these properties close to reality remains a challenge due to the lack of experimental data with regard to the spatial distribution of loadings at a photovoltaic module. However, since photovoltaic modules are mostly inclined, tangential and orthogonal non-uniform loads are acting on the surface. Until today, there is no general agreement on how such loads could be represented, neither conceptually nor experimentally. Nevertheless, we try to provide a brief insight into the subject within the following two subsections.

4.2.1.1 Snow Loads

Snow loads act in vertical direction because of gravity. It is obvious that they will occur at sub-zero temperatures. The mass density of snow is usually heterogeneous and highly dependent on climatic conditions as well as the thickness and age of the snow layer (Stoffel, 2005). The mechanics of snow, which influences subsequent loadings strongly, is topic of special research efforts (Mellor, 1975). For these reasons, only a few attempts have been made in order to approximate the distribution of snow loads. In a first try, the influence of the spatial distribution has been investigated experimentally, cf. Reil et al (2012). Another approach is the use of statistical distribution functions Ellingwood and O'Rourke (1985). Both methods are based on experimental results and restricted to specific climate zones. In addition, the standard EN 1991-1-3 is used by industry to determine the maximum snow load. In the case of fresh snow, the load distribution can be approximated with a sine wave since the load amplitude is decreasing near the edges because of snow slipping off. In addition, the surface properties of the photovoltaic module determine the adhesive strength such that snow rests at the module under inclination angle. Since photovoltaic modules are normally framed peripherally, this frame inhibits snow from slipping off additionally (Reil et al, 2012). However, the standard IEC 61215 indicates a maximum snow load of $5.4 \cdot 10^{-3}\,\mathrm{N}/\mathrm{mm}^2$.

4.2.1.2 Wind Loads

In contrast to snow loads, wind loads act almost parallel to Earth's surface such that their direction is assumed to be horizontal. First studies on wind loading can be found in Chevalier and Norton (1979) from 1979, where wind tunnel tests were used to analyse solar thermal collectors. This study gave averaged loading conditions of loaded surfaces. In Kopp et al (2002), wind loads on photovoltaic arrays were studied experimentally, where the outcome was also restricted to averaged values. In Ruscheweyh and Windhövel (2013), it is attempted to introduce pressure distributions due to experience, and approximations for practical reasons are formed. However, these load distributions are one dimensional only. Further experimental results examining the spatial dependence on a single module are provided by Aßmus and Köhl (2012). Investigations using computational fluid dynamics seem to be more fruitful gaining results of load distributions as shown in Shademan et al (2014). There, pressure distributions over the module surface are given for quasi-steady flows. Further numerical simulation of flow around modules and resulting pressure fields is done in Jubayer and Hangan (2014, 2016). Experimental studies from Abiola-Ogedengbe et al (2015) provide additional information affirming aforementioned statements. The resulting load distribution can be described by a standing wave in both plane directions, which does not need to be symmetric. As already shown in Aßmus and Köhl (2012), the loading is strongly dependent on the angles of inclination and incidence. However, all these investigations indicate that the wind direction and installation situation are crucial for resulting load distributions. Furthermore, similar to snow loads, wind loads are strongly dependent on the geographical location. The standard EN 1991-1-4 provides reference values for maximum wind loads. Thereby, extreme wind loads occur particularly near the coast or at high geographical locations (Jordan et al, 2012). As extreme conditions, storms lead to increased loading on photovoltaic modules. The standard IEC 61215 is indicating maximum wind load amplitudes of $2.4 \cdot 10^{-3}\,\mathrm{N/mm^2}$.

4.2.2 Mathematical Description of Mechanical Loads

4.2.2.1 Load Vector

As described in the previous section, climatic loads induce in-plane and out-of-plane loads on photovoltaic modules. Figure 4.2 shows a photovoltaic module mounted at an inclination angle θ. Let us consider a load vector $\boldsymbol{q}(X_1,X_2)$ acting at the outer surface $\mathfrak{O}$ of the module in the following form

$$\boldsymbol{q}(X_1,X_2) = \boldsymbol{s}(X_1,X_2) + \boldsymbol{p}(X_1,X_2) \tag{4.12}$$

with the in-plane load vector $\boldsymbol{s}(X_1,X_2) = -s_\alpha(X_1,X_2)\,\boldsymbol{e}_\alpha$ and the out-of-plane load vector $\boldsymbol{p}(X_1,X_2) = p(X_1,X_2)\,\boldsymbol{n}$.

4.2.2.2 Direction of Loads

This section focuses on the proper description of loads taking the inclination angle θ of a module into account. As a starting point, we introduce a fixed COOS $\{\boldsymbol{g}_i\}$ as well as another COOS $\{\boldsymbol{e}_i\}$ (with $\boldsymbol{e}_3 \equiv \boldsymbol{n}$), which is connected to the photovoltaic module, cf. also Fig. 4.2. Considering a rotation around $\boldsymbol{e}_2$ by the angle θ, the following relations hold true

$$\boldsymbol{g}_1 = \cos(\theta)\,\boldsymbol{e}_1 + \sin(\theta)\,\boldsymbol{e}_3\,, \tag{4.13}$$

$$\boldsymbol{g}_2 = \boldsymbol{e}_2\,, \tag{4.14}$$

$$\boldsymbol{g}_3 = -\sin(\theta)\,\boldsymbol{e}_1 + \cos(\theta)\,\boldsymbol{e}_3 \tag{4.15}$$

or vice versa

$$\boldsymbol{e}_1 = \cos(\theta)\,\boldsymbol{g}_1 - \sin(\theta)\,\boldsymbol{g}_3\,, \tag{4.16}$$

$$\boldsymbol{e}_2 = \boldsymbol{g}_2\,, \tag{4.17}$$

$$\boldsymbol{e}_3 = \sin(\theta)\,\boldsymbol{g}_1 + \cos(\theta)\,\boldsymbol{g}_3\,. \tag{4.18}$$

The proper orthogonal tensor $\boldsymbol{Q}$ maps the basis $\{\boldsymbol{g}_i\}$ onto the rotated basis $\{\boldsymbol{e}_i\}$ such that $\boldsymbol{Q}$ can be represented as follows

$$\boldsymbol{Q} = \boldsymbol{e}_i \otimes \boldsymbol{g}_i \,. \tag{4.19}$$

If one takes Eqs. (4.13)-(4.18) into account, one obtains the following representation for the rotation tensor

$$\boldsymbol{Q} = Q_{ij}\,\boldsymbol{e}_i \otimes \boldsymbol{e}_j = Q_{ij}\boldsymbol{g}_i \otimes \boldsymbol{g}_j \tag{4.20}$$

with

$$Q_{ij} = \begin{bmatrix} \cos(\theta) & 0 & \sin(\theta) \\ 0 & 1 & 0 \\ -\sin(\theta) & 0 & \cos(\theta) \end{bmatrix} . \tag{4.21}$$

Next, we assume that a load vector $\boldsymbol{a} = a_i \boldsymbol{g}_i$ is given with respect to the basis $\{\boldsymbol{g}_i\}$. The components a_i' of the vector $\boldsymbol{a} = a_i' \boldsymbol{e}_i$ with respect to the rotated basis $\{\boldsymbol{e}_i\}$ are computed as follows, cf. Bertram and Glüge (2015)

$$a_i' = Q_{ji} a_j \,. \tag{4.22}$$

In the following, above formulas are applied to loads resulting from natural weathering. We suppose that wind loads $\boldsymbol{q}_{\mathrm{W}}(X_1, X_2)$ are horizontal, while snow loads $\boldsymbol{q}_{\mathrm{S}}(X_1, X_2)$ refer to the vertical direction, cf. Fig. 4.2

$$\boldsymbol{q}_m(X_1, X_2) = \begin{cases} q_{\mathrm{W}}(X_1, X_2)\,\boldsymbol{g}_1, & \text{at wind loading}, \\ q_{\mathrm{S}}(X_1, X_2)\,\boldsymbol{g}_3, & \text{at snow loading}, \end{cases} \qquad \forall m = \{\mathrm{W}, \mathrm{S}\} \tag{4.23}$$

where the amplitude functions $q_m(X_1, X_2)$ need to be defined. With Eq. (4.22), one obtains the following representations for the load vectors

$$\boldsymbol{q}_m(X_1, X_2) = \begin{cases} q_{\mathrm{W}}(X_1, X_2)\cos(\theta)\,\boldsymbol{e}_1 + q_{\mathrm{W}}(X_1, X_2)\sin(\theta)\,\boldsymbol{n} & \text{at wind loading}, \\ -q_{\mathrm{S}}(X_1, X_2)\sin(\theta)\,\boldsymbol{e}_1 + q_{\mathrm{S}}(X_1, X_2)\cos(\theta)\,\boldsymbol{n} & \text{at snow loading}. \end{cases} \tag{4.24}$$

4.2.2.3 Amplitude and Spatial Distribution of Loads

In order to analyse the influence of spatially heterogeneous loads, the load distribution w.r.t. $\{\boldsymbol{g}_i\}$ is described by Eq. (4.23). In general, arbitrary amplitude functions $q_m(X_1, X_2)$ can be taken into account by the finite element, which is presented in the next sections. As one example, this paper focuses on a double sine function

$$q_m(X_1, X_2) = q_m^0 \sin\left(\frac{\pi}{L_1} X_1\right) \sin\left(\frac{\pi}{L_2} X_2\right) . \tag{4.25}$$

Herein, $q_m^0\ \forall m = \{\mathrm{W}, \mathrm{S}\}$ is the maximum amplitude. This concept can be easily generalised with a FOURIER series to approximate experimental data which, however, is not available for the present study.

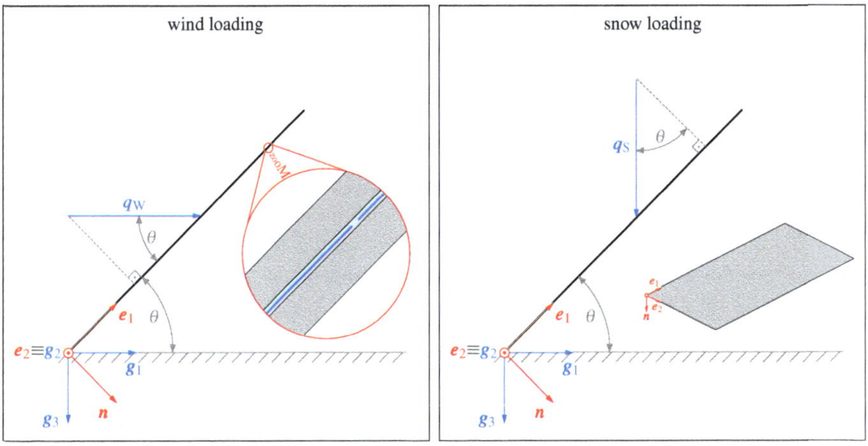

Fig. 4.2: Inclination angle θ with bases $\{\boldsymbol{g}_1, \boldsymbol{g}_2, \boldsymbol{g}_3\}$ and $\{\boldsymbol{e}_1, \boldsymbol{e}_2, \boldsymbol{n}\}$ for parametrisation of impact of snow and wind loads on a photovoltaic module

$$q_m(X_1, X_2) = \sum_{i=1}^{k} \sum_{j=1}^{k} \left(q_m^0\right)_{ij} \sin\left(\frac{\pi i}{L_1} X_1\right) \sin\left(\frac{\pi j}{L_2} X_2\right) \quad \forall m = \{\mathrm{W}, \mathrm{S}\} \wedge k \in \mathbb{Z}^+ . \tag{4.26}$$

Due to the lack of experimental findings concerning distribution functions, this investigations are restricted to FOURIER series with only one element ($k = 1 \Rightarrow$ biharmonic distribution). The load constructed in Eq. (4.25) in the context of the fixed COOS $\{\boldsymbol{g}_i\}$ used in Eq. (4.23) must now be rotated into the photovoltaic module COOS $\{\boldsymbol{e}_i\}$. Considering Eqs. (4.24), the approximations for wind and snow loads result in the succeeding representations for the loading components

$$\boldsymbol{q}_\mathrm{W} = -s_1 \boldsymbol{e}_1 + p\boldsymbol{n} \quad \text{with} \begin{cases} s_1(X_1, X_2) = -q_\mathrm{W}^0 \cos(\theta) \sin\left(\dfrac{\pi}{L_1} X_1\right) \sin\left(\dfrac{\pi}{L_2} X_2\right), \\ p(X_1, X_2) = \quad q_\mathrm{W}^0 \sin(\theta) \sin\left(\dfrac{\pi}{L_1} X_1\right) \sin\left(\dfrac{\pi}{L_2} X_2\right), \end{cases} \tag{4.27}$$

$$\boldsymbol{q}_\mathrm{S} = -s_1 \boldsymbol{e}_1 + p\boldsymbol{n} \quad \text{with} \begin{cases} s_1(X_1, X_2) = q_\mathrm{S}^0 \sin(\theta) \sin\left(\dfrac{\pi}{L_1} X_1\right) \sin\left(\dfrac{\pi}{L_2} X_2\right), \\ p(X_1, X_2) = q_\mathrm{S}^0 \cos(\theta) \sin\left(\dfrac{\pi}{L_1} X_1\right) \sin\left(\dfrac{\pi}{L_2} X_2\right). \end{cases} \tag{4.28}$$

Both loads can be combined, what results in an expression for the total load $\boldsymbol{q}_\Sigma$

$$\boldsymbol{q}_\Sigma = \boldsymbol{q}_\mathrm{W} + \boldsymbol{q}_\mathrm{S} . \tag{4.29}$$

4.3 Solution Approach with eXtended LayerWise Theory

4.3.1 Prerequisites

In XLWT, all considerations are restricted to the mid surface $\mathfrak{S}$ of the individual layers. The coordinates X_1, X_2 describe points in the mid surface, which is transversally positioned at $X_3 = 0$. Contemplations presented here are related to a planar mid surface, i.e. the structure is uncurved in the reference placement. It is equipped with a tripod COOS, while the unit vector perpendicular to the surface is called normal vector $\boldsymbol{n}$. In-plane unit vectors are designated with $\{\boldsymbol{e}_\alpha\}$. The surface boundary $\partial\mathfrak{S}$ in these directions is indicated by the boundary normal $\boldsymbol{\nu}$. Therefore $\boldsymbol{\nu} \cdot \boldsymbol{n} = 0$ holds. The surface continuum is equipped with five DOF's: two in-plane translational, one out-of-plane translational, and two out-of-plane rotational. These are in analogy with the kinematics suggested by Mindlin (1951). In-plane and out-of-plane loads s_α and p can be applied. In-plane loads are applied to the outer surfaces $\mathfrak{O}$. Therefore, they act with a lever of $\mp h/2$, while h is the thickness of the layer. Due to line volatility of out-of-plane loads, they can be applied to the surface $\mathfrak{S}$ directly. These assumptions are visualised in Fig. 4.3.

For structural analysis of composite structures, an integer multiple of the described surface continuum is required. For XLWT, three ($NK = 3$) mid surfaces are coupled, as depicted in Fig. 4.4. We designate these surfaces with the superscript indices

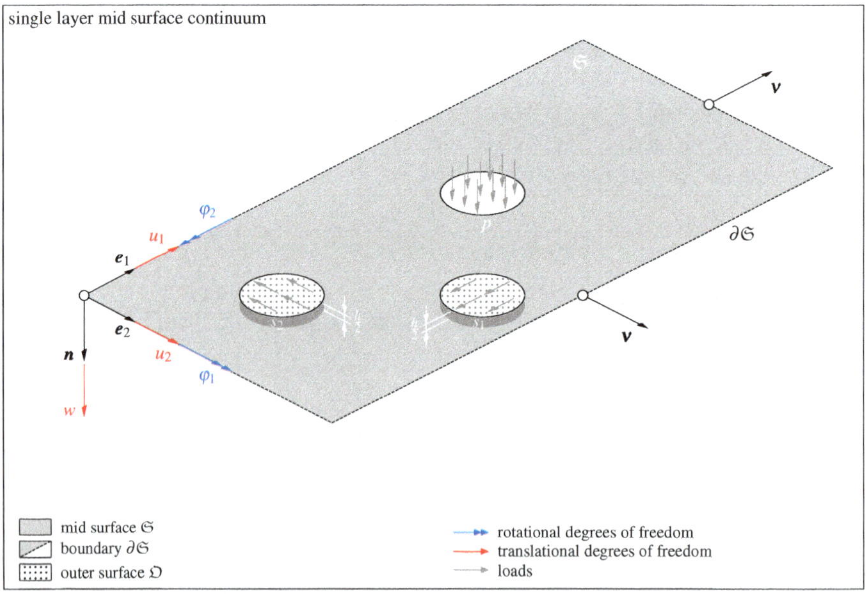

Fig. 4.3: Mechanical model of a surface continuum $\mathfrak{S}$ with boundary $\partial\mathfrak{S}$, COOS $\{\boldsymbol{e}_1, \boldsymbol{e}_2, \boldsymbol{n}\}$, DOF's $u_1, u_2, w, \varphi_1, \varphi_2$, and loads s_1, s_2, p

top (t), core (c), and bottom (b). All considerations are restricted to deformable plane surfaces, which are the mid planes ($\mathfrak{S}^t$, $\mathfrak{S}^c$, $\mathfrak{S}^b$) of every individual layer. At the layer interfaces ($\mathfrak{I}^t$, $\mathfrak{I}^b$), they are connected via constraints, kinematically and kinetically. All interacting forces act at these interfaces. External loading is applied at the outer surface $\mathfrak{O}$, while the back surface $\mathfrak{B}$ remains unloaded. The following points have been assumed concerning the physical structure.

- Uniform thickness of the corresponding layers.
- Perfect connection of the layers at their interfaces.

This leads to the subsequent properties for the surfaces.

- All material points per surface continuum are coplanar in the reference placement.
- All surfaces are plane-parallel in the reference placement and at constant distance to each other for any admissible placement.
- Mid surfaces, interfaces, as well as outer front and back surfaces are kinematically coupled via the straight normal hypothesis of the MINDLIN theory.

Specifications and consequences of these properties will be presented in the sequel, while we waive the derivation of the surface contiuum from 3D CAUCHY continuum theory, given for example in Aßmus et al (2016) or Naghdi (1973) in detail, based on a projection of kinematic, kinetic, and constitutive quantities.

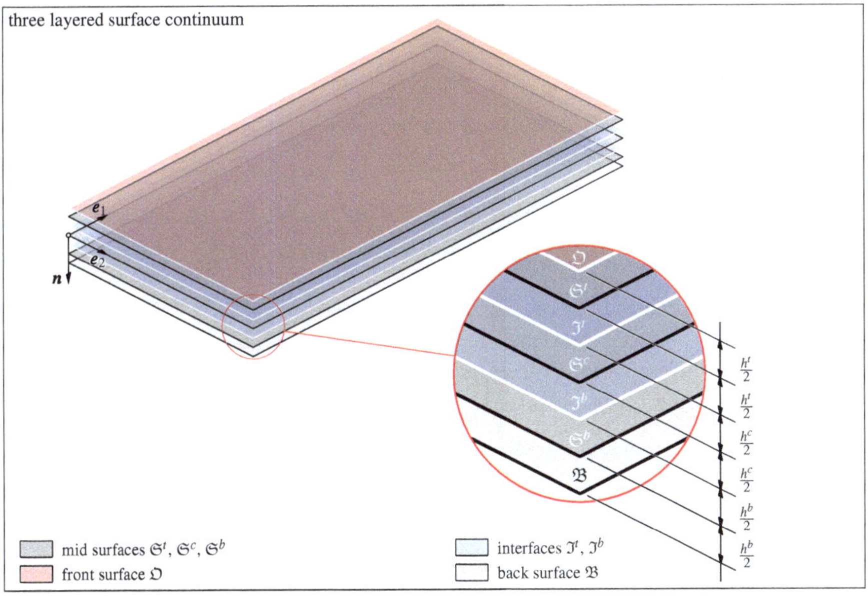

Fig. 4.4: Composition of the three layered surface continuum with interfaces, outer surface, and distances to each other assigned

4.3.2 Degrees of Freedom

In order to determine the position of a material point on a surface, two components are completely sufficient so that $\boldsymbol{x}^K_\ominus = X_\alpha \boldsymbol{e}^K_\alpha$ holds for every surface separately, while the superscript $K = \{t, c, b\}$ is the layer index. Since we operate on single surfaces only, it is possible to divide the DOF's into in-plane and out-of-plane measures. This is expressed by the in-plane displacements u^K_α and the deflections w^K. Additionally, rotational DOF's are introduced. While translational DOF's possess components in all three spatial directions of the Euclidean space $\mathbb{E}^3$, the rotational DOF's possess out-of-plane components only, as introduced by Reissner (1947). This introduction of rotations is therewith contrary to that of Cosserat and Cosserat (1909) or micropolar plates (Altenbach and Eremeyev, 2009) for practical reasons, cf. Zhilin (1976). The rotation about the normal to the surface is not considered as independent variable. However, the DOF's can thus be indicated as follows

$$\left.\begin{aligned} \text{in-plane displacements: } \boldsymbol{u}^K_\ominus &= u^K_\alpha \boldsymbol{e}_\alpha \\ \text{deflections: } w^K &\boldsymbol{n} \\ \text{out-of-plane rotations: } \boldsymbol{\varphi}^K &= -\varphi^K_2 \boldsymbol{e}_1 + \varphi^K_1 \boldsymbol{e}_2 \end{aligned}\right\} \forall \boldsymbol{x}^K_\ominus \in \mathfrak{S}^K . \tag{4.30}$$

Since we consider a three layered structure with $K = \{t, c, b\}$, a total of 15 DOF's results at this point.

4.3.3 Kinematical Measures

Under the assumption that strains and curvatures remain small, linear kinematical measures are introduced based on the aforementioned DOF's

$$\text{membrane strains: } \boldsymbol{E}^K = \nabla^{\text{sym}} \boldsymbol{u}^K_\ominus = E^K_{\alpha\beta} \boldsymbol{e}_\alpha \otimes \boldsymbol{e}_\beta , \tag{4.31}$$

$$\text{curvatures: } \boldsymbol{X}^K = \nabla^{\text{sym}} \boldsymbol{\varphi}^K = \chi^K_{\alpha\beta} \boldsymbol{e}_\alpha \otimes \boldsymbol{e}_\beta , \tag{4.32}$$

$$\text{transverse shear strains: } \boldsymbol{\gamma}^K = \nabla w^K + \boldsymbol{\varphi}^K = \gamma^K_\alpha \boldsymbol{e}_\alpha . \tag{4.33}$$

Herein, $\boldsymbol{E}^K$ is the membrane strain tensor, $\boldsymbol{X}^K$ is the curvature tensor, and $\boldsymbol{\gamma}^K$ is the transverse shear strain vector. The second rank tensors $\boldsymbol{E}^K$ and $\boldsymbol{X}^K$ are both symmetric.

4.3.4 Balance Equations and Kinetic Measures

The balance equations for the surface continuum, known as EULER's first and second law of motion, are given as follows for every layer separately

$$\left.\begin{array}{rlll} \text{balance of momentum: } \nabla \cdot \boldsymbol{F}^K & & +\boldsymbol{f}^K & = \mathbf{o} \\ \text{balance of moment of momentum: } \nabla \cdot \boldsymbol{M}^K & +\boldsymbol{F}^K \boxtimes \mathbf{1} & +\boldsymbol{m}^K & = \mathbf{o} \end{array}\right\} \forall \boldsymbol{x}_{\ominus}^K \in \mathfrak{S}^K . \tag{4.34}$$

Herin, $\boldsymbol{F}^K$ is the force tensor (also known as force-stress tensor), $\boldsymbol{F}^K \boxtimes \mathbf{1}$ is the vectorial invariant of $\boldsymbol{F}^K$ coupling both balance equations, $\boldsymbol{M}^K$ is the axial tensor of moments (also known as moment-stress tensor), $\boldsymbol{f}^K$ is the vector of interacting and external forces, and $\boldsymbol{m}^K$ is the vector of interacting as well as external moments. The tensors $\boldsymbol{F}^K$ and $\boldsymbol{M}^K$ have the following properties, which are in analogy to CAUCHY's theorem

$$\boldsymbol{\nu} \cdot \boldsymbol{F}^K = \boldsymbol{f}_\nu^K, \qquad \boldsymbol{n} \cdot \boldsymbol{F}^K = \mathbf{o}, \tag{4.35}$$

$$\boldsymbol{\nu} \cdot \boldsymbol{F}^K = \boldsymbol{m}_\nu^K, \qquad \boldsymbol{n} \cdot \boldsymbol{M}^K = \mathbf{o}. \tag{4.36}$$

Here $\boldsymbol{f}_\nu^K$ and $\boldsymbol{m}_\nu^K$ are forces and moments at the boundary with normal $\boldsymbol{\nu}$. Force and moment tensor are both non-symmetric

$$\boldsymbol{F}^K = F_{\alpha\beta}^K \boldsymbol{e}_\alpha \otimes \boldsymbol{e}_\beta + F_{\alpha 3}^K \boldsymbol{e}_\alpha \otimes \boldsymbol{n}, \tag{4.37}$$

$$\boldsymbol{M}^K = M_{\alpha\beta}^K \boldsymbol{e}_\alpha \otimes \boldsymbol{n} \times \boldsymbol{e}_\beta = -M_{\alpha\beta}^K \boldsymbol{e}_\alpha \otimes \boldsymbol{e}_\beta \times \boldsymbol{n}. \tag{4.38}$$

Since the force tensor is non-symmetric in general and distributed torques are not vanishing, the surface continum is considered as a polar medium, even if the physical three-dimensional structure of a photovoltaic module is non-polar, cf. Aßmus et al (2016). The force tensor can be decomposed into a symmetric and a skew part

$$\boldsymbol{F}^K = \left(\boldsymbol{F}^K\right)^{\mathrm{sym}} + \left(\boldsymbol{F}^K\right)^{\mathrm{skw}} \tag{4.39a}$$

$$= \frac{1}{2}\left[\boldsymbol{F}^K + \left(\boldsymbol{F}^K\right)^{\top}\right] + \frac{1}{2}\left[\boldsymbol{F}^K - \left(\boldsymbol{F}^K\right)^{\top}\right] \tag{4.39b}$$

$$= \boldsymbol{F}^K \cdot \boldsymbol{P} + \boldsymbol{F}^K \cdot \boldsymbol{n} \otimes \boldsymbol{n} \tag{4.39c}$$

$$= \boldsymbol{N}^K + \boldsymbol{q}_Q^K \otimes \boldsymbol{n} \tag{4.39d}$$

$$= N_{\alpha\beta}^K \boldsymbol{e}_\alpha \otimes \boldsymbol{e}_\beta + Q_{\alpha 3}^K \boldsymbol{e}_\alpha \otimes \boldsymbol{n}. \tag{4.39e}$$

Herein, $\boldsymbol{N}^K$ is the membrane force tensor, which is symmetric, and $\boldsymbol{q}_Q^K$ is the vector of transverse shear forces. Since the second index in $Q_{\alpha 3}^K$ is not varying, Q_α^K is introduced here, what allows us to describe the shear force terms as a vector $\boldsymbol{q}_Q^K$

$$\boldsymbol{q}_Q^K = \boldsymbol{F}^K \cdot \boldsymbol{n} = Q_\alpha^K \boldsymbol{e}_\alpha . \tag{4.40}$$

The second rank tensor $\boldsymbol{P} = \mathbf{1} - \boldsymbol{n} \otimes \boldsymbol{n}$ used in Eq. (4.39c) is the perpendicular projector, cf. Gurtin and Murdoch (1975). The vectorial invariant of the force tensor is determined by $\boldsymbol{F}^K \boxtimes \mathbf{1} = \boldsymbol{q}_Q^K \times \boldsymbol{n}$, so that $\left(\boldsymbol{F}^K\right)^{\mathrm{skw}} = -1/2\left(\boldsymbol{F}^K \boxtimes \mathbf{1}\right) \times \mathbf{1}$ holds true. Figure 4.5 provides free-body diagrams for the individual layers of the three layered composite structure where stress resultants are depicted at positive edges. Concerning the interacting and external forces, the following relations can be deduced

$$
\boldsymbol{f}^K = \begin{cases} (q^t + p\,)\,\boldsymbol{n} + \boldsymbol{s}^t - \boldsymbol{s} & \text{if } K = t\,, \\ (q^b - q^t)\,\boldsymbol{n} + \boldsymbol{s}^b - \boldsymbol{s}^t & \text{if } K = c\,, \\ \quad q^b\,\boldsymbol{n} - \boldsymbol{s}^b & \text{if } K = b\,. \end{cases} \qquad \forall\, \boldsymbol{f}^K \in \mathfrak{S}^K\,, \tag{4.41}
$$

In analogy to the decomposition of the force tensor $\boldsymbol{F}^K$, it is also possible to split $\boldsymbol{f}^K$ additively into an in-plane $\boldsymbol{f}^K_\ominus$ and out-of-plane part $f^K_\oplus$

$$
\boldsymbol{f}^K = \boldsymbol{f}^K \cdot (\boldsymbol{P} + \boldsymbol{n} \otimes \boldsymbol{n}) = \boldsymbol{f}^K \cdot \boldsymbol{P} + \boldsymbol{f}^K \cdot \boldsymbol{n} \otimes \boldsymbol{n} = \boldsymbol{f}^K_\ominus + f^K_\oplus \boldsymbol{n}\,, \tag{4.42}
$$

The interacting as well as the external moments are determined by the following set of equations, where we restrict ourselves to moments generated by forces only

$$
\boldsymbol{m}^K = \begin{cases} \frac{h^t}{2} \boldsymbol{n} \times (\boldsymbol{s} + \boldsymbol{s}^t) & \text{if } K = t\,, \\ \frac{h^c}{2} \boldsymbol{n} \times (\boldsymbol{s}^t + \boldsymbol{s}^b) & \text{if } K = c\,, \\ \frac{h^b}{2} \boldsymbol{n} \times \boldsymbol{s}^b & \text{if } K = b\,. \end{cases} \qquad \forall\, \boldsymbol{m}^K \in \mathfrak{S}^K\,, \tag{4.43}
$$

Variables used in Eqs. (4.41) and (4.43) are the in-plane interacting forces $\boldsymbol{s}^K = s^K_\alpha \boldsymbol{e}_\alpha$ and the out-of-plane interacting forces q^K. Inserting Relations (4.39)–(4.43) in Eqs. (4.34), we can determine the balances for the individual material surfaces of

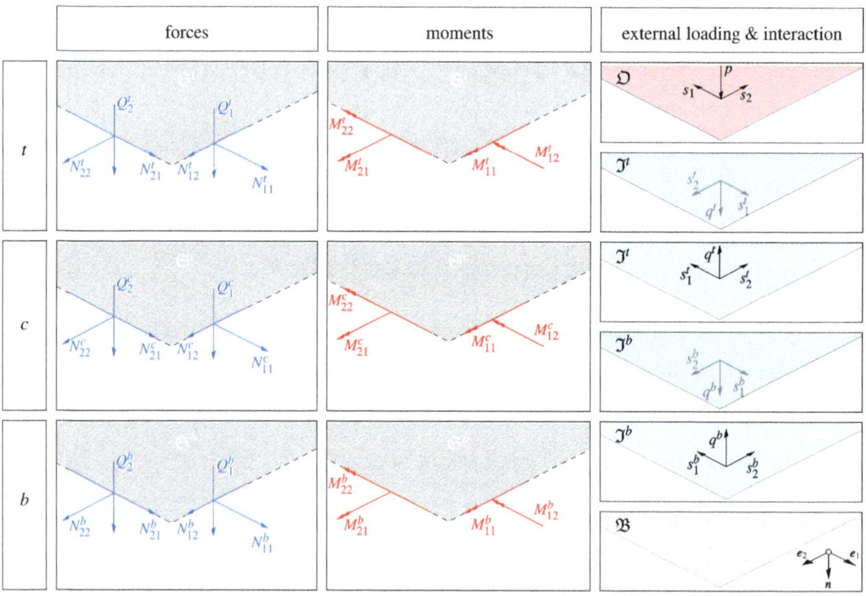

Fig. 4.5: Stress resultants at mid surfaces ($X^K_3 = 0$) as well as external and interacting forces at interfaces ($X_3 = \pm h^c/2$), front ($X_3 = -h^c/2 - h^t$), and back surface ($X_3 = h^c/2 + h^b$)

the composite structure. The balance of membrane forces takes the following form

$$\nabla \cdot \boldsymbol{N}^K + \boldsymbol{f}_{\ominus}^K = \boldsymbol{o} \quad \boldsymbol{f}_{\ominus}^K = \begin{cases} \boldsymbol{s}^t - \boldsymbol{s} & \text{if } K = t\,, \\ \boldsymbol{s}^b - \boldsymbol{s}^t & \text{if } K = c\,, \\ -\boldsymbol{s}^b & \text{if } K = b\,. \end{cases} \quad \forall\,\{\boldsymbol{s}^K, \boldsymbol{s}\} \in \{\mathfrak{I}^K, \mathfrak{O}\}\,, \tag{4.44}$$

The balance of transverse shear forces is indicated as follows

$$\nabla \cdot \boldsymbol{q}_Q^K + f_{\oplus}^K = 0 \quad f_{\oplus}^K = \begin{cases} q^t + p & \text{if } K = t\,, \\ q^b - q^t & \text{if } K = c\,, \\ -q^b & \text{if } K = b\,. \end{cases} \quad \forall\,\{q^K, p\} \in \{\mathfrak{I}^K, \mathfrak{O}\}\,, \tag{4.45}$$

Instead of the axial tensor of moments $\boldsymbol{M}^K$, the polar tensor of moments $\boldsymbol{L}^K = M_{\alpha\beta}^K \boldsymbol{e}_\alpha \otimes \boldsymbol{e}_\beta$ is introduced by the symmetrisation $\boldsymbol{L}^K = \boldsymbol{M}^K \times \boldsymbol{n}$. The balance of moment of momentum now reads as follows

$$\nabla \cdot \boldsymbol{L}^K - \boldsymbol{q}_Q^K + \boldsymbol{m}^K \times \boldsymbol{n} = \boldsymbol{o} \quad \boldsymbol{m}^K \times \boldsymbol{n} = \begin{cases} \frac{h^t}{2}(\boldsymbol{s} + \boldsymbol{s}^t) & \text{if } K = t\,, \\ \frac{h^c}{2}(\boldsymbol{s}^t + \boldsymbol{s}^b) & \text{if } K = c\,, \\ \frac{h^b}{2}\,\boldsymbol{s}^b & \text{if } K = b\,. \end{cases} \quad \forall\,\{\boldsymbol{s}^K, \boldsymbol{s}\} \in \{\mathfrak{I}^K, \mathfrak{O}\}\,, \tag{4.46}$$

4.3.5 Constitutive Equations

Considering a geometrically and physically linear theory, the constitutive equations for isotropic-homogeneous elastic materials (material symmetry, invariance under rotations) can be formulated layerwise as linear mappings

$$\text{membrane force - membrane strain relation: } \boldsymbol{N}^K = \boldsymbol{\mathcal{C}}^K : \boldsymbol{E}^K\,, \tag{4.47}$$

$$\text{bending moment - curvature relation: } \boldsymbol{L}^K = \boldsymbol{\mathcal{D}}^K : \boldsymbol{X}^K\,, \tag{4.48}$$

$$\text{transverse shear force - transverse shear strain relation: } \boldsymbol{q}_Q^K = \boldsymbol{Z}^K \cdot \boldsymbol{\gamma}^K\,. \tag{4.49}$$

The measures $\boldsymbol{\mathcal{C}}^K$, $\boldsymbol{\mathcal{D}}^K$, and $\boldsymbol{Z}^K$ used in Eqs. (4.47), (4.48), and (4.49) are the fourth rank membrane stiffness tensor, the fourth rank bending stiffness tensor, and the second rank transverse shear stiffness tensor, respectively. As is apparent, coupling stiffnesses are not considered. This decoupling of membrane, bending and shear state, i.e. decoupling of stretching, bending, twisting, and shearing is legit since every single surface $\mathfrak{S}^K$ is halfway up between the nearest interfaces $\mathfrak{I}^K$ respectively front $\mathfrak{O}$ or back surface $\mathfrak{B}$ in both transverse directions (geometrical symmetry). This fact coincides with the assumption of a physical plane layer which is symmetric to its mid plane at $X_3^K = 0$. In the case of isotropic material behaviour, the constitutive tensors read as follows (Naumenko and Eremeyev, 2014) membrane stiffness tensor: $$\boldsymbol{\mathcal{C}}^K = D_{\mathrm{M}}^K \, \nu^K \boldsymbol{P} \otimes \boldsymbol{P} + D_{\mathrm{M}}^K \frac{1-\nu^K}{2} \left(\boldsymbol{\mathcal{P}}_1 + \boldsymbol{\mathcal{P}}_2 \right) , \tag{4.50}$$

bending stiffness tensor: $$\boldsymbol{\mathcal{D}}^K = D_{\mathrm{B}}^K \, \nu^K \boldsymbol{P} \otimes \boldsymbol{P} + D_{\mathrm{B}}^K \frac{1-\nu^K}{2} \left(\boldsymbol{\mathcal{P}}_1 + \boldsymbol{\mathcal{P}}_2 \right) , \tag{4.51}$$

transverse shear stiffness tensor: $$\boldsymbol{Z}^K = D_{\mathrm{S}}^K \boldsymbol{P} . \tag{4.52}$$

Here, E^K, G^K, and ν^K are the YOUNG's modulus, the shear modulus, and the POISSON's ratio of the corresponding layer material. The membrane rigidity D_{M}^K, the bending rigidity D_{B}^K, and the shear rigidity D_{S}^K can be determined as follows

$$D_{\mathrm{M}}^K = \frac{E^K h^K}{1-(\nu^K)^2} , \tag{4.53a}$$

$$D_{\mathrm{B}}^K = \frac{E^K (h^K)^3}{12\left[1-(\nu^K)^2\right]} , \tag{4.53b}$$

$$D_{\mathrm{S}}^K = \kappa^K G^K h^K . \tag{4.53c}$$

These representations are in accordance with classical thin-walled structural theories known from Kirchhoff (1850); Reissner (1944), and Mindlin (1951). $\boldsymbol{\mathcal{P}}_1$ and $\boldsymbol{\mathcal{P}}_2$ are fourth rank tensors

$$\boldsymbol{\mathcal{P}}_1 = \boldsymbol{e}_\alpha \otimes \boldsymbol{e}_\beta \otimes \boldsymbol{e}_\beta \otimes \boldsymbol{e}_\alpha , \qquad \boldsymbol{\mathcal{P}}_2 = \boldsymbol{e}_\alpha \otimes \boldsymbol{e}_\beta \otimes \boldsymbol{e}_\alpha \otimes \boldsymbol{e}_\beta . \tag{4.54}$$

The stiffness tensors given in the compact representations (4.50), (4.51), and (4.52) are completely determined by two material parameters (E^K, ν^K, when considering isotropic materials $G^K = E^K/2(1+\nu^K)$ holds true) and one geometry parameter (h^K), respectively, up to an undetermined parameter κ^K, cf. Eq. (4.52). This so-called shear correction factor is a parameter to set the shear energy contribution at deformation processes with shear soft layers. A value of $\kappa^K = 1 \; \forall K = \{t, c, b\}$ is applied here since this setting showed best agreements with experimental validations, cf. Weps et al (2013). $\kappa^c = 1$ is manifested by the fact, that shear stresses acting on $\mathfrak{I}^t$ and $\mathfrak{I}^b$ are contrary but equal in magnitude for transverse symmetric structures while the core layer is comparatively thin, what results in constant shear stresses along X_3 with the limits $\mp h^K/2$. In contrast, $\kappa^K \; \forall K = \{t, b\}$ is not decisive due to the high shear rigidity of these layers.

4.3.6 Boundary Conditions

With regard to the practical implementation of the balance equations, the definition of boundary conditions is inevitable. Hereby, we introduce homogeneous DIRICHLET boundary conditions

$$\boldsymbol{u}(\boldsymbol{x}_{\ominus}^{K}) = \mathbf{o} \quad \wedge\vee \quad w(\boldsymbol{x}_{\ominus}^{K}) = 0 \quad \wedge\vee \quad \boldsymbol{\varphi}(\boldsymbol{x}_{\ominus}^{K}) = \mathbf{o} \qquad \forall \boldsymbol{x}_{\ominus}^{K} \in \partial\mathfrak{S}_{\mathrm{D}}^{K}, \tag{4.55}$$

and NEUMANN boundary conditions which are built on CAUCHY's theorem, cf. Eqs. (4.35) and (4.36)

$$\boldsymbol{\nu} \cdot \boldsymbol{N}^{K} = \boldsymbol{n}_{\star}^{K} \quad \wedge\vee \quad \boldsymbol{\nu} \cdot \boldsymbol{L}^{K} = \boldsymbol{m}_{\star}^{K} \quad \wedge\vee \quad \boldsymbol{\nu} \cdot \boldsymbol{q}_{Q}^{K} = q_{\star}^{K} \qquad \forall \boldsymbol{x}_{\ominus}^{K} \in \partial\mathfrak{S}_{\mathrm{N}}^{K}, \tag{4.56}$$

whereby $\boldsymbol{n}_{\star}^{K}$, $\boldsymbol{m}_{\star}^{K}$, and $q_{\star}^{K}$ are membrane forces, moments, and transverse shear forces prescribed at the boundaries, while $\partial\mathfrak{S}_{\mathrm{D}}^{K}$ and $\partial\mathfrak{S}_{\mathrm{N}}^{K}$ denote DIRICHLET and NEUMANN boundaries, respectively. Loads (in-plane $\boldsymbol{s}$ and out-of-plane p) at the outer surface of the top layer are implemented directly in the balance equations (4.44), (4.45), and (4.46).

4.3.7 Kinematical Constraints

The following expression holds true

$$\frac{\partial h^{K}}{\partial X_1} = \frac{\partial h^{K}}{\partial X_2} = 0 \tag{4.57}$$

since compression or dilatation in the transverse direction is neglected, what results in equal deflections for all layers

$$w^{t} = w^{c} = w^{b} = w \qquad \forall \boldsymbol{x}_{\ominus}^{K} = X_{\alpha}^{K} \boldsymbol{e}_{\alpha}. \tag{4.58}$$

This stipulates that the reduced number of independent DOF's for our composite is 13 at this point. As stated, we assume that all layers are rigidly connected at their interfaces for any admissible deformation process. The following kinematical constraints can therefore be deduced since the straight line hypothesis of Mindlin (1951) is used

$$\boldsymbol{u}_{\ominus}^{t} + \frac{h^{t}}{2}\boldsymbol{\varphi}^{t} = \boldsymbol{u}_{\ominus}^{c} - \frac{h^{c}}{2}\boldsymbol{\varphi}^{c}, \tag{4.59a}$$

$$\boldsymbol{u}_{\ominus}^{b} - \frac{h^{b}}{2}\boldsymbol{\varphi}^{b} = \boldsymbol{u}_{\ominus}^{c} + \frac{h^{c}}{2}\boldsymbol{\varphi}^{c}. \tag{4.59b}$$

Above constraints are used to substitute quantities of the core layer by the quantitites of the skin layers (t, b). As conseqence of Eqs. (4.59a) and (4.59b), the repeatedly reduced number of independent DOF's is 9.

4.3.8 Introduction of Mean and Relative Measures

For the sake of brevity, mean (superscript $\circ$) and relative (superscript Δ) displacements and rotations are introduced, cf. Eisenträger et al (2015b); Aßmus et al (2016)

$$\boldsymbol{u}^{\circ} = \frac{1}{2}\left(\boldsymbol{u}^{t}_{\ominus} + \boldsymbol{u}^{b}_{\ominus}\right), \qquad \boldsymbol{u}^{\Delta} = \frac{1}{2}\left(\boldsymbol{u}^{t}_{\ominus} - \boldsymbol{u}^{b}_{\ominus}\right), \tag{4.60a}$$

$$\boldsymbol{\varphi}^{\circ} = \frac{1}{2}\left(\boldsymbol{\varphi}^{t} + \boldsymbol{\varphi}^{b}\right), \qquad \boldsymbol{\varphi}^{\Delta} = \frac{1}{2}\left(\boldsymbol{\varphi}^{t} - \boldsymbol{\varphi}^{b}\right). \tag{4.60b}$$

Considering Eqs. (4.60a) and (4.60b), the set of independent DOF's is redefined as $\left\{u_1^{\circ}, u_2^{\circ}, u_1^{\Delta}, u_2^{\Delta}, w, \varphi_1^{\circ}, \varphi_2^{\circ}, \varphi_1^{\Delta}, \varphi_2^{\Delta}\right\}$. The corresponding membrane strain tensors, curvature tensors, and transverse shear strain vectors read as follows

$$\boldsymbol{E}^{\circ} = \nabla^{\text{sym}}\boldsymbol{u}^{\circ} = \frac{1}{2}\left(\boldsymbol{E}^{t} + \boldsymbol{E}^{b}\right), \tag{4.61a}$$

$$\boldsymbol{E}^{\Delta} = \nabla^{\text{sym}}\boldsymbol{u}^{\Delta} = \frac{1}{2}\left(\boldsymbol{E}^{t} - \boldsymbol{E}^{b}\right), \tag{4.61b}$$

$$\boldsymbol{X}^{\circ} = \nabla^{\text{sym}}\boldsymbol{\varphi}^{\circ} = \frac{1}{2}\left(\boldsymbol{X}^{t} + \boldsymbol{X}^{b}\right), \tag{4.61c}$$

$$\boldsymbol{X}^{\Delta} = \nabla^{\text{sym}}\boldsymbol{\varphi}^{\Delta} = \frac{1}{2}\left(\boldsymbol{X}^{t} - \boldsymbol{X}^{b}\right), \tag{4.61d}$$

$$\boldsymbol{\gamma}^{\circ} = \nabla w + \boldsymbol{\varphi}^{\circ} = \frac{1}{2}\left(\boldsymbol{\gamma}^{t} + \boldsymbol{\gamma}^{b}\right), \tag{4.61e}$$

$$\boldsymbol{\gamma}^{\Delta} = \boldsymbol{\varphi}^{\Delta} = \frac{1}{2}\left(\boldsymbol{\gamma}^{t} - \boldsymbol{\gamma}^{b}\right). \tag{4.61f}$$

In addition, mean and relative stress resultants are defined

$$\boldsymbol{N}^{\circ} = \boldsymbol{N}^{t} + \boldsymbol{N}^{c} + \boldsymbol{N}^{b}, \tag{4.62a}$$

$$\boldsymbol{N}^{\Delta} = \boldsymbol{N}^{t} - \boldsymbol{N}^{b}, \tag{4.62b}$$

$$\boldsymbol{L}^{\circ} = \boldsymbol{L}^{t} + \boldsymbol{L}^{c} + \boldsymbol{L}^{b} + \frac{1}{2}\left(h^{b} + h^{c}\right)\boldsymbol{N}^{b} - \frac{1}{2}\left(h^{t} + h^{c}\right)\boldsymbol{N}^{t}, \tag{4.62c}$$

$$\boldsymbol{L}^{\Delta} = \boldsymbol{L}^{t} - \boldsymbol{L}^{b}, \tag{4.62d}$$

$$\boldsymbol{q}_{Q}^{\circ} = \boldsymbol{q}_{Q}^{t} + \boldsymbol{q}_{Q}^{c} + \boldsymbol{q}_{Q}^{b}, \tag{4.62e}$$

$$\boldsymbol{q}_{Q}^{\Delta} = \boldsymbol{q}_{Q}^{t} - \boldsymbol{q}_{Q}^{b}. \tag{4.62f}$$

Constitutive equations with respect to the new stress resultants are formulated in analogy to Eqs. (4.47), (4.48), and (4.49)

$$\boldsymbol{N}^{\circ} = \mathcal{C}^{\circ} : \boldsymbol{E}^{\circ} + \mathcal{C}^{\Delta} : \boldsymbol{E}^{\Delta} + \mathcal{C}^{c} : \boldsymbol{E}^{c}, \tag{4.63a}$$

$$\boldsymbol{N}^{\Delta} = \mathcal{C}^{\Delta} : \boldsymbol{E}^{\circ} + \mathcal{C}^{\circ} : \boldsymbol{E}^{\Delta}, \tag{4.63b}$$

$$\boldsymbol{L}^{\circ} = \mathcal{D}^{\circ} : \boldsymbol{X}^{\circ} + \mathcal{D}^{\Delta} : \boldsymbol{X}^{\Delta} + \mathcal{D}^{c} : \boldsymbol{X}^{c}$$

$$
\begin{aligned}
&-\frac{1}{2}\left[h^{\Delta}\boldsymbol{\mathcal{C}}^{\circ}+(h^{c}+h^{\circ})\boldsymbol{\mathcal{C}}^{\Delta}\right]:\boldsymbol{E}^{\circ}\\
&-\frac{1}{2}\left[(h^{c}+h^{\circ})\boldsymbol{\mathcal{C}}^{\circ}+h^{\Delta}\boldsymbol{\mathcal{C}}^{\Delta}\right]:\boldsymbol{E}^{\Delta}, &(4.63c)
\end{aligned}
$$

$$\boldsymbol{L}^{\Delta}=\boldsymbol{\mathcal{D}}^{\Delta}:\boldsymbol{X}^{\circ}+\boldsymbol{\mathcal{D}}^{\circ}:\boldsymbol{X}^{\Delta}, \tag{4.63d}$$

$$\boldsymbol{q}_{Q}^{\circ}=\boldsymbol{Z}^{\circ}\cdot\boldsymbol{\gamma}^{\circ}+\boldsymbol{Z}^{\Delta}\cdot\boldsymbol{\gamma}^{\Delta}+\boldsymbol{Z}^{c}\cdot\boldsymbol{\gamma}^{c}, \tag{4.63e}$$

$$\boldsymbol{q}_{Q}^{\Delta}=\boldsymbol{Z}^{\Delta}\cdot\boldsymbol{\gamma}^{\circ}+\boldsymbol{Z}^{\circ}\cdot\boldsymbol{\gamma}^{\Delta}. \tag{4.63f}$$

In above equations, new stiffness tensors have been introduced

$$\boldsymbol{\mathcal{C}}^{\circ}=\boldsymbol{\mathcal{C}}^{t}+\boldsymbol{\mathcal{C}}^{b}, \qquad \boldsymbol{\mathcal{C}}^{\Delta}=\boldsymbol{\mathcal{C}}^{t}-\boldsymbol{\mathcal{C}}^{b}, \tag{4.64a}$$

$$\boldsymbol{\mathcal{D}}^{\circ}=\boldsymbol{\mathcal{D}}^{t}+\boldsymbol{\mathcal{D}}^{b}, \qquad \boldsymbol{\mathcal{D}}^{\Delta}=\boldsymbol{\mathcal{D}}^{t}-\boldsymbol{\mathcal{D}}^{b}, \tag{4.64b}$$

$$\boldsymbol{Z}^{\circ}=\boldsymbol{Z}^{t}+\boldsymbol{Z}^{b}, \qquad \boldsymbol{Z}^{\Delta}=\boldsymbol{Z}^{t}-\boldsymbol{Z}^{b}. \tag{4.64c}$$

Furthermore, new thickness measures are defined

$$h^{\circ}=\frac{1}{2}\left(h^{t}+h^{b}\right), \qquad h^{\Delta}=\frac{1}{2}\left(h^{t}-h^{b}\right). \tag{4.65}$$

4.3.9 Principle of Virtual Work

As basis for the numerical implementation, this section presents the principle of virtual work for the XLWT. The principle of virtual work is applied to the XLWT in the usual way, cf. also Eisenträger et al (2015b); Oñate (2013); Bathe and Zimmermann (2002). In a first step, the balance equations are combined with the corresponding virtual DOF's, i.e. Eqs. (4.44) are multiplied with the virtual in-plane displacements, Eqs. (4.45) are multiplied with the virtual deflections, and Eqs. (4.46) are multiplied with the virtual cross-section rotations. All balance equations are summed up and integrated over the surface. One introduces the mean and relative measures from the previous subsection and uses the kinematical constraints in order to replace the DOF's of the core layer by the mean and relative DOF's. After considering the constitutive equations, one obtains the following expressions for the internal and external work

$$
\begin{aligned}
\delta W_{\text{int}}=\int_{\mathfrak{S}}\Big\{&\delta\boldsymbol{E}^{\circ}:\boldsymbol{\mathcal{C}}^{\circ}:\boldsymbol{E}^{\circ}+\delta\boldsymbol{E}^{\Delta}:\boldsymbol{\mathcal{C}}^{\circ}:\boldsymbol{E}^{\Delta}+\delta\boldsymbol{E}^{\circ}:\boldsymbol{\mathcal{C}}^{\Delta}:\boldsymbol{E}^{\Delta}\\
&+\delta\boldsymbol{E}^{\Delta}:\boldsymbol{\mathcal{C}}^{\Delta}:\boldsymbol{E}^{\circ}+\left(\delta\boldsymbol{E}^{\circ}+\frac{1}{2}h^{\Delta}\delta\boldsymbol{X}^{\circ}+\frac{1}{2}h^{\circ}\delta\boldsymbol{X}^{\Delta}\right):\boldsymbol{\mathcal{C}}^{c}\\
&:\left(\boldsymbol{E}^{\circ}+\frac{1}{2}h^{\Delta}\boldsymbol{X}^{\circ}+\frac{1}{2}h^{\circ}\boldsymbol{X}^{\Delta}\right)+\delta\boldsymbol{\gamma}^{\circ}\cdot\boldsymbol{Z}^{\circ}\cdot\boldsymbol{\gamma}^{\circ}\\
&+\delta\boldsymbol{\gamma}^{\Delta}\cdot\boldsymbol{Z}^{\circ}\cdot\boldsymbol{\gamma}^{\Delta}+\delta\boldsymbol{\gamma}^{\circ}\cdot\boldsymbol{Z}^{\Delta}\cdot\boldsymbol{\gamma}^{\Delta}+\delta\boldsymbol{\gamma}^{\Delta}\cdot\boldsymbol{Z}^{\circ}\cdot\boldsymbol{\gamma}^{\circ}
\end{aligned}
$$

$$
\begin{aligned}
&+\left[\delta\boldsymbol{\gamma}^{\circ}-\frac{1}{h^{c}}\left(2\delta\boldsymbol{u}^{\Delta}+(h^{c}+h^{\circ})\,\delta\boldsymbol{\varphi}^{\circ}+h^{\Delta}\delta\boldsymbol{\varphi}^{\Delta}\right)\right]\cdot\boldsymbol{Z}^{c}\\
&\cdot\left[\boldsymbol{\gamma}^{\circ}-\frac{1}{h^{c}}\left(2\boldsymbol{u}^{\Delta}+(h^{c}+h^{\circ})\,\boldsymbol{\varphi}^{\circ}+h^{\Delta}\boldsymbol{\varphi}^{\Delta}\right)\right]\\
&+\delta\boldsymbol{X}^{\circ}:\boldsymbol{\mathcal{D}}^{\circ}:\boldsymbol{X}^{\circ}+\delta\boldsymbol{X}^{\Delta}:\boldsymbol{\mathcal{D}}^{\circ}:\boldsymbol{X}^{\Delta}+\delta\boldsymbol{X}^{\circ}:\boldsymbol{\mathcal{D}}^{\Delta}:\boldsymbol{X}^{\Delta}\\
&+\delta\boldsymbol{X}^{\Delta}:\boldsymbol{\mathcal{D}}^{\Delta}:\boldsymbol{X}^{\circ}+\frac{1}{(h^{c})^{2}}\left(2\delta\boldsymbol{E}^{\Delta}+h^{\circ}\delta\boldsymbol{X}^{\circ}+h^{\Delta}\delta\boldsymbol{X}^{\Delta}\right):\boldsymbol{\mathcal{D}}^{c}\\
&:\left(2\boldsymbol{E}^{\Delta}+h^{\circ}\boldsymbol{X}^{\circ}+h^{\Delta}\boldsymbol{X}^{\Delta}\right)\Bigg\}\mathrm{d}\mathfrak{S}\,,
\end{aligned}
\tag{4.66}
$$

$$
\begin{aligned}
\delta W_{\text{ext}}=&\int\limits_{\partial\mathfrak{S}_{\text{p}}}\Bigg\{\left(\delta\boldsymbol{u}^{\circ}+\frac{1}{2}h^{\Delta}\delta\boldsymbol{\varphi}^{\circ}\right)\cdot\boldsymbol{n}_{\nu}^{\circ}+\left[\delta\boldsymbol{u}^{\Delta}+\frac{1}{2}(h^{c}+h^{\circ})\,\delta\boldsymbol{\varphi}^{\circ}\right]\cdot\boldsymbol{n}_{\nu}^{\Delta}\\
&+\frac{1}{2}h^{\circ}\delta\boldsymbol{\varphi}^{\Delta}\cdot\boldsymbol{n}_{\nu}^{c}+\delta w q_{\nu}^{\circ}+\delta\boldsymbol{\varphi}^{\circ}\cdot\boldsymbol{m}_{\nu}^{\circ}+\delta\boldsymbol{\varphi}^{\Delta}\cdot\boldsymbol{m}_{\nu}^{\Delta}\\
&-\frac{1}{h^{c}}\left[2\delta\boldsymbol{u}^{\Delta}+(h^{c}+h^{\circ})\,\delta\boldsymbol{\varphi}^{\circ}+h^{\Delta}\delta\boldsymbol{\varphi}^{\Delta}\right]\cdot\boldsymbol{m}_{\nu}^{c}\Bigg\}\mathrm{d}\,\partial\mathfrak{S}_{\text{p}}\\
&+\int\limits_{\mathfrak{S}_{\text{p}}}\Bigg[\frac{h^{t}}{2}\left(\delta\boldsymbol{\varphi}^{\circ}+\delta\boldsymbol{\varphi}^{\Delta}\right)\cdot\boldsymbol{s}-\left(\delta\boldsymbol{u}^{\circ}+\delta\boldsymbol{u}^{\Delta}\right)\cdot\boldsymbol{s}\\
&+\delta w p\Bigg]\mathrm{d}\,\mathfrak{S}_{\text{p}}\,.
\end{aligned}
\tag{4.67}
$$

Hereby, $\mathfrak{S}_{\text{p}}$ and $\partial\mathfrak{S}_{\text{p}}$ denote the surface and boundary of the continuum, where boundary conditions with respect to the stress resultants are prescribed

$$
\left.\begin{aligned}
\boldsymbol{N}_{\text{p}}^{K}&=\boldsymbol{N}^{K}\big|_{\partial\mathfrak{S}_{\text{p}}}\\
\boldsymbol{L}_{\text{p}}^{K}&=\boldsymbol{L}^{K}\big|_{\partial\mathfrak{S}_{\text{p}}}\\
\boldsymbol{q}_{Q_{\text{p}}}^{K}&=\boldsymbol{q}_{Q}^{K}\big|_{\partial\mathfrak{S}_{\text{p}}}
\end{aligned}\right\}\forall\,K=\{\circ,\Delta,c\}\,.
\tag{4.68}
$$

Furthermore, stress resultants on the boundary of the surface continuum have been introduced

$$
\left.\begin{aligned}
\boldsymbol{n}_{\nu}^{K}&=\boldsymbol{\nu}\cdot\boldsymbol{N}_{\text{p}}^{K}\\
\boldsymbol{m}_{\nu}^{K}&=\boldsymbol{\nu}\cdot\boldsymbol{L}_{\text{p}}^{K}\\
q_{\nu}^{K}&=\boldsymbol{\nu}\cdot\boldsymbol{q}_{Q_{\text{p}}}^{K}
\end{aligned}\right\}\forall\,K=\{\circ,\Delta,c\}\,.
\tag{4.69}
$$

Equations (4.66) and (4.67) are inserted into the principle of virtual work, which states that the balance equations of a body are fulfilled if the virtual work of the internal forces equals the virtual work of the external forces (Bathe and Zimmermann, 2002).

$$
\delta W_{\text{int}}=\delta W_{\text{ext}}\,.
\tag{4.70}
$$

4.4 Numerical Implementation

4.4.1 Basic Procedure in Finite Element Method

Typically, quadrilateral finite elements are used to model thin-walled structures with single layer, while for structures with $NK > 1$ different approaches exist, cf. Carrera (2002, 2003b). The main idea of the present work is to represent all layers with one element in transverse direction. While in Eisenträger et al (2015b) a complete description of this approach with finite elements is given, the principles of numerical solution strategy are presented here.

The FEM is based on a strict separation of structural and element scales. The whole two-dimensional domain Ω is divided into subdomains Ω^e, i.e. the finite elements. Thereby, the domain Ω comprises the subdomains Ω^e, which must not overlap

$$\Omega = \bigcup_{e=1}^{NE} \Omega^e \qquad \Omega^i \cap \Omega^j = \emptyset \quad \text{if } i \neq j. \tag{4.71}$$

NE is the number of elements in the domain Ω. The principle of virtual work has to be fulfilled on the whole domain Ω, cf. also Eq. (4.70), and on each subdomain Ω^e

$$\delta W^e_{\text{int}} = \delta W^e_{\text{ext}}, \tag{4.72}$$

while the virtual work of the structure is obtained by adding the virtual work of all finite elements

$$\delta W_{\text{int}} = \sum_{e=1}^{NE} \delta W^e_{\text{int}} \qquad \delta W_{\text{ext}} = \sum_{e=1}^{NE} \delta W^e_{\text{ext}}. \tag{4.73}$$

4.4.2 Shape Functions

Since the exact values of the DOF's on the domain Ω and on the subdomains Ω^e are unknown, one uses the shape functions N^i in order to approximate the solution. For the finite layerwise element, SERENDIPITY type shape functions are applied, cf. also Fig. 4.6. The shape functions are defined with respect to the natural coordinates $-1 \leq \xi_j \leq 1, \forall\, j = \{1,2\}$ Szabó and Babuška (1991); Zienkiewicz and Taylor (2000)

$$\begin{aligned}
N^i(\boldsymbol{\xi}) &= \frac{1}{4}\left[1+\xi_1^i\xi_1\right]\left[1+\xi_2^i\xi_2\right]\left[\xi_1^i\xi_1+\xi_2^i\xi_2-1\right] && i=\{1,\dots,4\},\\
N^i(\boldsymbol{\xi}) &= \frac{1}{2}\left[1+\xi_1^i\xi_1\right]\left[1-\xi_2^2\right] && i=\{6,8\},\\
N^i(\boldsymbol{\xi}) &= \frac{1}{2}\left[1+\xi_2^i\xi_2\right]\left[1-\xi_1^2\right] && i=\{5,7\}.
\end{aligned}$$

The index i represents the node number, while the node numbering is depicted in the centre of Fig. 4.6.

4.4.3 JACOBI *Transformation*

The transformation of differential line elements $\mathrm{d}X_i$ and $\mathrm{d}\xi_i$ is based on the derivatives of the physical coordinates with respect to the natural coordinates. This transformation is performed via the JACOBI matrix $\mathbf{J}(\boldsymbol{\xi})$ and its inverse $\mathbf{J}(\boldsymbol{\xi})^{-1}$. In two dimensions, it holds true (Oñate, 2009)

$$\frac{\partial}{\partial \boldsymbol{\xi}} = \mathbf{J}(\boldsymbol{\xi})\frac{\partial}{\partial \mathbf{x}} \quad \Leftrightarrow \quad \frac{\partial}{\partial \mathbf{x}} = \mathbf{J}(\boldsymbol{\xi})^{-1}\frac{\partial}{\partial \boldsymbol{\xi}} \tag{4.74}$$

with

$$\mathbf{J}(\boldsymbol{\xi}) = \begin{bmatrix} \dfrac{\partial X_1}{\partial \xi_1} & \dfrac{\partial X_2}{\partial \xi_1} \\ \dfrac{\partial X_1}{\partial \xi_2} & \dfrac{\partial X_2}{\partial \xi_2} \end{bmatrix}, \qquad \mathbf{J}(\boldsymbol{\xi})^{-1} = \frac{1}{|\mathbf{J}(\boldsymbol{\xi})|}\begin{bmatrix} \dfrac{\partial X_2}{\partial \xi_2} & -\dfrac{\partial X_2}{\partial \xi_1} \\ -\dfrac{\partial X_1}{\partial \xi_2} & \dfrac{\partial X_1}{\partial \xi_1} \end{bmatrix}, \tag{4.75}$$

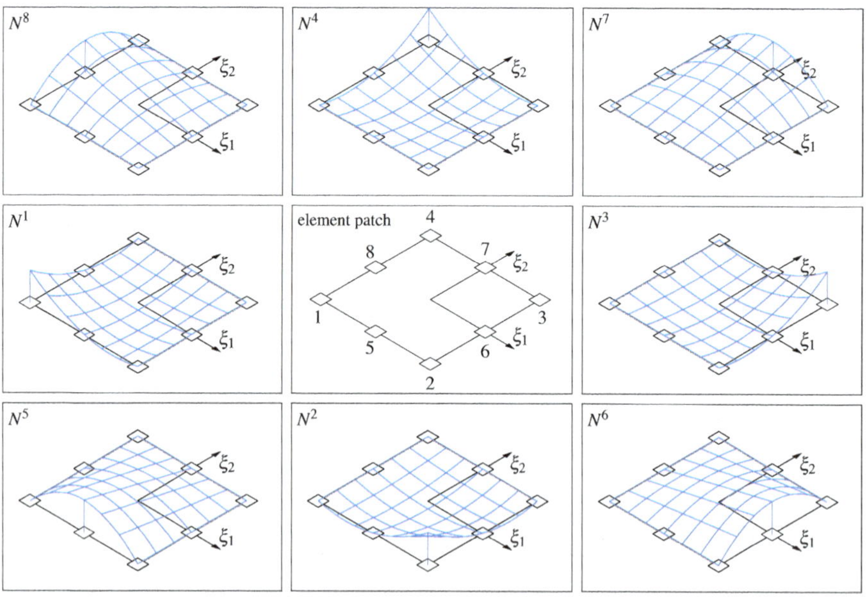

Fig. 4.6: Approximation of deformations with SERENDIPITY type shape functions of quadratic order of interpolation

$$\frac{\partial}{\partial \boldsymbol{\xi}} = \left[\frac{\partial}{\partial \xi_1} \; \frac{\partial}{\partial \xi_2} \right]^\top , \qquad \frac{\partial}{\partial \mathbf{x}} = \left[\frac{\partial}{\partial X_1} \; \frac{\partial}{\partial X_2} \right]^\top . \tag{4.76}$$

The determinant of the JACOBI matrix $\det[\mathbf{J}(\boldsymbol{\xi})] = |\mathbf{J}(\boldsymbol{\xi})|$ is utilised to transform an infinitesimal surface element $\mathrm{d}\Omega$ in physical coordinates into an infinitesimal surface element in natural coordinates

$$\mathrm{d}\Omega = \mathrm{d}X_1\, \mathrm{d}X_2 = |\mathbf{J}(\boldsymbol{\xi})|\; \mathrm{d}\xi_1\, \mathrm{d}\xi_2 . \tag{4.77}$$

4.4.4 Discretisation

4.4.4.1 Degrees of Freedom

In order to discretise equations, the vector of DOF's at every node i is specified as follows (recall DOF definitions of sections 4.3.7 and 4.3.8)

$$\mathbf{a}^i = \left[u_1^{\circ^i}\; u_2^{\circ^i}\; u_1^{\Delta^i}\; u_2^{\Delta^i}\; w^i\; \varphi_1^{\circ^i}\; \varphi_2^{\circ^i}\; \varphi_1^{\Delta^i}\; \varphi_2^{\Delta^i} \right]^\top \qquad \forall\, i = \{1, \ldots, N\!N\} \tag{4.78}$$

while $N\!N = 8$ is the number of nodes per element. All nodal vectors of DOF's are assembled to the element vector of DOF's referring to the nodal values of DOF's

$$\mathbf{a}^e = \left[\mathbf{a}^1\; \mathbf{a}^2\; \ldots\; \mathbf{a}^{N\!N} \right]^\top . \tag{4.79}$$

In order to obtain the fields of DOF's over the element with respect to the natural coordinates $\boldsymbol{\xi}$, the DOF's are interpolated into the shape functions, applying the isoparametric element concept

$$\begin{aligned} \mathbf{a}(\boldsymbol{\xi}) &= \left[u_1^\circ(\boldsymbol{\xi})\; u_2^\circ(\boldsymbol{\xi})\; u_1^\Delta(\boldsymbol{\xi})\; u_2^\Delta(\boldsymbol{\xi})\; w(\boldsymbol{\xi})\; \varphi_1^\circ(\boldsymbol{\xi})\; \varphi_2^\circ(\boldsymbol{\xi})\; \varphi_1^\Delta(\boldsymbol{\xi})\; \varphi_2^\Delta(\boldsymbol{\xi}) \right]^\top \\ &\approx \mathbf{N}(\boldsymbol{\xi})\, \mathbf{a}^e \end{aligned} \tag{4.80}$$

with the matrix of shape functions $\mathbf{N}(\boldsymbol{\xi})$

$$\mathbf{N}(\boldsymbol{\xi}) = \left[\mathbf{N}^1(\boldsymbol{\xi})\; \mathbf{N}^2(\boldsymbol{\xi})\; \ldots\; \mathbf{N}^{N\!N}(\boldsymbol{\xi}) \right] \tag{4.81}$$

and the matrix of shape functions at node i

$$\mathbf{N}^i(\boldsymbol{\xi}) = N^i(\boldsymbol{\xi})\, \mathbf{I} , \tag{4.82}$$

while $\mathbf{I}$ is a quadratic identity matrix, whose number of columns and rows is equal to the number of DOF's per node.

4.4.4.2 Kinematical Measures

As a next step, the kinematical equations, cf. also Eqs (4.61), are discretised. For this purpose, we introduce kinematical vectors with respect to the element

$$\mathbf{e}_{\mathrm{MB}} = \left[\mathbf{e}_{\mathrm{M}}^{\circ}\ \mathbf{e}_{\mathrm{M}}^{\Delta}\ \mathbf{e}_{\mathrm{B}}^{\circ}\ \mathbf{e}_{\mathrm{B}}^{\Delta}\right]^{\top} \qquad \mathbf{e}_{\mathrm{S}} = \left[\mathbf{e}_{\mathrm{S}}^{\circ}\ \mathbf{e}_{\mathrm{S}}^{\Delta}\right]^{\top} \tag{4.83}$$

with the auxiliary vectors for membrane and transverse strains as well as curvatures

$$\mathbf{e}_{\mathrm{M}}^{\circ} = \left[E_{11}^{\circ}\ E_{22}^{\circ}\ 2E_{12}^{\circ}\right]^{\top} = \left[u_{1,1}^{\circ}\ u_{2,2}^{\circ}\ u_{1,2}^{\circ}+u_{2,1}^{\circ}\right]^{\top}, \tag{4.84}$$

$$\mathbf{e}_{\mathrm{M}}^{\Delta} = \left[E_{11}^{\Delta}\ E_{22}^{\Delta}\ 2E_{12}^{\Delta}\right]^{\top} = \left[u_{1,1}^{\Delta}\ u_{2,2}^{\Delta}\ u_{1,2}^{\Delta}+u_{2,1}^{\Delta}\right]^{\top}, \tag{4.85}$$

$$\mathbf{e}_{\mathrm{B}}^{\circ} = \left[\chi_{11}^{\circ}\ \chi_{22}^{\circ}\ 2\chi_{12}^{\circ}\right]^{\top} = \left[\varphi_{1,1}^{\circ}\ \varphi_{2,2}^{\circ}\ \varphi_{1,2}^{\circ}+\varphi_{2,1}^{\circ}\right]^{\top}, \tag{4.86}$$

$$\mathbf{e}_{\mathrm{B}}^{\Delta} = \left[\chi_{11}^{\Delta}\ \chi_{22}^{\Delta}\ 2\chi_{12}^{\Delta}\right]^{\top} = \left[\varphi_{1,1}^{\Delta}\ \varphi_{2,2}^{\Delta}\ \varphi_{1,2}^{\Delta}+\varphi_{2,1}^{\Delta}\right]^{\top}, \tag{4.87}$$

$$\mathbf{e}_{\mathrm{S}}^{\circ} = \left[\gamma_1^{\circ}\ \gamma_2^{\circ}\right]^{\top} = \left[w_{,1}+\varphi_1^{\circ}\ w_{,2}+\varphi_2^{\circ}\right]^{\top}, \tag{4.88}$$

$$\mathbf{e}_{\mathrm{S}}^{\Delta} = \left[\gamma_1^{\Delta}\ \gamma_2^{\Delta}\right]^{\top} = \left[\varphi_1^{\Delta}\ \varphi_2^{\Delta}\right]^{\top}. \tag{4.89}$$

Please note that above kinematical vectors represent fields, i.e. these vectors depend on the natural coordinates $\boldsymbol{\xi}$. For the sake of brevity, this has not been written explicitly. Now, the kinematical fields are approximated in analogy to Eq. (4.80) by introducing the strain matrices $\mathbf{B}_{\mathrm{MB}}$ and $\mathbf{B}_{\mathrm{S}}$ for the membrane and bending as well as for the shear part, respectively

$$\mathbf{e}_{\mathrm{MB}}(\boldsymbol{\xi}) \approx \mathbf{B}_{\mathrm{MB}}(\boldsymbol{\xi})\,\mathbf{a}^e, \qquad \mathbf{e}_{\mathrm{S}}(\boldsymbol{\xi}) \approx \mathbf{B}_{\mathrm{S}}(\boldsymbol{\xi})\,\mathbf{a}^e. \tag{4.90}$$

The kinematical matrices are the products of the differential operators $\mathbf{D}$ and the matrix of shape functions $\mathbf{N}$

$$\mathbf{B}_{\mathrm{MB}}(\boldsymbol{\xi}) = \mathbf{D}_{\mathrm{MB}}\mathbf{N}(\boldsymbol{\xi}), \qquad \mathbf{B}_{\mathrm{S}}(\boldsymbol{\xi}) = \mathbf{D}_{\mathrm{S}}\mathbf{N}(\boldsymbol{\xi}) \tag{4.91}$$

with the differential operators for membrane, bending, and shear state

$$\mathbf{D}_{\mathrm{MB}} = \left[\mathbf{D}_{\mathrm{M}}^{\circ}\ \mathbf{D}_{\mathrm{M}}^{\Delta}\ \mathbf{D}_{\mathrm{B}}^{\circ}\ \mathbf{D}_{\mathrm{B}}^{\Delta}\right]^{\top}, \tag{4.92}$$

$$\mathbf{D}_{\mathrm{S}} = \left[\mathbf{D}_{\mathrm{S}}^{\circ}\ \mathbf{D}_{\mathrm{S}}^{\Delta}\right]^{\top} \tag{4.93}$$

and the auxiliary differential operators

$$\mathbf{D}_{\mathrm{M}}^{\circ} = \begin{bmatrix} \frac{\partial}{\partial X_1} & 0 & 0\,0\,0\,0\,0\,0\,0 \\ 0 & \frac{\partial}{\partial X_2} & 0\,0\,0\,0\,0\,0\,0 \\ \frac{\partial}{\partial X_2} & \frac{\partial}{\partial X_1} & 0\,0\,0\,0\,0\,0\,0 \end{bmatrix}, \tag{4.94}$$

$$\mathbf{D}_{\mathrm{M}}^{\Delta} = \begin{bmatrix} 0 & 0 & \frac{\partial}{\partial X_1} & 0 & 0 & 0 & 0 & 0 & 0 \\ 0 & 0 & 0 & \frac{\partial}{\partial X_2} & 0 & 0 & 0 & 0 & 0 \\ 0 & 0 & \frac{\partial}{\partial X_2} & \frac{\partial}{\partial X_1} & 0 & 0 & 0 & 0 & 0 \end{bmatrix} , \tag{4.95}$$

$$\mathbf{D}_{\mathrm{B}}^{\circ} = \begin{bmatrix} 0 & 0 & 0 & 0 & 0 & \frac{\partial}{\partial X_1} & 0 & 0 & 0 \\ 0 & 0 & 0 & 0 & 0 & 0 & \frac{\partial}{\partial X_2} & 0 & 0 \\ 0 & 0 & 0 & 0 & 0 & \frac{\partial}{\partial X_2} & \frac{\partial}{\partial X_1} & 0 & 0 \end{bmatrix} , \tag{4.96}$$

$$\mathbf{D}_{\mathrm{B}}^{\Delta} = \begin{bmatrix} 0 & 0 & 0 & 0 & 0 & 0 & 0 & \frac{\partial}{\partial X_1} & 0 \\ 0 & 0 & 0 & 0 & 0 & 0 & 0 & 0 & \frac{\partial}{\partial X_2} \\ 0 & 0 & 0 & 0 & 0 & 0 & 0 & \frac{\partial}{\partial X_2} & \frac{\partial}{\partial X_1} \end{bmatrix} , \tag{4.97}$$

$$\mathbf{D}_{\mathrm{S}}^{\circ} = \begin{bmatrix} 0 & 0 & 0 & 0 & \frac{\partial}{\partial X_1} & 1 & 0 & 0 & 0 \\ 0 & 0 & 0 & 0 & \frac{\partial}{\partial X_2} & 0 & 1 & 0 & 0 \end{bmatrix} , \tag{4.98}$$

$$\mathbf{D}_{\mathrm{S}}^{\Delta} = \begin{bmatrix} 0 & 0 & 0 & 0 & 0 & 0 & 0 & 1 & 0 \\ 0 & 0 & 0 & 0 & 0 & 0 & 0 & 0 & 1 \end{bmatrix} . \tag{4.99}$$

4.4.5 Constitutive Equations for FEM

In order to implement the finite element, the expressions for the virtual internal and external work, cf. Eqs. (4.66) and (4.67), need to be transferred to a matrix notation, what is discussed in the next section. For this reason, this section presents the corresponding constitutive equations in a matrix notation. First, vectors for the stress resultants are introduced in analogy to Eqs. (4.84)-(4.89)

$$\mathbf{s}_{\mathrm{M}}^{K} = \left[N_{11}^{K}\; N_{22}^{K}\; N_{12}^{K}\right]^{\top} \qquad \forall\, K = \{\circ, \Delta, c\} , \tag{4.100}$$

$$\mathbf{s}_{\mathrm{S}}^{K} = \left[Q_{1}^{K}\; Q_{2}^{K}\right]^{\top} \qquad \forall\, K = \{\circ, \Delta, c\} , \tag{4.101}$$

$$\mathbf{s}_{\mathrm{B}}^{K} = \left[M_{11}^{K}\; M_{22}^{K}\; M_{12}^{K}\right]^{\top} \qquad \forall\, K = \{\circ, \Delta, c\} . \tag{4.102}$$

Now, the constitutive equations are reformulated in a matrix notation, while Eqs. (4.63) serve as basis for the mean and relative variables, and Eqs. (4.47), (4.48), and (4.49) are evaluated for the core layer

$$\mathbf{s}_{\mathrm{M}}^{\circ} = \left(\hat{\mathbf{C}}_{\mathrm{M}}^{\circ} + \hat{\mathbf{C}}_{\mathrm{M}}^{c}\right) \mathbf{e}_{\mathrm{M}}^{\circ} + \hat{\mathbf{C}}_{\mathrm{M}}^{\Delta} \mathbf{e}_{\mathrm{M}}^{\Delta} + \frac{1}{2} \hat{\mathbf{C}}_{\mathrm{M}}^{c} \left(h^{\Delta} \mathbf{e}_{\mathrm{B}}^{\circ} + h^{\circ} \mathbf{e}_{\mathrm{B}}^{\Delta}\right) , \tag{4.103}$$

$$\mathbf{s}_{\mathrm{M}}^{\Delta} = \hat{\mathbf{C}}_{\mathrm{M}}^{\circ} \mathbf{e}_{\mathrm{M}}^{\Delta} + \hat{\mathbf{C}}_{\mathrm{M}}^{\Delta} \mathbf{e}_{\mathrm{M}}^{\circ} , \tag{4.104}$$

$$\mathbf{s}_{\mathrm{M}}^{c} = \hat{\mathbf{C}}_{\mathrm{M}}^{c} \left[\mathbf{e}_{\mathrm{M}}^{\circ} + \frac{1}{2} \left(h^{\Delta} \mathbf{e}_{\mathrm{B}}^{\circ} + h^{\circ} \mathbf{e}_{\mathrm{B}}^{\Delta}\right)\right] , \tag{4.105}$$

$$\mathbf{s}_{\mathrm{S}}^{\circ} = \hat{\mathbf{C}}_{\mathrm{S}}^{\circ} \mathbf{e}_{\mathrm{S}}^{\circ} + \hat{\mathbf{C}}_{\mathrm{S}}^{\Delta} \mathbf{e}_{\mathrm{S}}^{\Delta} + \hat{\mathbf{C}}_{\mathrm{S}}^{c} \left(\mathbf{e}_{\mathrm{S}}^{\circ} + \mathbf{A}_{1} \mathbf{a}\right) , \tag{4.106}$$

$$\mathbf{s}_{\mathrm{S}}^{\Delta} = \hat{\mathbf{C}}_{\mathrm{S}}^{\circ} \mathbf{e}_{\mathrm{S}}^{\Delta} + \hat{\mathbf{C}}_{\mathrm{S}}^{\Delta} \mathbf{e}_{\mathrm{S}}^{\circ} , \tag{4.107}$$

$$\mathbf{s}_{\mathrm{S}}^{c} = \hat{\mathbf{C}}_{\mathrm{S}}^{c} \left(\mathbf{e}_{\mathrm{S}}^{\circ} + \mathbf{A}_{1} \mathbf{a} \right) , \tag{4.108}$$

$$\begin{aligned} \mathbf{s}_{\mathrm{B}}^{\circ} &= \hat{\mathbf{C}}_{\mathrm{B}}^{\circ} \mathbf{e}_{\mathrm{B}}^{\circ} + \hat{\mathbf{C}}_{\mathrm{B}}^{\Delta} \mathbf{e}_{\mathrm{B}}^{\Delta} - \frac{1}{h^{c}} \hat{\mathbf{C}}_{\mathrm{B}}^{c} \left(2\mathbf{e}_{\mathrm{M}}^{\Delta} + h^{\circ} \mathbf{e}_{\mathrm{B}}^{\circ} + h^{\Delta} \mathbf{e}_{\mathrm{B}}^{\Delta} \right) \\ &\quad - \frac{1}{2} \hat{\mathbf{C}}_{\mathrm{M}}^{\circ} \left[h^{\Delta} \mathbf{e}_{\mathrm{M}}^{\circ} + (h^{c} + h^{\circ}) \, \mathbf{e}_{\mathrm{M}}^{\Delta} \right] \\ &\quad - \frac{1}{2} \hat{\mathbf{C}}_{\mathrm{M}}^{\Delta} \left[h^{\Delta} \mathbf{e}_{\mathrm{M}}^{\Delta} + (h^{c} + h^{\circ}) \, \mathbf{e}_{\mathrm{M}}^{\circ} \right] , \end{aligned} \tag{4.109}$$

$$\mathbf{s}_{\mathrm{B}}^{\Delta} = \hat{\mathbf{C}}_{\mathrm{B}}^{\circ} \mathbf{e}_{\mathrm{B}}^{\Delta} + \hat{\mathbf{C}}_{\mathrm{B}}^{\Delta} \mathbf{e}_{\mathrm{B}}^{\circ} , \tag{4.110}$$

$$\mathbf{s}_{\mathrm{B}}^{c} = -\frac{1}{h^{c}} \hat{\mathbf{C}}_{\mathrm{B}}^{c} \left(2\mathbf{e}_{\mathrm{M}}^{\Delta} + h^{\circ} \mathbf{e}_{\mathrm{B}}^{\circ} + h^{\Delta} \mathbf{e}_{\mathrm{B}}^{\Delta} \right) . \tag{4.111}$$

The constitutive matrices $\hat{\mathbf{C}}_{\mathrm{M}}^{K}$, $\hat{\mathbf{C}}_{\mathrm{S}}^{K}$, $\hat{\mathbf{C}}_{\mathrm{B}}^{K}$ $\forall\, K = \{\circ, \Delta, c\}$ as well as the auxiliary matrix $\mathbf{A}_1$ are defined in the appendix.

4.4.6 Element Stiffness Relation

This section derives the element stiffness relation based on the principle of virtual work. As a first step, the virtual internal and external work are presented in a matrix notation. First, Eq. (4.66) is transferred to a matrix notation considering the vectors of strains and stress resultants as well as the constitutive relations introduced in Sections 4.4.4.2 and 4.4.5. Then, Eqs. (4.80) and (4.90) are inserted such that one obtains

$$\begin{aligned} \delta W_{\mathrm{int}} = \int\limits_{\Omega^e} \delta \mathbf{a}^{e\top} \Big[& \mathbf{B}_{\mathrm{S}}^{\top} \left(\mathbf{C}_{\mathrm{S}}^{\circ} + \mathbf{C}_{\mathrm{S}}^{\Delta} + \mathbf{C}_{\mathrm{S}}^{\Delta\,\top} + \mathbf{A}_2^{\top} \hat{\mathbf{C}}_{\mathrm{S}}^{c} \mathbf{A}_2 \right) \mathbf{B}_{\mathrm{S}} \\ & + \mathbf{B}_{\mathrm{S}}^{\top} \mathbf{A}_2^{\top} \hat{\mathbf{C}}_{\mathrm{S}}^{c} \mathbf{A}_1 \mathbf{N} + \left(\mathbf{B}_{\mathrm{S}}^{\top} \mathbf{A}_2^{\top} \hat{\mathbf{C}}_{\mathrm{S}}^{c} \mathbf{A}_1 \mathbf{N} \right)^{\top} + \mathbf{N}^{\top} \mathbf{A}_1^{\top} \hat{\mathbf{C}}_{\mathrm{S}}^{c} \mathbf{A}_1 \mathbf{N} \\ & + \mathbf{B}_{\mathrm{MB}}^{\top} (\mathbf{C}_{\mathrm{MB}}^{\circ} + \mathbf{C}_{\mathrm{MB}}^{\Delta} + \mathbf{C}_{\mathrm{MB}}^{\Delta\,\top} + \mathbf{A}_3^{\top} \hat{\mathbf{C}}_{\mathrm{M}}^{c} \mathbf{A}_3 \\ & + \mathbf{A}_4^{\top} \hat{\mathbf{C}}_{\mathrm{B}}^{c} \mathbf{A}_4) \mathbf{B}_{\mathrm{MB}} \Big] \mathbf{a}^{e} \, \mathrm{d}\Omega^{e} . \end{aligned} \tag{4.112}$$

The introduced auxiliary matrices $\mathbf{A}_i\ \forall\, i = \{1, \ldots, 5\}$ are defined in the appendix. In order to shorten above expression, the generalised constitutive matrices $\mathbf{C}_{\mathrm{MB}}^{\circ}$, $\mathbf{C}_{\mathrm{MB}}^{\Delta}$, $\mathbf{C}_{\mathrm{S}}^{\circ}$, and $\mathbf{C}_{\mathrm{S}}^{\Delta}$ have been introduced. Their definitions can be found in the appendix, too. Furthermore, all integrals are performed with respect to the element surface Ω^e or the element boundary $\partial\Omega^e$, respectively. Now, Eq. (4.67) is proceeded similarly, what results in the following expression

$$\delta W_{\mathrm{ext}} = \int\limits_{\partial\Omega_{\mathrm{p}}^{e}} \delta \mathbf{a}^{e\top} \mathbf{N}^{\top} \mathbf{A}_5 \mathbf{t} \, \mathrm{d}\partial\Omega_{\mathrm{p}}^{e} + \int\limits_{\Omega^e} \delta \mathbf{a}^{e\top} \mathbf{N}^{\top} \mathbf{q} \, \mathrm{d}\Omega^{e} . \tag{4.113}$$

The vectors $\mathbf{t}$ and $\mathbf{q}$ contain loads that are distributed over a curve or a surface of the plate, respectively

$$\mathbf{t} = \left[\mathbf{n}_\nu^\circ \; \mathbf{n}_\nu^\Delta \; \mathbf{n}_\nu^c \; q_\nu^\circ \; \mathbf{m}_\nu^\circ \; \mathbf{m}_\nu^\Delta \; \mathbf{m}_\nu^c\right]^\top \tag{4.114}$$

$$\mathbf{q} = \left[-s_1 \; -s_2 \; -s_1 \; -s_2 \; p \; \tfrac{h^t}{2}s_1 \; \tfrac{h^t}{2}s_2 \; \tfrac{h^t}{2}s_1 \; \tfrac{h^t}{2}s_2\right]^\top . \tag{4.115}$$

The vectors $\mathbf{n}_\nu^\circ$, $\mathbf{n}_\nu^\Delta$, $\mathbf{n}_\nu^c$, $\mathbf{m}_\nu^\circ$, $\mathbf{m}_\nu^\Delta$, and $\mathbf{m}_\nu^c$ refer to the corresponding vectors in tensor notation defined in Eq. (4.69).

In order to derive the element stiffness relation, Eqs. (4.112) and (4.113) are inserted into Eq. (4.70), and coefficients are equated with respect to the vector $\delta\mathbf{a}^{e\top}$. This yields the element stiffness relation

$$(\mathbf{K}_{\mathrm{MB}}^e + \mathbf{K}_{\mathrm{S}}^e)\,\mathbf{a}^e = \mathbf{r}^e . \tag{4.116}$$

Hence, the stiffness matrix for the membrane and bending state $\mathbf{K}_{\mathrm{MB}}^e$ and the stiffness matrix for the shear state $\mathbf{K}_{\mathrm{S}}^e$ are determined as follows

$$\mathbf{K}_{\mathrm{MB}}^e = \int_{\Omega^e} \mathbf{B}_{\mathrm{MB}}^\top \big(\mathbf{C}_{\mathrm{MB}}^\circ + \mathbf{C}_{\mathrm{MB}}^\Delta + \mathbf{C}_{\mathrm{MB}}^{\Delta\,\top} + \mathbf{A}_3^\top \hat{\mathbf{C}}_{\mathrm{M}}^c \mathbf{A}_3 + \mathbf{A}_4^\top \hat{\mathbf{C}}_{\mathrm{B}}^c \mathbf{A}_4\big)\mathbf{B}_{\mathrm{MB}}\,\mathrm{d}\Omega^e , \tag{4.117}$$

$$\begin{aligned}\mathbf{K}_{\mathrm{S}}^e = \int_{\Omega^e} \Big[& \mathbf{B}_{\mathrm{S}}^\top \big(\mathbf{C}_{\mathrm{S}}^\circ + \mathbf{C}_{\mathrm{S}}^\Delta + \mathbf{C}_{\mathrm{S}}^{\Delta\,\top} + \mathbf{A}_2^\top \hat{\mathbf{C}}_{\mathrm{S}}^c \mathbf{A}_2\big)\mathbf{B}_{\mathrm{S}} \\ & + \mathbf{B}_{\mathrm{S}}^\top \mathbf{A}_2^\top \hat{\mathbf{C}}_{\mathrm{S}}^c \mathbf{A}_1 \mathbf{N} + \big(\mathbf{B}_{\mathrm{S}}^\top \mathbf{A}_2^\top \hat{\mathbf{C}}_{\mathrm{S}}^c \mathbf{A}_1 \mathbf{N}\big)^\top \\ & + \mathbf{N}^\top \mathbf{A}_1^\top \hat{\mathbf{C}}_{\mathrm{S}}^c \mathbf{A}_1 \mathbf{N} \Big]\,\mathrm{d}\Omega^e .\end{aligned} \tag{4.118}$$

The right-hand-side vector $\mathbf{r}^e$ comprises all the line loads $\mathbf{r}_1^e$ and the surface loads $\mathbf{r}_2^e$ on the element

$$\mathbf{r}^e = \mathbf{r}_1^e + \mathbf{r}_2^e \tag{4.119}$$

with

$$\mathbf{r}_1^e = \int_{\partial\Omega_{\mathrm{p}}^e} \mathbf{N}^\top \mathbf{A}_5 \mathbf{t}\,\mathrm{d}\partial\Omega_{\mathrm{p}}^e , \qquad \mathbf{r}_2^e = \int_{\Omega_{\mathrm{p}}^e} \mathbf{N}^\top \mathbf{q}\,\mathrm{d}\Omega_{\mathrm{p}}^e . \tag{4.120}$$

4.4.7 Surface Load Vector

Since this paper aims to analyse surface loads on photovoltaic modules, we focus on the surface load vector $\mathbf{r}_2^e$. For the numerical procedure, the load vector $\mathbf{q}$ is approximated via the isoparametric element concept, in analogy to Eq. (4.80)

$$\mathbf{q}(\boldsymbol{\xi}) \approx \mathbf{N}(\boldsymbol{\xi})\,\mathbf{q}^e, \tag{4.121}$$

with the element load vector $\mathbf{q}^e$, which comprises all nodal load vectors $\mathbf{q}^i$

$$\mathbf{q}^e = \left[\mathbf{q}^1\ \mathbf{q}^2 \ldots \mathbf{q}^{NN}\right]^\top, \tag{4.122}$$

$$\mathbf{q}^i = \left[-s_1^i\ -s_2^i\ -s_1^i\ -s_2^i\ p^i\ \tfrac{h^t}{2}s_1^i\ \tfrac{h^t}{2}s_2^i\ \tfrac{h^t}{2}s_1^i\ \tfrac{h^t}{2}s_2^i\right]^\top. \tag{4.123}$$

The variables s_1^i, s_2^i, and p^i are determined by evaluating the load distribution functions at the corresponding node coordinates $\left(X_1^i, X_2^i\right)$

$$s_1^i = s_1\left(X_1^i, X_2^i\right), \qquad s_2^i = s_2\left(X_1^i, X_2^i\right), \qquad p^i = p\left(X_1^i, X_2^i\right). \tag{4.124}$$

Equation (4.121) is inserted into Eq. (4.120) such that the surface load vector is approximated as follows

$$\mathbf{r}_2^e \approx \int\limits_{\Omega_\mathrm{p}^e} \mathbf{N}(\boldsymbol{\xi})^\top \mathbf{N}(\boldsymbol{\xi})\,\mathbf{q}^e\,\mathrm{d}\Omega_\mathrm{p}^e. \tag{4.125}$$

4.4.8 Assembling

In order to obtain numerical equations for the whole structure, the virtual energies of all finite elements belonging to the structure, i.e. $e \in [1, NE]$, are summed up, considering the relations between the DOF's of every element and structural DOF's. This procedure transfers variables on elemental scale to the structure. Based on the displacement vectors, stiffness matrices, and right-hand-side vectors of the element, one obtains the corresponding structural vectors and matrices.

The symbolic operator $\bigcup$ is used to ensemble the finite elements (Zienkiewicz and Taylor, 2000)

$$\mathbf{K}_\mathrm{MB} = \bigcup_{e=1}^{NE} \mathbf{K}_\mathrm{MB}^e, \qquad \mathbf{K}_\mathrm{S} = \bigcup_{e=1}^{NE} \mathbf{K}_\mathrm{S}^e, \tag{4.126}$$

$$\mathbf{a} = \bigcup_{e=1}^{NE} \mathbf{a}^e, \qquad \mathbf{r} = \bigcup_{e=1}^{NE} \mathbf{r}^e. \tag{4.127}$$

Finally, the structural stiffness relation is formulated.

$$\mathbf{K}\mathbf{a} = \mathbf{r}. \tag{4.128}$$

The structural stiffness matrix is composed by a membrane-bending $\mathbf{K}_\mathrm{MB}$ and a shear portion $\mathbf{K}_\mathrm{S}$

$$\mathbf{K} = \mathbf{K}_\mathrm{MB} + \mathbf{K}_\mathrm{S}. \tag{4.129}$$

4.4.9 Numerical Integration and Artificial Stiffening Effects

To determine the stiffness matrices and load vectors, it is necessary to calculate integrals over the element surface. The method to solve these integrations numerically used here is the so called GAUSS-LEGENDRE quadrature (Zienkiewicz and Taylor, 2000), which is established in FEM. In order to solve the exact analytical integration I of a function $f(\boldsymbol{\xi})$ over the element surface, Eq. (4.77) is used:

$$I = \int_{\Omega^e} f(\boldsymbol{\xi})\, \mathrm{d}\Omega^e = \int_{-1}^{1}\int_{-1}^{1} f(\boldsymbol{\xi})\, |\mathbf{J}(\boldsymbol{\xi})|\, \mathrm{d}\xi_1\, \mathrm{d}\xi_2 . \tag{4.130}$$

The integral is approximated by the weighted summation of function values (Schwarz and Köckler, 2011)

$$I \approx \sum_{i=1}^{NG_1} \sum_{j=1}^{NG_2} \alpha_1^i \alpha_2^j f(\xi_1^i, \xi_2^j) . \tag{4.131}$$

The function to be integrated is evaluated at the integration or GAUSS points (ξ_1^i, ξ_2^i) and multiplied with the weighting factors α_1^i, α_2^i, while NG is the number of integration points. The position of the integration points is shown in Fig. 4.7 at the SERENDIPITY element.

Typical problems with finite elements are related to artificial stiffening effects. The probably best known phenomenon of this kind is the so-called shear locking. This stiffening effect becomes particularly relevant when bending states at slender structures are under investigation (Babuška and Suri, 1992). In that case, the shear rigidity is parasitic. Due to that, in addition to complete integration, two additional modes of integration are introduced to prevent shear locking (Stolarski and Belytschko, 1982). The modes providing a remedy are the reduced and the selective integration. In this case, a reduced number of GAUSS points is addressed for integration, or a reduced or the complete number of GAUSS points is selected according to the rigidity portion. The principle procedures of all three integration modes are explained in Fig. 4.7, while coordinates of GAUSS points and corresponding weights for complete and reduced integration are given in Table 4.1.

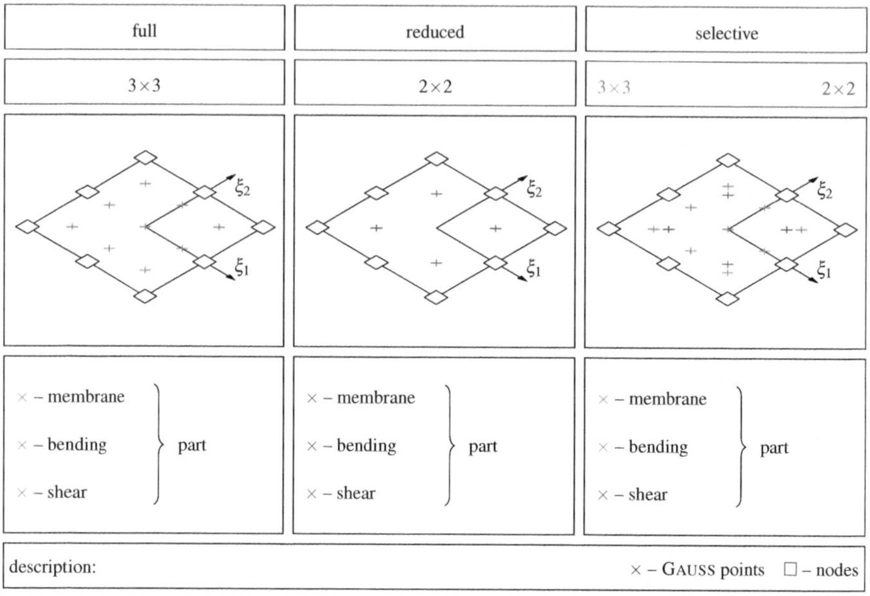

Fig. 4.7: Implemented integration modes for SERENDIPITY element with quadratic shape functions: full, reduced, and selective integration

Table 4.1: Coordinates of GAUSS points and assigned weights of GAUSS-LEGENDRE quadrature (Schwarz and Köckler, 2011)

$m \times m$	ξ_1^i	ξ_2^j	α^i, α^j
3×3	$\xi_1^1 = -\sqrt{\frac{3}{5}}$	$\xi_2^1 = -\sqrt{\frac{3}{5}}$	$\alpha^1 = \frac{5}{9}$
	$\xi_1^2 = 0$	$\xi_2^2 = 0$	$\alpha^2 = \frac{8}{9}$
	$\xi_1^3 = +\sqrt{\frac{3}{5}}$	$\xi_2^3 = +\sqrt{\frac{3}{5}}$	$\alpha^3 = \frac{5}{9}$
2×2	$\xi_1^1 = -\frac{1}{\sqrt{3}}$	$\xi_2^1 = -\frac{1}{\sqrt{3}}$	$\alpha^1 = 1$
	$\xi_1^2 = +\frac{1}{\sqrt{3}}$	$\xi_2^2 = +\frac{1}{\sqrt{3}}$	$\alpha^2 = 1$

4.5 Structural Analysis

4.5.1 Test Structure

As also done in different studies (Eisenträger et al, 2015b; Aßmus et al, 2016), a commercial, 72 cell photovoltaic module with glass front and back cover and ethylene-vinyl acetate (EVA) as core layer material is used for the case studies in the present contribution. Nevertheless, there is a large variation of possible geometries and materials at the market, as reported in Aßmus et al (2016). However, lengths and thicknesses used here are exemplaric. Geometrical measures used for this study are presented in Table 4.2. Hence, a length ratio $\mathit{LR} = {}^{L_2}/_{L_1} = 0.5$ results. For the sake of simplicity, a structure, whose geometrical and material properties are symmetric in transverse direction, is used. The thickness ratio is $\mathit{TR} = {}^{h^c}/_{(h^t+h^b)} \approx 15.6 \cdot 10^{-2}$, and the thickness-to-length ratio is $\mathit{TLR} = {}^{H}/_{L_{\min}} = 9.1 \cdot 10^{-3}$. Material parameters at room temperature (23°C) of the photovoltaic module under consideration are also given in Table 4.2, where isotropic elastic material behaviour is assumed. Glasses for the skin layers are stiff brittle materials, while the EVA for the core layer is a soft rubber-like material. Due to the extreme differences in material properties, the shear modulus ratio is $\mathit{GR} = {}^{G^c}/_{G^s} = 9.98 \cdot 10^{-5}\ \forall s = \{t, b\}$, and the normalised shear rigidity parameter is $\beta = 9.40$ according to Naumenko and Eremeyev (2014). While GR confirms that classical composite theories cannot be applied here due to the highly differing material properties, the shear rigidity parameter indicates the application range of XLWT. Compared to studies presented in Naumenko and Eremeyev (2014); Eisenträger et al (2015a,b); Aßmus et al (2016), β with respect to room temperature is in the range, where the first order shear deformation theory is also applicable to obtain useful results. Since the aim of this study is also the consideration of realistic temperatures at loading, β will vary such that the necessity of XLWT will be shown in the sequel.

For the mounting of a photovoltaic module, a peripheral frame surrounding the composite structure as shown in Aßmus et al (2016) is considered here. Since it is al-

Table 4.2: Geometrical and material parameters of the test structure used for convergence analysis (room temperature $\vartheta = 23$°C)

K	L_1^K [mm]	L_2^K [mm]	h^K [mm]	E^K [N/mm²]	ν^K [−]
t/b	1620	810	3.2	$73.0 \cdot 10^3$ (Cahn et al, 1991)	0.30 (Cahn et al, 1991)
c			1.0	7.9 (Eitner, 2011)	0.41 (Eitner, 2011)

Table 4.3: Definition of boundaries at the layerwise structure and chosen boundary conditions for convergence study ($H = \sum h^K$)

boundary	Γ_1^1	Γ_1^2	Γ_2^1	Γ_2^2	Γ_N
$\boldsymbol{e}_1$	0	L_1	$0 \ldots L_1$	$0 \ldots L_1$	$0 \ldots L_1$
$\boldsymbol{e}_2$	$0 \ldots L_2$	$0 \ldots L_2$	0	L_2	$0 \ldots L_2$
$\boldsymbol{n}$	$-H/2 \ldots H/2$	$-H/2 \ldots H/2$	$-H/2 \ldots H/2$	$-H/2 \ldots H/2$	$-H/2$
condition	$w = 0$	$w = 0,\ u_1^K = 0$	$w = 0$	$w = 0$	$q_\mathrm{W} = 500\,\mathrm{N/mm^2}$

ready shown in Schicker et al (2014); Beinert et al (2016) that framing all-round the photovoltaic composite is favourable due to lower mechanical stresses compared to mounting at specific points at the sides along L_α, this concept is also applied here. It is assumed that the bedding material in the groove of the frame can transmit pressure, but not tension. In the absence of data on the compliance of the bedding material, an elastic bedding is renounced here. For a clear definition of the mounting, the boundaries Γ_α^β are introduced in Table 4.3. Additionally, the boundary Γ_N is defined as loaded area at the photovoltaic module.

4.5.2 Discretisation and Convergence

For verification of the numerical implementation of XLWT, a convergence study is conducted here via discretisation variation. Used boundary conditions are given in Table 4.3. A constant wind load is applied to the outer surface of the photovoltaic module under inclination angle of 35° such that the loading is uniform, but non-orthogonal. This loading scenario corresponds to an orthogonal portion with $p = 409.58\,\mathrm{N/mm^2}$ and a transverse portion of $s_1 = 286.79\,\mathrm{N/mm^2}$, where Eq. (4.24) holds true for the component notation at Γ_N.

The spatial discretisation of the structure presented in the previous section is done via the finite element described in Sect. 4.4.4. A structured mesh is used, where all interior angles of the finite elements account for 90°. The discretisation density has been varied within a convergence study by changes of the element edge lengths. As can be seen in Table 4.4, the aspect ratio of the finite elements is kept constant at $AR = h^e_\mathrm{max}/h^e_\mathrm{min} = 1$ for all discretisation variations, i.e. $h_1^e = h_2^e\ \forall\ \Omega^e \subset \Omega$ holds true. Therein, h_α^e is the element edge length. Furthermore, $NN = NE \times NNE$ is the number of nodes and NEQ is the total number of degrees of freedom ($NEQ = NN \times NDOF$) in corresponding discretised domain, where NNE is the number of nodes per element (8 in present case) and $NDOF$ is the number of degrees of freedoms per node (9), while NE is the total number of elements. Considering the number of integration points

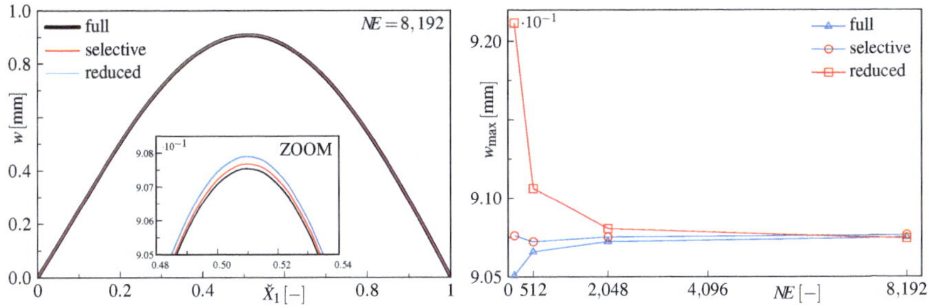

Fig. 4.8: Discretisation variation through h^e adaption: deflection path along plate bisecting line and convergence of maximum deflection

(*NG*) given in Table 4.4, we distinguish between full and reduced (red.) integration scheme according to Fig. 4.7 resulting from Sect. 4.4.9. In summary, three integration schemes at four discretisation variants are used for the convergence study.

Results of the convergence analysis are shown in Fig. 4.8. On the left-hand side, the deflection is depicted along a path representing the bisecting line of the plate. $\check{X}_1 = {}^{X_1}/_{L_1}$ is a normalised measure. On the right-hand side, the maximum deflection of the plate versus the number of elements is given. It is conspicuous that it is sufficient to calculate with a maximum of $8,192$ elements in the present case, compared to conventional structural analysis with 3D elements, where 10^6 to 10^8 elements are needed to gain convergence (Dietrich, 2014). By varying the number of elements, it is found that selective integration provides fastest convergence. Maximum calculation times were clearly below one minute for the highest disretisation density, while all simulations were performed on an eight core Intel i7-5960X CPU and 64 GB RAM by using the solver of ABAQUS 6.14. It should be noted that deflections decrease with increasing number of elements applying the reduced integration scheme, which seems unphysical. However, this phenomenon is shown in Eisenträger et al (2015b) also and will not be discussed here. As is apparent, the deformation field along $\check{X}_1$ is unsymmetric (see magnification in left plot of Fig. 4.8) due to the non-orthogonal loading. For the maximum element number used here, reduced integration results in the largest deflections, while full integration leads to the stiffening of the elements. However, difference in the maximum deflection of all three integration schemes is around 10^{-3} mm, while the maximum deflection is $9.08 \cdot 10^{-1}$ mm.

In the sequel, all calculations are performed by using $8,192$ elements, where the selective integration is favoured since this scheme seems to be insensitive to artificial stiffening, also at low discretisation densities.

Table 4.4: Characteristic parameters of the discretisation variation resulting through h^e_α adaptation at the composite structure

mesh		1	2	3	4
NE		128	512	2,048	8,192
NN		433	1,633	6,337	24,961
NEQ		3,897	14,697	57,033	224,649
NG	full	2,304	4,608	18,432	73,728
	red.	1,024	2,048	8,192	32,768
AR		1	1	1	1

4.5.3 Case Studies

As already mentioned, snow loads act at sub-zero temperatures. In contrast, wind loads are independent of temperature. Due to the lack of data for material parameters of the core layer at any admissible temperature with respect to the range of application, cf. Aßmus et al (2016), these case studies are restricted to snow loading at $-40°$C and to wind loading at $+80°$C as two extrema. In contrast to the temperature dependent material behaviour of the core layer, we assume that the material parameters of the skin layers made of glass are insensitive with respect to temperature changes. Temperature dependent material parameters of the core layer for both case studies are given in Table 4.5. The high value of $\beta = 106.74$ at decreased temperatures indicates that also the KIRCHHOFF theory can be used. Contrary to that, $\beta = 2.41$ at elevated temperatures requires the XLWT (Naumenko and Eremeyev, 2014; Eisenträger et al, 2015a,b). This behaviour is also confirmed by the shear modulus ratio *GR*.

Boundary conditions for edge support and loading at the front surface are given in Table 4.6 for snow and wind load. In both cases, the load distribution at Γ_{N} is

Table 4.5: Temperature dependent material parameters of core layer material (based on Eitner (2011))

ϑ [°C]	E^c [N/mm²]	ν^c [−]	G^c [N/mm²]	GR [−]	β [−]
−40	1019.04	0.41	361.36	$1.29 \cdot 10^{-2}$	106.74
+80	0.52	0.41	0.18	$6.57 \cdot 10^{-6}$	2.41

assumed to be well described by a double sine function with respect to COOS $\{\boldsymbol{g}_i\}$ as given in Eq. (4.25). This double sine function is rotated in the COOS $\{\boldsymbol{e}_i\}$ by the angle θ using Eqs. (4.27) or (4.28). For both load cases, the angle is varied, i.e. $\theta = \{25°, 35°, 45°, 55°\}$. Compared to load intensities in natural weathering discussed in Sect. 4.2.1, the loads used here are kept small since just the procedure is in the foreground. At both cases, the maximum amplitude of loading is $500\,{}^{\mathrm{N}}/_{\mathrm{mm}^2}$.
Since the transverse loads of both loading scenarios act opposite when considering the loaded surface inclined with $0 < \theta < 90°$, different boundary conditions for the photovoltaic composite are needed at Γ_1^1 and Γ_1^2. These are presented in Table 4.6.
In the sequel, the evaluations are divided in two case studies. Figures 4.9 and 4.10 show the DOF's for all layers along two paths with respect to a varying inclination angle. The paths are A-A ($X_1 = 0 \ldots L_1$, $X_2 = {}^{L_2}/_2$) and B-B ($X_1 = {}^{L_1}/_2$, $X_2 = 0 \ldots L_2$), while normalised length measures $\breve{X}_\alpha = {}^{X_\alpha}/_{L_\alpha}$ are used. In addition, kinetic and kinematical quantities for the individual layers are juxtaposed comparatively in Figs. 4.11, 4.12, 4.13, and 4.14, while only an inclination angle of $\theta = 35°$ is taken into account for the sake of clarity. This value represents a typical average for modules mounted in the area of Germany (around $47° \ldots 52°$ of latitude), see Huld et al (2008), since the optimum setting is always perpendicular to the sun. Due to the symmetry of the second rank tensors $\boldsymbol{N}^K$, $\boldsymbol{L}^K$, $\boldsymbol{E}^K$ and $\boldsymbol{X}^K$, the evaluation is restricted to the elements $\square_{11}^K$, $\square_{22}^K$, and $\square_{12}^K$ $\forall \square = \{N, M, E, \chi\}$. All results are discussed in Sect. 4.5.4.

4.5.4 Results and Discussion

4.5.4.1 Degrees of Freedom

Figure 4.9 displays the deflection, the in-plane displacements, and the rotations along the Paths A-A and B-B for wind loading at elevated temperatures. It becomes obvious that the deflection increases with growing inclination angle. In contrast to pure orthogonal loading, cf. Aßmus et al (2016); Eisenträger et al (2015b), the deflection is non-symmetric due to the introduction of tangential loads. The maximum deflection is not located at the plate centre anymore, i.e. $w_{\max} \neq w({}^{L_1}/_2, {}^{L_2}/_2)$, and equals $w_{\max} \approx 1.95\,\mathrm{mm}$ for $\theta = 55°$. The displacement u_1^K approaches its extremum

Table 4.6: Boundary conditions at the layerwise structure used for case studies at wind and snow loading

boundary	Γ_1^1	Γ_1^2	Γ_2^1	Γ_2^2	Γ_N
wind loading	$w = 0$	$w = 0$, $u_1^K = 0$	$w = 0$	$w = 0$	$q_\mathrm{W}^0 = 500\,{}^{\mathrm{N}}/_{\mathrm{mm}^2}$
snow loading	$w = 0$, $u_1^K = 0$	$w = 0$	$w = 0$	$w = 0$	$q_\mathrm{S}^0 = 500\,{}^{\mathrm{N}}/_{\mathrm{mm}^2}$

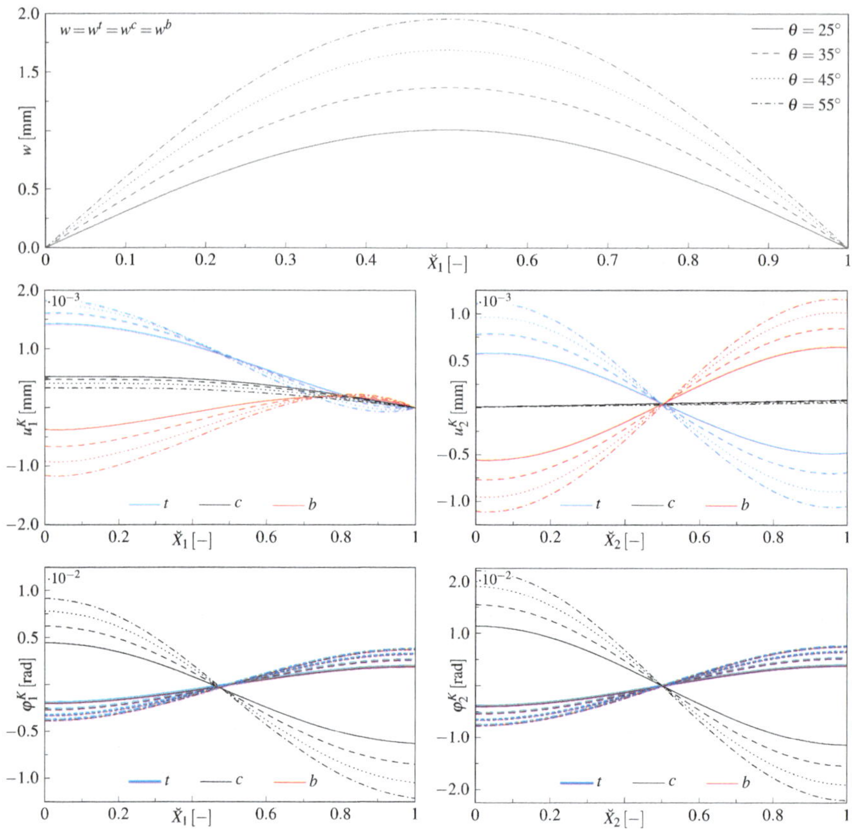

Fig. 4.9: Angle-dependent layerwise kinematics of the photovoltaic composite structure at wind loading

near the edge $\check{X}_1 = 0$, while $u_1^K(\check{X}_1 = 1) = 0$ holds true due to the boundary conditions in Table 4.6. The absolute displacement $|u_2^K|$ is equal in top and core layer along the Path B-B with opposite signs and the displacement in the core layer tends to zero. Contrarily, the rotations φ_1^K and φ_2^K are equal in the skin layers, while the core layer rotates in opposite direction.

For comparison, Fig. 4.10 displays the deflection, the in-plane displacements, and the rotations along the Paths A-A and B-B for snow loading at reduced temperatures. Contrary to wind loading, the deflection increases with decreasing inclination angle, such that the maximum deflection occurs at $\theta = 25°$ and equals $w_{\max} \approx 0.45\,\text{mm}$. At $\check{X}_1 = 0$, the derivative of the deflection with respect to $\check{X}_1$ tends to zero. This is due to the constrained displacements u_1^K. Subsequently, the mutual sliding between the skin layers is decreased significantly because of the high shear rigidity of the core layer material at reduced temperatures. For this reason, the rotations φ_1^K are

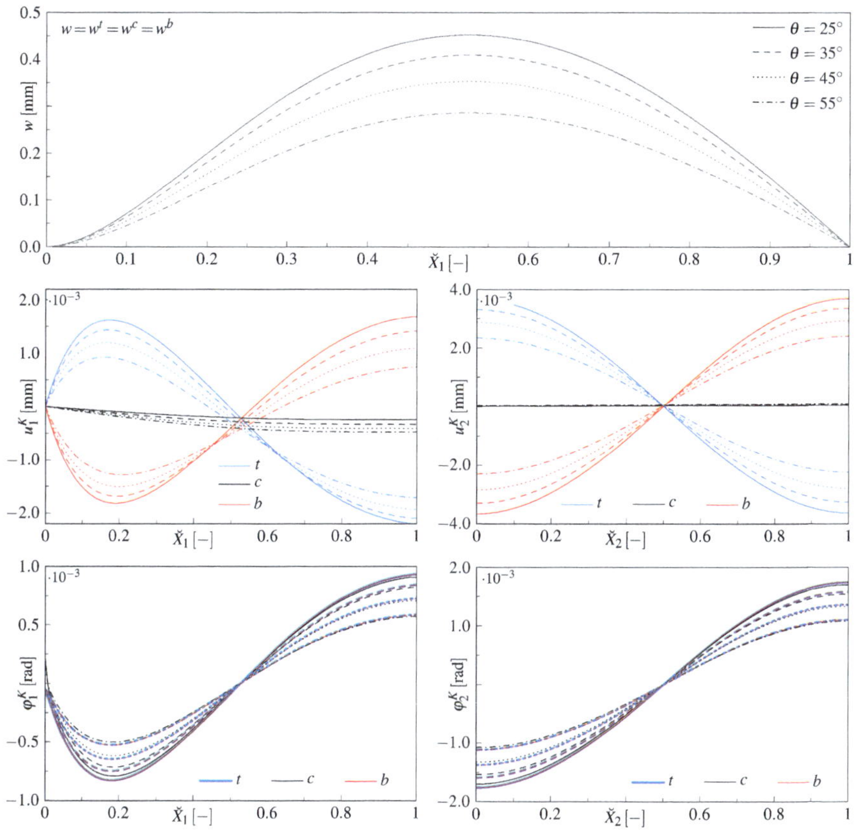

Fig. 4.10: Angle-dependent layerwise kinematics of the photovoltaic composite structure at snow loading

vanishing at this point. Also in the case of snow loading, the maximum deflection is shifted from the plate centre, so that $w_{\max} \neq w(L_1/2, L_2/2)$ holds true. The in-plane displacements u_1^K approach their extrema at $\check{X}_1 = 1$. Due to the symmetry of boundary conditions with respect to $\check{X}_2$, the displacements u_2^K approach the maximum or minimum, respectively, at the edges of the module and are zero in the centre. The displacements in the front cover are contrary to those in the back cover, whereas the displacements in the core layer are approximately zero. A relevant displacement in the core layer cannot be observed. Due to the increased shear rigidity of the core layer, all layers rotate in the same direction. The skin layers rotate around the same angles, while the rotation in the core layer differs slightly. The rotations φ_2^K are zero at $\check{X}_2 = 0.5$ and reach their maximum or minimum, respectively, at the edges.

For present geometry and material parameters when restricting to 55° inclination, the deflections at wind loading and elevated temperatures are around one order of magnitude higher than at snow loading at reduced temperatures. Thereby, it should be remembered that the load intensities at both loading scenarios are equal. Comparing snow and wind loading, the in-plane displacements are in the same order of magnitude. Basically, it should be noted that the rotations are about one order of magnitude higher at elevated temperatures than at reduced temperatures. This fact can become critical when loads and/or inclination angle at elevated temperatures are rising, what results in high shear loads in the core layer.

4.5.4.2 Kinetic and Kinematic Quantities

Figure 4.11 presents the stress resultants (both upper rows) and the conjugate kinematical measures (both lower rows) along the Path A-A for wind loading at elevated temperatures. For the membrane force, $N_{11}^K(\check{X}_1 = 0) = 0$ holds true because the edge Γ_1^1 is unconstrained with respect to the in-plane displacements. Furthermore, the membrane force N_{12}^K equals zero approximately, considering the scale of the ordinate. One should note that no membrane forces are present in the core layer. Since the deflections are constrained at the edges, the transverse shear forces show typical boundary layer effects Brank (2008). Refining the mesh near the edges would reduce the influence of these effects. The bending moments $M_{\alpha\alpha}^K$ equal zero at the edges $\check{X}_1 = 0$ and $\check{X}_1 = 1$ since the rotations φ_α^K are unconstrained along the edges Γ_1^1 and Γ_1^2. Similar to the membrane force N_{12}^K, the twisting moment M_{12}^K is zero. All moments $M_{\alpha\beta}^c$ are zero, i.e. the core layer is moments-free. The membrane strains behave similarly to the membrane forces, and the shear strains are also influenced by the boundary layer effects. In the shear-rigid skin layers, the shear strains γ_2^K are zero. In addition, $\chi_{11}^K \approx 0$ is valid for all layers.

Additionally, Fig. 4.12 depicts the stress resultants (both upper rows) and the conjugate kinematical measures (both lower rows) along the Path A-A for snow loading at reduced temperatures. Because of the different boundary conditions with respect to the in-plane displacements for wind and snow loading, cf. Table 4.6, the membrane forces at snow loading show a different behaviour compared to wind loading. In analogy to wind loading, the membrane forces N_{12}^K are approximately zero. Boundary layer effects are also observable at the transverse shear forces Q_α^K. The behaviour of the bending moments at snow loading differs strongly from the bending moments at wind loading near the edge $\check{X}_1 = 0$. As in the case of wind loading, the core layer is moments-free and the twisting moment approximates zero.

The membrane strains E_{11}^K reach their extrema at $\check{X}_1 = 0$ and become zero at the opposite edge. Overall, the membrane strains present a similar behaviour as the corresponding membrane forces. Once again, the membrane strain E_{12}^K is zero, approximately, and the shear strain γ_2^K disappears in the skin layers. Also in the case of snow loading, the boundary layer effects are reflected by the shear strains, and the curvature χ_{11}^K is zero in all layers. The curvature χ_{22}^c of the core layer reaches its minimum at $\check{X}_1 = 0$, while it tends to zero at $\check{X}_1 = 1$. Comparing the magnitudes of the quantities for snow and wind loading, it becomes obvious that the membrane forces

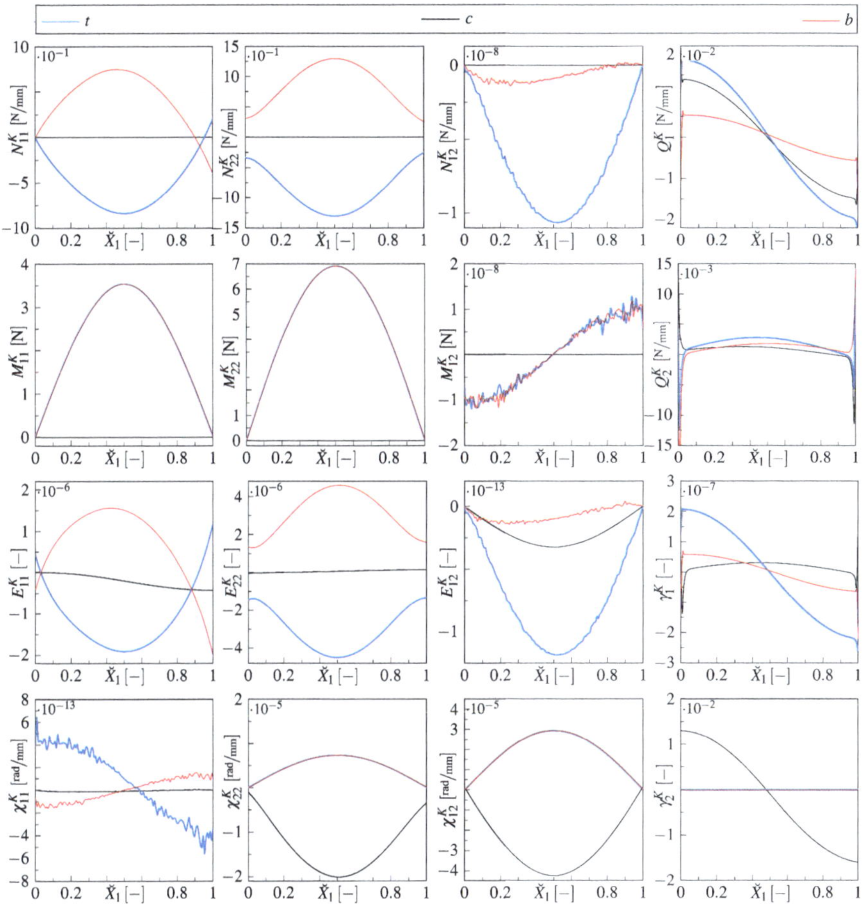

Fig. 4.11: Layerwise stress resultants, strains, and curvatures of the photovoltaic composite structure at wind loading for path A-A

and strains are higher at snow loading than at wind loading, whereas the behaviour of the moments and curvatures is inverse.

In analogy to Figs. 4.11 and 4.12, Figs. 4.13 and 4.14 present the corresponding quantities along the Path B-B. The membrane forces in the skin layers reach their extremum in the centre of the plate, i.e. $\check{X}_2 = 0.5$, while the membrane forces in the core layer are zero. In both case studies, the transverse shear force Q_1^K equals zero in the interior of the plate and reaches its extrema at the edges. Once again, a boundary layer effect is observable with respect to the transverse shear force Q_2^K. For the bending moments, $M_{\alpha\alpha}^K(\check{X}_1 = 0) = M_{\alpha\alpha}^K(\check{X}_1 = 1) = 0$ holds true. In both case studies, the moments show the same behaviour, while the moments at wind loading are approximately one order of magnitude higher than the corresponding quantities

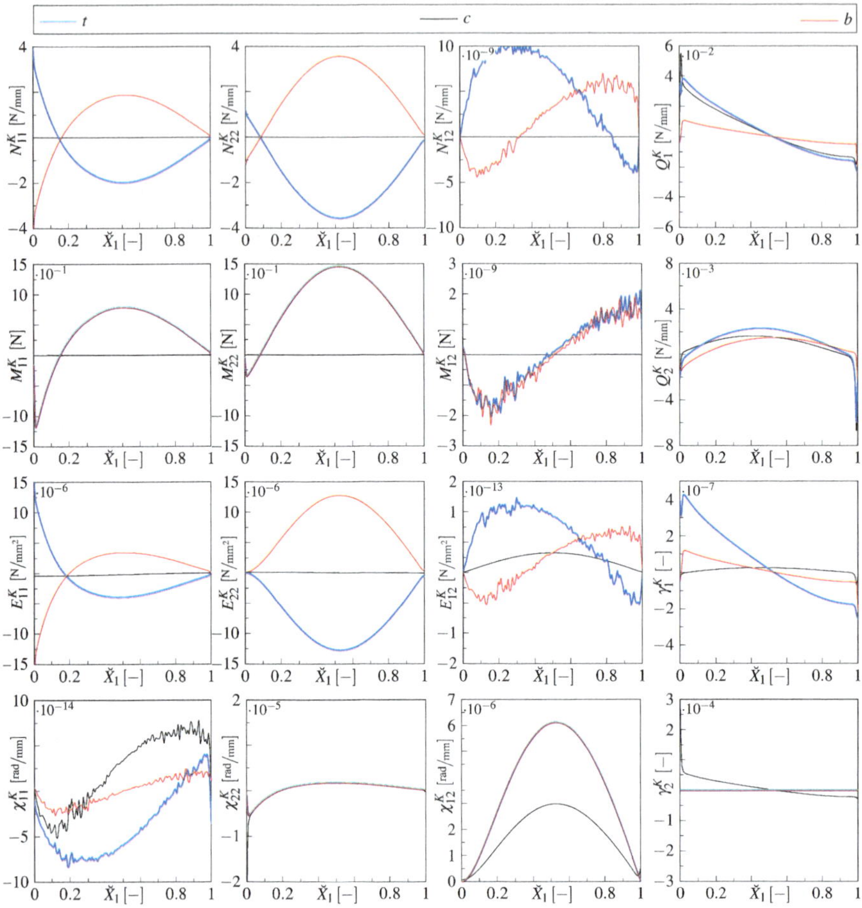

Fig. 4.12: Layerwise stress resultants, strains, and curvatures of the photovoltaic composite structure at snow loading for path A-A

at snow loading. The twisting moment M_{12}^K is influenced by a boundary layer effect, and the core layer is moments-free.

At wind loading, the membrane strain E_{11}^K is approximately constant in the interior of the photovoltaic module. In general, the curves of the membrane strains have a similar shape as the curves of the corresponding membrane forces. The membrane strains at snow loading are significantly higher than the ones at wind loading. As one would expect, the shear strains at elevated temperatures, i.e. at wind loading, exceed the corresponding shear strains at reduced temperatures (snow loading) by several orders of magnitude. In both case studies, the transverse shear strains γ_α^K in the skin layers are negligible compared to the shear strains in the core layer. In contrast to the other kinematical measures, the curves for the curvatures at wind loading differ strongly from the corresponding curves at snow loading.

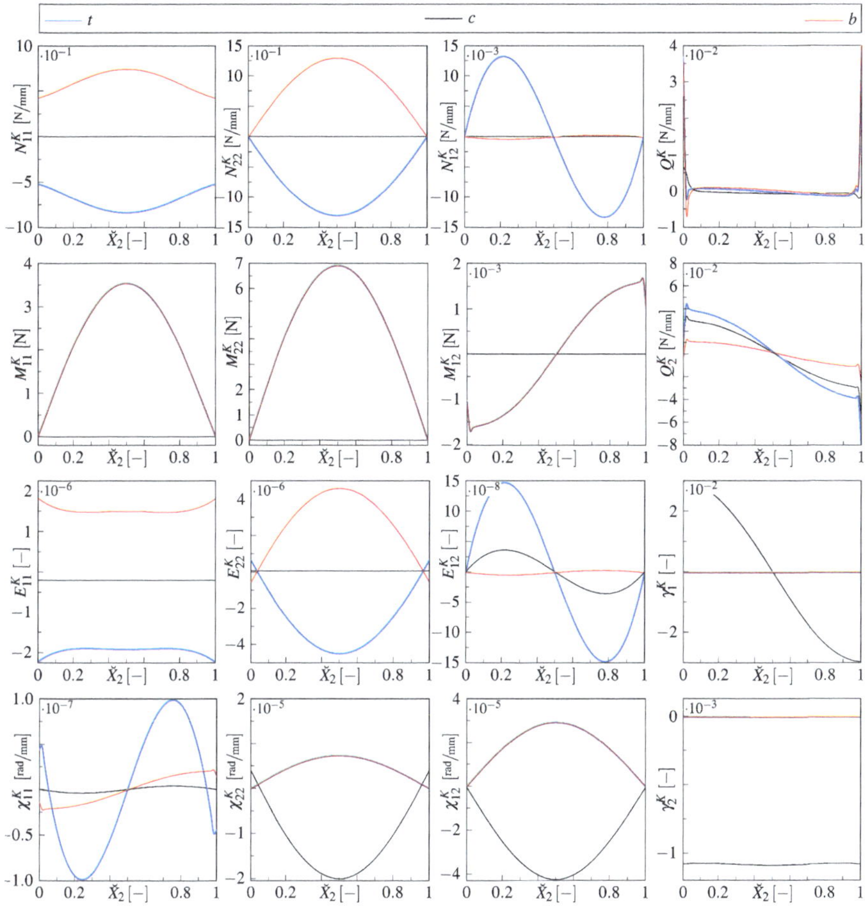

Fig. 4.13: Layerwise stress resultants, strains, and curvatures of the photovoltaic composite structure at wind loading for path B-B

4.6 Conclusion

In the foregoing investigations, we have introduced an approach for approximating natural mechanical loading at photovoltaic modules and evaluating its effect on structural stresses. By exploiting the XLWT, a framework for the mechanics of photovoltaic modules is derived, while a computational solution approach provided by the FEM is used.

For the presence of wind and snow loading, we have obtained adequate approximations of load distributions by introducing FOURIER series. Additionally, we have introduced a structural inclination to represent realistic operating and mounting conditions, and to account for directions of loads applied. By this procedure, we have

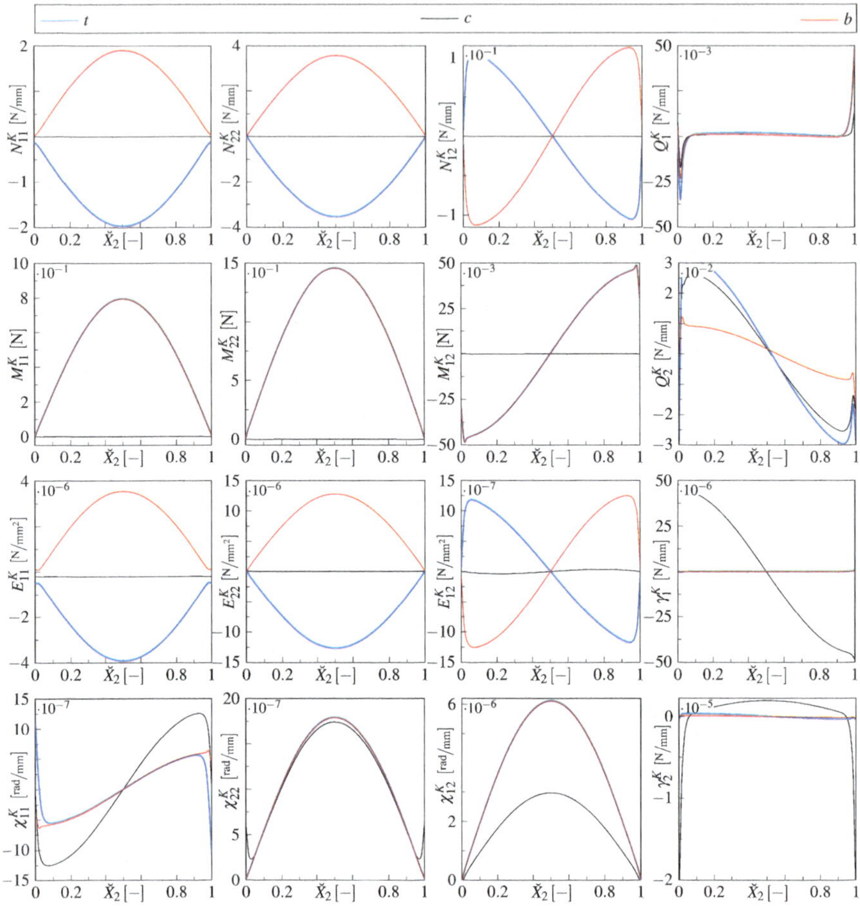

Fig. 4.14: Layerwise stress resultants, strains, and curvatures of the photovoltaic composite structure at snow loading for path B-B

generated a catalogue of different possibilities for load approximations at inclined surfaces. This allows us to consider a broad class of mechanical problems on structural analysis of photovoltaic modules. However, this framework can be easily enlarged by considering an azimuthal angle, or by using general periodic functions to describe location dependent load distributions at a surface. The description of non-uniform and heterogeneous loads is not restricted to the use with the XLWT. It is applicable to any procedure of structural mechanics at photovoltaic modules.
For structural analysis, we have introduced the concept of deformable elastic (directed) surfaces, as these form the basis of the multilayered systems used to describe the present composite structure. As basis, an extension of the five parameter thin-walled structural theory based on the direct approach, expanded to a three layered system is carried out. Thereby, kinematic and kinetic constraints on the inter-

faces between the layers are introduced, such that it is possible to describe the structure completely with only 9 DOF's. The attraction of this procedure is to involve the membrane, bending, and shear state independently and layerwise. In contrast to Naumenko and Eremeyev (2014); Eisenträger et al (2015b), and Aßmus et al (2016), tangential loading is taken into account by using the XLWT. Afterwards a computational solution strategy is introduced based on FEM. For this reason, the governing equations of XLWT are embedded in a variational framework, namely the principle of virtual work. By exploiting these equations, one obtains the element stiffness relation, where C^0 continuous elements of SERENDIPITY type are introduced. Since shear locking is a well known problem within the context of theories considering transverse shear strains, two alternative arrangements to avoid such artificial stiffening effects are introduced.

Finally, the effects of the load approximation are investigated on two characteristic loading scenarios. The case studies presented here use inclination angles typically for latitudes in middle Europe. Concerning maximum load intensities used in the present investigations, we restrict ourselves to the validity of the fully linear XLWT. With regard to our conceptual framework for load approximations, load intensities with arbitrary values can be applied. The results reveal the significance of direction dependent loading when analysing photovoltaic composite structures, in addition to temperature dependent analyses as already emphasised in Aßmus et al (2016). This is particularly evident comparing the achieved results with the evaluations in Eisenträger et al (2015b).

There is a fundamental agreement in the scientific community that theories for slender structures exhibit only an approximate character, even if the direct approach is known as geometrically exact. However, experimental evidence of the performance of the XLWT will remain an open problem. Since manufacturers of photovoltaic modules typically do not provide any information on the mechanical properties of the materials used, experimental comparisons are still a venture. Therefore, prototypical structures are needed, where the materials of components have been analysed in foregoing material tests. With regard to the theoretical background based on the direct approach, hypotheses have then to be introduced to incorporate three dimensional information for the underlying physical structure.

One of the shortcomings of the approach using the XLWT is the missing information regarding the loads at components embedded in the core layer of the photovoltaic module. If one is interested in concrete data for stresses and strains occurring at solar cells, encapsulations, or further ingredients during the loading processes, one is referred to Aßmus et al (2016), where a multi-scale approach is proposed to generate such local data from the present results.

However, as it is the main goal of every design engineer to develop structures that do not deform inelastic during operation, the whole procedure presented here constitutes a distinguished methodology to optimise photovoltaic module structures or to invent new structural design concepts. By using the computational solution strategy for the governing equations of XLWT, the present approach provides a powerful tool for structural analysis considering any admissible mounting and loading condition. This opens up the question for optimum geometrical and material parameters of photovoltaic composites. For this purpose, an optimisation problem can be formulated, where kinematic and kinetic quantities are control variables, while aforementioned structural parameters are manipulating variables. Nevertheless, there are also constraints since the costs of photovoltaic module production represent a decisive factor. Besides, not all dimensions can be realised technologically and not all materials can be selected due to requirements for further physical properties, e.g. transparency, conductivity, moisture absorption etc.

Acknowledgements The financial support rendered by the German Research Foundation (DFG) in context of the research training group 'Micro-Macro-Interactions in Structured Media and Particle Systems' (GRK 1554) is gratefully acknowledged.

4.A Appendix

4.A.1 Constitutive Matrices

The constitutive matrices, introduced in Eqs. (4.103)-(4.111), are defined as follows

$$\hat{\mathbf{C}}_L^K = \begin{bmatrix} a_L^K + 2b_L^K & b_L^K & 0 \\ a_L^K & a_L^K + 2b_L^K & 0 \\ 0 & 0 & b_L^K \end{bmatrix} \qquad \forall\, K = \{\circ, \Delta, c\} \wedge L = \{\mathrm{M}, \mathrm{B}\}\,, \tag{4.132}$$

$$\hat{\mathbf{C}}_\mathrm{S}^K = a_\mathrm{S}^K \begin{bmatrix} 1 & 0 \\ 0 & 1 \end{bmatrix} \qquad \forall\, K = \{\circ, \Delta, c\} \tag{4.133}$$

with the abbreviations

$$a_L^K = \begin{cases} D_L^t \nu^t + D_L^b \nu^b & \text{if } K = \circ \\ D_L^t \nu^t - D_L^b \nu^b & \text{if } K = \Delta \\ D_L^c \nu^c & \text{if } K = c \end{cases} \qquad \forall L = \{\mathrm{M}, \mathrm{B}\}\,, \tag{4.134}$$

$$b_L^K = \begin{cases} \frac{1-\nu^t}{2} D_L^t + \frac{1-\nu^t}{2} D_L^b & \text{if } K = \circ \\ \frac{1-\nu^t}{2} D_L^t - \frac{1-\nu^t}{2} D_L^b & \text{if } K = \Delta \\ \frac{1-\nu^t}{2} D_L^c & \text{if } K = c \end{cases} \qquad \forall L = \{\mathrm{M}, \mathrm{B}\}\,, \tag{4.135}$$

$$a_\mathrm{S}^K = \begin{cases} D_\mathrm{S}^t + D_\mathrm{S}^b & \text{if } K = \circ \\ D_\mathrm{S}^t - D_\mathrm{S}^b & \text{if } K = \Delta \\ D_\mathrm{S}^c & \text{if } K = c \end{cases} \tag{4.136}$$

With the matrices above, one can define the generalised constitutive matrices

$$\mathbf{C}^{\circ}_{\mathrm{MB}} = \begin{bmatrix} \hat{\mathbf{C}}^{\circ}_{\mathrm{M}} & \mathbf{0} & \mathbf{0} & \mathbf{0} \\ \mathbf{0} & \hat{\mathbf{C}}^{\circ}_{\mathrm{M}} & \mathbf{0} & \mathbf{0} \\ \mathbf{0} & \mathbf{0} & \hat{\mathbf{C}}^{\circ}_{\mathrm{B}} & \mathbf{0} \\ \mathbf{0} & \mathbf{0} & \mathbf{0} & \hat{\mathbf{C}}^{\circ}_{\mathrm{B}} \end{bmatrix}, \tag{4.137}$$

$$\mathbf{C}^{\Delta}_{\mathrm{MB}} = \begin{bmatrix} \mathbf{0} & \hat{\mathbf{C}}^{\Delta}_{\mathrm{M}} & \mathbf{0} & \mathbf{0} \\ \mathbf{0} & \mathbf{0} & \mathbf{0} & \mathbf{0} \\ \mathbf{0} & \mathbf{0} & \mathbf{0} & \hat{\mathbf{C}}^{\Delta}_{\mathrm{B}} \\ \mathbf{0} & \mathbf{0} & \mathbf{0} & \mathbf{0} \end{bmatrix}, \tag{4.138}$$

$$\mathbf{C}^{\circ}_{\mathrm{S}} = \begin{bmatrix} \hat{\mathbf{C}}^{\circ}_{\mathrm{S}} & \mathbf{0} \\ \mathbf{0} & \hat{\mathbf{C}}^{\circ}_{\mathrm{S}} \end{bmatrix}, \tag{4.139}$$

$$\mathbf{C}^{\Delta}_{\mathrm{S}} = \begin{bmatrix} \mathbf{0} & \hat{\mathbf{C}}^{\Delta}_{\mathrm{S}} \\ \mathbf{0} & \mathbf{0} \end{bmatrix}. \tag{4.140}$$

The zero matrices in Eqs. (4.137) and (4.138) have three rows and columns, while the zero matrices in Eqs. (4.139) and (4.140) possess only two rows and columns.

4.A.2 Auxiliary Matrices

Furthermore, the auxiliary matrices $\mathbf{A}_i \; \forall \, i = \{1, \ldots, 5\}$ are defined as follows:

$$\mathbf{A}_1 = \frac{1}{h^c} \begin{bmatrix} 0 & 0 & -2 & 0 & 0 & -(h^{\circ}+h^c) & 0 & -h^{\Delta} & 0 \\ 0 & 0 & 0 & -2 & 0 & 0 & -(h^{\circ}+h^c) & 0 & -h^{\Delta} \end{bmatrix}, \tag{4.141}$$

$$\mathbf{A}_2 = \begin{bmatrix} 1 & 0 & 0 & 0 \\ 0 & 1 & 0 & 0 \end{bmatrix}, \tag{4.142}$$

$$\mathbf{A}_3 = \begin{bmatrix} \mathbf{I} & \mathbf{0} & \frac{1}{2}h^{\Delta}\mathbf{I} & \frac{1}{2}h^{\circ}\mathbf{I} \end{bmatrix}, \tag{4.143}$$

$$\mathbf{A}_4 = \frac{1}{h^c} \begin{bmatrix} \mathbf{0} & 2\mathbf{I} & h^{\circ}\mathbf{I} & h^{\Delta}\mathbf{I} \end{bmatrix}, \tag{4.144}$$

$$\mathbf{A}_5 = \begin{bmatrix} 1 & 0 & 0 & 0 & 0 & 0 & 0 & 0 & 0 & 0 & 0 & 0 & 0 \\ 0 & 1 & 0 & 0 & 0 & 0 & 0 & 0 & 0 & 0 & 0 & 0 & 0 \\ 0 & 0 & 1 & 0 & 0 & 0 & 0 & 0 & 0 & 0 & 0 & -\frac{2}{h^c} & 0 \\ 0 & 0 & 0 & 1 & 0 & 0 & 0 & 0 & 0 & 0 & 0 & 0 & -\frac{2}{h^c} \\ 0 & 0 & 0 & 0 & 0 & 0 & 1 & 0 & 0 & 0 & 0 & 0 & 0 \\ \frac{1}{2}h^{\Delta} & 0 & \frac{1}{2}(h^{\circ}+h^c) & 0 & 0 & 0 & 0 & 0 & -1 & 0 & 0 & -\frac{h^{\circ}+h^c}{h^c} & 0 \\ 0 & \frac{1}{2}h^{\Delta} & 0 & \frac{1}{2}(h^{\circ}+h^c) & 0 & 0 & 0 & 1 & 0 & 0 & 0 & 0 & -\frac{h^{\circ}+h^c}{h^c} \\ 0 & 0 & 0 & 0 & \frac{1}{2}h^{\circ} & 0 & 0 & 0 & 0 & 0 & -1 & -\frac{h^{\Delta}}{h^c} & 0 \\ 0 & 0 & 0 & 0 & 0 & \frac{1}{2}h^{\circ} & 0 & 0 & 0 & 1 & 0 & 0 & -\frac{h^{\Delta}}{h^c} \end{bmatrix}. \tag{4.145}$$

The identity matrices $\mathbf{I}$ and the zero matrices $\mathbf{0}$ used in Eqs. (4.143)-(4.144) have three rows and columns each.

References

Abiola-Ogedengbe A, Hangan H, Siddiqui K (2015) Experimental investigation of wind effects on a standalone photovoltaic (PV) module. Renewable Energy 78:657–665, DOI 10.1016/j.renene.2015.01.037

Altenbach H (2015) Kontinuumsmechanik: Einführung in die materialunabhängigen und materialabhängigen Gleichungen, 3rd edn. Springer Vieweg, Berlin · Heidelberg, DOI 10.1007/978-3-662-47070-1

Altenbach H, Eremeyev VA (2009) On the linear theory of micropolar plates. Zeitschrift für Angewandte Mathematik und Mechanik 89(4):242–256, DOI 10.1002/zamm.200800207

Aßmus M, Köhl M (2012) Experimental investigation of the mechanical behavior of photovoltaic modules at defined inflow conditions. Journal of Photonics for Energy 2(1):1–11, DOI 10.1117/1.JPE.2.022002

Aßmus M, Jack S, Weiß KA, Köhl M (2011) Measurement and simulation of vibrations of PV-modules induced by dynamic mechanical loads. Progress in Photovoltaics: Research and Applications 19(6):688–694, DOI 10.1002/pip.1087

Aßmus M, Naumenko K, Altenbach H (2016) A multiscale projection approach for the coupled global-local structural analysis of photovoltaic modules. Composite Structures 158:340–358, DOI 10.1016/j.compstruct.2016.09.036

Babuška I, Suri M (1992) On Locking and Robustness in the Finite Element Method. SIAM Journal on Numerical Analysis 29(5):1261–1293, DOI 10.1137/0729075

Bathe KJ, Zimmermann P (2002) Finite-Elemente-Methoden. Springer

Beinert A, Ebert M, Eitner U, Aktaa J (2016) Influence of Photovoltaic Module Mounting Systems on the Thermo-Mechanical Stresses in Solar Cells by FEM Modelling. In: Proceedings of the 32nd European Photovoltaic Solar Energy Conference and Exhibition (EU PVSEC), WIP Munich, pp 1833–1836, DOI 10.4229/EUPVSEC20162016-5BV.1.14

Bertram A (2012) Elasticity and Plasticity of Large Deformations: An Introduction, 3rd edn. Springer, Berlin · Heidelberg, DOI 10.1007/978-3-642-24615-9

Bertram A, Glüge R (2015) Solid Mechanics. Theory, Modeling, and Problems. Springer, DOI 10.1007/978-3-319-19566-7

Brank B (2008) On boundary layer in the Mindlin plate model: Levy plates. Thin-Walled Structures 46(5):451–465, DOI http://dx.doi.org/10.1016/j.tws.2007.11.003

Cahn R, Haasen P, Kramer E (1991) Materials science & technology. A comprehensive treatment. Vol. 9: Glasses and amorphous materials. Wiley-VCH, Weinheim

Carrera E (2002) Theories and finite elements for multilayered, anisotropic, composite plates and shells. Archives of Computational Methods in Engineering 9(2):87–140, DOI 10.1007/BF02736649

Carrera E (2003a) Historical review of Zig-Zag theories for multilayered plates and shells. Applied Mechanics Reviews 56(3):287–308, DOI 10.1115/1.1557614

Carrera E (2003b) Theories and finite elements for multilayered plates and shells: A unified compact formulation with numerical assessment and benchmarking. Archives of Computational Methods in Engineering 10(3):215–296, DOI 10.1007/BF02736224

Chevalier HL, Norton DJ (1979) Wind loads on solar collector panels and support structure. Tech. rep., Aerospace Engineering Department, Texas Agricultural & Mechanical University, Houston, Texas (USA), URL www.osti.gov/scitech/servlets/purl/5350425, Technical Report, Aerospace Engineering Department, Texas Agricultural & Mechanical University, Houston, Texas (USA)

Cosserat E, Cosserat F (1909) Théorie des corps déformables. A. Hermann et fils, Paris, URL http://jhir.library.jhu.edu/handle/1774.2/34209, (english translation by D.H. Delphenich available at http://www.neo-classical-physics.info/uploads/3/4/3/6/34363841/cosserat_chap_i-iii.pdf and http://www.neo-classical-physics.info/uploads/3/4/3/6/34363841/cosserat_chap_iv-vi.pdf)

Dietrich S (2014) Numerische Untersuchungen zur mechanischen Zuverlässigkeit verkapselter Siliziumsolarzellen. PhD thesis, Otto von Guericke University Magdeburg, URL http://d-nb.info/1066224463

Eisenträger J, Naumenko K, Altenbach H, Köppe H (2015a) Application of the first-order shear deformation theory to the analysis of laminated glasses and photovoltaic panels. International Journal of Mechanical Sciences 96–97:163–171, DOI 10.1016/j.ijmecsci.2015.03.012

Eisenträger J, Naumenko K, Altenbach H, Meenen J (2015b) A user-defined finite element for laminated glass panels and photovoltaic modules based on a layer-wise theory. Composite Structures 133, DOI 10.1016/j.compstruct.2015.07.049

Eitner U (2011) Thermomechanics of photovoltaic modules. PhD thesis, Martin Luther University Halle-Wittenberg, URL http://nbn-resolving.de/urn:nbn:de:gbv:3:4-5812

Ellingwood B, O'Rourke M (1985) Probabilistic models of snow loads on structures. Structural safety 2(4):291–299, DOI 10.1016/0167-4730(85)90015-3

Ferrara C, Philipp D (2012) Why Do PV Modules Fail? Energy Procedia 15:379–387, DOI 10.1016/j.egypro.2012.02.046

Gurtin ME, Murdoch AI (1975) A continuum theory of elastic material surfaces. Archive for Rational Mechanics and Analysis 57(4):291–323, DOI 10.1007/BF00261375

Huld T, Šúri M, Dunlop ED (2008) Comparison of potential solar electricity output from fixed-inclined and two-axis tracking photovoltaic modules in europe. Progress in Photovoltaics: Research and Applications 16(1):47–59, DOI 10.1002/pip.773

Jordan DC, Wohlgemuth JH, Kurtz SR (2012) Technology and climate trends in pv module degradation. 27th European Photovoltaic Solar Energy Conference and Exhibition (PVSEC):3118–3124, DOI 10.4229/27thEUPVSEC2012-4DO.5.1

Jubayer CM, Hangan H (2014) Numerical simulation of wind effects on a stand-alone ground mounted photovoltaic (PV) system. Journal of Wind Engineering and Industrial Aerodynamics 134:56–4, DOI 10.1016/j.jweia.2014.08.008

Jubayer CM, Hangan H (2016) A numerical approach to the investigation of wind loading on an array of ground mounted solar photovoltaic (PV) panels. Journal of Wind Engineering and Industrial Aerodynamics 153:60–70, DOI 10.1016/j.jweia.2016.03.009

Kajari-Schröder S, Kunze I, Eitner U, Köntges M (2011) Spatial and orientational distribution of cracks in crystalline photovoltaic modules generated by mechanical load tests. Solar Energy Materials and Solar Cells 95(11):3054–3059, DOI 10.1016/j.solmat.2011.06.032

Kirchhoff GR (1850) Über das Gleichgewicht und die Bewegung einer elastischen Scheibe. Crelles Journal für die reine und angewandte Mathematik 40:51–88, DOI 10.1515/crll.1850.40.51

Köntges M, Kurtz S, Packard C, Jahn U, Berger KA, Kato K, Friesen T, Liu H, Iseghem MV (2014) Review of Failures of Photovoltaic Modules. Tech. rep., International Energy Agency (IEA), URL http://iea-pvps.org/index.php?id=57&eID=dam_frontend_push&docID=2064, Report, IEA-PVPS T13-01:2014, , Photovoltaic Power Systems Programme

Kopp GA, Surry D, Chen K (2002) Wind loads on a solar array. Wind and Structures: An International Journal 5(5):393–406, DOI 10.12989/was.2002.5.5.393

Lai W, Rubin D, Krempl E (2009) Introduction to Continuum Mechanics, 4th edn. Butterworth-Heinemann, Oxford

Lebedev LP, Cloud MJ, Eremeyev VA (2010) Tensor Analysis with Applications in Mechanics. World Scientific, New Jersey · London · Singapore · Beijing · Shanghai · Hong Kong · Taipei · Chennai, DOI 10.1142/9789814313995

Mellor M (1975) A review of basic snow mechanics. Tech. rep., US Army Cold Regions Research and Engineering Laboratory, uS Army Cold Regions Research and Engineering Laboratory

Mindlin RD (1951) Influence of rotatory inertia and shear on flexural motions of isotropic, elastic plates. Journal of Applied Mechanics 18:31–38

Moore D (1977) Low-Cost Silicon Solar Array Project: cyclic pressure-load development testing of solar panels. Tech. rep., Jet Propulsion Laboratory, California Institute of Technology,

Pasadena, California (USA), Technical Report, Jet Propulsion Laboratory, California Institute of Technology, Pasadena, California (USA)
Naghdi PM (1973) The theory of shells and plates. In: Truesdell C (ed) Linear Theories of Elasticity and Thermoelasticity: Linear and Nonlinear Theories of Rods, Plates, and Shells, Springer, Berlin, Heidelberg, pp 425–640, DOI 10.1007/978-3-662-39776-3_5
Naumenko K, Eremeyev VA (2014) A layer-wise theory for laminated glass and photovoltaic panels. Composite Structures 112:283–291, DOI 10.1016/j.compstruct.2014.02.009
Oñate E (2009) Structural Analysis with the Finite Element Method. Linear Statics. Volume 1: Basis and Solids. Springer, DOI 10.1007/978-1-4020-8733-2
Oñate E (2013) Structural Analysis with the Finite Element Method. Linear Statics. Volume 2: Beams, Plates and Shells. Springer, DOI 10.1007/978-1-4020-8743-1
Reil F, Mathiak G, Strohkendl K, Raubach S, Schloth C, v Wangenheim B, Fotak K (2012) Experimental testing of PV modules under inhomogeneous snow loads. 27th European Photovoltaic Solar Energy Conference and Exhibition (PVSEC):3414–3417, DOI 10.4229/27thEUPVSEC2012-4BV.2.53
Reissner E (1944) On the theory of bending of elastic plates. Journal of Mathematics and Physics 23(1–4):184–191, DOI 10.1002/sapm1944231184
Reissner E (1945) The effect of transverse shear deformation on the bending of elastic plates. Journal of Applied Mechanics 12:69–77
Reissner E (1947) On the bending of elastic plates. Quarterly of Applied Mathematics 5(1):55–68, URL http://www.jstor.org/stable/43633584
Ruscheweyh H, Windhövel R (2013) Wind loads at solar and photovoltaic modules for large plants. In: 13th International Conference on Wind Engineering (ICWE13), Amsterdam, The Netherlands
Sander M, Dietrich S, Pander M, Ebert M, Bagdahn J (2013) Systematic investigation of cracks in encapsulated solar cells after mechanical loading. Solar Energy Materials and Solar Cells 111:82–89, DOI 10.1016/j.solmat.2012.12.031
Schicker J, Hirschl C, Leidl R (2014) Effect of PV module frame boundaries on cell stresses in solar cells. Journal of Energy Challenges and Mechanics 1:155–162, DOI 10.13140/2.1.3374.1441
Schwarz HR, Köckler N (2011) Numerische Mathematik, 8th edn. Vieweg+Teubner, Wiesbaden, DOI 10.1007/978-3-8348-8166-3
Shademan M, Barron RM, Balachandar R, Hangan H (2014) Numerical simulation of wind loading on ground-mounted solar panels at different flow configurations. Canadian Journal of Civil Engineering 41(8):728–738, DOI 10.1139/cjce-2013-0537
Stoffel M (2005) Numerical Modelling of Snow using Finite Elements. PhD thesis, ETH Zurich - Swiss Federal Institute of Technology, DOI 10.3929/ethz-a-005189531
Stolarski H, Belytschko T (1982) Membrane locking and reduced integration for curved elements. Journal of Applied Mechanics, Transactions ASME 49(1):172–176, DOI 10.1115/1.3161961
Szabó B, Babuška I (1991) Finite Element Analysis. John Wiley & Sons, Inc.
Weps M, Naumenko K, Altenbach H (2013) Unsymmetric three-layer laminate with soft core for photovoltaic modules. Composite Structures 105:332–339, DOI 10.1016/j.compstruct.2013.05.029
Zhilin PA (1976) Mechanics of deformable directed surfaces. Int J Solids Struct 12(9):635–648, DOI 10.1016/0020-7683(76)90010-X
Zienkiewicz O, Taylor R (2000) The Finite Element Method. Volume 1: The Basis, 5th edn. Butterworth-Heinemann

Chapter 5
Block Element Method for the Stamps of the no Classical Form

Vladimir A. Babeshko, Olga V. Evdokimova & Olga M. Babeshko

Abstract We present a block element method for solving 3D integral equations with difference kernel arising in boundary value problems of continuum mechanics and in mathematical physics. The approach has been induced by the Wiener–Hopf method, and its extension to the 3D case is called the integral factorization method and is mostly used in applications to domains with smooth boundary. In the present paper, the method is applied to domains with piecewise smooth boundary and corner points, which necessitates its generalization to the case of functions of two variables. Mixed boundary value problems have numerous applications in mechanics as well as in theoretical and technical physics. The method was tested on a vector contact problem for a wedge-shaped punch occupying the first quadrant. Techniques for obtaining various characteristics of solutions are described in detail. They are based on the inversion of a system of 1D linear integral equations typical of dynamic and static contact problems for strip punches.

Key words: Vector contact problem · System of integral equations · Wedge-shaped area · Block element · Factorization · Approximate values · Singular characteristics

---

Vladimir A. Babeshko
Southern Scientific Center RAS, Chehov st. 41, 344006 Rostov-on-Don & Kuban State University, Stavropolckia st. 149, 350040 Krasnodar, Russia,
e-mail: babeshko41@mail.ru

Olga V. Evdokimova
Southern Scientific Center RAS, Chehov st. 41, 344006 Rostov-on-Don, Russia,
e-mail: evdokimova.olga@mail.ru

Olga M. Babeshko
Kuban State University, Stavropolckia st. 149, 350040 Krasnodar, Russia,
e-mail: babeshko49@mail.ru

H. Altenbach et al. (eds.), *Mechanics for Materials and Technologies*,
Advanced Structured Materials 46, DOI 10.1007/978-3-319-56050-2_5

5.1 Introduction

The integral factorization method, which originates from Wiener and Hopf (1932), permits studying numerous mixed boundary value problems in continuum mechanics as well as in mathematical and theoretical physics numerically and analytically (Arutyunan and Manzhirov, 1999; Manzhirov, 2001, 1987; Noble, 1958; Vorovich and Babeshko, 1974; Babeshko et al, 2015). This method, which is closely related to the factorization of functions and matrix functions of one complex variable, was successfully adapted to the case of 2D domains with smooth boundary. Several versions of the factorization method for integral equations are described in Noble (1958). However, in the present paper the method is only applied to 1D integral equations or to equations that can readily be reduced to 1D equations. A generalization of the method for 3D domains with smooth boundary is presented in Vorovich and Babeshko (1974); Babeshko et al (2015); Babeshko (1984); Babeshko et al (1989); Eskin (1980). At the same time, this method has not been applied to problems with non-smooth boundary. Apparently, the present paper is the first to use this method for solving static and dynamic boundary value problems for wedge-type domains. The simplest case is that of a rectangular wedge, that is, of a quadrant in a rectangular coordinate system. An important theoretical application of the factorization methods is their role in the development of the block element method for solving boundary value problems for partial differential equations (Babeshko et al, 2006). This topologically based method permits solving such boundary value problems block by block, thus covering the entire domain where the boundary value problem is posed. Some advantages of the block element method, which is used to solve the integral equations arising in 3D mixed boundary value problems, can be illustrated by several examples of contact problems. As a rule, the boundary conditions are changed on the interface in the contact problem under study. For example, in the contact problems for a rigid punch, this is the transition from the domain of displacements under the punch, where the integral equation is posed, to the domain of stresses, and the opposite transition occurs in the crack theory. In contrast to other approaches, including numerical ones, the block element method permits simultaneously solving the integral equations and calculating the physical parameters of the mixed problem even outside the domain where the integral equations are posed. In contact problems, this is the behavior of displacements outside the punch area. In what follows, the system of integral equations of the mixed boundary value problem is considered in the wedge area occupying the first quadrant in the rectangular coordinate system. Such integral equations arise in problems on the action of a wedge-shaped punch on a multilayer medium under the condition of rigid adhesion between the punch and the medium or in the case where there are only tangential stresses in the contact region; in particular, this is the case for contact problems with creep Arutyunan and Manzhirov (1999); Manzhirov (2001, 1987). Such integral equations also arise in problems on the behavior of a crack, wedge-shaped in the horizontal projection, in a multilayer medium, on a thin-walled covering of such a shape on a multilayer substratum, on the behavior of a drifting ice floe wedge-shaped in the horizontal projection, in mixed boundary value problems describing the cold plasma behavior in such a domain, and in other boundary value problems.

5.2 Statement of the Problem

In what follows, the block method is used to study systems of integral equations of the following form, which has not yet been considered:

$$\begin{aligned}
\mathrm{K}q &= \int_0^\infty\int_0^\infty k(x-\xi,y-\eta)q(\xi,\eta)d\xi d\eta = f(x,y), \quad 0\le x\le\infty, \quad 0\le y\le\infty,\\
k(x,y) &= \frac{1}{4\pi^2}\int_{\Gamma_1}\int_{\Gamma_2} K(\alpha,\beta)e^{-i(\alpha x+\beta y)}d\alpha d\beta,\\
Q(\alpha,\beta) &= \int_0^\infty\int_0^\infty q(\xi,\eta)e^{i(\alpha\xi+\beta\eta)}d\xi d\eta
\end{aligned} \tag{5.1}$$

The integral equation can be written as

$$\begin{aligned}
&\frac{1}{4\pi^2}\int_{\Gamma_1}\int_{\Gamma_2} K(\alpha,\beta)Q(\alpha,\beta)e^{-i(\alpha x+\beta y)}d\alpha d\beta = f(x,y), \quad 0\le x\le\infty, \quad 0\le y\le\infty,\\
&F(\alpha,\beta) = \int_0^\infty\int_0^\infty f(\xi,\eta)e^{i(\alpha\xi+\beta\eta)}d\xi d\eta
\end{aligned}$$

The entries of the matrix function $K(\alpha,\beta)$ are assumed to be meromorphic functions in both complex variables and behave at infinity as

$$\begin{aligned}
K(\alpha,\beta) &= O(\alpha^{-1}), \quad \beta = \text{const};\\
K(\alpha,\beta) &= O(\beta^{-1}), \quad \alpha = \text{const}, \quad \alpha,\beta\to\infty.
\end{aligned}$$

Further, it is assumed that the zero and polar sets of these entries are analytic curves in the space of two complex variables, the polar sets being solvable for each of the parameters and representable in the form $\alpha_n = \alpha_n(\beta)$, $\beta_n = \beta_n(\alpha)$. The same is true for the determinant of the matrix function. Finitely many poles can be real-valued in dynamic mixed boundary value problems (Vorovich and Babeshko, 1974). The contours Γ_1 and Γ_2 lie on the real axis of the complex α- and β-planes everywhere except for the regions where the real poles of entries of the matrix function $K(\alpha,\beta)$ appear on the real axis. In this case, the contours deviate into the complex plane in sufficiently narrow domains of the real axes to bypass the poles. The contours are chosen so as to justify the statement of the boundary value problem physically in the case of time-harmonic mixed boundary value problems, namely, so as to satisfy the radiation principle at infinity, which means that the phase velocities are directed to infinity for normal materials and from infinity for mesomaterials in electronics (Vorovich and Babeshko, 1974). In the first case, the real poles of entries of the symbol $K(\alpha,\beta)$ in the first quadrant are bypassed by the contours from above if, in the transition to complex amplitudes, the function $\exp(-i\omega t)$, where ω is the oscillation frequency and t is time, is accepted. In the second case, the poles in the first quadrant are bypassed from below. The case of reversed phase velocities, which is described in Vorovich and Babeshko (1974), takes a special place. The contours in the second quadrant are in the opposite position. It should be noted that there are many papers where the specific characteristics of the solution at the corner point are studied (Babeshko et al, 1989; Glushkov and Glushkova, 1992). The solutions of systems of integral equations of this and similar types have important applications in the foundation engineering, seismology, element base devices in electronics, in studying the mixed boundary value problems of plasma behavior in various states, and in several other important technical areas.

5.3 Properties of the Integral Equations

To study the system of integral equations by using the block element method, we introduce the following set of block elements $\Omega_{mn}, m=1,2, n=1,2$. The second and fourth quadrants are denoted by Ω_{12} and Ω_{21}, respectively, and the third quadrant, by Ω_{22}. The same subscripts are used to denote vector functions supported in these domains. The first quadrant is denoted by Ω. We use the following notation. To avoid multiple use of the symbols of direct and inverse Fourier transform, the Fourier transforms of vector functions denoted by lowercase letters are denoted by the corresponding uppercase letters; e.g., the Fourier transforms of $k(x,y), f(x,y), \phi(x,y)$ are $K(\alpha,\beta), F(\alpha,\beta), \Phi(\alpha,\beta)$, respectively. Further, we need to factorize vector functions into sums and products. These operations are carried out for vector functions of two complex variables α and β. To perform the factorization into a sum, we use the notation introduced in Vorovich and Babeshko (1974) applying the curly brackets. For example, if a vector function $G(\alpha,\beta)$ is factorized as a sum over the parameterα with respect to the contour Γ_1, then the vector functions obtained by this operation are denoted by the formulas

$$G(\alpha,\beta) = \{G(\alpha,\beta)\}_\alpha^+ + \{G(\alpha,\beta)\}_\alpha^- . \tag{5.2}$$

The first vector function on the right is regular in the complex plane above the contour Γ_1, and the second one is regular in the complex plane below this contour. In the case where the vector functions are also factorized with respect to the other parameter β, the factorized vector functions are denoted by

$$G_\alpha^+ = \{G_\alpha^+\}_\beta^+ + \{G_\alpha^+\}_\beta^- ,$$
$$\{G_\alpha^+\}_\beta^+ \equiv G_{\alpha\beta}^{++} \equiv \{\{G(\alpha,\beta)\}_\alpha^+\}_\beta^+ ,$$
$$\{G_\alpha^+\}_\beta^- \equiv G_{\alpha\beta}^{+-} \equiv \{\{G(\alpha,\beta)\}_\alpha^+\}_\beta^- .$$

For vector functions with subscripts, say, indicating the quadrants where these vector functions are supported, we use the convention that these subscripts follow the subscripts referring to factorization; e.g.,

$$G^{++}_{\alpha\beta mn} \equiv \{G^{+}_{\alpha mn}\}^{+}_{\beta} \equiv \{\{G_{mn}(\alpha,\beta)\}^{+}_{\alpha}\}^{+}_{\beta},$$
$$G^{+-}_{\alpha\beta mn} \equiv \{G^{+}_{\alpha mn}\}^{-}_{\beta} \equiv \{\{G_{mn}(\alpha,\beta)\}^{+}_{\alpha}\}^{-}_{\beta}.$$

The operator given below carries out the factorization into a sum (Vorovich and Babeshko, 1974),

$$\{G(\alpha,\beta)\}^{+}_{\alpha} = \frac{1}{2\pi i}\int_{\Gamma_1} \frac{G(\xi,\beta)}{\xi-\alpha}d\xi, \quad \alpha > \Gamma_1,$$
$$\{G(\alpha,\beta)\}^{-}_{\alpha} = -\frac{1}{2\pi i}\int_{\Gamma_1} \frac{G(\xi,\beta)}{\xi-\alpha}d\xi, \quad \alpha < \Gamma_1$$

The abstract inequalities following the integral expressions denote the position of the parameter α above the contour Γ_1 in the first case or below the contour Γ_1 in the second case.

The factorization of matrix functions into a product has the form

$$K(\alpha,\beta) = K_{+\alpha}(\alpha,\beta)K_{-\alpha}(\alpha,\beta), \qquad K(\alpha,\beta) = K_{+\beta}(\alpha,\beta)K_{-\beta}(\alpha,\beta). \quad (5.3)$$

The methods used to factorize matrix functions into products are more complicated. Some methods for factorizing matrix functions were developed and published in Vorovich and Babeshko (1974). Triangular and functionally commutative matrix functions can be factorized by well-known formulas. If the matrix functions are functionally commutative in the first parameter, i.e., $K(\alpha,\beta)K(\gamma,\beta) = K(\gamma,\beta)K(\alpha,\beta)$, then the factorization formulas become

$$K(\alpha,\beta)_{+\alpha}(\alpha,\beta) = \exp\frac{1}{2\pi i}\int_{\Gamma_1} \frac{\ln K(\alpha,\beta)(\xi,\beta)}{\xi-\alpha}d\xi, \quad \alpha > \Gamma_1,$$
$$K(\alpha,\beta)_{-\alpha}(\alpha,\beta) = \exp\left(-\frac{1}{2\pi i}\int_{\Gamma_1} \frac{\ln K(\alpha,\beta)(\xi,\beta)}{\xi-\alpha}d\xi\right), \quad \alpha < \Gamma_1$$

The other conditions imposed on the matrix functions for the formulas of factorization into a sum or product to apply are described in detail in the literature (e.g., see Noble (1958); Vorovich and Babeshko (1974)), and we do not discuss them here.

5.4 The Block Element Method for a System of Integral Equations

The system of integral equations (5.1) is studied by an approach traditional for the integral factorization method Vorovich and Babeshko (1974); Babeshko (1984); Babeshko et al (2006). To this end, we equivalently embed the system of integral equations in the compact topological space of analytic manifolds and simply complete it with new unknown functions to cover the whole plane homeomorphic to the sphere and apply the double Fourier transform. For this, we rewrite the integral system (5.1) as

$$\begin{aligned} &\frac{1}{4\pi^2}\int\limits_{\Gamma_1}\int\limits_{\Gamma_2} K(\alpha,\beta)Q(\alpha,\beta)e^{-i(\alpha x+\beta y)}d\alpha d\beta \\ &= f(x,y)+\phi_{12}(x,y)+\phi_{21}(x,y)+\phi_{22}(x,y) \end{aligned} \tag{5.4}$$

Here the new unknown vector function $\varphi_{12}(x,y)$ is supported in the second quadrant, $\varphi_{21}(x,y)$, in the fourth, and $\varphi_{22}(x,y)$, in the third.

After an application of the Fourier transform, according the discussion in the first section, the continued system of integral equations (5.4) becomes

$$K(\alpha,\beta)Q(\alpha,\beta)-\Phi_{12}(\alpha,\beta)-\Phi_{21}(\alpha,\beta)-\Phi_{22}(\alpha,\beta)-F(\alpha,\beta)=0 \tag{5.5}$$

We obtain a certain relation between the analytic manifolds and, by equating the sums of similar components of these manifolds to zero, we construct relations that allow us to obtain the solution $Q(\alpha,\beta)$ of the system of integral equations. The instruments for solving this problem are factorizations, which allow us to distinguish similar components. After several transformations, where the factorization relations (5.2) and (5.3) are used, we obtain the main result of our study, which states that the solution $q(x,y)$ can be represented as

$$q(x,y)=\frac{1}{4\pi^2}\int\limits_{\Gamma_1}\int\limits_{\Gamma_2} Q(\alpha,\beta)e^{-i(\alpha x+\beta y)}d\alpha d\beta$$

where the function $Q(\alpha,\beta)$has the form (which follows from (5))

$$Q(\alpha,\beta)=K^{-1}(\alpha,\beta)\left[\,F(\alpha,\beta)+\Phi^{+-}_{\beta\alpha 12}(\alpha,\beta)+\Phi^{+-}_{\alpha\beta 21}(\alpha,\beta)+\Phi^{--}_{\alpha\beta 22}(\alpha,\beta)\,\right] \tag{5.6}$$

The functions $\Phi^{+-}_{\beta\alpha 12}(\alpha,\beta)$, $\Phi^{+-}_{\alpha\beta 21}(\alpha,\beta)$, are determined from the system of two 1D equations like those obtained for the strip punch (Vorovich and Babeshko, 1974; Babeshko, 1984; Babeshko et al, 1989),

$$\Phi^{+-}_{\beta\alpha 12}+\{\,K_{-\alpha}\left\{K^{-1}_{-\alpha}\Phi^{+-}_{\alpha\beta 21}\right\}^{-}_{\alpha}\,\}^{+}_{\beta}=-\{\,K_{-\alpha}\left\{K^{-1}_{-\alpha}F(\alpha,\beta)\right\}^{-}_{\alpha}\,\}^{+}_{\beta} \tag{5.7}$$

$$\Phi^{+-}_{\alpha\beta 21} + \{ K_{-\beta} \left\{ K^{-1}_{-\beta} \Phi^{+-}_{\beta\alpha 12} \right\}^{-}_{\beta} \}^{+}_{\alpha} = - \{ K_{-\beta} \left\{ K^{-1}_{-\beta} F(\alpha,\beta) \right\}^{-}_{\beta} \}^{+}_{\alpha} \tag{5.8}$$

After this, the function $\Phi^{--}_{\alpha\beta 22}(\alpha,\beta)$ is taken in one of equivalent forms and expressed via the already obtained functions as

$$\Phi^{--}_{\beta\alpha 22} = - \{ K_{-\alpha} \{ K^{-1}_{-\alpha} F(\alpha,\beta) \}^{-}_{\alpha} \}^{-}_{\beta} - \{ K_{-\alpha} \left\{ K^{-1}_{-\alpha} \Phi^{+-}_{\alpha\beta 21} \right\}^{-}_{\alpha} \}^{-}_{\beta}$$

$$\Phi^{--}_{\alpha\beta 22} = - \{ K_{-\beta} \left\{ K^{-1}_{-\beta} F(\alpha,\beta) \right\}^{-}_{\beta} \}^{-}_{\alpha} - \{ K_{-\beta} \left\{ K^{-1}_{-\beta} \Phi^{+-}_{\beta\alpha 12} \right\}^{-}_{\beta} \}^{-}_{\alpha}$$

These formulas with the factorization integrals taken into account give an illusion that it is very difficult to calculate the integrals which are required to solve the system of equations and to determine the unknown functions. But in many mixed boundary value problems, the functions participating in the representation of the solutions are meromorphic, which allows us to use residue theory to calculate the integrals. Moreover, by calculating the integrals, we obtain systems of linear algebraic equations which permit numerical calculations if we know the zeros and poles of meromorphic functions similar to those studied in Vorovich and Babeshko (1974); Babeshko (1984); Babeshko et al (1989).

5.5 Study of the Properties of the Solution of the System of Integral Equations and a Boundary Value Problem

We assume that the characteristic properties of the symbol $K(\alpha,\beta)$of the system of integral equations occurring in many boundary value problems for multilayer isotropic and anisotropic media and for the punch contact with friction or without it have the form

$$K_{-\alpha}(\alpha,\beta) = O(\alpha^{-\theta_1}), \quad K_{-\beta}(\alpha,\beta) = O(\alpha^{-\theta_2}), \quad 0 < \theta_m < 1, \quad |\alpha| \to \infty$$

and that the vector function $f(x,y)$ absolutely integrable in the first quadrant is twice continuously differentiable. We see that the operator of system (5.7), (5.8) is bounded in the space of functions continuous with a power-law weight on the contours $\Gamma_m, m = 1,2$ (Vorovich and Babeshko, 1974).
This allows us to investigate the properties of the solution of the system of integral equations in the first block (in the first quadrant) and the behavior of the surface in the other blocks (quadrants). Let us show that the constructed solution has characteristic properties typical of the solutions of contact problems obtained in simpler domains similar to the quadrant under study. Such regions can be the zone remote from the boundaries of contact region and the zone on the boundary of the contact region.

1. Let us rewrite relation (5.6) as

$$q(x,y) = \frac{1}{4\pi^2} \int\limits_{\Gamma_1} \int\limits_{\Gamma_2} K^{-1}(\alpha,\beta) [\, F(\alpha,\beta) + \Phi^{+-}_{\beta\alpha 12}(\alpha,\beta) +$$
$$+ \Phi^{+-}_{\alpha\beta 21}(\alpha,\beta) + \Phi^{--}_{\alpha\beta 22}(\alpha,\beta)\,] e^{-i(\alpha x + \beta y)} d\alpha d\beta, \quad 0 < x,y < \infty$$

On the right-hand side, we let x and y tend to infinity in a neighborhood of the bisector of the first quadrant angle and, by the Riemannian–Lebesgue theorem, obtain the following estimate for Fourier integrals:

$$q(x,y) \approx \left(\frac{1}{4\pi^2} \int\limits_{\Gamma_1} \int\limits_{\Gamma_2} K^{-1}(\alpha,\beta) F(\alpha,\beta) e^{-i(\alpha x + \beta y)} d\alpha d\beta \right) [1 + o(1)]$$

Thus, far from the boundaries, the solution of the integral equations tends to a degenerate solution given by the solution of convolution equation under the assumption that the equation holds in the entire plane (Vorovich and Babeshko, 1974).

2. To study the properties of the solution on the punch boundaries by using the factorization formulas and the formulas representing the operators in system (5.7), (5.8), we can use the estimates

$$\Phi^{+-}_{\beta\alpha 12} + \left\{ K_{-\alpha} \left\{ K^{-1}_{-\alpha} \Phi^{+-}_{\alpha\beta 21} \right\}^{-}_{\alpha} \right\}^{+}_{\beta} = - \left\{ K_{-\alpha} \left\{ K^{-1}_{-\alpha} F(\alpha,\beta) \right\}^{-}_{\alpha} \right\}^{+}_{\beta},$$
$$\left\{ K^{-1}_{-\alpha} \Phi^{+-}_{\alpha\beta 21} \right\}^{-}_{\alpha} = G^{-}_{\alpha}(\alpha,\xi),$$
$$\left\{ K_{-\alpha} \left\{ K^{-1}_{-\alpha} \Phi^{+-}_{\alpha\beta 21} \right\}^{-}_{\alpha} \right\}^{+}_{\beta} = \frac{1}{2\pi i} \int\limits_{\Gamma_2} K_{-\alpha}(\alpha,\xi) \frac{G^{-}_{\alpha}(\alpha,\xi)}{\xi - \beta} d\xi = O(\alpha^{-\theta_1})$$
$$K^{-1}(\alpha,\beta) \left\{ K_{-\alpha} \left\{ K^{-1}_{-\alpha} \Phi^{+-}_{\alpha\beta 21} \right\}^{-}_{\alpha} \right\}^{+}_{\beta} = O(\alpha^{1-\theta_1}),$$
$$q(x,y) \sim \frac{1}{2\pi} \int\limits_{\Gamma_1} K^{-1}(\alpha,\beta) \left\{ K_{-\alpha} \left\{ K^{-1}_{-\alpha} \Phi^{+-}_{\alpha\beta 21} \right\}^{-}_{\alpha} \right\}^{+}_{\beta} e^{-i\alpha x} d\alpha = O(x^{-\theta_1}),$$
$$x \to 0, \quad y > 0$$
$$\Phi^{+-}_{\alpha\beta 21} + \left\{ K_{-\beta} \left\{ K^{-1}_{-\beta} \Phi^{+-}_{\beta\alpha 12} \right\}^{-}_{\beta} \right\}^{+}_{\alpha} = - \left\{ K_{-\beta} \left\{ K^{-1}_{-\beta} F(\alpha,\beta) \right\}_{\beta} \right\}^{+}_{\alpha},$$
$$\left\{ K^{-1}_{-\beta} \Phi^{+-}_{\beta\alpha 12} \right\}^{-}_{\beta} = G^{-}_{\beta}(\xi,\beta)$$
$$\left\{ K_{-\beta} \left\{ K^{-1}_{-\beta} \Phi^{+-}_{\beta\alpha 12} \right\}^{-}_{\beta} \right\}^{+}_{\alpha} = \frac{1}{2\pi i} \int\limits_{\Gamma_1} K_{-\beta}(\xi,\beta) \frac{G^{-}_{\beta}(\xi,\beta)}{\xi - \alpha} d\xi = O(\beta^{-\theta_2}), \quad \alpha > \Gamma_1,$$
$$K^{-1}(\alpha,\beta) \left\{ K_{-\beta} \left\{ K^{-1}_{-\beta} \Phi^{+-}_{\beta\alpha 12} \right\}^{-}_{\beta} \right\}^{+}_{\alpha} = O(\beta^{1-\theta_2})$$
$$q(x,y) \sim \frac{1}{2\pi} \int\limits_{\Gamma_2} K^{-1}(\alpha,\beta) \left\{ K_{-\beta} \left\{ K^{-1}_{-\beta} \Phi^{+-}_{\beta\alpha 12} \right\}^{-}_{\beta} \right\}^{+}_{\alpha} e^{-i\beta y} d\beta = O(y^{-\theta_2}),$$
$$y \to 0, \quad x > 0$$

It follows from these estimates that, on the boundaries of the wedge-shape punch, there is an unbounded increase in the contact stresses, which were studied in detail for various conditions on the interface between the punch and the medium in Babeshko et al (1989); Glushkov and Glushkova (1992). It was shown in these papers that the behavior of the contact stresses near the corner point is described by the formula $(x^2+y^2)^{-0,5\gamma}, \quad x,y \to 0$.

3. To study the properties of the solution of the mixed boundary value problem on the surface outside the punch, we estimate the behavior of the functions in the other blocks. To this end, the following constructions are used:

$$\phi_{12}(x,y) \sim \frac{1}{8\pi^3 i} \int_{\Gamma_1}\int_{\Gamma_2}\int_{-\infty}^{\infty} K_{-\alpha}(\alpha,\xi)\frac{G_{\alpha}^{-}(\alpha,\xi)}{\xi-\beta} d\xi e^{-i(\alpha x+\beta y)} d\alpha d\beta,$$
$$K_{-\alpha}(\alpha,\xi) = \sum_n \frac{C_n(\xi)}{\alpha-\alpha_n(\xi)}, \quad \Im\alpha_n \geq 0$$
$$\phi_{12}(x,y) = O(e^{-i\alpha_{n0}x}), \quad \Re(-i\alpha_{n0}) \geq 0, \quad \Im(-i\alpha_{n0}) \leq 0,$$
$$-\infty < x < 0, \quad y > 0, \quad x \to -\infty$$
$$\phi_{21}(x,y) \sim \frac{1}{8\pi^3 i} \int_{\Gamma_1}\int_{\Gamma_2}\int_{-\infty}^{\infty} K_{-\beta}(\xi,\beta)\frac{G_{\beta}^{-}(\xi,\beta)}{\xi-\alpha} d\xi e^{-i(\alpha x+\beta y)} d\alpha d\beta,$$
$$K_{-\beta}(\xi,\beta) = \sum_n \frac{S_n(\xi)}{\beta-\beta_n(\xi)}, \quad \Im\beta_n \geq 0$$
$$\phi_{21}(x,y) = O(e^{-i\beta_{n0}y}), \quad \Re(-i\beta_{n0}) \geq 0, \quad \Im(-i\beta_{n0}) \leq 0,$$
$$-\infty < y < 0, \quad x > 0, \quad y \to -\infty$$

The obtained estimates imply the following results. If a static boundary value problem is considered, then, as the punch is indented, the surface in the second and forth quadrants exponentially tends to the unperturbed state when moving away from the boundaries of the first quadrant to infinity in the second and the forth quadrants. In the case of a dynamical boundary value problem about the punch vibration, in the second and the fourth quadrants, there arise waves that move away from the punch boundaries to infinity with respective phase velocities $-\omega\alpha_n^{-1}$ and $-\omega\beta_n^{-1}$ in the cases $\Im\alpha_n = \Im\beta_n = 0$ and of normal material of the medium, that is, in the cases of coinciding directions of the group and phase velocities. For the other, complex, poles of the meromorphic functions, the additions to the waves running away exponentially tends to zero as the coordinates tend to infinity. In the third quadrant, as $x,y \to \infty$, there are both types of waves described above.

5.6 Acknowledgments

This work was supported in part by the Russian Foundation for Basic Research, projects nos. (15-01-01379), (15-08-01377), (16-41-230214), (16-41-230218), (16-48-230216), and (17-08-00323), by the Program no. 1-33P of the Presidium of the Russian Academy of Sciences, projects from (0256-2015-0088) to (0256-2015-0093), and by the Federal Target Program for 2017, projects nos. (9.8753.2017/BCh) and (0256-2014-0006).

References

Arutyunan NK, Manzhirov AV (1999) Contact Problems in the Theory of Creep (in Russ.). Izd-vo NAN RA, Erevan

Babeshko BA (1984) Generalized Factorization Method in Spatial Dynamical Mixed Problems of Elasticity Theory (in Russ.). Nauka, Moscow

Babeshko BA, Glushkov EV, Zinchenko ZF (1989) Dynamics of Nonhomogeneous Linear Elastic Medium (in Russ.). Nauka, Moscow

Babeshko VA, Babeshko OM, Evdokimova OV (2006) About integral and differential methods of factorization. DAN 410(2):168 – 172

Babeshko VA, Evdokimova OV, Babeshko OM (2015) Analogy between an engineering heat-conduction problem and one climatic event. Journal of Applied Physics and Technical Mechanics 56(6):959 – 965

Eskin GI (1980) Boundary Value Problems for Elliptic Pseudodifferential Equations, Math. Monographs, vol 52. American Mathematical Society, Providence, RI

Glushkov EV, Glushkova NV (1992) On the singularities of the elastic stress field near the tip of a wedge-shaped spatial crack. Isv AN MTT [Mechanics of Solids] (4):82 – 86

Manzhirov A (1987) Contact problems of the interaction between viscoelastic foundations subjected to ageing and systems of stamps not applied simultaneously. Journal of Applied Mathematics and Mechanics 51(4):523 – 535

Manzhirov AV (2001) Contact problems for inhomogeneous aging viscoelastic solids. In: Mechanics of Contact Interactions, Fismatlit, Moscow, pp 549–565

Noble B (1958) Methods Based on the Wiener-Hopf Technique for the Solution of Partial Differential Equations. International series of monographs on pure and applied mathematics, Pergamon Press

Vorovich II, Babeshko BA (1974) Dynamic Mixed Problems of Elasticity for Non-Classical Domains (in Russ.). Nauka, Moscow

Wiener N, Hopf E (1932) Über eine Klasse singularer Integralgleichungen. Sitz Ber Preuss Akad Wiss pp 696–706

Chapter 6
On the Irreversible Deformations Growth in the Material with Elastic, Viscous, and Plastic Properties and Additional Requirements to Yield Criteria

Alexandra S. Begun, Alexander A. Burenin, Larisa V. Kovtanyuk, Galina L. Panchenko, Alexander N. Prokudin

Abstract The closed mathematical model of the deformation processes of materials with elastic, viscous and plastic properties under large deformations is developed. The reversible and irreversible determinations of components are given by the differential equations of change (transfer). The stress-strain equations are derived within the frameworks of the hypothesis of the thermodynamic potential independence on irreversible deformation. The irreversible deformations accumulation is modelled without separation of creep and plastic parts. During unloading process the reverse consequence of the dissipative mechanisms is assumed. The irreversible deformations and rates continuity conditions are provided by the plastic potential specification (yield criterion). The developed approach is illustrated by examples. In virtue of proposed approach the boundary value problems for cylinder under pressure gradient and viscometric deformations are solved.

Key words: large deformations, elasticity, plasticity, viscosity, large creep deformations.

6.1 Introduction

There are technological processes of intensive materials forming with the specifications not allowing high process velocity. In such processes (for example cold forming) the deforming is provided by the slow creep process. The irreversible creep strain rates are determined by the level and the distribution of stresses. However,

Alexandra S. Begun · Larisa V. Kovtanyuk
Institute of automation and control processes of FEB RAS, 690041 Vladivostok, Russia
e-mail: asustinova@dvo.ru, lk@dvo.ru

Alexander A. Burenin · Galina L. Panchenko · Alexander N. Prokudin
Institute of Machine Science and metallurgy of FEB RAS, 681005 Komsomol'sk-on-Amur, Russia,
e-mail: burenin@dvo.ru, sunbeam_85@mail.ru

H. Altenbach et al. (eds.), *Mechanics for Materials and Technologies*,
Advanced Structured Materials 46, DOI 10.1007/978-3-319-56050-2_6

the stresses essentially depend on the plastic domain presence, which vanishing is impossible. Consequently the design simulation of similar technological processes need to be based on a mathematical model of large deformations when both processes (the slow(creep) and the fast(plastic) flow are presented. The modern mechanics does not have such theories. In order to develop a phenomenological model we present here a number of fundamental reasons for this. As the basis of developed approach we choose a mathematical model of large elastic-plastic deformation (Burenin et al, 1996). The advantage of the approach developed in Burenin et al (1996), is the satisfaction of the classical elastic-plastic theory principles. Firstly the irreversible strain tensor under elastic deformation and unloading isn't changed. Secondly the stress is fully calculated by the reversible strain tensor. These lead to necessity of the assumption of thermodynamic potentials (internal energy, free energy) independence on the irreversible deformations attributing to the dissipation. The assumptions allow us to state and solve boundary value problems (Burenin et al, 2003; Kovtanyuk, 2005; Burenin et al, 2010) including analytical solutions, which are insoluble within frameworks of the other models. The theory generalization on non-isothermal cases see in Burenin et al (2015).

Geometrically and thermodynamically consistent theory of large deformation accounting the simultaneous change of the reversible and the irreversible deformations has been proposed by Lee (1969). The multiplicative separation of the total strain tensor on the reversible and the irreversible parts was widely discussed and generalised (Clifton, 1972; Kondaurov, 1982; Nemat-Nasser, 1982; Naghdi, 1985; Levitas, 1996).

Perfect geometrical formulations in this approach are based on the hypothesis of an unique unloading state corresponding to each actual state of the solids with vanishing reversible deformation. The inaccessibility of this state leads to the theoretical and numerical difficulties in the formulation and solution of boundary value problems. Therefore attempts were made to develop a mathematical model (Rogovoi, 2005) with kinematics different from the Lee principles.

The method of solving elastic-plastic problems proposed by Hill (1958) is widely used. One is based on the procedure of the constitutive equations differentiation. Some finite element method computation packages are attributed to this assumption. These large deformation theories represent the strain rate tensor as the sum of the reversible (elastic) and the irreversible (plastic) strain rate tensors due to one of the objective derivative. The thermodynamical and geometrical incorrentness was shown in Khan and Huang (1995); Simo and Pister (1984). Nevertheless it is commonly used in numerical calculations (Xiao et al, 2006) even in metal processing computing packages (Firat et al, 2008). Note also that one can obtain wrong solution by virtue of discontinues of stress and displacements.

6.2 Large Deformations Kinematics

The displacement vector in a rectangular Cartesian coordinate system we define as

$$a_i = a_i(x_1, x_2, x_3, t),\\ u_i = x_i - a_i = u_i(x_1, x_2, x_3, t). \tag{6.1}$$

Herein t is the time, a_i are material point coordinates at the referential state, x_i are material point coordinates at the actual state (Euler coordinates), u_i are displacements vector components.

For distortion tensor $a_{i,j} = \partial a_i/\partial x_j$ and metric tensor $g_{ij} = a_{k,i}a_{k,j}$ one can derive the differential equations of change (transfer) as follows

$$\frac{da_{i,j}}{dt} + a_{i,k}v_{k,j} = 0,\\ \frac{dg_{i,j}}{dt} + g_{ik}v_{k,j} + v_{k,j}g_{kj} = 0,\\ v_i = \frac{dx_i}{dt} = \frac{du_i}{dt} = \frac{\partial u_i}{\partial t} + v_j u_{i,j}. \tag{6.2}$$

The Almansi strain tensor $d_{ij} = \frac{1}{2}(\delta_{ij} - g_{ij})$ according to Eqs. (6.1, 6.2) is calculated by

$$\frac{dd_{ij}}{dt} + d_{ik}v_{k,j} + v_{k,j}g_{kj} = \varepsilon_{ij},\\ \varepsilon_{ij} = \frac{1}{2}(v_{i,j} + v_{j,i}). \tag{6.3}$$

The transfer equation for the strain tensor is fundamentally different from ones for the distortion and metric tensor due to the source tensor, as is the Euler strain rate tensor. If the distortion tensor and the metric tensor are conserved then the deformations are produced by the solids motion at a point along its trajectory. Deformation does not change when the $\varepsilon_{ij} = 0$. In this case, the body moves as a rigid one. The transfer equations of the strain tensor and the metric tensor can be transformed as

$$\frac{\overline{D}d_{ij}}{\overline{D}t} = \varepsilon_{ij}; \frac{\overline{D}g_{ij}}{\overline{D}t} = 0, \tag{6.4}$$

wherein $\overline{D}/\overline{D}t$ denotes the Cotter–Rivlin objective derivative becoming Jaumann one under condition $\varepsilon_{ij} = 0$

$$\frac{\hat{D}g_{ij}}{\hat{D}t} = \frac{dg_{ij}}{dt} + g_{ik}\omega_{kj} + \omega_{ki}g_{kj} = 0,\\ \frac{\hat{D}d_{ij}}{\hat{D}t} = \frac{dd_{ij}}{dt} + d_{ik}\omega_{kj} + \omega_{ki}d_{kj} = 0,\\ \omega_{ij} = v_{i,j} - \varepsilon_{ij}. \tag{6.5}$$

The rotation tensor ω_{ij} defines the angular velocity of the solids under unchanging deformations $\varepsilon_{ij} = 0$. At the basis of the Lee elastic-plastic kinematics (Lee, 1969) the unloading configurations with zero reversible deformations along with the referential and actual ones were introduced. However this configuration is reachable if we divide the solids on infinite number of infinitesimal parts with zero loads. Otherwise after the removal of the external loads the reversible deformation will be always present along with irreversible ones residual stresses playing a positive role providing the construction tightness. But often we need to reduce residual stresses by the special technological operations (tempering, annealing). The representation E. Lee of the metric tensor in our notation reads by equation

$$g_{ij} = g^p_{ik} g^e_{kj} = \frac{\partial a_m}{\partial b_i}\frac{\partial a_m}{\partial b_k}\frac{\partial b_n}{\partial x_k}\frac{\partial b_n}{\partial x_j}. \tag{6.6}$$

Here (6.6) b_i are the point coordinates in unloading state with zero elastic deformations. Then the total strain can be splitted by

$$g_{ij} = (\delta_{ik} - 2e^p_{ik})(\delta_{kj} - 2e^e_{kj}). \tag{6.7}$$

Equations (6.7) follow from the Eq. (6.6) where the tensors e^e_{ij} and e^p_{ij} are the reversible and irreversible parts. Note the Clifton theory (see Clifton, 1972) is based on permutations of multipliers in Eqs. (6.7).

For distortion and metric tensor g_{ij} we accept the following representation

$$\begin{aligned} a_{i,j} &= Y_{i,k}(\delta_{kj} - e_{kj}), \\ g_{ij} = a_{s,i}a_{s,j} &= (\delta_{ik} - e_{ik})(\delta_{km} - 2p_{km})(\delta_{mj} - e_{mj}). \end{aligned} \tag{6.8}$$

Herein δ_{ij} are the Kronecker symbols.

Then for tensor p_{ij} in Eq. (6.2) one can obtain

$$p_{ij} = \frac{1}{2}(\delta_{ij} - Y_{m,i}Y_{m,j})$$

The tensor p_{ij} introduced in Eqs. (6.8) is the symmetric one $p_{ij} = p_{ji}$. But the tensor e_{ij} in general case presented by Eqs. (6.8) may not be a symmetric. First equation in (6.8) is not orthogonal decomposition of the distortion tensor. Taking into account Eqs. (6.1) and (6.8) the transfer equations for tensors (6.8) can be derived in form

$$\begin{aligned} \frac{de_{ij}}{dt} &= v_{i,j} - b_{ij} - e_{ik}v_{k,j} + b_{ik}e_{kj}, \\ \frac{dp_{ij}}{dt} &= \frac{1}{2}(b_{ij} + b_{ji}) - p_{ik}b_{kj} + b_{ki}p_{kj}, \\ b_{ij} &= -Y^{-1}_{ik}\frac{dY_{kj}}{dt}. \end{aligned} \tag{6.9}$$

Let us assume the symmetry of the tensor e_{ij}. It possible only under conditions manifesting by (6.9)

$$b_{km}(\delta_{mj} - e_{mj}) - (\delta_{km} - e_{km})b_{mj} = (\delta_{km} - e_{km})v_{m,j} - v_{m,k}(\delta_{mj} - e_{mj}). \tag{6.10}$$

Consequently the symmetry condition $e_{ij} = e_{ji}$ for tensor in Eq. (6.9) is assured by Eq. (6.10). Differently, the tensor b_{km} is not an arbitrary and satisfies to Eq. (6.10) and only if $e_{ij} = e_{ji}$. The Eq. (6.10) in this case is the tensor one for b_{km}. The solution of Eq. (6.10) is written by

$$b_{ij} = r_{ij} + (\delta_{ik} - e_{ik})t_{kj}. \tag{6.11}$$

Herein (6.11) t_{kj} is an arbitrary symmetric tensor ($t_{kj} = t_{jk}$). The tensor r_{ij} is skew-symmetric $r_{ij} = -r_{ji}$ and represented by equation

$$\begin{aligned} r_{ij} = w_{ij} + A^{-1}[B^2(\varepsilon_{ik}e_{kj} - e_{ik}\varepsilon_{kj}) + B(\varepsilon_{ik}e_{km}e_{mj} - e_{ik}e_{km}\varepsilon_{mj}) + \\ + e_{ik}\varepsilon_{km}e_{mn}e_{nj} - e_{ik}e_{km}\varepsilon_{mn}e_{nj}] \\ v_{i,j} = \frac{1}{2}(v_{i,j} + v_{j,i}) - \frac{1}{2}(v_{i,j} - v_{j,i}) = \varepsilon_{ij} + w_{ij}, \\ A = 8 - 8E_1 + 3E_1^2 - E_2 - \frac{1}{3}E_1^3 + \frac{1}{3}E_3,\ B = 2 - E_1, \\ E_1 = e_{jj}, \quad E_2 = e_{ij}e_{ji}, \quad E_3 = e_{ij}e_{jk}e_{ki}. \end{aligned} \tag{6.12}$$

Taking into account Eqs. (6.11) and (6.12) we can transform the transfer equations (6.9) for the tensors e_{ij} and p_{ij} as follows

$$\begin{aligned} \frac{de_{ij}}{dt} &= \varepsilon_{ij} - t_{ij} - r_{ij} + w_{ij} - e_{ik}(\varepsilon_{kj} + w_{kj} - t_{kj}) + (r_{ik} + t_{ik})e_{kj} - e_{im}t_{mk}e_{kj}, \\ \frac{dp_{ij}}{dt} &= t_{ij} - \frac{1}{2}(e_{ik}t_{kj} + t_{ik}e_{kj}) + (r_{ik} - t_{ik})p_{kj} - p_{ik}(r_{kj} + t_{kj}) + p_{ik}e_{km}t_{mj} + t_{im}e_{mk}p_{kj}. \end{aligned} \tag{6.13}$$

The components of an arbitrary tensor t_{ij} in Eq. (6.11) are initially vanished. Thus we can simplify the transfer equations (6.13)

$$\begin{aligned} \frac{de_{ij}}{dt} &= \varepsilon_{ij} + w_{ik}e_{kj} - e_{ik}w_{kj} - \frac{1}{2}(e_{ik}\varepsilon_{kj} + \varepsilon_{ik}e_{kj}) + \frac{1}{2}(z_{ik}e_{kj} + e_{ik}z_{kj}), \\ \frac{dp_{ij}}{dt} &= w_{ik}p_{kj} - p_{ik}w_{kj} + z_{ik}p_{kj} - p_{ik}z_{kj}, \\ z_{ij} &= r_{ij} - w_{ij}. \end{aligned} \tag{6.14}$$

If the medium is undeformed, then $\varepsilon_{ij} \equiv 0$ and Eq. (6.12) involve $z_{ij} = 0$. Moreover the second equation in (6.14) suggests the zero Jaumann derivative on tensor p_{ij} and it transformation as a rigid motion. But the two terms in second equation (6.14) distinguishing its from the Jaumann derivative on the tensor p_{ij} do not alter the change of the objective derivative

$$\frac{Dp_{ij}}{Dt} = \frac{dp_{ij}}{dt} - r_{ik}p_{kj} + p_{ik}r_{kj} = 0. \tag{6.15}$$

Contrary to the Jaumann derivative in Eq. (6.15) instead of the rotation tensor w_{ij} the skew-symmetric tensor ($r_{ij} = -r_{ji}$) is used. The components of the tensor r_{ij} equal w_{ij} in the its principal linear part. The tensor p_{ij} describes by the Eq. (6.15). This situation provides an opportunity to identify the tensor p_{ij} with irreversible strain. Moreover the case $t_{ij} \equiv 0$ is the elastic deformation process. The irreversible deformations may occur in the solid and satisfy to Eq. (6.15) i.e. if the medium is moved as a rigid solid without irreversible deformation tensor changes. The two last terms in second equation (6.14) connecting with geometrical correctness of the choice of objective derivative and irreversible strain invariance p_{ij} during elastic deforming $t_{ij} \equiv 0$. Note that elastic deforming is the kinematically determinate state while $t_{ij} \equiv 0$. The tensor t_{ij} must be computed by the special conditions under increasing irreversible deformations. The Almansi strain tensor d_{ij} according to Eq. (6.2) is calculated by the following decomposition

$$\begin{aligned} d_{ij} &= \frac{1}{2}(\delta_{ij} - g_{ij}) = \frac{1}{2}(u_{i,j} + u_{j,i} - u_{k,i}u_{k,j}) = \\ &= e_{ij} + p_{ij} - \frac{1}{2}e_{ik}e_{kj} - e_{ik}p_{kj} - p_{ik}e_{kj} + e_{ik}p_{km}e_{mj} \end{aligned} \tag{6.16}$$

Finally if the p_{ij} is the irreversible strains then the equation for the reversible strains reads according to Equ. (6.16) in form

$$s_{ij} = e_{ij} - 0.5e_{ik}e_{kj}.$$

6.3 Governing Equations

Initially we consider reversible deforming process when the irreversible deformation isn't increased. In this case the energy conservation law is formulated as

$$\rho\frac{de}{dt} + q_{j,j} = \sigma_{ij}\varepsilon_{ij}. \tag{6.17}$$

Herein ρ is the mass density, e is the entire energy density, q_i are the components of the heat flux vector, σ_{ij} are the components of the stress tensor.

Considering the slow processes, as the thermodynamic potential we use the free energy Ψ. For its density one can get equation as follows

$$\begin{aligned} \Psi(e_{ij},T) &= e(e_{ij},s) - sT, \\ \frac{\partial\Psi}{\partial T} &= -s, \\ \frac{\partial e}{\partial s} &= T, \end{aligned} \tag{6.18}$$

wherein T is the temperature, s is the entropy density. Equation (6.18) gives the thermodynamic potential satisfying to the hypothesis of the conservative deforming mechanism. Otherwise we assume the independence of Ψ on the irreversible deformation.
Substitution of Eq. (6.18) into (6.19) leads to following formula

$$\rho\left(\frac{\partial\Psi}{\partial e_{ij}}\frac{de_{ij}}{dt}+T\frac{ds}{dt}\right)+q_{i,j}-\sigma_{ij}\varepsilon_{ij}=0. \tag{6.19}$$

Eliminating from Eq. (6.19) the strains derivative by the first equation of (6.14) one can derive

$$\begin{aligned}&\left(\rho\left(\frac{\partial\Psi}{\partial e_{ij}}-\frac{\partial\Psi}{\partial e_{ik}}e_{kj}+A^{-1}B^2\left(e_{ik}\frac{\partial\Psi}{\partial e_{km}}e_{mj}-\frac{\partial\Psi}{\partial e_{ik}}e_{km}e_{mj}\right)+\right.\right.\\&A^{-1}B\left(e_{ik}e_{km}\frac{\partial\Psi}{\partial e_{mn}}e_{nj}-\frac{\partial\Psi}{\partial e_{ik}}e_{km}e_{mn}e_{nj}\right)+\\&\left.\left.+A^{-1}\left(e_{ik}e_{km}\frac{\partial\Psi}{\partial e_{mn}}e_{nt}e_{tj}-e_{ik}\frac{\partial\Psi}{\partial e_{km}}e_{mn}e_{nt}e_{tj}\right)\right)\sigma_{ij}\right)\varepsilon_{ij}+\\&+\rho r_{ij}\left(e_{ik}\frac{\partial\Psi}{\partial e_{kj}}-\frac{\partial\Psi}{\partial e_{ik}}e_{kj}\right)+\rho T\frac{ds}{dt}+q_{i,j}=0.\end{aligned} \tag{6.20}$$

Hence, by virtue of the processes independence given by e_{ij}, r_{ij}, T and symmetry condition for stress tensor the constitutive equations read by

$$\begin{aligned}\sigma_{ij}&=\rho\frac{\partial\Psi}{\partial e_{ik}}(\delta_{kj}-e_{kj}),\\ \rho T\frac{ds}{dt}&+q_{i,j}=0.\end{aligned} \tag{6.21}$$

The first equation of (6.21) is the analogue of the well-known in the nonlinear theory of elasticity (Lurie, 1990) as Murnaghan constitutive equations, the second one is the entropy balance equation during reversible deforming. Inside the domains with zero irreversible deformation ($p_{ij}\equiv 0$) the Murnaghan law takes the form

$$\begin{aligned}\sigma_{ij}&=\rho\frac{\partial\Psi}{\partial d_{ik}}(\delta_{kj}-2d_{kj}),\\ d_{ij}&=e_{ij}-\frac{1}{2}e_{ik}e_{kj}.\end{aligned} \tag{6.22}$$

Consequently we have two type of the strain-stress relations. First type (6.22) for the reversible deforming domain. Second type for irreversible deforming domain (6.21). The only potential function Ψ is needed for irreversible deforming domain $\Psi=\Psi(e_{ij},T)$ and for reversible deforming domain $\Psi=\Psi(d_{ij},T)$. The both functions must be equal under p_{ij} approaching to zero. This situation is quite similar to the classic Prandtl-Reuss elastic-plastic theory. At present study the stresses depend only on the reversible deformations as consequence of the hypothesis of the thermodynamic potentials (free energy, internal energy) independence on irreversible deformations.

Suppose now that $t_{ij} \neq 0$ during the deforming. Consequently the irreversible deformations are increased. After substituting of the first equation in (6.13) into Eq. (6.19) we can derive

$$\begin{gathered}\left(\rho\frac{\partial\Psi}{\partial e_{ik}}(\delta_{ij}-e_{kj})-\sigma_{ij}\right)\varepsilon_{ij}+\rho T\frac{ds}{dt}+q_{j,i}=\tau_{ij}t_{ij},\\ \tau_{ij}=\rho\frac{\partial\Psi}{\partial e_{ik}}(\delta_{kj}-2e_{kj}+e_{km}e_{mj})=\rho\frac{\partial\Psi}{\partial e_{ik}}(\delta_{kj}-s_{kj}).\end{gathered} \tag{6.23}$$

As expected the Murnaghan formula (6.21) and the entropy balance equation ensure from (6.23). The canonical form of balance equation can rewritten as

$$\frac{\partial(\rho s)}{\partial t}=-(T^{-1}q_j+\rho s v_j)_j-T^{-2}q_jT_{,j}+T^{-1}\tau_{ij}t_{ji}. \tag{6.24}$$

In Eq. (6.24) the first term is the entropy flux vector, the second one is the entropy production due to the heat conduction and the third one is the entropy production due to irreversible deformations. Hereafter we will consider the isothermal deformation processes. The generalization to the non-isothermal case isn't complex. In the case of isothermal entropy production is carried out only by the irreversible deformation process that is due to plastic flow, or due to the viscous (creep) deformation. Entropy production due to plasticity and viscosity occurs in different ways, but it can be generalized by the equation

$$D=\sigma_{ij}\gamma_{ij}. \tag{6.25}$$

Indeed, we have the well known representation of the entropy source (de Groot and Mazur, 1984) for non-plastic material or unloading and pre-plastic processes

$$D=\sigma_{ij}e^{v}_{ij}. \tag{6.26}$$

In the case of perfect plasticity for entropy production we also have the equation

$$D=\sigma_{ij}e^{p}_{ij}. \tag{6.27}$$

In Eqs. (6.26) and (6.27) the tensors e^{v}_{ij} and e^{p}_{ij} are the creep strain rate and the plastic strain rate respectively. Consequently $\gamma_{ij}=\varepsilon^{v}_{ij}$ inside the non-plastic domain and $\gamma_{ij}=e^{p}_{ij}$ during the plastic flow. The representation of the γ_{ij} is more complex for non-perfectly plastic domains. The viscous properties of the solids can inhibit the plastic flow being often modeled by adding appropriate terms in the plastic potential (yield surface). However, the representation (6.25) will be valid in this case; for γ_{ij} is only to be finding a more precise definition. According to (6.24), (6.25) in the general case of isothermal deformation we obtain

$$\sigma_{ij}\gamma_{ij}=T^{-1}\tau_{ij}t_{ij}. \tag{6.28}$$

Equation (6.28) is transformed by Eqs. (6.21), (6.23)

$$\begin{aligned}\tau_{ij} &= T\sigma_{ik}(\delta_{kj} - e_{kj}),\\ \gamma_{ij} &= t_{ik}(\delta_{kj} - e_{kj}).\end{aligned} \tag{6.29}$$

Last equation in (6.29) relates the unknown symmetric tensors t_{ik} and γ_{ij}. The transfer equations (6.13) are rearranged by the unknown tensor excluding into Eqs. (6.29)

$$\begin{aligned}\frac{De_{ij}}{Dt} &= \varepsilon_{ij} - \gamma_{ij} - \frac{1}{2}\left((\varepsilon_{ik} - \gamma_{ik} + z_{ik})e_{kj} + e_{ik}(\varepsilon_{kj} - \gamma_{kj} - z_{kj})\right),\\ \frac{Dp_{ij}}{Dt} &= \gamma_{ij} - p_{ik}\gamma_{kj} - \gamma_{ik}p_{kj},\\ \frac{Dn_{ij}}{Dt} &= \frac{dn_{ij}}{dt} - r_{ik}n_{kj} + n_{ik}r_{kj}.\end{aligned} \tag{6.30}$$

According to Eq. (6.30) we have the reversible deformation only if $\gamma_{ij} = 0$. Otherwise, under condition $\gamma_{ij} = 0$ the irreversible strain p_{ij} is changed similar to the rigid motion. The special objective derivative (6.15) ensures the geometrical correctness of kinematics.

In previous sections the irreversible deformation is not splitted into the plastic and creep parts. The difference of the rate tensors is associated with the dissipative mechanisms. The creep deformation rates ε^{v}_{ij} are specified by the non-zero stresses inside solids. The plastic deformation rates ε^{p}_{ij} arise if the stress state reaches the yield surface (when the plastic flow conditions are satisfied). Thus the deformation domain is separated into parts in dependence on the yield criterion satisfaction. The irreversible deformation is rapidly increase in the plastic flow domain and slowly increase in the creep one. The moving elastic-plastic borders separating the domain are the place of the exchange of irreversible deformation accumulation mechanisms. Note that the specification of the mechanisms of creep, plastic flow, elastic deformation potential can be experimentally determined. It is necessary to make sure that creep and plastic deformations are continuous at the elastic-plastic borders.

Assume that the viscous deformations occur before plastic flow and during unloading. These properties also resist the plastic flow. In the simplest case, under such conditions it is necessary to assume that the yield surface is defined by the equation: $f(\sigma_{ij}, \varepsilon^{p}_{ij}) = 0$. Then we can derive the von Mises maximum principle and the associated plastic flow rule is obtained in the form

$$\gamma_{ij} = \varepsilon^{p}_{ij} - \varepsilon^{v_0}_{ij} = \lambda\frac{\partial f}{\partial \sigma_{ij}}, \quad \lambda(\varepsilon^{p}_{ij}, \quad \varepsilon^{v_0}_{ij}) > 0. \tag{6.31}$$

Herein $\varepsilon^{v_0}_{ij}$ is the creep strain rate tensor at the time of the given material point reaches the elastic-plastic border, i.e. at the beginning of plastic flow. This irreversible strain rate is the initial value for the subsequent plastic flow. In this case the maximum shear stress condition takes the form

$$\max|\sigma_i - \sigma_j| = 2k + 2\eta\max|\varepsilon^{p}_{k} - \varepsilon^{v_0}_{k}|. \tag{6.32}$$

In Eq. (6.32) $\sigma_i, \varepsilon_k^p$ denote the principle values of the stress tensor and the plastic strain rate one, $\varepsilon_k^{v_0}$ the principle value of the tensor $\varepsilon_{ij}^{v_0}$, k yield stress, η is the coefficient of viscous resistance to plastic flow.

In the case of the linear piecewise plastic potential as in virtue of the Tresca yield criterion (6.32) it is convenient for creep potential also assume the piecewise-linear form

$$\gamma_{ij} = \varepsilon_{ij}^v = \frac{\partial V(\Sigma)}{\partial \sigma_{ij}}, \quad \Sigma = \max|\sigma_i - \sigma_j|, \quad V = B\Sigma^n. \tag{6.33}$$

Equation (6.33) contains the two-constant (B, n) Norton power creep law. We emphasize once again that (6.32) and (6.33) are the simplest case of the dissipation potentials. Specification of the elasticity law can also be assumed as for an incompressible medium in form (Lurie, 1990)

$$\begin{gathered} W = W(J_1, J_2) = \rho^{-1} = -2\mu J_1 - \mu J_2 + bJ_1^2 + (b-\mu)J_1J_2 - \chi J_1^3 + \ldots \\ J_1 = s_{jj}, \quad J_2 = s_{ij}s_{ji}, \quad s_{ij} = e_{ij} - \frac{1}{2}e_{ik}e_{kj} \\ \sigma_{ij} = -P_1\delta_{ij} + \frac{\partial W}{\partial e_{ik}}(\delta_{kj} - e_{kj}). \end{gathered} \tag{6.34}$$

Herein P_1 is the additional hydrostatic pressure, μ shear module, b, χ are the high order elastic modulus.

6.4 The Flow of Elastic-Viscous-Plastic Solids Inside the Cylindrical Tube

The problems in the frameworks of the large deformation model described above is solved in this section. Let continuum takes a form of cylindrical tube with radius R. Introduce the cylindrical coordinates ρ, φ, z. The deforming process under the pressure gradient is considered

$$\frac{\partial P_1(\rho,\, z,\, t)}{\partial z} = -\psi(t), \quad \psi(0) = 0. \tag{6.35}$$

The initial and boundary conditions are given by

$$\begin{gathered} e_{ij}\big|_{t=0} = p_{ij}\big|_{t=0} = 0, \\ \vec{u}\Big|_{\rho=R} = \vec{v}\Big|_{\rho=R} = 0. \end{gathered} \tag{6.36}$$

The unknown displacement and velocity vectors have the only non-zero vertical component. In view of axial symmetry we obtain

$$
\begin{aligned}
u &= u_z(\rho, t), \quad v = v_z(\rho, t), \\
d_{\rho\rho} &= -\frac{1}{2}\left(\frac{\partial u}{\partial \rho}\right)^2, \quad d_{\rho z} = d_{z\rho} = \frac{1}{2}\frac{\partial u}{\partial \rho}, \\
\varepsilon_{\rho z} &= \varepsilon_{z\rho} = \frac{1}{2}\frac{\partial v}{\partial \rho}, \quad w_{z\rho} = -w_{\rho z} = \frac{1}{2}\frac{\partial v}{\partial \rho}.
\end{aligned}
\tag{6.37}
$$

Hereafter the Helmholtz free energy is used as the thermodynamic potential. Assuming that the free energy density ψ is independent on irreversible deformations we get equation

$$W = \rho_0 \psi.$$

Then, the isotropic elastic potential for an incompressible solids in the form of a Taylor series expansion with respect to the referential state can be written in following form

$$
\begin{aligned}
W = W(I_1, I_2) &= (\alpha - \mu) I_1 + \alpha I_2 + \beta I_1^2 - \xi I_1 I_2 - \zeta I_1^3 + \ldots \\
J_1 = s_{jj}, \quad J_2 &= s_{ij} s_{ji}, \quad s_{ij} = e_{ij} - \frac{1}{2} e_{ik} e_{kj}.
\end{aligned}
\tag{6.38}
$$

Here $\mu,\ \alpha,\ \beta,\ \xi, \chi$ are the material constants.

After substituting Eq (6.38) into Eq. (6.22) the stress-strain relations can be formulated by

$$
\begin{aligned}
\sigma_{\rho\rho} &= -(P_1 + 2\mu) + 2b\left(e_{\rho\rho} + e_{zz} + e_{\varphi\varphi}\right) + 2\mu e_{\rho\rho} + \mu e_{\rho z}^2 + \ldots \\
\sigma_{\varphi\varphi} &= -(P_1 + 2\mu) + 2b\left(e_{\rho\rho} + e_{zz} + e_{\varphi\varphi}\right) + 2\mu e_{\varphi\varphi} - 2\mu e_{\rho z}^2 + \ldots \\
\sigma_{zz} &= -(P_1 + 2\mu) + 2b\left(e_{\rho\rho} + e_{zz} + e_{\varphi\varphi}\right) + 2\mu e_{zz} + \mu e_{\rho z}^2 + \ldots \\
\sigma_{\rho z} &= 2\mu e_{\rho z} + \ldots
\end{aligned}
\tag{6.39}
$$

The equations of motion in cylindrical coordinate taking into account the axial symmetry and zero volume forces and inertion can be derived

$$
\begin{aligned}
&\frac{\partial \sigma_{\rho\rho}}{\partial \rho} + \frac{\partial \sigma_{\rho z}}{\partial z} + \frac{\sigma_{\rho\rho} - \sigma_{\varphi\varphi}}{\rho} = 0, \\
&\frac{\partial \sigma_{\rho z}}{\partial \rho} + \frac{\partial \sigma_{zz}}{\partial z} + \frac{\sigma_{\rho z}}{\rho} = 0.
\end{aligned}
\tag{6.40}
$$

The transfer equations for the reversible and irreversible strains read

$$
\begin{aligned}
\gamma_{\rho\rho} &= \frac{dp_{\rho\rho}}{dt} + 2\left(p_{\rho\rho}\gamma_{\rho\rho} + p_{\rho z}\left(r_{z\rho} + \gamma_{\rho z}\right)\right),\\
\gamma_{zz} &= \frac{dp_{zz}}{dt} + 2\left(p_{zz}\gamma_{zz} + p_{\rho z}\left(r_{\rho z} + \gamma_{\rho z}\right)\right),\\
\gamma_{\rho z} &= \frac{dp_{\rho z}}{dt} + r_{z\rho}\left(p_{\rho\rho} - p_{zz}\right) + \left(p_{\rho z}\left(\gamma_{zz} + \gamma_{\rho\rho}\right) + \gamma_{\rho z}\left(p_{zz} + p_{\rho\rho}\right)\right),\\
\gamma_{\varphi\varphi} &= \frac{dp_{\varphi\varphi}}{dt} + 2p_{\varphi\varphi}\gamma_{\varphi\varphi},\\
-\gamma_{\rho\rho} &= \frac{de_{\rho\rho}}{dt} + 2r_{z\rho}e_{\rho z},\\
-\gamma_{zz} &= \frac{de_{zz}}{dt} + 2r_{\rho z}e_{\rho z},\\
\varepsilon_{\rho z} - \gamma_{\rho z} &= \frac{de_{\rho z}}{dt} + r_{\rho z}\left(e_{\rho\rho} - e_{zz}\right) +\\
&+ \frac{1}{2}\left(e_{\rho\rho}\left(\gamma_{\rho z} - \varepsilon_{\rho z}\right) + e_{zz}\left(\varepsilon_{\rho z} - \gamma_{\rho z}\right) + e_{\rho z}\left(\gamma_{zz} - \gamma_{\rho\rho}\right)\right),\\
-\gamma_{\varphi\varphi} &= \frac{de_{\varphi\varphi}}{dt}.
\end{aligned}
\tag{6.41}
$$

Let us going to the definition of the irreversible deformation source. The creep potential V according to the Norton power creep law reads

$$
\begin{aligned}
V\left(\sigma_{ij}\right) &= B\Sigma^n\left(\sigma_1,\ \sigma_2, \sigma_3\right),\\
\Sigma &= \max\left|\sigma_i - \sigma_j\right|,
\end{aligned}
\tag{6.42}
$$

wherein σ_i is the principle values of the stress tensor, B, n are the creep modulus. The value Σ is expressed by the stresses as follows

$$
\Sigma = \sqrt{4\sigma_{\rho z}^2 + \left(\sigma_{\rho\rho} - \sigma_{zz}\right)^2}. \tag{6.43}
$$

The equation for the creep deformation source can be found in form

$$
\begin{aligned}
\varepsilon_{\rho z}^c &= (-1)^n 2^n Bn\mu^{n-1} e_{\rho z}^{n-1},\\
\varepsilon_{\rho\rho}^c &= -\varepsilon_{zz}^c = \frac{\varepsilon_{\rho z}^c}{2}\left(\frac{e_{\rho\rho} - e_{zz}}{e_{\rho z}}\right).
\end{aligned}
\tag{6.44}
$$

The modified von Mises plastic potential with hardening and viscosity properties is derived by

$$
f\left(\tau_{ij},\quad p_{ij}, \varepsilon_{ij}^p\right) = \left(\tau_{ij} - cp_{ij} - \eta\varepsilon'^p_{ij}\right)\left(\tau_{ji} - cp_{ji} - \eta\varepsilon'^p_{ij}\right) - \frac{8}{3}k^2, \tag{6.45}
$$

wherein $\tau_{ij} = \sigma_{ij} - \frac{1}{3}\sigma_{kk}\delta_{ij}$, $\varepsilon'^p_{ij} = \varepsilon^p_{ij} - \frac{1}{3}\varepsilon^p_{kk}\delta_{ij}$, c is the material parameter describing the Baushinger effect, η is the plastic flow viscosity, k is the yield stress.
The following equation we obtain due to the associated plastic flow rule

$$\varepsilon_{ij}^{p} = \frac{2\lambda\left(\tau_{ij} - cp_{ij}\right)}{1 + 2\lambda\eta}. \tag{6.46}$$

Substituting Eq. (6.46) to the plastic potential $f(\tau_{ij},\ p_{ij}, \varepsilon_{ij}^{p})$ we can find unknown function λ and calculate the plastic strain rate. Further it can be expressed by virtue of Eq. (6.39) and finally

$$\begin{gathered}
\varepsilon_{\rho\rho}^{p} = \frac{1}{3\eta}\frac{Q - \sqrt{\frac{8}{3}}k}{Q}\left(\mu\left(4e_{\rho\rho} - 2e_{\varphi\varphi} - 2e_{zz} + 3e_{\rho z}^{2}\right) - 3cp_{\rho\rho}\right),\\
\varepsilon_{\varphi\varphi}^{p} = \frac{1}{3\eta}\frac{Q - \sqrt{\frac{8}{3}}k}{Q}\left(\mu\left(4e_{\varphi\varphi} - 2e_{\rho\rho} - 2e_{zz} - 6e_{\rho z}^{2}\right) - 3cp_{\varphi\varphi}\right),\\
\varepsilon_{zz}^{p} = \frac{1}{3\eta}\frac{Q - \sqrt{\frac{8}{3}}k}{Q}\left(\mu\left(4e_{zz} - 2e_{\rho\rho} - 2e_{\varphi\varphi} + 3e_{\rho z}^{2}\right) - 3cp_{zz}\right),\\
\varepsilon_{rz}^{p} = \frac{1}{\eta}\frac{Q - \sqrt{\frac{8}{3}}k}{Q}\left(2\mu e_{rz} - cp_{\rho z}\right).
\end{gathered} \tag{6.47}$$

The function Q is computed by

$$\begin{gathered}
Q = \sqrt{\frac{1}{3}\left(3c^{2}B_{1} + 2c\mu B_{2} + 2\mu^{2}B_{3}\right)},\\
B_{1} = p_{\rho\rho}^{2} + p_{\varphi\varphi}^{2} + p_{zz}^{2} + 2p_{\rho z}^{2},\\
B_{2} = 3e_{\rho z}^{2}\left(-p_{\rho\rho} + 2p_{\varphi\varphi} - p_{zz}\right) + 2e_{rr}\left(-2p_{\rho\rho} + p_{\varphi\varphi} + p_{zz}\right) +\\
+2e_{\varphi\varphi}\left(p_{\rho\rho} - 2p_{\varphi\varphi} + p_{zz}\right) + 2e_{zz}\left(p_{\rho\rho} + p_{\varphi\varphi} - 2p_{zz}\right) - 12e_{\rho z}^{2}p_{\rho z},\\
B_{3} = 12e_{\rho z}^{2} + 6e_{\rho z}^{2}\left(e_{rr} - 2e_{\varphi\varphi} + e_{zz}\right) + 4e_{\rho\rho}^{2} +\\
+4e_{\varphi\varphi}^{2} + 4e_{zz}^{2} - 4e_{rr}e_{zz} - 4e_{rr}e_{\varphi\varphi} - 4e_{rr}e_{zz} + 9e_{\rho z}^{4}.
\end{gathered} \tag{6.48}$$

The general solution for second equilibrium equation (6.40) taking into account the (6.35) is obtained in following form

$$\sigma_{\rho z}(\rho,\ t) = -\frac{\psi(t)}{2}\rho \tag{6.49}$$

Hereafter the following dimensionless notation is used

$$\begin{gathered}
\hat{\rho} = \frac{\rho}{R},\quad \hat{z} = \frac{z}{R},\quad \hat{u} = \frac{u}{R},\quad \tau = \frac{t}{R}\sqrt{\frac{\mu}{\rho_0}},\\
\hat{\sigma}_{ij} = \frac{\sigma_{ij}}{\mu},\quad \hat{b} = \frac{b}{\mu},\quad \hat{k} = \frac{k}{\mu},\quad \hat{c} = \frac{c}{\mu},
\end{gathered} \tag{6.50}$$

wherein ρ_0 is the referential mass density.

Material and process parameters are given by

$$
\begin{gathered}
n=3, \quad \frac{BnR\mu^{n-1}\sqrt{\rho_0}}{\sqrt{\mu}}=3.5, \quad \hat{\eta}=\frac{2\sqrt{\mu}R\sqrt{\rho_0}}{\eta}=1.0, \\
\hat{b}=4, \quad \hat{k}=0.00125, \quad \hat{c}=0.05, \\
\tau_1=16, \quad \tau_2=32, \quad \tau_3=48, \quad \psi_{max}=5\cdot 10^{-3}.
\end{gathered} \tag{6.51}
$$

The pressure gradient is assumed as the following piecewise function

$$
\psi(\tau)=\begin{cases}
\frac{\psi_{max}}{2}\cdot\left(1+\sin\left(\frac{\pi}{t_1}\tau-\frac{\pi}{2}\right)\right), & 0\le\tau\le\tau_1, \\
\psi_{max}, \ \tau_1<\tau\le\tau_2 \\
\frac{\psi_{max}}{2}\cdot\left(1+\sin\left(\frac{\pi}{(t_3-t_2)}(\tau-\tau_2)+\frac{\pi}{2}\right)\right), & \tau_2<\tau\le\tau_3.
\end{cases} \tag{6.52}
$$

The resulting differential equations system (6.41), (6.40) in virtue of dimensionless description (6.50) is numerically solved under the initial and boundary conditions (6.36). The central difference scheme is used for spatial derivatives approximation and the explicit deference scheme is used for time derivatives approximation. The elastic-plastic border is computed by the following equation

$$
Q-\sqrt{\frac{8}{3}}k=0. \tag{6.53}
$$

The Fig. 6.1 shows the graphical results for irreversible deformations $p_{\rho z}, p_{\rho\rho}, p_{zz}$, displacement u at the times τ_1, τ_2, τ_3.

6.5 Viscometric Deformation of the Incompressible Cylindrical Layer

As another problem considering in this section is viscometric deformation of the cylindrical layer $r_0\le r\le R$ of the incompressible elastic-plastic-creep material. The viscometric walls $r=r_0$, $r=R$ are rigid under adhesion conditions. The deformation is carried out by rotation of the inner rigid cylinder wall. The cylindrical coordinate system (r,ϕ,z) is used. Thus for the non-zero components of displacements and deformations in this case we can obtain

$$
\begin{gathered}
u_r=r(1-\cos\theta); u_\phi=r\sin\theta; d_{r\phi}=\frac{1}{2}r\theta_{,r}; d_{rr}=-\frac{1}{2}(r\theta_{,r})^2, \\
\theta|_{r=R}=0, \quad \sigma_{r\phi}|_{r=r_0}=\frac{c(t)}{r_0^2}, \quad c(0)=0,
\end{gathered} \tag{6.54}
$$

wherein the $c(t)$ is the given function.

The stresses during reversible deforming can be computed by (6.54) as follows

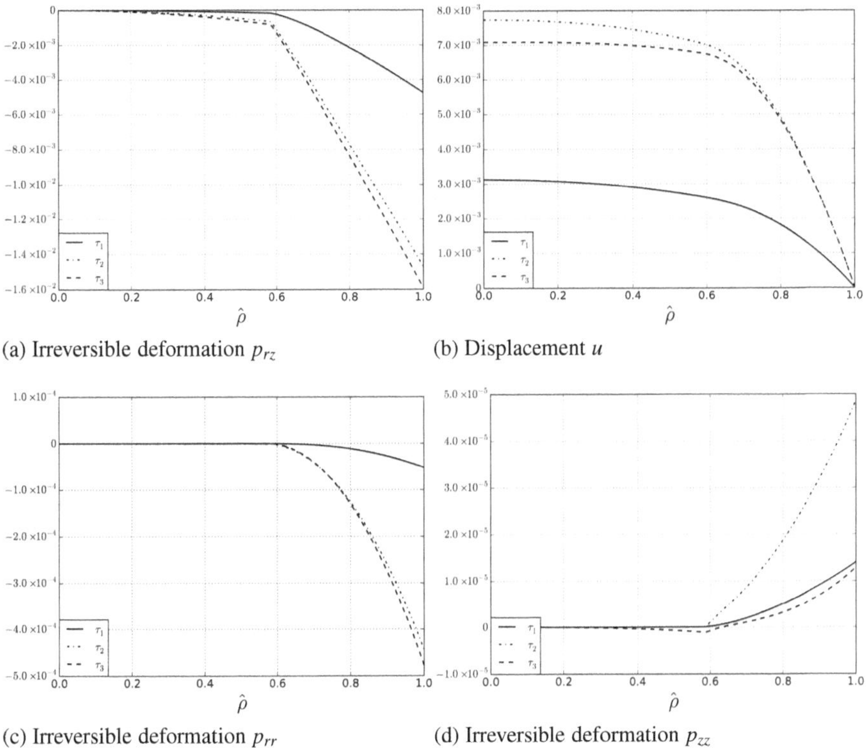

(a) Irreversible deformation p_{rz}

(b) Displacement u

(c) Irreversible deformation p_{rr}

(d) Irreversible deformation p_{zz}

Fig. 6.1: Numerical results at times τ_1, τ_2, τ_3

$$
\begin{aligned}
\sigma_{rr} &= -P_1 + 2(b+\mu)e_{rr} + 2be_{\phi\phi} - (2b+3\chi)e_{rr}^2 + \mu e_{r\phi}^2 - \\
&\quad -2(b+\mu+3\chi)e_{rr}e_{\phi\phi} - (\mu+3\chi)e_{\phi\phi}^2 + \dots \\
\sigma_{\phi\phi} &= -P_1 + 2(b+\mu)e_{\phi\phi} + 2be_{rr} - (\mu+3\chi)e_{rr}^2 + \mu e_{r\phi}^2 - \\
&\quad -2(b+\mu+3\chi)e_{rr}e_{\phi\phi} - (2b+3\chi)e_{\phi\phi}^2 + \dots \\
\sigma_{zz} &= -P_1 + 2b(e_{rr}+e_{\phi\phi}) - (\mu+3\chi)(e_{rr}^2 + e_{\phi\phi}^2) - 2\mu e_{r\phi}^2 - 6\chi e_{rr}e_{\phi\phi} + \dots \\
\sigma_{r\phi} &= 2\mu e_{r\phi} - (\mu - 2b)e_{r\phi}(e_{rr}+e_{\phi\phi}) + \dots
\end{aligned}
\tag{6.55}
$$

Herein the dots denote higher order terms than the second component of reversible deformation (in the numerical calculations we take into account sixth order terms). The following equation is derived by (6.55)

$$
\frac{\sigma_{rr} - \sigma_{\phi\phi}}{\sigma_{r\phi}} = \frac{e_{rr} - e_{\phi\phi}}{e_{r\phi}}
\tag{6.56}
$$

The non-zero components of the kinematical variables in this case read

$$v_\phi = v = r\dot{\theta}; \varepsilon_{r\phi} = \frac{1}{2}r\dot{\theta}_{,r}; \omega_{r\phi} = -\dot{\theta} - \frac{1}{2}r\dot{\theta}_{,r};$$
$$r_{r\phi} = -\dot{\theta} + \frac{2\varepsilon_{r\phi}(1-e_{\phi\phi})}{e_{rr}+e_{\phi\phi}-2}; \dot{\theta} = \frac{\partial\theta}{\partial t}; \dot{\theta}_{,r} = \frac{\partial^2\theta}{\partial t\partial r}. \tag{6.57}$$

Consequently one can substitute Eq. (6.57) in Eq. (6.33) in view of the (6.56) and find

$$\varepsilon^v_{r\phi} = 2Bn\sigma_{r\phi}\Phi^{\frac{n}{2}-1}; \varepsilon^v_{rr} = -\varepsilon^v_{\phi\phi} = Bn\sigma_{r\phi}\frac{e_{rr}-e_{\phi\phi}}{e_{r\phi}}\Phi^{\frac{n}{2}-1};$$
$$\Phi = \sigma^2_{r\phi}(4+(\frac{e_{rr}-e_{\phi\phi}}{e_{r\phi}})^2) \tag{6.58}$$

The resulting system consists of the Eqs. (6.30) after rearranging (6.57), (6.58) and adding the latter equation from (6.55)

$$\begin{aligned}
\dot{p}_{rr} &= 2p_{r\phi}(S-X)+Y(1-2p_{rr}),\\
\dot{p}_{\phi\phi} &= -2p_{r\phi}(S+X)-Y(1-2p_{\phi\phi}),\\
\dot{p}_{r\phi} &= (p_{\phi\phi}-p_{rr})S+X(1-p_{rr}-p_{\phi\phi}),\\
\dot{e}_{rr} &= e_{r\phi}(3S+X\frac{e_{\phi\phi}-e_{rr}}{e_{rr}+e_{\phi\phi}-2})+\Upsilon(e_{rr}-1),\\
\dot{e}_{\phi\phi} &= e_{r\phi}(X-S)-\Upsilon(e_{\phi\phi}-1),\\
\dot{e}_{r\phi} &= (e_{\phi\phi}-1)S+\frac{1}{2}X(e_{rr}+e\phi\phi-2),\\
c(t)r^{-2} &= 2\mu e_{r\phi}-(\mu-2b)e_{r\phi}(e_{rr}+e\phi\phi).
\end{aligned} \tag{6.59}$$

In Eq. (6.59) the following notation is used

$$S = \dot{\theta}_{,r}\frac{r(1-e_{\phi\phi})}{e_{rr}+e_{\phi\phi}-2}, \quad X = \frac{2Bnc(t)}{r^2}(\frac{c^2(t)}{r^4}(4+\frac{e_{rr}-e_{\phi\phi}}{e_{r\phi}})^2)^{\frac{n}{2}-1}, \Upsilon = X\frac{e_{rr}-e_{\phi\phi}}{2e_{r\phi}}. \tag{6.60}$$

The stress component $\sigma_{r\phi}$ according to Eq. (6.59), the equilibrium equation and boundary condition on the inner cylinder is

$$\sigma_{r\phi} = c(t)r^{-2}. \tag{6.61}$$

The system of equations (6.59) is the system of seven partial differential equations with respect to seven unknown functions $e_{rr}, e_{\phi\phi}, e_{r\phi}, p_{rr}, p_{\phi\phi}, p_{r\phi}, \theta$ which is numerically integrated by finite deference scheme. The initial conditions are the free state conditions. The boundary conditions are given by kinematics of solids (6.54). The hydrostatic pressure $p_{r,t}$ is found by integrating of the equilibrium equation and simultaneously the stress tensor components $\sigma_{rr}, \sigma_{\phi\phi}, \sigma_{zz}$ are determined. But stress, strain tensor and displacements calculating by the latter approach are valid until the time $t = t_1$ under increasing pressure. At the time the yield criterion (6.32) is satisfied on cylindrical surface $r = r_0$ in form

$$\sqrt{\Phi} = 2k + 2\eta\sqrt{(\varepsilon^p_{r\phi})^2 + (\varepsilon^p_{rr})^2}. \tag{6.62}$$

Since the time t_1 the elastic-plastic border $r = r_1(t)$ is propagated from the surface $r = r_0$. Inside the domain $r_1(t) \neq r \neq R$ the irreversible deformations increase according to creep law. Inside the domain $r_0 \neq r \neq r_1(t)$ the ones increase due to plastic flow. The following equations of plastic strain rates are derived in virtue of associated plastic flow rule

$$\begin{gathered}\varepsilon^p_{r\phi} = 2H\sigma_{r\phi} + \varepsilon^{v_0}_{r\phi}; \varepsilon^p_{rr} = -\varepsilon^p_{\phi\phi} = H(\sigma_{rr} - \sigma_{\phi\phi}) + \varepsilon^{v_0}_{rr}, \\ H = \frac{1}{\eta}(\frac{1}{2} - \frac{k}{\sqrt{\Phi}}).\end{gathered} \tag{6.63}$$

The final system inside plastic domain is obtained by exchanging of the variables X, Υ to X_1, Υ_1 in Eqs. (6.59)

$$\begin{gathered}X_1 = \frac{c(t)}{\eta r^2} - \frac{2ke_{r\phi}}{\eta\sqrt{(e_{rr} - e_{/phi\phi})^2 + 4(e_{r\phi})^2}} + \varepsilon^{v_0}_{r\phi}, \\ \Upsilon_1 = X_1\frac{e_{rr} - e_{\phi\phi}}{2e_{r\phi}} + \varepsilon^{v_0}_{rr}.\end{gathered} \tag{6.64}$$

Thus reversible and irreversible deformations are determined by the corresponding systems integration inside two domains using the same boundary conditions and continuity conditions of strain and function $\theta(r,t)$ on elastic-plastic border $r = r_1(t)$.

Suppose that, starting from the time $t = t_2 > t_1$ the load is decreased. Whereupon at the time the unloading elastic-plastic border is separated from $r_1(t_2)$ and moves to one $r = r_0$. Consequently inside undeforming domain $r_1(t_2) \leq r \leq R$ and plastic domain $r_2(t) \leq r \leq r_2(t_2)$ the irreversible deformations is increased by virtue of creep. The reversible and irreversible deformations are found by integrating of Eqs. (6.59). Inside plastic domain $r_0 \leq r \leq r_2(t)$ in Eq. (6.59) we need to replace X, Υ to X_1, Υ_1. The stresses is calculated as above. The implementation of the boundary conditions and continuity conditions on the elastic-plastic borders allows us to specify at each time the elastic-plastic border spatial position.

The calculations are carried out by dimensionless variables

$$\widetilde{r} = r/R, \tau = \sqrt{\mu/\rho_0 t}/R, \widetilde{\sigma}_{ij} = \sigma_{ij}/\mu$$

and material constants

$$k/\mu = 0.003, b/\eta = 4, \chi/\mu = 80, r_0/R = 0.5, n = 3, BnR\mu^{n-1}\sqrt{\rho_0/\mu} = 3.5.$$

The Fig. 6.2 shows the rotation angle evolution θ. The Figs. 6.3 demonstrate the irreversible deformations evolution in respect to time on surface $r = r_0$.

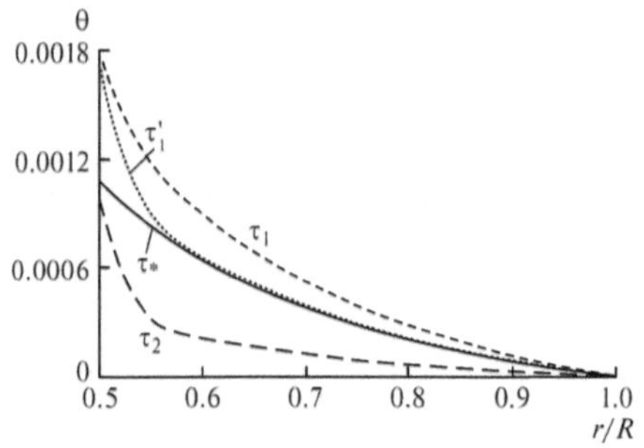

Fig. 6.2: Rotation angle

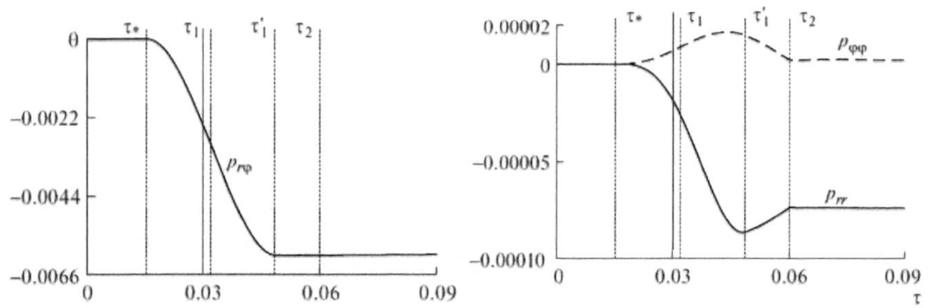

(a) Irreversible deformation $p_{r\phi}$ (b) Irreversible deformation $p_{\phi\phi}$

Fig. 6.3: Irreversible deformation on surface $r = r_0$

6.6 Conclusion

The mathematical model of large deformation of materials with viscous, elastic and plastic properties has been developed. The model is based on differential transfer equations for the reversible and irreversible strains. The proposed model has several advantages compared to alternative approaches. Firstly, the governing equations derivation isn't use the terms of the intermediate and unloading configurations. Secondly, the objective derivative is derived by the thermodynamic laws, and is independent on a researcher choice. In virtue of proposed approach the boundary value problems for cylinder under pressure gradient and viscometric deformations have been solved. In solved problems the irreversible deformations have been accumulated by creep and plastic flow. These solutions can be used for testing of numerical algorithms for the calculation of large deformations of elastic-viscous-plastic material.

References

Burenin AA, Bykovtsev GI, Kovtanyuk LV (1996) A simple model of finite strain in an elastoplastic medium. Doklady Physics 41(3):127–129

Burenin AA, Kovtanyuk LV, Polonik MV (2003) The formation of a one-dimensional residual stress field in the neighbourhood of a cylindrical defect in the continuity of an elastoplastic medium. Journal of Applied Mathematics and Mechanics 67(2):283 – 292

Burenin AA, Kovtanyuk LV, Mazelis AL (2010) Development of a rectilinear axisymmetric viscoplastic flow and elastic aftereffect after its stop. Journal of Applied Mechanics and Technical Physics 51(2):261–268

Burenin AA, Kovtanyuk LV, Panchenko GL (2015) Nonisothermal motion of an elastoviscoplastic medium through a pipe under a changing pressure drop. Doklady Physics 60(9):419–422

Clifton RJ (1972) On the equivalence of $\boldsymbol{f}^e \boldsymbol{f}^p$ and $\bar{\boldsymbol{f}}^p \bar{\boldsymbol{f}}^e$. ASME J Appl Mech 39(1):287–289

de Groot SR, Mazur P (1984) Non-Equilibrium Thermodynamics. Dover, New York

Firat M, Kaftanoglu B, Eser O (2008) Sheet metal forming analyses with an emphasis on the springback deformation. Journal of Materials Processing Technology 196(1-3):135 – 148

Hill R (1958) A general theory of uniqueness and stability in elastic-plastic solids. Journal of the Mechanics and Physics of Solids 6(3):236 – 249

Khan AS, Huang S (1995) Continuum Theory of Plasticity. Wiley, New York

Kondaurov VI (1982) Equations of elastoviscoplastic medium with finite deformations. Journal of Applied Mechanics and Technical Physics 23(4):584–591

Kovtanyuk LV (2005) On the forcing of an elastoviscoplastic material through an inflexible circular cylindrical die. Doklady Physics 50(2):112–114

Lee EH (1969) Elastic-plastic deformation at finite strains. ASME J Appl Mech 36(1):1–6

Levitas VI (1996) Large Deformation of Materials with Complex Rheological Properties at Normal and High Pressure. Nova Science Publishers, Commack, N.Y

Lurie AI (1990) Non-Linear Theory of Elasticity. North-Holland

Naghdi PM (1985) Recent development in finite deformation plasticity. In: Sawczuk A, Bianchi G (eds) Plasticity Today: Modeling, Methods and Applications, Springer Netherland, pp 75–83

Nemat-Nasser S (1982) On finite deformation elasto-plasticity. International Journal of Solids and Structures 18(10):857 – 872

Rogovoi AA (2005) Constitutive relations for finite elastic-inelastic strains. Journal of Applied Mechanics and Technical Physics 46(5):730–739

Simo J, Pister K (1984) Remarks on rate constitutive equations for finite deformation problems: computational implications. Computer Methods in Applied Mechanics and Engineering 46(2):201 – 215

Xiao H, Bruhns OT, Meyers A (2006) Elastoplasticity beyond small deformations. Acta Mechanica 182(1):31–111

Chapter 7
On Nonlocal Surface Elasticity and Propagation of Surface Anti-Plane Waves

Victor A. Eremeyev

Abstract We discuss the nonlocal model of surface elasticity that is the model with surface energy density which depends on first and higher gradients of displacements. To demonstrate the peculiarities of the model we consider the propagation of anti-plane surface waves in an elastic halfspace with the surface energy. Using the least action principle we derive the governing equations for the problem. Analyzing the anti-plane deformations we obtain the dispersion relation and analyze its dependence on surface elastic moduli and on the order of considered gradients.

Key words: anti-plane waves · Surface elasticity · Strain gradient elasticity · Surface waves · *N*th-order strain gradient elasticity

7.1 Introduction

Recent developments in the nanotechnologies involve in common practise enriched models of continua such as surface elasticity, strain gradient elasticity and other generalized models of the continuum mechanics, see for example recent comparison of the popular models in Liebold and Müller (2015). The surface elasticity model proposed by Gurtin and Murdoch (1975, 1978) found many applications in micro- and nanomechanics, see Duan et al (2008); Wang et al (2011); Javili et al (2012); Eremeyev (2016) and reference therein. The Gurtin-Murdoch model requires constitutive equations for surface strain energy density and for kinetic surface energy density in addition to the constitutive equations in the bulk. This model is general-

Victor A. Eremeyev
Southern Federal University, Institute of Mathematics, Mechanics and Computer Science, Milchakova Street 8a, Rostov-on-Don, 344090, Russia and Rzeszów University of Technology, Faculty of Mechanical Engineering and Avionics, al. Powstańców Warszawy 8, 35959 Rzeszów, Poland,
e-mail: `eremeyev.victor@gmail.com; veremeyev@prz.edu.pl`

H. Altenbach et al. (eds.), *Mechanics for Materials and Technologies*,
Advanced Structured Materials 46, DOI 10.1007/978-3-319-56050-2_7

ized in Steigmann and Ogden (1997, 1999) where bending stiffness related to the surface was taken into account. Further extensions for surface and interface models can be found in Placidi et al (2014); Lurie et al (2009); dell'Isola et al (2009a); Javili et al (2013); Eremeyev (2016); Eremeyev et al (2016). Within the various extensions of surface elasticity more complex constitutive equations are usually introduced.

The strain gradient models were applied for nanostructured materials in Aifantis (1999, 2016); Forest et al (2011). Let us note that there some similarities between strain gradient elasticity and surface stresses theory known from the theory of capillarity (Rowlinson and Widom, 2003; de Gennes et al, 2004). After Mindlin (1965) modelling of surface stresses relates also with gradient theories of elasticity, see for example de Gennes (1981); dell'Isola et al (2009b); dell'Isola and Seppecher (1997); Askes and Aifantis (2011). Both theories may change dramatically the behaviour of solutions of boundary-value problem. In particular, the influence of surface stresses may be crucial in the vicinity of geometrical singularities, holes, notches, crack tips, etc. Let us only note here the existence of surface and interfacial anti-plane waves decaying with the distance from the surface or interface observed in elastic solids modelled using the surface elasticity (Fan et al, 2006; Xu et al, 2015; Eremeyev et al, 2016) and modelled with the strain gradient elasticity (Vardoulakis and Georgiadis, 1997; Yerofeyev and Sheshenina, 2005; Gourgiotis and Georgiadis, 2015; Georgiadis et al, 2000). It is worse to note that such surface anti-plane waves do not exist in the classical linear elasticity (Achenbach, 1973).

The aim of this paper is to discuss the possible extension of a surface elasticity considering nonlocal interactions (long range forces). Such interactions lead to nonlocal form of constitutive equations for the surface stresses. For simplicity we are restricted ourselves by infinitesimal deformations. Considering this model we analyze the propagation of surface anti-plane waves localized near plane boundary of an elastic half space and to derive the form of a dispersion equation for these waves. We search the anti-plane waves whose amplitude decays exponentially with distance from the surface. Here we extend the recent results (Eremeyev et al, 2016) considering most general model of surface strain energy for small deformations. We assumed that the surface strain energy is a quadratic function of surface strains and surface gradients of displacements up to Nth order. In other words we consider weak nonlocal model of surface elasticity which includes also the Gurtin–Murdoch model. The considered deformations with nonlocal surface energy in some sense can be treated as continuum limit of discreet models such as Nieves et al (2016); Gorbushin and Mishuris (2016a,b); Carcaterra et al (2015); Rahali et al (2015); Porubov (2014); Gerasimov et al (2016, 2017).

The paper is organized as follows. In Sect. 7.2 we present the basic equations of the considered model. Here we introduce the strain energy and kinetic surface densities. The strain energy density depends on first and higher order of surface gradients of displacements (up to N-order, $N > 1$). Using the least action principle we derive the dynamic boundary conditions. The derived boundary conditions contain the tangent partial derivatives of the displacements up to $2N$-order. In Sect. 7.3 we reduce the general statement to the case of anti-plane deformations. As a result we obtain the classical wave equation in the half space and non-classic boundary equation at the half space plane. Finally we derived the dispersion equation that is the relation between phase velocity and wave number.

7.2 Governing Equations

Let an elastic solid occupies the region V in $\mathbb{R}$ with the smooth boundary $A = \partial V$. We are restricted ourselves by linear elastic isotropic solids for the description of the material behaviour in the bulk. So the infinitesimal deformations of an elastic solid are described by the displacement field

$$\boldsymbol{u} = \boldsymbol{u}(\boldsymbol{x},t), \tag{7.1}$$

where $\boldsymbol{u}$ is twice differentiable vector-function of $\boldsymbol{x}$, $\boldsymbol{x}$ is the position vector and t is time, whereas the strain energy W is given by

$$W = \mu \boldsymbol{e} : \boldsymbol{e} + \frac{1}{2}\lambda (\mathrm{tr}\boldsymbol{e})^2, \quad \boldsymbol{e} = \frac{1}{2}\left(\nabla \boldsymbol{u} + (\nabla \boldsymbol{u})^{\mathrm{T}}\right), \tag{7.2}$$

where $\boldsymbol{e}$ is the strain tensor, the double dot stands for scalar (inner) product of two second-order tensors, λ and μ are Lamé moduli, $\mu > 0$, $3\lambda + 2\mu > 0$, and ∇ is the 3D nabla operator. The corresponding stress tensor is determined as follows

$$\boldsymbol{\sigma} \equiv \frac{\partial \mathscr{W}}{\partial \boldsymbol{e}} = 2\mu \boldsymbol{e} + \lambda \boldsymbol{I}\,\mathrm{tr}\boldsymbol{e}.$$

Here $\boldsymbol{I}$ is the three-dimensional unit tensor.

In addition to (7.2) we introduce the kinetic energy density by the formula

$$T = \frac{1}{2}\rho \dot{\boldsymbol{u}} \cdot \dot{\boldsymbol{u}}, \tag{7.3}$$

where ρ is the mass density and overdot denotes the derivative with respect to t.

Within the framework of the surface elasticity we introduce the surface kinetic energy T_s and the surface strain energy density W_s. Taking into account the surface mass density m we use the following formula for surface kinetic energy density (Gurtin and Murdoch, 1978)

$$\mathscr{K}_s = \frac{1}{2} m \dot{\boldsymbol{u}} \cdot \dot{\boldsymbol{u}}\Big|_{\boldsymbol{x}\in A}. \tag{7.4}$$

For the surface strain energy density we assume the following dependence

$$W_s = W_s(\varepsilon, \nabla_s^2 \boldsymbol{u}, \ldots, \nabla_s^N \boldsymbol{u}), \tag{7.5}$$

where $\nabla_s \equiv \nabla - \boldsymbol{n}\partial/\partial n$ is the surface nabla operator (Lebedev et al, 2010),

$$\varepsilon = \frac{1}{2}\left((\nabla_s \boldsymbol{u}) \cdot \boldsymbol{A} + \boldsymbol{A} \cdot (\nabla_s \boldsymbol{u})^{\mathrm{T}}\right)$$

is the surface strain tensor, $\mathbf{A} \equiv \mathbf{I} - \mathbf{n} \otimes \mathbf{n}$ is the surface unit second-order tensor, $\boldsymbol{n}$ is the unit vector of outer normal to A, $\otimes$ denotes the tensorial product of two vectors. Constitutive equation (7.5) satisfies the principle of the material frame indifference in the case of small deformations. Indeed, it is obviously that (7.5) is invariant under transformations of superimposed rigid body motion

$$\boldsymbol{u} \to \boldsymbol{u} + \boldsymbol{a} + \boldsymbol{b} \times \boldsymbol{x},$$

where $\boldsymbol{a}$ and $\boldsymbol{b}$ are constant vectors and $\times$ is the cross product. Constitutive equation (7.5) gives an example of nonlocal surface strain energy density. Indeed, here W_s depends on derivatives up to Nth order like Nth order strain gradient elasticity (dell'Isola et al, 2012). For simplicity we assume W_s in the form

$$W_s = \mu_s \varepsilon : \varepsilon + \frac{1}{2}\lambda_s (\operatorname{tr} \varepsilon)^2 + \frac{1}{2}\sum_{i=2}^{N} \mu_i \|\nabla_s^i \boldsymbol{u}\|^2, \tag{7.6}$$

where λ_s and μ_s are the surface elastic moduli called also surface Lamé moduli used in the Gurtin–Murdoch approach (Gurtin and Murdoch, 1975, 1978), and $\mu_i \geq 0$ are additional surface stiffness moduli. Here $\|\cdot\|$ denotes the Euclidian norm in the space of ith-order tensors (Lebedev et al, 2010). Thus, (7.6) contains the Gurtin–Murdoch model of surface elasticity as a special case.

For derivation of governing equations we use the Hamilton variational principle (the principle of least action). Functional of action is defined as follows

$$\mathscr{H} = \int_0^{\mathrm{T}} \int_V (T - W)\, \mathrm{d}V\, \mathrm{d}t + \int_0^{\mathrm{T}} \int_A (T_s - W_s)\, \mathrm{d}A\, \mathrm{d}t. \tag{7.7}$$

After standard technique of calculus of variations the variational equation $\delta \mathscr{H} = 0$ results in the following motion equation

$$\nabla \cdot \sigma = \rho \ddot{\boldsymbol{u}}, \quad \forall \boldsymbol{x} \in V, \tag{7.8}$$

and the dynamic boundary conditions

$$\boldsymbol{n} \cdot \sigma = \nabla_s \cdot \tau - m\ddot{\boldsymbol{u}}, \quad \forall \boldsymbol{x} \in A, \tag{7.9}$$

where τ is the surface stress tensor. For nonlocal model (7.6) it takes the form

$$\begin{aligned}
\tau &\equiv \frac{\partial \mathscr{W}_s}{\partial \varepsilon} - \sum_{i=2}^{2N} (-1)^i \nabla_s \cdot \left(\ldots \nabla_s \cdot \left(\frac{\partial \mathscr{W}_s}{\partial \nabla_s^i \boldsymbol{u}}\right) \ldots\right) \\
&= \mu_s \varepsilon + \lambda_s \mathbf{A} \operatorname{tr} \varepsilon - \sum_{i=2}^{N} (-1)^i \mu_i \underbrace{\nabla_s \cdot (\ldots \nabla_s \cdot}_{2i \text{ times}} (\nabla_s^i \boldsymbol{u}) \ldots),
\end{aligned} \tag{7.10}$$

Nonclassical boundary conditions (7.9) with (7.10) generalizes the classic Young-Laplace equation used in the theory of capillarity and its analogues known from the Gurtin–Murdoch and Steigmann-Ogden models.
For example, in the case $N = 2$ Eq. (7.10) takes the form

$$\boldsymbol{\tau} = \mu_s \boldsymbol{\varepsilon} + \lambda_s \mathbf{A} \operatorname{tr} \boldsymbol{\varepsilon} - \mu_2 \nabla_s \cdot \left(\nabla_s \cdot \nabla_s^2 \boldsymbol{u}\right).$$

7.3 Anti-Plane Surface Waves in an Elastic Half-Space

Let us consider stationary waves of an elastic half-space $x_1 \leq 0$, see Fig. 7.1. Here x_1, x_2 and x_3 are the Cartesian coordinates and $\boldsymbol{i}_k$ are corresponding unit base vectors.

For anti-plane motion the vector of displacement takes the form (Achenbach, 1973)

$$\boldsymbol{u} = u(x_1, x_2, t)\boldsymbol{i}_3. \tag{7.11}$$

Now we have more simple formulae for gradients and strains

$$\nabla \boldsymbol{u} = u_{,\alpha} \boldsymbol{i}_\alpha \otimes \boldsymbol{i}_3 = \nabla u \otimes \boldsymbol{i}_3, \quad \nabla_s \boldsymbol{u} = u_{,2} \boldsymbol{i}_2 \otimes \boldsymbol{i}_3, \quad \nabla_s^i \boldsymbol{u} = \frac{\partial^i u}{\partial x_2^i} \underbrace{\boldsymbol{i}_2 \otimes \ldots \otimes \boldsymbol{i}_2}_{i \text{ times}} \otimes \boldsymbol{i}_3,$$

$$\boldsymbol{e} = \frac{1}{2}(\nabla u \otimes \boldsymbol{i}_3 + \boldsymbol{i}_3 \otimes \nabla u), \quad \boldsymbol{\varepsilon} = \frac{1}{2} u_{,2} (\boldsymbol{i}_2 \otimes \boldsymbol{i}_3 + \boldsymbol{i}_2 \otimes \boldsymbol{i}_3).$$

Hereafter Greek indices take values $1, 2$ and for brevity we use the notation

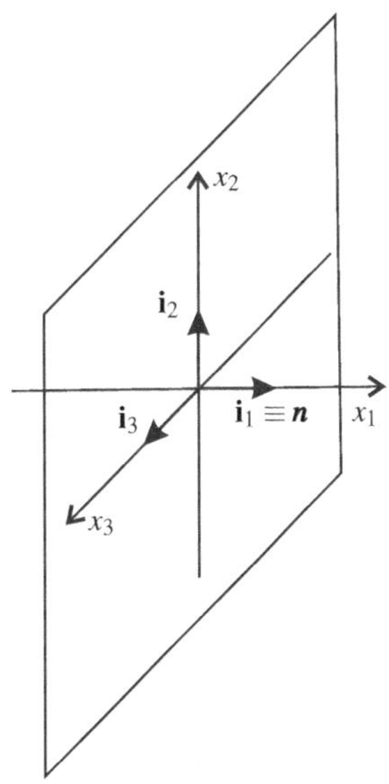

Fig. 7.1 Elastic half-space and used Cartesian coordinates and base vectors

$$u_{,\alpha} = \frac{\partial u}{\partial x_\alpha}.$$

For the anti-plane deformation (7.11) the motion equations reduce to the wave equation

$$\rho \ddot{u} = \mu \Delta u, \tag{7.12}$$

where $\Delta u = u_{,11} + u_{,22}$.
Assuming a steady state and looking for solution of (7.12) in the form

$$u = U(x_1) e^{i(kx_2 - \omega t)}, \tag{7.13}$$

we obtain the ordinary differential equation with respect to U

$$\mu[U'' - k^2 U] + \rho \omega^2 U = 0, \tag{7.14}$$

where k is the wavenumber, ω is the circular velocity, $i = \sqrt{-1}$, and the prime stands for differentiation with respect to x_1. Assuming that $U(x_1)$ decays exponentially with distance from the half-space surface, we find the solution of (7.14) in form

$$U = U_0 \exp \sqrt{k^2 - \omega^2 / c_{\mathrm{T}}^2} x_1,$$

where U_0 is an amplitude and $c_{\mathrm{T}} = \sqrt{\mu / \rho}$ is the phase velocity of transverse waves (Achenbach, 1973). Thus, the general stationary solution of (7.12) takes the form

$$u = U_0 \exp \left[\sqrt{k^2 - \omega^2 / c_{\mathrm{T}}^2} x_1 + i(kx_2 - \omega t) \right]. \tag{7.15}$$

For anti-plane deformations the boundary condition (7.9) reduces to scalar equation

$$-m\ddot{u} + \mu_{\mathrm{s}} u_{,22} - \sum_{i=2}^{N} (-1)^i \mu_i \frac{\partial^{2i} u}{\partial x_2^{2i}} = \mu u_{,1}. \tag{7.16}$$

Substituting (7.15) into (7.16) we obtain the dispersion equation

$$m\omega^2 - \mu_{\mathrm{s}} k^2 - \sum_{i=2}^{N} \mu_i k^{2i} = \mu \sqrt{k^2 - \frac{\omega^2}{c_{\mathrm{T}}^2}}. \tag{7.17}$$

Introducing the phase velocity $c = \omega / k$ we transform (7.17) into

$$c^2 = c_{\mathrm{s}}^2 + \frac{1}{m} \sum_{i=2}^{N} \mu_i k^{2i-2} + \frac{\mu}{m} \frac{1}{|k|} \sqrt{1 - \frac{c^2}{c_{\mathrm{T}}^2}}, \tag{7.18}$$

where $c_{\mathrm{s}} = \sqrt{\mu_{\mathrm{s}} / m}$ is the shear wave velocity in the thin film associated with the Gurtin–Murdoch model (Eremeyev et al, 2016). Clearly, the solution of (7.18) exists if $c \leq c_{\mathrm{T}}$.

Let us consider few cases. For $\mu_i = 0$ Eq. (7.18) transforms to

$$c^2 = c_s^2 + \frac{\mu}{m}\frac{1}{|k|}\sqrt{1 - \frac{c^2}{c_T^2}}. \qquad (7.19)$$

Detailed analysis of (7.19) for the Gurtin–Murdoch model is performed in Eremeyev et al (2016). The solution of (7.19) exists if and only if $c_S < c \le c_T$. It was shown that $c \approx c_s$ and $\omega \approx c_s k$ for $k \to \infty$, that is for short waves, and dispersion curve starts from the point $k = 0$, $c = c_T$. So, for long waves that is for $k \approx 0$ the propagation of waves is determined almost by the bulk properties of material. The dependence c vs. k is given in Fig. 7.2, see the dashed curve. Here the relation $c_S = c_T/4$ is assumed. Comparison of (7.18) with Eq. (20) in Eremeyev et al (2016) shows that the difference consist of polynomial

$$\frac{1}{m}\sum_{i=2}^{N}\mu_i k^{2i-2} = \mu_2 k^2 + \mu_3 k^4 + \ldots + \mu_N k^{2N-2}$$

with positive coefficients. Again, when $k \to 0$ $c \to c_T$. But now we can conclude that for $k \to \infty$ now the asymptotic behaviour of c is not constant, it polynomially depends on the highes term. For $N = 2$ Eq. (7.18) transforms to

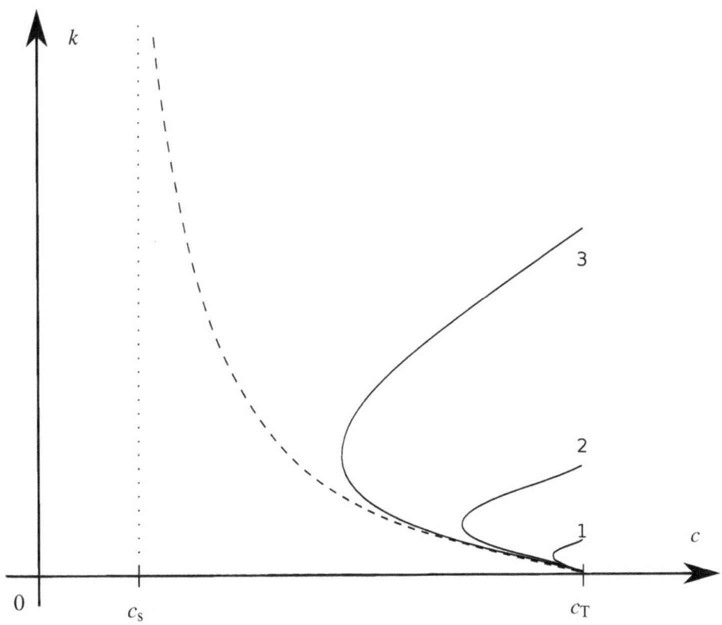

Fig. 7.2: Dispersion curves. The dashed curve corresponds to the Gurtin-Murdoch model, curves 1, 2, 3 are for $\bar{\mu}_2 = 0.01, 0.001, 0.0001$, respectively.

$$c^2 = c_s^2 + \frac{\mu_2}{m}k^2 + \frac{\mu}{m}\frac{1}{|k|}\sqrt{1 - \frac{c^2}{c_T^2}}. \tag{7.20}$$

Now the dispersion curves change dramatically. In Fig. 7.2 three curves corresponding to different values of μ_2 is given. Here $\bar{\mu}_2$ is the normalized (dimensionless) elastic modulus. Even for small μ_2 that is when $\mu_2 \to 0$ there is no limit to the previous model. So one can conclude that even for small nonlocality we have significant difference in the dispersion curve behaviour. In particular, the the anti-plane waves exist in different range of phase velocities and wave numbers.

Thus, increasing the level of non-locality that is increasing N we obtain different behavior of solutions in the case of short waves, in general. In other words, there is no limit at $N \to \infty$ and analysis of strong nonlocal models may be required.

7.4 Conclusions

We present here the analysis of anti-plane surface waves propagation in the isotropic linear elastic medium with surface energy which depends on surface gradients of displacements up to Nth-order. The dispersion relation is derived and analyzed. We shown that for long waves the phase velocity is almost determined by the bulk properties of material whereas for short waves the phase velocity is determined by highest order member in the surface strain energy density. Thus, considering further gradient terms in the surface energy density, that is increasing N, we can change the dynamical properties of material for short waves.

Acknowledgements Author acknowledges the support by the Russian Science Foundation (grant number 15-19-10008).

References

Achenbach J (1973) Wave Propagation in Elastic Solids. North Holland, Amsterdam

Aifantis E (2016) Chapter one – Internal length gradient (ilg) material mechanics across scales and disciplines. Advances in Applied Mechanics 49:1–110

Aifantis EC (1999) Gradient deformation models at nano, micro, and macro scales. Journal of Engineering Materials and Technology 121(2):189–202

Askes H, Aifantis EC (2011) Gradient elasticity in statics and dynamics: An overview of formulations, length scale identification procedures, finite element implementations and new results. Int J Solids Struct 48(13):1962–1990

Carcaterra A, dell'Isola F, Esposito R, Pulvirenti M (2015) Macroscopic description of microscopically strongly inhomogenous systems: A mathematical basis for the synthesis of higher gradients metamaterials. Archive for Rational Mechanics and Analysis 218(3):1239–1262

dell'Isola F, Seppecher P (1997) Edge contact forces and quasi-balanced power. Meccanica 32(1):33–52 dell'Isola F, Madeo A, Seppecher P (2009a) Boundary conditions at fluid-permeable interfaces in porous media: A variational approach. Int J Solids Struct 46(17):3150–3164

dell'Isola F, Sciarra G, Vidoli S (2009b) Generalized Hooke's law for isotropic second gradient materials. Proc Roy Soc London A 465(2107):2177–2196

dell'Isola F, Seppecher P, Madeo A (2012) How contact interactions may depend on the shape of Cauchy cuts in *n*th gradient continua: approach "á la d'Alembert". ZAMP 63(6):1119–1141

Duan HL, Wang J, Karihaloo BL (2008) Theory of elasticity at the nanoscale. In: Adv. Appl. Mech., vol 42, Elsevier, pp 1–68

Eremeyev VA (2016) On effective properties of materials at the nano-and microscales considering surface effects. Acta Mech 227(1):29–42

Eremeyev VA, Rosi G, Naili S (2016) Surface/interfacial anti-plane waves in solids with surface energy. Mechanics Research Communications 74:8–13

Fan H, Yang J, Xu L (2006) Piezoelectric waves near an imperfectly bonded interface between two half-spaces. Appl Phys Lett 88(20):203509

Forest S, Cordero N, Busso EP (2011) First vs. second gradient of strain theory for capillarity effects in an elastic fluid at small length scales. Computational Materials Science 50(4):1299–1304

de Gennes PG (1981) Some effects of long range forces on interfacial phenomena. J Physique Lettres 42(16):377–379

de Gennes PG, Brochard-Wyart F, Quéré D (2004) Capillarity and Wetting Phenomena: Drops, Bubbles, Pearls, Waves. Springer, New York

Georgiadis H, Vardoulakis I, Lykotrafitis G (2000) Torsional surface waves in a gradient-elastic half-space. Wave Motion 31(4):333–348

Gerasimov RA, Eremeyev VA, Petrova TO, Egorov VI, Maksimova OG, Maksimov AV (2016) Computer simulation of the mechanical properties of metamaterials. Journal of Physics: Conference Series 738(1):012,100

Gerasimov RA, Maksimova OG, Petrova TO, Eremeyev VA, Egorov VI, Maksimov AV (2017) Analytical and computer methods to evaluate mechanical properties of the metamaterials based on various models of polymeric chains. In: Sumbatyan M (ed) Wave Dynamics and Composite Mechanics for Microstructured Materials and Metamaterials, Advanced Structured Materials, vol 59, p 39

Gorbushin N, Mishuris G (2016a) Analysis of dynamic failure of the discrete chain structure with non-local interactions. Mathematical Methods in the Applied Sciences

Gorbushin N, Mishuris G (2016b) Dynamic crack propagation along the interface with non-local interactions. Journal of the European Ceramic Society 36(9):2241–2244

Gourgiotis P, Georgiadis H (2015) Torsional and {SH} surface waves in an isotropic and homogenous elastic half-space characterized by the Toupin–Mindlin gradient theory. International Journal of Solids and Structures 62(0):217 – 228

Gurtin ME, Murdoch AI (1975) A continuum theory of elastic material surfaces. Arch Ration Mech An 57(4):291–323

Gurtin ME, Murdoch AI (1978) Surface stress in solids. Int J Sol Struct 14(6):431–440

Javili A, McBride A, Steinmann P (2012) Thermomechanics of solids with lower-dimensional energetics: On the importance of surface, interface, and curve structures at the nanoscale. A unifying review. Appl Mech Rev 65:010,802–1–31

Javili A, dell'Isola F, Steinmann P (2013) Geometrically nonlinear higher-gradient elasticity with energetic boundaries. J Mech Phys Solids 61(12):2381–2401

Lebedev LP, Cloud MJ, Eremeyev VA (2010) Tensor Analysis with Applications in Mechanics. World Scientific, New Jersey

Liebold C, Müller WH (2015) Are microcontinuum field theories of elasticity amenable to experiments? A review of some recent results. In: Differential Geometry and Continuum Mechanics, Springer, pp 255–278

Lurie S, Volkov-Bogorodsky D, Zubov V, Tuchkova N (2009) Advanced theoretical and numerical multiscale modeling of cohesion/adhesion interactions in continuum mechanics and its applications for filled nanocomposites. Comp Mater Sci 45(3):709–714

Mindlin RD (1965) Second gradient of strain and surface-tension in linear elasticity. Int J Solids Struct 1(4):417–438

Nieves M, Mishuris G, Slepyan L (2016) Analysis of dynamic damage propagation in discrete beam structures. Int Journal Solids Struct

Placidi L, Rosi G, Giorgio I, Madeo A (2014) Reflection and transmission of plane waves at surfaces carrying material properties and embedded in second-gradient materials. Math Mech Solids 19(5):555–578

Porubov A (2014) Modeling of strongly nonlinear effects in diatomic lattices. Archive of Applied Mechanics 84(9-11):1533–1538

Rahali Y, Giorgio I, Ganghoffer J, dell'Isola F (2015) Homogenization à la Piola produces second gradient continuum models for linear pantographic lattices. International Journal of Engineering Science 97:148–172

Rowlinson JS, Widom B (2003) Molecular Theory of Capillarity. Dover, New York

Steigmann DJ, Ogden RW (1997) Plane deformations of elastic solids with intrinsic boundary elasticity. Proceedings of the Royal Society A 453(1959):853–877

Steigmann DJ, Ogden RW (1999) Elastic surface-substrate interactions. Proceedings of the Royal Society A 455(1982):437–474

Vardoulakis I, Georgiadis HG (1997) SH surface waves in a homogeneous gradient-elastic half-space with surface energy. Journal of Elasticity 47(2):147–165

Wang J, Huang Z, Duan H, Yu S, Feng X, Wang G, Zhang W, Wang T (2011) Surface stress effect in mechanics of nanostructured materials. Acta Mech Solida Sin 24:52–82

Xu L, Wang X, Fan H (2015) Anti-plane waves near an interface between two piezoelectric half-spaces. Mechanics Research Communications 67:8–12

Yerofeyev VI, Sheshenina OA (2005) Waves in a gradient-elastic medium with surface energy. Journal of Applied Mathematics and Mechanics 69(1):57 – 69

Chapter 8
Deformation of Spherical Inclusion in an Elastic Body with Account for Influence of Interface Considered as Infinitesimal Layer with Abnormal Properties

Robert V. Goldstein, Valentin A. Gorodtsov & Konstantin B. Ustinov

Abstract The model for surface (interface) elasticity accounting for the influence of bulk and interface eigenstrains as well as the influence of not only in-plane but also out-of-plane stresses on the surface deformation, was proposed by the authors. Definition of all interface values as integrals of the excesses of the corresponding bulk values over the normal to the interface, and procedure of energy variation resulted in constitutive equations for the interface of more general type then the Shuttleworth equations (Shuttleworth, 1950). Here the model is added with the boundary conditions at the interface. The model is used to describe deformation of a spherical inclusion in elastic media.

Key words: Deep reservoirs · Rock · Well · Test facility · Deformations · Stresses · Strength · Elastic moduli

8.1 Introduction

With reducing space dimensions under consideration down to nanometers, due to the changes in the character of molecular interaction and increasing role of defects in the vicinity of materials the classical theory of elasticity has been needed some generalization for adequate description of the observed mechanical behavior. The generalization leads to appearance of a scale effect absent in the traditional elasticity. To describe such a mechanical behavior two approaches are known: to describe the abnormal behavior in the frame of continuum mechanics (Shuttleworth, 1950; Gurtin and Ian Murdoch, 1975; Murdoch, 2005; Podstrigach and Povstenko, 1985; Ibach, 1997; Hashin, 1991), which results in various variants of surface elasticity;

Robert V. Goldstein · Valentin A. Gorodtsov · Konstantin B. Ustinov
Ishlinsky Institute for Problems in Mechanics RAS, Vernadskiy Ave, 101 bldg 1, 119526 Moscow, Russia,
e-mail: goldst@ipmnet.ru, gorod@ipmnet.ru, ustinov@ipmnet.ru

H. Altenbach et al. (eds.), *Mechanics for Materials and Technologies*,
Advanced Structured Materials 46, DOI 10.1007/978-3-319-56050-2_8

and to use discrete molecular modeling (Altenbach et al, 2010; Ivanova et al, 2002; Goldstein and Chentsov, 2005; Odegard, 2006; Golovnev et al, 2015; Korobeynikov et al, 2015; Berinskii and Krivtsov, 2010; Zubko, 2016). Both approaches have been used successfully for solving particular problems, related to mechanics of nanoscale objects (Eremeyev et al, 2009; Girchenko et al, 2011; Cammarata, 1994; Duan et al, 2005, 2009; Goldstein et al, 2010; Bochkarev and Grekov, 2015; Grekov and Kostyrko, 2016). In the works of authors (Ustinov et al, 2013; Goldstein et al, 2014) the generalization of the traditional model of surface elasticity (Shuttleworth, 1950; Gurtin and Ian Murdoch, 1975; Murdoch, 2005; Podstrigach and Povstenko, 1985; Ibach, 1997) has been suggested, for which the governing equations for the interfaces include not only the stresses and strains acting in the plane of the interface, but also the stresses and strains having components coinciding with the normal of this plane. In the present work this model is generalized for curved surfaces, and is applied for the problem of spherically symmetrical deformation of the body with a spherical inclusion.

8.2 Model of the Interface Elasticity

According to Ustinov et al (2013); Goldstein et al (2014) consider a layer of material of thickness h, between the layers of other materials, so that the total thickness of the package is $H > h$ (Fig. 8.1a). All materials are supposed linear elastic and possessing transverse anisotropy with the isotropy plane being parallel to the layer boundaries. In Cartesian coordinates with x_3 directed normally to the layers, the elastic properties are described by five constants: $C^i_{11}, C^i_{12}, C^i_{13}, C^i_{33}, C^i_{44}, C^i_{66} = (C^i_{11} - C^i_{22})/2$ ($i = A, B, C$ for the lower, upper and intermediate layers, respectively).

Consider another configuration where the intermediate layer is replaced with an interface of zero thickness, and the upper and lower layers expanded so that the total package thickness conserve (Fig. 8.1b). The properties of the introduced interface are determined so that the elastic energy of both systems be equal for arbitrary uniform boundary conditions. This is achieved by as-suming the relation between volume and surface stresses $(\sigma_{ij}, \sigma^s_{ij})$ and volume and surface strains $(\varepsilon_{ij}, \varepsilon^s_{ij})$ as follows (Shuttleworth, 1950; Gurtin and Ian Murdoch, 1975)

$$\begin{aligned}
\sigma^s_{11} &= A_{1111}\varepsilon_{11} + A_{1122}\varepsilon_{22} + A_{1133}\varepsilon^s_{33}, \\
\sigma^s_{22} &= A_{1122}\varepsilon_{11} + A_{1111}\varepsilon_{22} + A_{1133}\varepsilon^s_{33}, \\
\sigma^s_{12} &= 2A_{1212}\varepsilon_{12}, \\
\sigma^s_{33} &= A_{1133}\varepsilon_{11} + A_{1133}\varepsilon_{22} + A_{3333}\varepsilon^s_{33}, \\
\sigma^s_{13} &= 2A_{1313}\varepsilon^s_{12}, \\
\sigma^s_{23} &= 2A_{1313}\varepsilon^s_{23}
\end{aligned} \tag{8.1}$$

Here the surface strains are understood as the difference between displacements of the upper and lower boundary of the interface, similarly to Winkler or spring model,

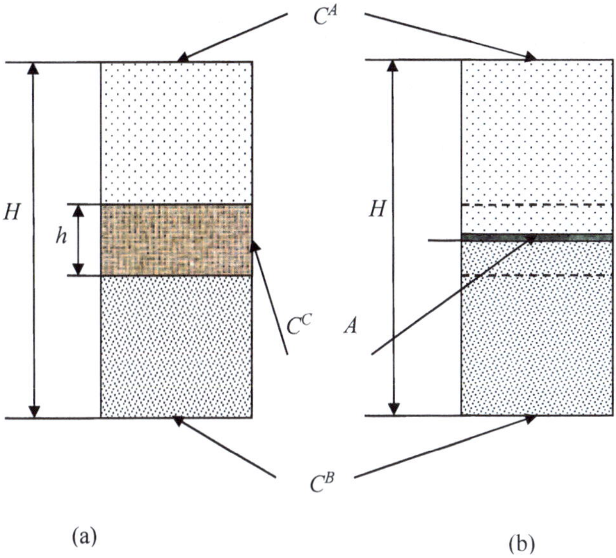

Fig. 8.1: Model of composed layer: (a) initial configuration; (b) equivalent configuration

$$\varepsilon_{33}^{s} = [u_3], 2\varepsilon_{13}^{s} = [u_1], 2\varepsilon_{23}^{s} = [u_2] \tag{8.2}$$

In general case the volume strain and stress may possess discontinuities across the surface, therefore in (8.1) the volumetric values should be under-stood as the average of the corresponding values for the upper and lower boundary

$$\begin{aligned} \varepsilon_{11} &= \frac{1}{2}(\varepsilon_{11}^{A} + \varepsilon_{11}^{B}), \;\; \varepsilon_{22} = \frac{1}{2}(\varepsilon_{22}^{A} + \varepsilon_{22}^{B}), \;\; \varepsilon_{12} = \frac{1}{2}(\varepsilon_{12}^{A} + \varepsilon_{12}^{B}) \\ \sigma_{33} &= \frac{1}{2}(\sigma_{33}^{A} + \sigma_{33}^{B}), \; \sigma_{13} = \frac{1}{2}(\sigma_{13}^{A} + \sigma_{13}^{B}), \; \sigma_{23} = \frac{1}{2}(\sigma_{23}^{A} + \sigma_{23}^{B}) \end{aligned} \tag{8.3}$$

The above formulae together with Young-Laplace equation

$$\boldsymbol{\sigma}^{A} - \boldsymbol{\sigma}^{B} = \boldsymbol{\nabla}_{s}\boldsymbol{\sigma}^{s} \tag{8.4}$$

may be considered as the boundary conditions for the internal interface. Here $\boldsymbol{\nabla}_s$ is 2-D gradient in the coordinate frame related to the surface.

The elastic coefficients are

$$
\begin{aligned}
A_{12}=A_{21} &= \frac{h}{4C_{33}^A C_{33}^B - 2C_{33}^C(C_{33}^A + C_{33}^B)}[C_{13}^{B\,2}(2C_{33}^A - C_{33}^C) + C_{13}^{A\,2}(2C_{33}^B - C_{33}^C) \\
&\quad + 2C_{13}^{C\,2}(2C_{33}^A + C_{33}^B) - 4C_{13}^B C_{13}^C C_{33}^A - 4C_{13}^A C_{13}^C C_{33}^B + 2C_{13}^A C_{13}^B C_{33}^C \\
&\quad - (2C_{12}^C - C_{12}^A - C_{12}^B)(C_{33}^C(C_{33}^A + C_{33}^B) - 2C_{33}^A C_{33}^B)], \\
A_{13}=A_{23} &= \frac{(C_{13}^B C_{33}^A + C_{13}^A C_{33}^B)C_{33}^C - 2C_{13}^C C_{33}^A (C_{33}^B}{C_{33}^C(C_{33}^A + C_{33}^B) - 2C_{33}^A C_{33}^B} = A_{31} = A_{32}, \\
A_{33} &= \frac{1}{h}\left(\frac{1}{C_{33}^C} - \frac{1}{2C_{33}^A} - \frac{1}{2C_{33}^B}\right)^{-1}, \\
A_{44}=A_{55} &= \frac{1}{h}\left(\frac{1}{C_{44}^C} - \frac{1}{2C_{44}^A} - \frac{1}{2C_{44}^B}\right)^{-1}, \\
A_{66} &= h\left(C_{66}^C - \frac{C_{66}^A + C_{66}^B}{2}\right)
\end{aligned} \tag{8.5}
$$

It is worth to emphasize that relations (8.1), (8.1) due to presence of cross-terms are not reduced to a combination of the traditional surface elasticity (Shuttleworth, 1950; Gurtin and Ian Murdoch, 1975; Murdoch, 2005; Podstrigach and Povstenko, 1985; Ibach, 1997) and spring theory, considered, e.g. in (Hashin, 1991). Table 8.1 illustrates the difference between the various types of interface descriptions.

Table 8.1: Variants interface conditions

Full contact	Winkler layer	Surface elasticity	Generalized surface elasticity
Kinematics			
$[u_i]=0$	$[u_i]=\varepsilon_{ni}^s$	$[u_i]=0$	$[u_i]=\varepsilon_{ni}^s$
Statics			
$[\sigma_{ni}^s]=0$	$[\sigma_{ni}^s]=0$	$[\sigma_{ni}^s]=\nabla_s\sigma_{\alpha\beta}^s$ Young-Laplace equation	
Additional variables			
	ε_{ni}^s - 3 displacement discontinuities	$\sigma_{\alpha\beta}^s$ - 3 components of surface stress	$\varepsilon_{ni}^s, \sigma_{\alpha\beta}^s$ - 6 units (3 displacement discontinuities and 3 components) of surface stress)
Additional equation for them			
	$\mu_{(i)}^s\varepsilon_{ni}^s=\sigma_{ni}$ Winkler equations	$\sigma_{\alpha\beta}^s=\Lambda_{\alpha\beta\gamma\delta}\varepsilon_{\gamma\delta}$ Shattleworth equations	suggested equations (8.5)

8.3 Problem of Spherical Inclusion. Various Solutions

The above theory allows generalization for curvilinear interfaces if the curvature radius remains much bigger than the thickness of the original inter-mediate layer. Consider an infinite media with a spherical inclusion of radius R with the above described contact conditions. Let us restrict ourselves for simplicity with the case of isotropic media and inclusion, and transverse isotropic interface with the isotropy axis at each point normal to the spherical surface. According to the general solution of the of the elasticity problem in case of spherical symmetry (Landau and Lifshitz, 1979) the displacements, strains and stresses inside and outside of the inclusion (reduced to convenient for our purposes notation) are

$$\begin{aligned} &u^M = C_1^M r + \frac{C_1^M}{r^2}, \quad \varepsilon_{rr}^M = C_1^M - \frac{2C_2^M}{r^3}, \quad \varepsilon_{\theta\theta}^M = \varepsilon_{\varphi\varphi}^M = C_1^M + \frac{2C_2^M}{r^3}, \\ &\sigma_{rr}^M = 3K^M C_1^M - \frac{4\mu^M C_2^M}{r^3}, \quad \sigma_{\theta\theta}^M = 3K^M C_1^M + \frac{2\mu^M C_2^M}{r^3}, \\ &M = A, B, \qquad C_2^A = 0 \end{aligned} \tag{8.6}$$

Here $M = A, B$ for the internal and external zones, respectively; $C_1^B = \sigma^\infty / 3K^B$ determines the field at infinity in case of hydrostatic stresses σ^∞, K, μ are volumetric and shear moduli. Constants C_1^A, C_2^B are determined with the help of (8.1), (8.3), (8.4), which for considered case are reduced at $r = R$ to

$$\begin{aligned} &\frac{R}{2}(\sigma_{rr}^B - \sigma_{rr}^A) - (A_{1111} + A_{1122})\frac{\varepsilon_{\theta\theta}^A + \varepsilon_{\theta\theta}^B}{2} - A_{1133}(u^B - u^A) = 0, \\ &\frac{\sigma_{rr}^B + \sigma_{rr}^A}{2} = A_{3333}(u^B - u^A) + A_{3311}(\varepsilon_{\theta\theta}^A + \varepsilon_{\theta\theta}^B) \end{aligned} \tag{8.7}$$

Consider the case where the constants of surface elasticity A_{ijkl} correspond to a layer of final thickness h and are determined by (8.5). Calculating any component of stress and displacement is straightforward. Thus the coefficient of strain concentration (ratio of strains within the inclusion and at infinity) is

$$\frac{\varepsilon^A}{\varepsilon^B} = \frac{3(\lambda^B + 2\mu^B)}{3K^A + 4\mu^B} - \frac{36(\lambda^B + 2\mu^B)(K^A - K^C)(\mu^B - \mu^C)}{(3K^A + 4\mu^B)^2(\lambda^C + 2\mu^C)}\frac{h}{R} + 0\left(\frac{h}{R}\right)^2 \tag{8.8}$$

Here $K^M, \lambda^M, \mu^M, M = A, B, C$ are volumetric moduli and Lamé's constants of the materials.

Compare the obtained solution with the solution of the problem of spherical inclusion surrounded by an intermediate layer of final thickness h. To do this let us again use formulae (8.6), setting $M = A, B, C, C_2^A = 0$, where index C corresponds to the intermediate layer. The boundary conditions in case of full contact for at $r = R$ are

$$\begin{aligned} &\sigma_{rr}^A - \sigma_{rr}^C = 0, \quad u^A - u^C = 0 \quad \text{for} \quad r = R - h/2, \\ &\sigma_{rr}^B - \sigma_{rr}^C = 0, \quad u^B - u^C = 0 \quad \text{for} \quad r = R + h/2 \end{aligned} \tag{8.9}$$

Solution of system (8.9) with account for the expressions for stresses and displacements (8.6) yields values of constants $C_1^A, C_2^B, C_1^C, C_2^C$, substitution of which into (8.6) yields the solution of the imposed problem. Coefficient of strain concentration, calculated in such a way, coincides with (8.8) for the leading term of expansion over h/R.

Consider the problem in the frame of the traditional surface elasticity. The boundary condition at the interface at $r = R$ in our notation are (Cammarata, 1994; Duan et al, 2005, 2009)

$$\frac{R}{2}(\sigma_{rr}^{B} - \sigma_{rr}^{A}) - (A_{1111} + A_{1122})\varepsilon_{\theta\theta}^{A} = 0, \quad (u^{B} - u^{A}) = 0 \Rightarrow \varepsilon_{\theta\theta}^{A} = \varepsilon_{\theta\theta}^{B} \tag{8.10}$$

Substitution (8.10) into (8.6) gives the solution coinciding with the result obtained in Cammarata (1994). Coefficient of strain concentration calculated this way is

$$\frac{\varepsilon^{A}}{\varepsilon^{\infty}} = \frac{3(\lambda^{B} + 2\mu^{B})}{3K^{A} + 4\mu^{B}} - \frac{12(\lambda^{B} + 2\mu^{B})(\lambda^{C} - \mu^{C})}{(3K^{A} + 4\mu^{B})^{2}}\frac{h}{R} + 0\left(\frac{h}{R}\right)^{2} \tag{8.11}$$

in more general case does not coincides with (8.8). The results yielded by (8.8) and (8.11) become close for relatively rigid intermediate layer, i.e. for λ^{C}, μ^{C} or λ^{A}, μ^{A} or λ^{B}, μ^{B}, however no exact coincidence is observed.

8.4 Conclusion

The boundary conditions at the interface separating two bodies with different elastic properties were formulated in the frame of the suggested by authors generalized model of surface (interface) elasticity. The problems of spherically symmetrical deformation of an infinite body with spherical inclusion under action of the externally applied stress field and eigenstrains both in the inclusion and interface are considered. The obtained solution coincides with the leading term (over the ratio of the interface thickness and the inclusion radius) of the solution obtained for the case of thin interface with the final thickness. No such a coincidence is observed for the solution obtained in the frame of the traditional surface elasticity.

Acknowledgements The work is executed at financial support of the RAS Presidium Program for Fundamental Research of Presidium RAS 1.1.

References

Altenbach H, Eremeev VA, Morozov NF (2010) On equations of the linear theory of shells with surface stresses taken into account. Mechanics of Solids 45(3):331–342

Berinskii IE, Krivtsov AM (2010) On using many-particle interatomic potentials to compute elastic properties of graphene and diamond. Mechanics of Solids 45(6):815–834

Bochkarev A, Grekov M (2015) On symmetrical and antisymmetrical buckling of a plate with circular nanohole under uniaxial tension. Appl Math Sci 9(125):125 – 128

Cammarata RC (1994) Surface and interface stress effects in thin films. Progress in Surface Science 46(1):1 – 38

Duan HL, Wang J, Huang ZP, Karihaloo BL (2005) Eshelby formalism for nano-inhomogeneities. Proceedings of the Royal Society of London A: Mathematical, Physical and Engineering Sciences 461(2062):3335–3353

Duan HL, Wang J, Karihaloo BL (2009) Theory of elasticity at the nanoscale. In: Aref H, van der Giessen E (eds) Advances in Applied Mechanics, vol 42, Elsevier, pp 1 – 68

Eremeyev VA, Altenbach H, Morozov NF (2009) The influence of surface tension on the effective stiffness of nanosize plates. Doklady Physics 54(2):98–100

Girchenko AA, Eremeyev VA, Morozov NF (2011) Modeling of spiral nanofilms with piezoelectric properties. Physical Mesomechanics 14(1):10 – 15

Goldstein RV, Chentsov AV (2005) Discrete-continuous model of a nanotube. Mechanics of Solids 40(4):45–59

Goldstein RV, Gorodtsov VA, Ustinov KB (2010) Effect of residual surface stress and surface elasticity on deformation of nanometer spherical inclusions in an elastic matrix. Physical Mesomechanics 13(5):318 – 328

Goldstein RV, Gorodtsov VA, Ustinov KB (2014) On surface elasticity theory for plane interfaces. Physical Mesomechanics 17(1):30–38

Golovnev IF, Golovneva EI, Fomin VM (2015) Molecular dynamics study into the role of the surface in fracture of nanostructures. Physical Mesomechanics 18(2):127–133

Grekov MA, Kostyrko SA (2016) Surface effects in an elastic solid with nanosized surface asperities. International Journal of Solids and Structures 96:153 – 161

Gurtin ME, Ian Murdoch A (1975) A continuum theory of elastic material surfaces. Archive for Rational Mechanics and Analysis 57(4):291–323

Hashin Z (1991) Thermoelastic properties of particulate composites with imperfect interface. Journal of the Mechanics and Physics of Solids 39(6):745 – 762

Ibach H (1997) The role of surface stress in reconstruction, epitaxial growth and stabilization of mesoscopic structures. Surface Science Reports 29(5):195 – 263

Ivanova EA, Krivtsov AM, Morozov NF (2002) Peculiarities of the bending-stiffness calculation for nanocrystals. Doklady Physics 47(8):620–622

Korobeynikov SN, Alyokhin VV, Annin BD, Babichev AV (2015) Quasi-static buckling simulation of single-layer graphene sheets by the molecular mechanics method. Mathematics and Mechanics of Solids 20(7):836–870

Landau LD, Lifshitz EM (1979) A Course of Theoretical Physics, vol 7 (Theory of Elasticity. Pergamon Press

Murdoch AI (2005) Some fundamental aspects of surface modelling. Journal of Elasticity 80(1):33–52

Odegard GM (2006) Equivalent-continuum modeling of nanostructured materials. In: Rieth M, Schommers W (eds) Handbook of Theoretical and Computational Nanotechnology, American Scientific Publishers

Podstrigach Y, Povstenko Y (1985) Introduction to mechanics of surface effects in solids (in Russ.). Naukova Dumka, Kiev

Shuttleworth R (1950) The surface tension of solids. Proc Phys Soc A63:444–457

Ustinov KB, Goldstein RV, Gorodtsov VA (2013) On the modeling of surface and interface elastic effects in case of eigenstrains. In: Altenbach H, Morozov NF (eds) Surface Effects in Solid Mechanics: Models, Simulations and Applications, Springer, Berlin, Heidelberg, pp 167–180

Zubko IY (2016) Computation of elastic moduli of graphene monolayer in nonsymmetric formulation using energy-based approach. Physical Mesomechanics 19(1):93–106

Chapter 9
Analysis of Internal Stresses in a Viscoelastic Layer in Sliding Contact

Irina G. Goryacheva, Feodor I. Stepanov & Elena V. Torskaya

Abstract Sliding contact of a smooth indenter with a linear viscoelastic layer adhered to a rigid half-space is under consideration. The quasistatic problem is investigated by constructing a solution for the case of a moving load, distributed in a rectangular element, which allows us to use the boundary element method and iterative procedure. The effect of sliding velocity, the layer thickness and viscoelastic properties on distribution of contact and internal stresses is analyzed based on the numerical solution of the problem.

Key words: Contact problem · Sliding · Viscoelasticity · Layer · Internal stresses

9.1 Introduction

Studying the contact problem for viscoelastic coating, which is much more compliant than the substrate material, it is possible to model it by viscoelastic layer adhered to a rigid half-space. 2-D contact problems for different types of indenters sliding over the viscoelastic layer are considered in a number of papers, for example Aleksandrov and Mark (2009); Mark (2008); Morozov and Makhovskaya (2007). Contact pressure and sliding resistance, associated with energy dissipation due to the rheological properties of the material, are analyzed based on the solutions of such problems.

For 3-D contact problem one-dimensional model of viscoelastic layer is widely used (Lyubicheva, 2008; Soldatenkov, 2015) to obtain analytical solution for a single slider or a system of sliders and to take into account mutual effect (Aleksandrov et al, 2010; Soldatenkov, 2015) and surface adhesion (Lyubicheva, 2008;

Irina G. Goryacheva · Feodor I. Stepanov · Elena V. Torskaya
Ishlinsky Institute for Problems in Mechanics RAS, Vernadskiy Ave, 101 bldg 1, 119526 Moscow, Russia,
e-mail: goryache@ipmnet.ru, stepanov_ipm@gmail.ru, torskaya@ipmnet.ru

H. Altenbach et al. (eds.), *Mechanics for Materials and Technologies*,
Advanced Structured Materials 46, DOI 10.1007/978-3-319-56050-2_9

Goryacheva et al, 2014). 3-D quasi-static contact problems are also developed for a slider and viscoelastic half-space (Goryacheva et al, 2015, 2016; Koumi et al, 2015; Stepanov and Torskaya, 2016; Stepanov, 2015; Kusche, 2016; Stepanov and Torskaya, 2017). Contact problem solutions are obtained for linear viscoelastic material models with one (Goryacheva et al, 2015, 2016; Koumi et al, 2015; Stepanov and Torskaya, 2016; Stepanov, 2015; Kusche, 2016) or three (Aleksandrov et al, 2010) retardation times for homogeneous materials (Aleksandrov et al, 2010; Goryacheva et al, 2015, 2016; Stepanov and Torskaya, 2016; Stepanov, 2015) and materials, containing inclusions Koumi et al (2015). Tangential stresses in contact area are taken into account in (Goryacheva et al, 2015, 2016; Stepanov and Torskaya, 2016; Stepanov, 2015; Kusche, 2016). Mutual effect is considered for the system of two sliders in Stepanov (2015). The contact problem solutions are used to calculate the contact stresses, contact area and to evaluate hysteretic losses during sliding. Internal stresses in viscoelastic material are analyzed in Stepanov and Torskaya (2016) for a homogeneous half-space loaded by normal and tangential contact stresses, and in Koumi et al (2015) for a half-space with inclusions loaded only by the normal contact stresses.

The method of contact problem solution for a slider moving with constant velocity over viscoelastic layer, which is bonded with a rigid half-space, is proposed in Stepanov and Torskaya (2017). In the current paper the method of calculation of internal stresses within the layer is developed and used to analyze the influence of the layer thickness, its rheological properties and sliding velocity on stress distribution under the slider.

9.2 Problem Formulation

Sliding contact of a rigid smooth indenter and viscoelastic layer of thickness h is considered under the assumption that the layer is in perfect adhesion with rigid half-space. The indenter, which is loaded by normal force Q, moves with constant velocity V in direction Ox (Fig. 9.1). The coordinate system is related to the indenter at the point of initial contact of the indenter and the layer. Note that in the moving system of coordinates all stresses and displacements do not depend on time. Boundary conditions at the surface $z = 0$ are the following:

$$\begin{aligned}
&w(x,y,0) = f(x,y) + D, \ \ (x,y) \in \Omega; \\
&\sigma_z(x,y,0) = 0, \ \ (x,y) \notin \Omega; \\
&\tau_{xz}(x,y,0) = \tau_{yx}(x,y,0) = 0, \ -\infty < x < +\infty, \ -\infty < y < +\infty
\end{aligned} \tag{9.1}$$

Here Ω is the contact region, $w(x,y,0)$ are surface normal displacements, D is penetration of the indenter, σ_z, τ_{xz}, τ_{yz} are normal and tangential stresses. The shape of the indenter is determined by the smooth function $f(x,y)$. Contact pressure $p(x,y) = -\sigma_z(x,y,0)$ and contact region Ω are unknown.

Equilibrium condition is

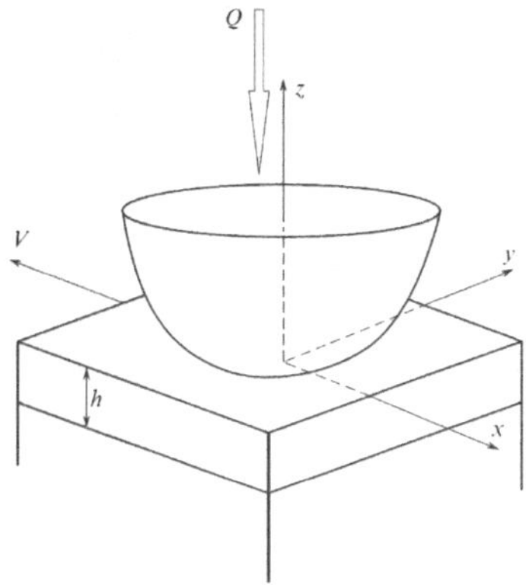

Fig. 9.1 Scheme of contact

$$Q = \iint_{\Omega} p(x,y)dxdy \tag{9.2}$$

and the condition of zero pressures at the boundary of the contact zone Ω are also satisfied. The boundary conditions at the layer-substrate interface ($z = h$) correspond to perfect adhesion:

$$w(x,y,h) = 0, \quad u_x(x,y,h) = 0, \quad u_y(x,y,h) = 0 \tag{9.3}$$

Here $u_x(x,y,z)$ and $u_y(x,y,z)$ are shear displacements of the layer material.
The linear viscoelastic material model is described by an integral operator (Aleksandrov et al, 2010), which defines the dependence of the shear deformation $\gamma(t)$ on the shear stress $\tau(t)$ by the following relationship:

$$\gamma(t) = \frac{1}{G}\tau_{xz}(t) + \frac{1}{G}\int_{-\infty}^{t} \tau(t)K(t-\tau)d\tau \tag{9.4}$$

Here G is instantaneous shear modulus, the creep kernel $K(t)$ is the following:

$$K(t) = k\exp\left(-\frac{t}{\omega}\right) \tag{9.5}$$

here ω is retardation time, k is the reciprocal of relaxation time. The Poisson ratio ν is assumed to be a constant value.

9.3 Method of Solution

At the first stage let us consider sliding of a constant pressure q distributed inside a square $2a\times2a$, over a viscoelastic layer bonded to the rigid half-space. Conditions at the upper layer bound are the following:

$$\begin{array}{ll} \sigma_z(x,y,0) = -q, & |x| \le a,\, |y| \le a; \\ \sigma_z(x,y,0) = 0, & |x| > a,\, |y| > a; \\ \tau_{xz}(x,y,0) = 0, & \tau_{yx}(x,y,0) = 0, \quad -\infty < x < \infty, \quad -\infty < y < \infty \end{array} \tag{9.6}$$

The solution of the similar boundary problem for a two-layered elastic half-space was obtained in Nikishin and Shapiro (1970), based on double Fourier transforms. For the particular case of the elastic layer bonded to the rigid base the expression for surface normal displacements has the form:

$$w'(x',y',0) = -\frac{q}{2G}\int_0^{\pi/2}\int_0^{\infty} \Delta(\nu,\gamma,\varphi,\lambda)\cos(x'\gamma\cos\varphi)\cos(y'\gamma\sin\varphi)d\gamma d\varphi \tag{9.7}$$

Here x', y', w' are dimensionless coordinates and normal displacements related to half-side of the square a, G is shear modulus of the layer material, ν is Poisson ratio, γ, φ are the coordinates for inverse double Fourier transforms, $\lambda = h/a$ is dimensionless layer thickness. The function $\Delta(\nu,\gamma,\varphi,\lambda)$ which is a result of solution of linear system of functional equations [16], derived from (9.3), (9.6) after using biharmonic function for representation of stresses and displacements, and also after double Fourier transform of constant pressure q, has the following form:

$$\begin{aligned} \Delta(\nu,\gamma,\varphi,\lambda) &= \frac{4}{\pi^2}\frac{\sin(\gamma\cos\varphi)\sin(\gamma\sin\varphi)}{\gamma^2\sin\varphi\cos\varphi}\times \\ &\times\left(\frac{-24\nu^2e^{-4\gamma\lambda}-8\nu^2+26\nu e^{-4\gamma\lambda}+0.4\nu\gamma\lambda(e^{-4\gamma\lambda}+1)}{-e^{-2\gamma\lambda}(16\nu^2+4\gamma^2\lambda^2+10)+e^{-4\gamma\lambda}(1+24\nu)+4\nu-3}+\right. \\ &\left.+\frac{8\nu e^{-2\gamma\lambda}(\gamma\lambda+1)+14\nu-6-0.3\gamma\lambda e^{-4\gamma\lambda}-0.2\gamma\lambda e^{-2\gamma\lambda}-0.3\gamma\lambda-6e^{-4\gamma\lambda}-4e^{-2\gamma\lambda}}{-e^{-2\gamma\lambda}(16\nu^2+4\gamma^2\lambda^2+10)+e^{-4\gamma\lambda}(1+24\nu)+4\nu-3}\right) \end{aligned} \tag{9.8}$$

Taking into account that for the case of the viscoelastic layer G in (9.7) is the operator (9.4), we reduce the following relation for normal displacements of the viscoelastic layer boundary:

$$\begin{aligned} w'(x',y',0) &= -\frac{q}{2G}\int_0^{\pi/2}\int_0^{\infty}\Delta(\gamma,\nu,\varphi,\lambda)\cos(y'\gamma\sin\varphi)\times \\ &\times\left(\cos(x'\gamma\cos\varphi)+\int_{-\infty}^{0}K(-\tau)\cos\left(\left(x'+\frac{V'\tau}{\omega}\right)\gamma\cos\varphi\right)d\tau\right)d\gamma d\varphi \end{aligned} \tag{9.9}$$

Here $V' = V\omega/a$ is dimensionless sliding velocity.

Substitution of the creep kernel (9.5) in (9.9) and integration over time leads to the following relation:

$$w'(x',y',0) = -\frac{q}{2G}\int_0^{\pi/2}\int_0^{\infty}\Delta(\gamma,\varphi,\nu,\lambda)\cos(y'\gamma\sin\varphi)\times \\ \times\left(\cos(x'\gamma\cos\varphi) + c\frac{V'\gamma\cos\varphi\sin(x'\gamma\cos\varphi)+\cos(x'\gamma\cos\varphi)}{1+(V'\gamma\cos\varphi)^2}\right)d\gamma d\varphi \tag{9.10}$$

Here $c = k\omega$ is the ratio of retardation to relaxation times.

The relation (9.10) together with the boundary conditions (9.1) and (9.2) are used for the solution of the contact problem. The hypothetical contact area is divided into N square elements with unknown constant pressures $p_i \quad (i = 1,...N)$ within each of them. Then the system of equations for determination of the contact pressures follows from (9.1) and (9.2):

$$\begin{pmatrix} 4a^2 & \cdots & 4a^2 & 0 \\ \kappa_1^1 & \cdots & \kappa_N^1 & -1 \\ \vdots & \ddots & \vdots & \vdots \\ \kappa_1^N & \cdots & \kappa_N^N & -1 \end{pmatrix} \times \begin{pmatrix} p_1 \\ \vdots \\ p_N \\ D \end{pmatrix} = \begin{pmatrix} Q \\ f_1 \\ \vdots \\ f_N \end{pmatrix} \tag{9.11}$$

here $f_1...f_N$ describe the indenter shape in each element. Coefficients κ_i^j are determined from (9.10):

$$\kappa_i^j = -\frac{1}{2G}\int_0^{\pi/2}\int_0^{\infty}\Delta(\gamma,\nu,\varphi,\lambda)\cos(y_{ij}\gamma\sin\varphi)\times \\ \times\left(\cos(x_{ij}\gamma\cos\varphi) + c\frac{V'\gamma\cos\varphi\sin(x_{ij}\gamma\cos\varphi)+\cos(x_{ij}\gamma\cos\varphi)}{1+(V'\gamma\cos\varphi)^2}\right)d\gamma d\varphi \tag{9.12}$$

Here x_{ij}, y_{ij} are distances between square elements over $0x$ and $0y$ respectively.

The solution of (9.11) in an arbitrary (a priori exceeding resulting) contact region includes, in general, the negative pressure in some cell-squares. On the next iteration the pressure in these elements is assumed zero, the rank of the system (9.11) is reduced due to the zero-point data elements, and the system is solved again. As a result, the iterative process leads to a positive contact pressure $p(x,y)$ within the definite contact area Ω. The procedure of numerical solution of the equation (9.11) is described in details in Goryacheva et al (2016).

The resulting pressure distribution is used to calculate the internal stresses in the viscoelastic layer. Internal stresses within the viscoelastic layer can be calculated based on the expressions for the elastic layer (Nikishin and Shapiro, 1970):

$$
\begin{aligned}
\sigma_x &= \int_0^{\pi/2}\int_0^{\infty} \left(\Delta_x(\nu,\gamma,\varphi,\lambda) - \cos^2\varphi\, \Delta_u(\nu,\gamma,\varphi,\lambda)\right) \times \\
&\times \cos(x'\gamma\cos\varphi)\cos(y'\gamma\sin\varphi)\gamma d\gamma d\varphi, \\
\sigma_y &= \int_0^{\pi/2}\int_0^{\infty} \left(\Delta_y(\nu,\gamma,\varphi,\lambda) - \sin^2\varphi\, \Delta_u(\nu,\gamma,\varphi,\lambda)\right) \times \\
&\times \cos(x'\gamma\cos\varphi)\cos(y'\gamma\sin\varphi)\gamma d\gamma d\varphi, \\
\sigma_z &= \int_0^{\pi/2}\int_0^{\infty} \Delta_z(\nu,\gamma,\varphi,\lambda)\cos(x'\gamma\cos\varphi)\cos(y'\gamma\sin\varphi)\gamma d\gamma d\varphi, \\
\tau_{xy} &= \int_0^{\pi/2}\int_0^{\infty} \Delta_{xy}(\nu,\gamma,\varphi,\lambda)\sin(x'\gamma\cos\varphi)\sin(y'\gamma\sin\varphi)\gamma d\gamma \sin\varphi\cos\varphi d\varphi, \\
\tau_{xz} &= \int_0^{\pi/2}\int_0^{\infty} \Delta_{xz}(\nu,\gamma,\varphi,\lambda)\sin(x'\gamma\cos\varphi)\cos(y'\gamma\sin\varphi)\gamma d\gamma \cos\varphi d\varphi, \\
\tau_{yz} &= \int_0^{\pi/2}\int_0^{\infty} \Delta_{yz}(\nu,\gamma,\varphi,\lambda)\cos(x'\gamma\cos\varphi)\sin(y'\gamma\sin\varphi)\gamma d\gamma \sin\varphi d\varphi
\end{aligned}
\tag{9.13}
$$

It is worth to note that the stresses (9.13) depend only on the Poisson ratio. Since for the viscoelastic model of the layer under consideration the Poisson ratio is constant, the relations (9.13) can be used to calculate the internal stresses in the viscoelastic layer.

9.4 Analysis of Internal Stresses

The method presented above is used to calculate and analyze stresses, which arise in sliding contact of a rigid spherical indenter with radius R and a viscoelastic layer. The following dimensionless parameters are used for analysis: coordinates $(x,y)/R$, velocity $V^* = V\omega/R$, layer thickness h/R, load $Q' = Q/R^2G_l$ (G_l is longitudinal shear modulus), and parameter$c = k\omega$. Dimensionless stresses σ_x', σ_y', σ_z', τ_{xy}', τ_{xz}', τ_{yz}' are related to the longitudinal shear modulus.

Figure 9.2 illustrates the distribution of the normal stresses in the layer for various values of parameters. Cross-sections by $z = 0$ plane present σ_z' stress distributions. All presented distributions are not symmetric, the fact is caused by the viscoelasticity of material, and the most asymmetric distribution occurs for material with relatively large ratio of instantaneous to longitudinal shear moduli (Fig. 9.2 d). The main features of the pressure curves (σ_z' stress at the surface) correspond to the results, which are obtained for viscoelastic half-space (Goryacheva et al, 2015, 2016; Koumi et al, 2015). The increase of a sliding velocity leads to the contact pressure increase (Figures 9.2 a and 9.2 c); the same effect takes place if the layer thickness

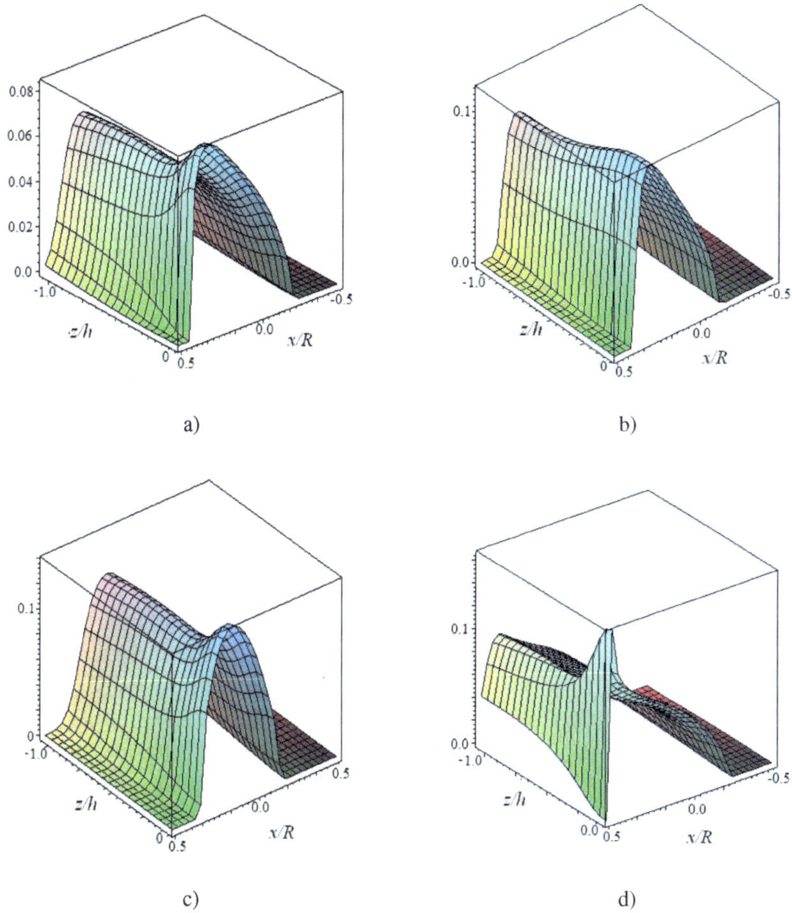

Fig. 9.2: Stresses σ_z' in the viscoelastic layer: $V^* = 0.333$ (a,b,d), $V^* = 1.666$ (c); c=5 (a,b,c), c=20 (d); h/R=0.1 (a,c,d), h/R=0.033 (b); $\nu = 0.3$; $Q' = 0.1$ (a,b,c), $Q' = 0.035$ (a,b,c).

decreases (Fig. 9.2 b). For the case of relatively thin layers the stresses weakly depend on coordinate z, it means that a 1-D model of viscoelastic material can be used for contact problem solution. For thicker layers the stresses decrease sharply from the surface, and then they tend to a constant value.

Figure 9.3 illustrates tensile-compressive stress distribution at the surface. The analysis of this stress component is important, because tensile stress concentration may lead to material damage. The stress maxima are located at the surface for all values of input parameters used for calculations. Positive values correspond to compression, and negative values – to tension. In any case we have maximal compression

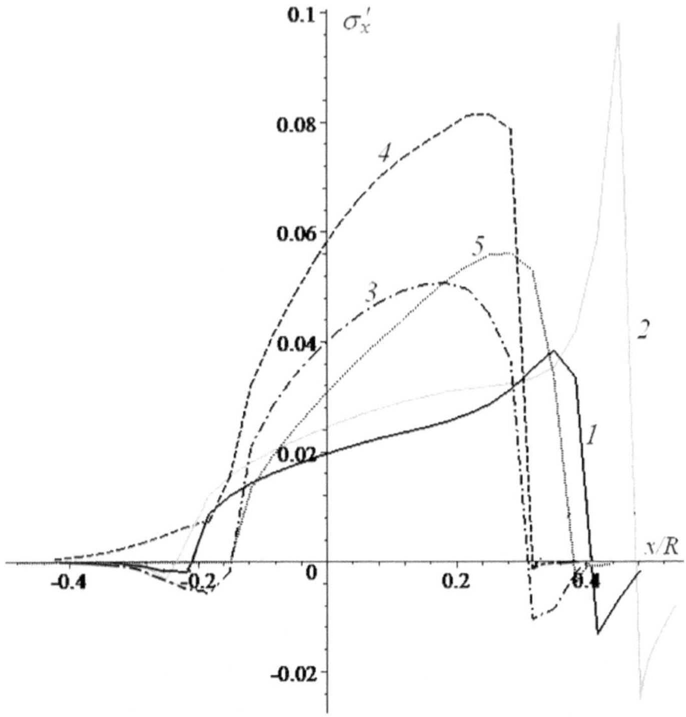

Fig. 9.3: Stresses σ'_x at the surface of viscoelastic layer: $V^* = 0.333$ (curves 1,2,4,5), $V^* = 1.666$ (curve 3) ; c=5 (curves 1, 3-5), c=20 (curve 2); h/R=0.1 (curves 1-4), h/R=0.033 (curve 5); $\nu = 0.3$ (curves 1-3, 5), $\nu = 0.45$ (curve 4); $Q' = 0.1$ (curves 1, 3-5), $Q' = 0.035$ (curve 2)

within the contact zone at the point of maximum pressure. At the front of the contact zone stresses have 'jump' from positive to negative or zero values. Tension may also occur behind the contact zone with lower values of local maxima. It is interesting to note, that tensile stresses are mostly sensitive to the changes of Poisson ratio and the layer thickness; increase of Poisson ratio and decrease of the layer thickness lead to essential decrease of tensile stresses at the front of contact zone. Simultaneously, normal and compressive stresses increase.

The analysis of the stress component τ'_{xz} distribution at the layer-substrate interface shows that the stresses are equal to zero at the surface and maximal at the interface (Fig. 9.4). The stress concentration may cause the layer delamination. Here we analyze absolute stress values, because positive or negative values correspond to different direction of their action. The maxima locate under the contact zone boundary , and the absolute maximum is under the front of the contact. The effect of Poisson ratio is most essential for shear stresses: curves 1 and 4 are calculated for ν=0.3 and 0.45 respectively, and the shear stress maxima are more than 2 times greater for large Poisson ratio.

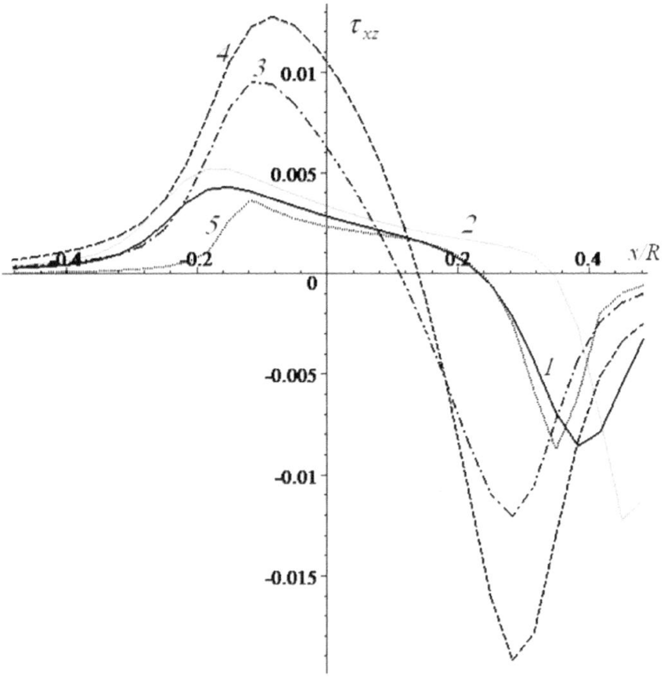

Fig. 9.4: Stresses τ'_{xz} at the layer-substrate interface: $V^* = 0.333$ (curves 1,2,4,5), $V^* = 1.666$ (curve 3) ; c=5 (curves 1, 3-5), c=20 (curve 2); h/R=0.1 (curves 1-4), h/R=0.033 (curve 5); $\nu = 0.3$ (curves 1-3, 5), $\nu = 0.45$ (curve 4); $Q' = 0.1$ (curves 1, 3-5), $Q' = 0.035$ (curve 2)

Principal shear stresses for the layers of different thickness are presented in Fig. 9.5. The stress distribution is almost symmetric for the thicker layer, it means that the layer thickness influences on detection of rheological properties of material in sliding contact. The 5 times increase of the layer thickness leads to almost 3 times decrease of maximum of the principal shear stress, which is localized under the surface.

9.5 Conclusions

The internal stresses within the viscoelastic layer bonded with a rigid half-space, in contact with the spherical slider were calculated. The dependence of the stress distribution on layer thickness and its rheological properties, as well as on sliding velocity was analyzed. It is shown, that

- the increase of viscoelastic layer thickness leads to the decrease of the maximal values of the internal stresses;

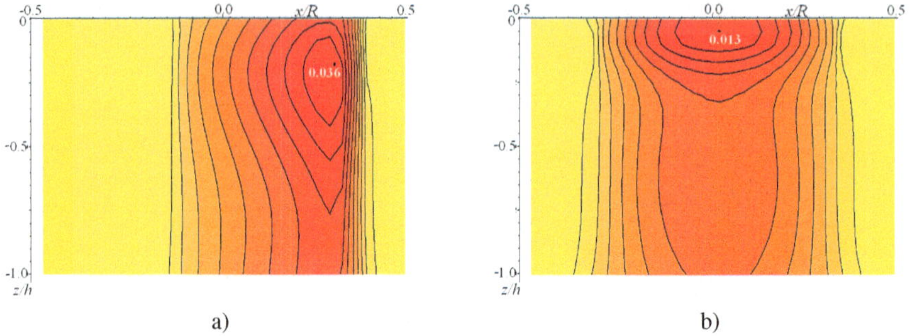

Fig. 9.5: Principal shear stresses in the viscoelastic layer: $V^* = 0.333$; c=5; $\nu = 0.3$; $Q' = 0.1$; h/R=0.033 (a), h/R=0.166

- for relatively thin layers the stress components weakly depend on coordinate z, it means that a 1-D model of viscoelastic material can be used for the contact problem solution for thin coatings;
- the increase of sliding velocity and parameter c, which determines the rheological properties of material, leads to increase of the maximal values of the internal stresses; for higher values of velocity and parameter c the principal shear stress maximum is larger and closer to the surface;
- the increase of Poisson ratio leads to the essential decrease of tensile stresses at the front of the contact region; the effect of Poisson ratio is also essential for the maximal values of shear stresses at the layer-substrate interface ;

The results can be used to predict the coating fracture in friction interaction.

Acknowledgements The work was supported by Russian Science Foundation (No. 14-29-00198).

References

Aleksandrov VM, Mark AV (2009) Quasistatic periodic contact problem for a viscoelastic layer, a cylinder, and a space with a cylindrical cavity. Journal of Applied Mechanics and Technical Physics 50(5):866–871

Aleksandrov VM, Goryacheva IG, Torskaya EV (2010) Sliding contact of a smooth indenter and a viscoelastic half-space (3d problem). Doklady Physics 55(2):77–80

Goryacheva IG, Gubenko MM, Makhovskaya YY (2014) Sliding of a spherical indenter on a viscoelastic foundation with the forces of molecular attraction taken into account. Journal of Applied Mechanics and Technical Physics 55(1):81–88

Goryacheva IG, Stepanov FI, Torskaya EV (2015) Sliding of a smooth indentor over a viscoelastic half-space when there is friction. Journal of Applied Mathematics and Mechanics 79(6):596–603

Goryacheva IG, Stepanov FI, Torskaya EV (2016) Effect of friction in sliding contact of a sphere over a viscoelastic half-space. In: Neittaanmäki P, Repin S, Tuovinen T (eds) Mathematical

Modeling and Optimization of Complex Structures, Computational Methods in Applied Sciences, pp 93–104

Koumi KE, Chaise T, Nelias D (2015) Rolling contact of a rigid sphere/sliding of a spherical indenter upon a viscoelastic half-space containing an ellipsoidal inhomogeneity. Journal of the Mechanics and Physics of Solids 80:1–25

Kusche S (2016) Frictional force between a rotationally symmetric indenter and a viscoelastic half-space. ZAMM - Journal of Applied Mathematics and Mechanics DOI 10.1002/zamm.201500169

Lyubicheva AN (2008) Analysis of the mutual influence of contact spots in sliding of the periodic system of asperities on a viscoelastic base of the Winkler type. Journal of Friction and Wear 29(2):92–98

Mark AV (2008) The uniform motion of rectangular and parabolic punches in a viscoelastic layer. Journal of Applied Mathematics and Mechanics 72(4):492–498

Morozov AV, Makhovskaya YY (2007) Experimental and theoretical evaluation of the deformation component of the coefficient of friction. Journal of Friction and Wear 28(4):331–337

Nikishin VS, Shapiro GS (1970) Space problems of the elasticity theory for multilayered media (in Russ. VTs AN SSSR, Moscow

Soldatenkov IA (2015) Calculation of friction for indenter with fractal roughness that slides against a viscoelastic foundation. Journal of Friction and Wear 36(3):193–196

Stepanov FI (2015) Sliding of two smooth indenters on a viscoelastic foundation in the presence of friction. Journal of Applied Mechanics and Technical Physics 56(6):1071–1077

Stepanov FI, Torskaya EV (2016) Study of stress state of viscoelastic half-space in sliding contact with smooth indenter. Journal of Friction and Wear 37(2):101–106

Stepanov FI, Torskaya EV (2017) Modeling of indenter sliding over viscoelastic layer bonded to rigid foundation. Mechanics of Solids 50(6):in print

Chapter 10
On the Problem of Diffusion in Materials Under Vibrations

Dmitry A. Indeitsev & Yulia A. Mochalova

Abstract We present a closed set of coupled equations of impurity transport during deformation of solids following from the general laws of continuum mechanics. Within a two-component continuum model, we derived the generalized diffusion equation. As an application, the phenomenon of impurity localization is considered in a rod under vibrations. The obtained numerical and analytical results are compared with experiments on the investigation of impurity redistribution during the deformation of samples.

Key words: Diffusion · Vibrations · Two-Component model · Elastic media · Analytic solution

10.1 Introduction

Vibration mechanics as a special section of dynamic problems of continuum mechanics holds a momentous place in the study of the mass transfer phenomenon in inhomogeneous media (in the most general case, there are diffusion of impurities in solids, degradation of metals, flow of liquids through the porous media or other liquid movement of special fluids). A detailed exposition of the main problems of vibrational mechanics and techniques is represented by Blekhman (2000). Our interest is the problem of mutual influence of impurity diffusion and deformation of elastic medium under vibrations. It is known that non-stationary mechanical loads lead to a redistribution of impurities in material, which can have a significant effect on the mechanical properties of a sample (Polyanskiy et al, 2005; Belyaev et al, 2012). In most studies on diffusion in solids, the phenomenological models are used. It allows

Dmitry A. Indeitsev · Yulia A. Mochalova
Institute for Problems in Mechanical Engineering RAS, V.O., Bolshoy pr., 61, 199178 St. Petersburg, Russia,
e-mail: dmitry.indeitsev@gmail.com, yumochalova@yandex.ru

H. Altenbach et al. (eds.), *Mechanics for Materials and Technologies*,
Advanced Structured Materials 46, DOI 10.1007/978-3-319-56050-2_10

us to take into account only the effect of stress state of the medium on impurity diffusion (Aifantis, 1980). In this case the question of how kinetic processes in the material (diffusion of impurities, the sedimentation of impurity particles on the main structure) influence on mechanical material properties of materials remains open. To describe the impurity diffusion in solids, a two-component continuum model is suggested in Indeitsev et al (2012); Indeitsev and Mochalova (2014). Interaction between the main stricture and the flow of impurity is determined by internal forces (which depend on the material deformation (stress state)) and mass transfer between the impurity and the main structure. Within this model governing equations are obtained by Indeitsev and Mochalova (2014) and static stress state of the material with an impurity is investigated. The paper by Indeitsev et al (2009) proposes a model of internal transformation in a sample under dynamical load. Simplified balance equations without the convective diffusion terms are used. It is shown that the change in mechanical properties of the material is due to internal transformation when the impurity particles are attached to the main structure. In the paper by Indeitsev and Osipova (2009) the stress-strain state of material with the impurity is researched within statistical physics. It good agreement between the two-component mechanical model and statistical physics approach is shown. The equation of impurity motion (the generalized diffusion equation) is obtained by Indeitsev et al (2012). The equation allows us to consider the effect of the main structure deformation on the impurity diffusion. A metal sample under the cyclic loading is examined and the localization of impurity in the central part of the sample is shown.

This paper is an extension of the researches started in Indeitsev et al (2009, 2012). Aim of this paper is the research of mutual influence of impurity diffusion and deformation within the two-component continual model. It is assumed that the free particles of the impurity can be attached to the main structure and it mechanical properties can be transformed. We restrict the study to particular cases, which qualitatively describe two important experimentally observed phenomena in the material under vibrations: the impurity localization and the drop of generalized stiffness in the sample. For each problems approximate analytical solutions are obtained, which have a good agreement with previously numerical studies and experimental data. These model problems allow us to investigate features of diffusion processes in solids under non-stationary external action and can be tests for the analysis of complex problems of diffusion processes in multi-component media.

10.2 The Equation of Impurity Motion

Within the two-component approach, we consider two interpenetrating environments: dynamically deformable main solid structure (lattice of the metal, for example) and a movable impurity in the body (Indeitsev and Mochalova, 2014). Joint movement and coupling of these two environments are described in the absolute coordinate system by the following system of equations:

- the equation of motion of the main structure

$$\nabla \cdot \boldsymbol{\sigma} = \rho_1 \frac{\partial \boldsymbol{v}_1}{\partial t} - J\boldsymbol{v}_1 - \boldsymbol{R}, \tag{10.1}$$

- the equation of motion of the mobile impurity

$$-\nabla p = \rho_2 \frac{\partial \boldsymbol{v}_2}{\partial t} + J\boldsymbol{v}_2 + \boldsymbol{R}, \tag{10.2}$$

- the continuity equation of the main structure

$$\frac{\partial \rho_1}{\partial t} + \nabla \cdot (\rho_1 \boldsymbol{v}_1) = -J, \tag{10.3}$$

- the continuity equation of the mobile impurity

$$\frac{\partial \rho_2}{\partial t} + \nabla \cdot (\rho_2 \boldsymbol{v}_2) = J, \tag{10.4}$$

- the state equation of the main structure

$$\boldsymbol{\sigma} = \boldsymbol{\sigma}(\boldsymbol{\varepsilon}), \tag{10.5}$$

- the state equation of mobile impurity

$$p = p(\rho_2). \tag{10.6}$$

Here ρ_1,$\boldsymbol{v}_1$ $\boldsymbol{\sigma}$, ε are the density, velocity vector, the stress tensor and strain of the basic structure of the material; ρ_2, $\boldsymbol{v}_2$, p are density, velocity vector and pressure of the mobile impurity. Reactive forces $J\boldsymbol{v}_1$ and $J\boldsymbol{v}_2$ in Eqs. (10.1), (10.2) are neglected. Interaction between the components is determined by Indeitsev and Mochalova (2014) and J is the source term, $\boldsymbol{R}$ is the internal force of the interaction between the main structure and the mobile impurity

$$\boldsymbol{R} = \mu(\varepsilon)\rho_2(\boldsymbol{v}_2 - \boldsymbol{v}_1). \tag{10.7}$$

Here $\mu(\varepsilon)$ is the drag coefficient.

We introduce the relative velocity of movement of the impurity $\boldsymbol{\vartheta} = \boldsymbol{v}_2 - \boldsymbol{v}_1$. Substituting $\boldsymbol{v}_2$ expressed in terms of the relative velocity into (10.2) and neglecting $\rho_2 \partial \boldsymbol{\vartheta}/\partial t$, we get

$$\boldsymbol{\vartheta} = -\frac{1}{\mu(\varepsilon)\rho_2}\left(\nabla p + \rho_2 \frac{\partial \boldsymbol{v}_1}{\partial t}\right). \tag{10.8}$$

Substituting $\boldsymbol{\vartheta}$ into expression (10.4), we obtain the diffusion equation which takes into account the effect of deformation and inertial forces:

$$\frac{\partial \rho_2}{\partial t} - \nabla \cdot \left[\frac{1}{\mu(\varepsilon)}\left(\nabla p + \rho_2 \frac{\partial \boldsymbol{v}_1}{\partial t}\right) - \rho_2 \boldsymbol{v}_1\right] = J. \tag{10.9}$$

Differentiating Eq. (10.6) with respect to ρ_2, and assuming that the speed of sound in the impurity $c_2^2 = \partial p/\partial \rho$ does not change much, Eq. (10.9) can be rewritten as

$$\frac{\partial \rho_2}{\partial t} - \nabla \cdot \left[D(\varepsilon) \left(\nabla \rho_2 + \frac{\rho_2}{c_2^2} \frac{\partial \mathbf{v}_1}{\partial t} \right) - \rho_2 \mathbf{v}_1 \right] = J, \tag{10.10}$$

where $D(\varepsilon) = c_2^2/\mu(\varepsilon)$ is the diffusion coefficient, which is proportional to the characteristic size of penetrability of the main structure and dependents on the spherical deformation (Indeitsev et al, 2012). Thus, the obtained equation determines the motion of impurity. The first term

$$\nabla \cdot \left[D(\varepsilon) \left(\nabla \rho_2 + \frac{\rho_2}{c_2^2} \frac{\partial \mathbf{v}_1}{\partial t} \right) \right] \tag{10.11}$$

gives diffusion of the impurity. The second term $\nabla \cdot \left[\mathbf{v}_s \rho_f \right]$ defines the impurity transfer due to the motion of main structure. Note that in (10.10) the diffusion coefficient depends on the deformation of main structure. In addition the member defined by vibration of the main structure appears in (10.11).

In the absence of deformation equation (10.10) reduces to the classical diffusion equation with diffusion coefficient $D_0 = D(0)$. Therefore, in the first approximation, D is determined by the experimental values obtained in the static theory (Indeitsev and Mochalova, 2014). For the small deformations we can assume that the value of the diffusion coefficient D is linearly dependent on ε and $D(\varepsilon) = D_0 - D_1 \varepsilon$, where $D_1 > 0$.

The system of equations (10.1), (10.3), (10.5) and (10.10) allows us to determine the dynamics of the impurity redistribution during deformation of the material.

10.3 Statement of the Problem: Governing Equations

As an example, we consider a one-dimensional rod with the impurity of length $2L$ under the cyclic load applied to its ends $F_0 \sin \omega t$, ω is the frequency of load. Let ρ_1^+ be the density of impurity particles attached to the main structure and $\rho_1 = \rho_1^0 + \rho_1^+$, where $\rho_1^0 = \text{const}$. Then, from Eq. (10.3) we get

$$\frac{\partial \rho_1^+}{\partial t} + \frac{\partial}{\partial x} \left(\rho_1^+ v_1 \right) = -J. \tag{10.12}$$

Note that $\rho_1^+ \ll \rho_1$ and in Eq. (10.1) $\rho_1 = \text{const}$ can be considered. Besides, in most cases the influence of internal interaction force R on the motion of main structure can be neglected in Eq. (10.1).

The stress of main structure depends on the density of attached impurity ρ_1^+ and the equation of main structure state (10.5) is written as

$$\sigma = E(\rho_1^+)\,\varepsilon, \tag{10.13}$$

where $E(\rho_1^+)$ is the effective Young's modulus (Indeitsev and Mochalova, 2014).

The source terms J has the following form $J = -\alpha\rho_2$. The positive parameter α defines the rate of impurity sedimentation on main stricture Indeitsev et al (2009).

Passing to displacement u ($v_1 = \partial u/\partial t$, $\varepsilon = \partial u/\partial x$) in the Eqs. (10.1), (10.10) and (10.12), we get the following problem:

$$\frac{\partial\sigma}{\partial x} = \rho_1\frac{\partial^2 u}{\partial t^2}, \quad \sigma = E(\rho_1^+)\varepsilon, \tag{10.14}$$

$$\frac{\partial\rho_1^+}{\partial t} + \frac{\partial}{\partial x}\left(\rho_1^+\frac{\partial u}{\partial t}\right) = \alpha\rho_2, \tag{10.15}$$

$$\frac{\partial\rho_2}{\partial t} - \frac{\partial}{\partial x}\left[\left(D_0 - D_1\frac{\partial u}{\partial x}\right)\left(\frac{\partial\rho_2}{\partial x} + \frac{\rho_2}{c_2^2}\frac{\partial^2 u}{\partial t^2}\right) - \rho_2\frac{\partial u}{\partial t}\right] = -\alpha\rho_2, \tag{10.16}$$

boundary and initial conditions are

$$u\big|_{x=0} = 0, \quad \sigma\big|_{x=L} = F_0\sin\omega t, \tag{10.17}$$

$$\frac{\partial\rho_2}{\partial x}\bigg|_{x=0} = 0, \quad \left(D_0 - D_1\frac{\partial u}{\partial x}\right)\left[\frac{\partial\rho_2}{\partial x} + \frac{\rho_2}{c_2^2}\frac{\partial^2 u}{\partial t^2}\right] - \rho_2\frac{\partial u}{\partial t}\bigg|_{x=L} = 0, \tag{10.18}$$

$$\rho_2\big|_{t=0} = \rho_0(x), \quad \rho_1^+\big|_{t=0} = 0.$$

Introducing the dimensionless parameters as follows:

$$t_1 = \frac{D_0}{L^2}t, \quad x_1 = \frac{x}{L}, \quad \overline{u} = \frac{u}{L}, \quad \overline{\rho}_2 = \frac{\rho_2}{\rho_1}, \quad \Omega = \frac{L^2}{D_0}\omega, \quad \overline{\sigma} = \frac{L^2}{\rho_1 D_0^2}\sigma,$$

$$\overline{E}_0 = \frac{L^2}{\rho_1 D_0^2}E_0, \quad \overline{\alpha} = \frac{\alpha L^2}{D_0}, \quad \beta = D_1/D_0, \quad \eta = \frac{D_0^2}{c^2 L^2},$$

we rewrite the set of equations (10.14)–(10.16) in the form

$$\frac{\partial\overline{\sigma}}{\partial x_1} = \frac{\partial^2\overline{u}}{\partial t_1^2}, \quad \overline{\sigma} = \overline{E}(\overline{\rho}_1^+)\varepsilon, \tag{10.19}$$

$$\frac{\partial\overline{\rho}_2}{\partial t_1} - \frac{\partial}{\partial x_1}\left[\left(1 - \beta\frac{\partial\overline{u}}{\partial x_1}\right)\left(\frac{\partial\overline{\rho}_2}{\partial x_1} + \eta\overline{\rho}_2\frac{\partial^2\overline{u}}{\partial t_1^2}\right) - \overline{\rho}_1\frac{\partial\overline{u}}{\partial t_1}\right] = -\overline{\alpha}\overline{\rho}_1, \tag{10.20}$$

$$\frac{\partial\overline{\rho}_1^+}{\partial t_1} + \frac{\partial}{\partial x_1}\left[\overline{\rho}_1^+\frac{\partial\overline{u}}{\partial t_1}\right] = \overline{\alpha}\overline{\rho}_2, \tag{10.21}$$

the boundary conditions are

$$\overline{u}\big|_{x_1=0} = 0, \quad \overline{\sigma}\big|_{x_1=1} = \overline{F}_0\sin\Omega t_1, \tag{10.22}$$

$$\left.\frac{\partial \overline{\rho}_2}{\partial x_1}\right|_{x_1=0}=0,\quad \left.\left(1-\beta\frac{\partial \overline{u}}{\partial x_1}\right)\left(\frac{\partial \overline{\rho}_2}{\partial x_1}+\eta\,\overline{\rho}_2\frac{\partial^2 \overline{u}}{\partial t_1^2}\right)-\overline{\rho}_2\frac{\partial \overline{u}}{\partial t_1}\right|_{x_1=1}=0,$$

where $\overline{F}_0=L^2F_0/\rho_1D_0^2$ and the initial conditions

$$\overline{\rho}_2|_{t_1=0}=\rho_0(x),\quad \overline{\rho}_1^+|_{t_1=0}=0. \tag{10.23}$$

The set of governing equations (10.19)–(10.23) is too complicated for a direct mathematical analysis as it describes the relationship of dynamic processes having different time scales: the wave processes in the basic structure, impurity diffusion, mass transfer between the components. Therefore, we consider a number of special cases. Further the overline is omitted.

10.4 Localization of Diffusion Process

Let $J=0$ and there is no mass transfer between the components. In this case, the impurity particles are not attached to the main structure, $\rho_1^+=0$, and the rod rigidity remains constant $E=E_0$. Thus, the system of equations (10.19)–(10.23) splits into two problems. The first problem defines oscillations of the rod under vibration load

$$\frac{\partial^2 u}{\partial t^2}=c_0^2\frac{\partial^2 u}{\partial x^2}, \tag{10.24}$$

$$u\big|_{x=0}=0,\quad \left.\frac{\partial u}{\partial x}\right|_{x=1}=F_0\sin\Omega t_1. \tag{10.25}$$

Here $c_0^2=(cL/D_0)^2$, $c=\sqrt{E_0/\rho_1}$ is the speed of sound in the rod. Naturally to consider the steady-state oscillations. Then, the solution of (10.24), (10.25) has the form

$$u(x,t)=\frac{c_0F_0}{\Omega\cos\left(\Omega/c_0\right)}\sin\Omega t\,\sin\left(\Omega x/c_0\right).$$

Note that $\Omega/c_0\ll 1$ and we can assume that

$$u(x,t)=F_0x\sin\Omega t. \tag{10.26}$$

The second problem describes the impurity redistribution in the rod under vibration load

$$\frac{\partial \rho_2}{\partial t}-\frac{\partial}{\partial x}\left[\left(1-\beta\frac{\partial u}{\partial x}\right)\left(\frac{\partial \rho_2}{\partial x}+\eta\,\rho_2\frac{\partial^2 u}{\partial t^2}\right)-\rho_2\frac{\partial u}{\partial t}\right]=0, \tag{10.27}$$

$$\left.\frac{\partial \rho_2}{\partial x}\right|_{x=0}=0,\quad \left.\left(1-\beta\frac{\partial u}{\partial x}\right)\left(\frac{\partial \rho_2}{\partial x}+\eta\,\rho_2\frac{\partial^2 u}{\partial t^2}\right)-\rho_2\frac{\partial u}{\partial t}\right|_{x=1}=0, \tag{10.28}$$

$$\rho_2|_{t=0}=\rho_0(x). \tag{10.29}$$

Let us proceed to the study of the generalized diffusion equation. Without the deformation of main structure, Eq. (10.27) becomes the classical diffusion equation and the solution of corresponding Cauchy problem is well known (see Polyanin and Manzhirov, 2007) and diffusion leads to a uniform distribution of impurity density along the length of rod

$$\rho_2 = \int_0^1 \rho_0 \, \mathrm{d}x \quad \text{as} \quad t \to \infty.$$

Assuming that the diffusion coefficient does not depend on the rod strain, Eq. (10.27) reduces to the advective-diffusion equation (Papanicolaou, 1995). In this case, the diffusion leads to a redistribution of the impurity followed by fluctuations in the impurity concentration around the mean value, which are caused by the vibrations.

Returning to the problem (10.27)–(10.29), we assume that the impurity density ρ_2 varies insignificantly over the period of the rod oscillations, that is

$$\langle \rho_2 \rangle = \frac{\Omega}{2\pi} \int_0^{\frac{2\pi}{\Omega}} \rho_2(x,t) \mathrm{d}x = \rho_2(x,t).$$

Substituting (10.26) into (10.27)–(10.29) and averaging obtained system over the interval $[t, t+2\pi/\Omega]$ leads to the following Cauchy problem:

$$\frac{\partial \rho_2}{\partial t} = \frac{\partial}{\partial x}\left[\frac{\partial \rho_2}{\partial x} + \gamma x \rho_2\right], \tag{10.30}$$

$$\frac{\partial \rho_2}{\partial x} + \gamma x \rho_f \Big|_{x=1} = 0, \quad \frac{\partial \rho_2}{\partial x}\Big|_{x=0} = 0, \tag{10.31}$$

$$\rho_2\big|_{t=0} = \rho_0, \tag{10.32}$$

where $\gamma = \beta \eta \Omega^2/2$. The solution of the problem (10.30)–(10.32) is sought by the Fourier method. Therefore, at first we solve the corresponding Sturm–Liouville problem, namely define the values of the parameter λ for which there are non-trivial solutions of the equation

$$L\varphi = -\lambda \varphi, \quad L\varphi = (\varphi_x + \gamma x \varphi)_x' \tag{10.33}$$

with the boundary conditions

$$\varphi_x\big|_{x=0} = 0, \quad \varphi_x + \gamma \varphi\big|_{x=1} = 0. \tag{10.34}$$

Note that after the transformations $\varphi = \exp(-\gamma \zeta^2/4)\,\psi$ and $x = \sqrt{\gamma}\zeta$, Eq. (10.33) reduces to the Weber equation $\psi_{\zeta\zeta} - [\frac{1}{4}\zeta^2 + \frac{1}{2} - \lambda]\psi = 0$. Its solution is conveniently written in terms of confluent hypergeometric functions (Abramowitz and

Stegun, 1972; Polyanin and Manzhirov, 2007). Then the solution of Eq. (10.33) can be written as

$$\varphi(x) = \left[a\,{}_1F_1\Big(-\frac{\lambda}{2\gamma}, \frac{1}{2}, \frac{\gamma x^2}{2}\Big) + bx\,{}_1F_1\Big(\frac{\gamma-\lambda}{2\gamma}, \frac{3}{2}, \frac{\gamma x^2}{2}\Big)\Big)\right] \exp(-\gamma x^2/2), \quad (10.35)$$

where ${}_1F_1$ is the confluent hypergeometric function, a, b are unknown constants. Differentiating (10.35) on x and substituting into the boundary conditions (10.34), we find that $b = 0$ and the equation for determining the eigenvalues λ has the form

$$\gamma\,{}_1F_1\Big(-\frac{\lambda}{2\gamma}, \frac{1}{2}, \frac{\gamma}{2}\Big) - (\gamma+\lambda)\,{}_1F_1\Big(-\frac{\lambda}{2\gamma}, \frac{3}{2}, \frac{\gamma}{2}\Big) = 0. \quad (10.36)$$

Investigating Eq. (10.36) for different values of the parameter γ, we find that the operator L has a simple discrete spectrum of eigenvalues starting with $\lambda_0 = 0$:

$$0 < \lambda_1 < \lambda_2 < \ldots$$

The corresponding eigenfunctions are defined by the following expression

$$\varphi_k(x) = {}_1F_1\Big(-\frac{\lambda_k}{2\gamma}, \frac{1}{2}, \frac{\gamma x^2}{2}\Big) \exp(-\gamma x^2/2)/N_k, \quad k = 0, 1, \ldots, \quad (10.37)$$

where $N_k = \Big[\int_0^1 {}_1F_1\big(-\frac{\lambda_k}{2\gamma}, \frac{1}{2}, \frac{\gamma x^2}{2}\big) \exp(-\gamma x^2/2)\mathrm{d}x\Big]^{1/2}$ $(k = 0, 1, \ldots)$.

The eigenvalues $\lambda_k \to (\pi k)^2$ for $\gamma \to 0$ and the corresponding eigenfunctions $\varphi_k \to \cos\pi k x$, $(k = 0, 1, \ldots)$. It gives a good approximation of eigenvalues and eigenfunctions of the operator L in the case $\gamma < 1$.

Now we seek the solution of the problem (10.30)–(10.32) as an expansion in eigenfunctions (10.37). Using the procedure of the Fourier method yields

$$\rho_2(x,t) = c_0 \exp(-\gamma x^2/2) + \sum_{k=1}^{\infty} c_k\,{}_1F_1\Big(-\frac{\lambda_k}{2\gamma}, \frac{1}{2}, \frac{\gamma x^2}{2}\Big) \exp(-\lambda_k t - \gamma\frac{x^2}{2}), \quad (10.38)$$

$$c_k = \frac{1}{N_k}\int_0^1 \rho_0(x)\varphi_k(x)\,\mathrm{d}x, \; k = 0, 1, \ldots.$$

It is noted that for $\rho_0 =$ const the impurity density $\rho_2(x,t)$ tends to the stationary solution of the problem (10.30):

$$\rho_2(x,t) \to \sqrt{\frac{2\gamma}{\pi}} \frac{\rho_0}{\operatorname{erf}\sqrt{\gamma/2}} \exp(-\gamma x^2/2), \quad t \to \infty.$$

Thus, the obtained approximate analytical solution of the problem shows that the effect of vibrations leads to the accumulation of impurity in the central part of the rod. Considering as impurity hydrogen atoms dissolved in the material, we can show that the solution has a good agreement with experimental data (the redistribution of hydrogen under long-time cyclic loading Polyanskiy et al, 2005). In addition, the solution is consistent with the numerical results obtained by Indeitsev et al (2012).

10.5 Structural Transformations of Materials

Consider the second particular case. Let us neglect the convective diffusion part of the mass balance equations (10.15), (10.16), and assume that the impurity densities are almost unchanged over the length of the rod $\rho_2(x,t) = \rho_2(t)$, $\rho_1^+(x,t) = \rho_1^+(t)$. Then the solution of Eqs. (10.20) and (10.21) has the form

$$\rho_2 = \rho_0 \exp\{-\alpha t\}, \quad \rho_1^+ = \rho_0\big(1 - \exp\{-\alpha t\}\big).$$

The modulus of elasticity takes the following form $E(\rho_1^+) = \big[1+\kappa\big(1-\exp\{-\alpha t\}\big)\big]^{-1}$ Indeitsev and Mochalova (2014) and the equation of main structure motion (10.19) can be written as

$$\frac{\partial^2 u}{\partial x^2} = \Big[1 + \kappa\big(1 - \exp\{-\alpha t\}\big)\Big]\frac{\partial^2 u}{\partial t^2},$$

$$u|_{x=0} = 0, \quad \left.\frac{\partial u}{\partial x}\right|_{x=1} = \overline{F}_0\big[1 + \kappa\big(1 - \exp\{-\alpha t\}\big)\big]\sin\Omega t,$$

where κ is dimensionless coefficient, which determines the ratio of the Young's modulus of pure material in the absence of impurity to Young's modulus of the material with attached particles of impurity.

Let $\alpha/\Omega \ll 1$ be a small parameter. Using the multiple scale method, we introduce two scales of time t and $\tau = \alpha t$. For the zero approximation we obtain

$$\frac{\partial^2 u_0}{\partial x^2} = \Big[1 + \kappa\big(1 - \exp\{-\tau\}\big)\Big]\frac{\partial^2 u_0}{\partial t^2},$$

$$u_0|_{x=0} = 0, \quad \left.\frac{\partial u_0}{\partial x}\right|_{x=1} = F_0\big[1 + \kappa_2\big(1 - \exp\{-\tau\}\big)\big]\sin\Omega t,$$

and

$$u_0(x,t,\tau) = \frac{F_0\Omega_0(\tau)}{\Omega^2\cos\Omega_0(\tau)}\sin\Omega_0(\tau)x\sin\Omega t,$$

where $\Omega_0(\tau) = \Omega\sqrt{1+\kappa\big(1-\mathrm{e}^{-\tau}\big)}$.

Thus, assuming that $\alpha \ll 1$, we obtain decrease of the structural rigidity and growth of the amplitude of oscillations. If $\alpha \gg 1$ the particles of impurity attach to the main structure with a speed, which is greater then the speed of sound in material, and structural changes in the main structure (change of generalized rigidity) occur almost instantaneously (Indeitsev et al, 2009).

Note that in this model, we assume that α is the constant, however α, as the diffusion coefficient, must depend on the main structure deformation, and perhaps, on the impurity concentration.

10.6 Conclusion

Considered two-component model allows us to describe diffusion in solids (in particular the changes of material structure) and outlines the interaction of the three competing processes:

- impurity diffusion, which can depend on the rate of deformation of the main structure;
- attachment of impurity to the main structure, which generally also depends on the deformation of the solid;
- non-stationary wave processes in the material.

These processes have different scales of time which is decisive for solving specific problems.

We have examined two particular cases. First, when the mass transfer between the impurity and the main structure can be neglected (source terms $J = 0$). In this case, the influence of cyclic loading has leaded to the localization of impurity in the central part of the sample. In the second case, it is assumed that the rate of impurity sedimentation on the main structure is much greater than the rate of diffusion and diffusion-convective terms in the balance equations can be neglected. It has led to structural changes in the material, namely, to the decrease of generalized stiffness sample.

In general, diffusion and mass transfer processes are interrelated. However, it is possible to understand for specific problems how the rates of these processes relate. For example, if the rate of impurity diffusion is greater then the rate of sedimentation on the main structure, then at first the impurity is localized in a region of the sample. Impurity particles will be attached to the main structure in this local region, it can lead to local changes in the structure of sample. If the rate of impurity sedimentation significantly exceeds the rate of diffusion, impurity is embedded in the main structure not having enough time for the localization.

Finally we note that the solution of specific problems can be investigated by using the method of multiple scales and the averaging method. The solution is suggested to be sought in the form (10.26) and (10.38), where coefficients are unknown functions of time.

Acknowledgements The work was carried out with support of the grant No. 14.607.21.0120, id RFMEFI60715X0120 of the Ministry of Education and Science of the Russian Federation.

References

Abramowitz M, Stegun IA (1972) Handbook of Mathematical Functions with Formulas, Graphs, and Mathematical Tables. Dover, New York

Aifantis EC (1980) On the problem of diffusions in solids. Acta Mechanica 37:265–296

Belyaev AK, Polyanskiy VA, Yakovlev YA (2012) Stresses in pipeline affected by hydrogen. Acta Mechanica 223:1611–1619

Blekhman II (2000) Vibrational Mechanics. World Scientific, Singapore

Indeitsev DA, Mochalova YA (2014) Mechanics of multi-component media with exchange of mass and non-classical supplies. In: Irschik H, Belyaev AK (eds) Dynamics of Mechanical Systems with Variable Mass, Springer, Wien, pp 165–194

Indeitsev DA, Osipova EV (2009) Hydrogen embrittlement under load as a first-order phase transition. Doklady Physics 51:1901–1906

Indeitsev DA, Naumov VN, Semenov BN, Belyaev AK (2009) Thermoelastic waves in a continuum with complex structure. ZAMM 89:279–287

Indeitsev DA, Semenov BN, Sterlin MD (2012) The phenomenon of localization of diffusion process in a dynamically deformed solid. Doklady Physics 57:171–173

Papanicolaou GP (1995) Diffusion in random media. In: Keller JB, McLaughlin DW, Papanicolaou GC (eds) Surveys in Applied Mathematics, Springer, Boston, pp 205–253

Polyanin AD, Manzhirov AV (2007) Handbook of Mathematics for Engineers and Scientists. Chapman and Hall/CRC Press, Boca Raton-London

Polyanskiy AM, Polyanskiy VA, Popov-Diumin DB (2005) Diagnostics of mechanical condition of structural material by method of high-temperature hydrogen vacuum-extraction. In: Ziegler F, Heuer R, Adam C (eds) Proceedings of the Sixth International Congress on Thermal Stresses, Vienna, Austria, Technischen Universität Wien, pp 589–592

Chapter 11
A Study of Objective Time Derivatives in Material and Spatial Description

Elena Ivanova, Elena Vilchevskaya & Wolfgang H. Müller

Abstract This paper presents an in-depth discussion of the transformation properties of total and material (substantial) time derivative operators during a change of the Frame of Reference (FoR). For this purpose it is first necessary to establish the transformation properties of gradients of scalar fields, which then leads to the transformation properties of the gradient itself, a.k.a. nabla operator. The analysis is based on the notion of so-called tensor and vector *image transfer* from one frame to another, as originally introduced in the scientific work of Zhilin. Emphasis is put on several issues, namely, first, the observation point considered in context with all operators must be the same for all FoRs at all times. Second, the effect of all operators on an invariant scalar must be investigated. This will then consecutively result in transformation properties of the operators themselves. Third, the arguments of the scalar field must be the same for all FoRs in order to guarantee a meaningful comparison. This in mind it will be shown (a) that the nabla operator in the current configuration is invariant, (b) that the nabla operator in the reference configuration is not, and (c) that the aforementioned time derivatives in material and in spatial description are invariant.

Elena Ivanova
Peter the Great Saint-Petersburg Polytechnic University, Politekhnicheskaja 29, 195251 St.-Petersburg, Russia,
e-mail: elenaivanova239@gmail.com

Elena Vilchevskaya
Institute for Problems in Mechanical Engineering of the Russian Academy of Sciences, Bol'shoy pr. 61, V.O., 199178 St. Petersburg & Peter the Great Saint-Petersburg Polytechnic University, Politekhnicheskaja 29, 195251 St.-Petersburg, Russia,
e-mail: vilchevska@gmail.com

Wolfgang H. Müller
Institute of Mechanics, Chair of Continuum Mechanics and Materials Theory, Technical University of Berlin, Einsteinufer 5, 10587 Berlin, Germany,
e-mail: whmueller1000@gmail.com

H. Altenbach et al. (eds.), *Mechanics for Materials and Technologies*,
Advanced Structured Materials 46, DOI 10.1007/978-3-319-56050-2_11

Key words: Frame indifference · Total time derivative · Substantial time derivative · Nabla operator · Images · Frame of reference

11.1 Introduction and Outline to the Paper

This paper presents a study of the frame indifference properties of operators for total and material (a.k.a. substantial) time derivatives within the context of material as well as spatial description of continuous fields. In papers devoted to different formulations of the concept of material objectivity (see, *e.g.*, Truesdell and Noll, 1965; Noll, 1958; Bressan, 1972; Bertram and Svendsen, 2001; Svendsen and Bertram, 2001; Edelen and McLennan, 1973; Ryskin, 1985; Matolcsi, 1986; Frewer, 2009, and references therein) the notion of frame-indifference is, in general, widely discussed. At the same time the question as to whether time derivatives show properties of objectivity and frame indifference is often not addressed comprehensively enough in the literature. It is conceivable that the authors of the corresponding papers believe that this issue is not directly relevant to their articles, or maybe they think that the answer to this question is obvious and/or has already been given. One way or the other, in most cases authors (without any comments) use the same symbol for the total and for the substantial time derivative in different reference systems, implying objectivity of this operation (see, *e.g.*, Massoudi, 2002; Liu, 2005; Rivlin, 2006).

Without questioning the objectivity of the total or of the material derivative, we believe that the proof of this fact is not obvious, and that statements found in the literature are not always satisfactory and often contradictory, to say the least. For example, in Muschik and Restuccia (2008) it is shown that the material time derivative is an observer-invariant operator, while in Bampi and Morro (1980) it is written:

> ... the material time derivative, when viewed as a rule assigning to every frame the time derivative relative to that frame, is nonobjective.

The intentions of this paper are manifold. Above all we shall attempt to clarify sometimes obscure and confusing terms appearing in the literature. This concerns, first, a precise introduction of the notion of a *Frame of Reference* (FoR). Second, based on physics arguments, we will investigate how a change between different FoRs should conceptually be properly performed. In this context the notion of *vector and tensor images* will be presented, which was originally introduced in the Russian literature by Zhilin, *e.g.*, in Zhilin (2001) and Zhilin (2003), a concept that proves to be of great value in classical continuum theory.

Moreover, in context with the kinematic relations required during the change of reference frames we will create awareness for the existence and for the use of the so-called *Poisson relation*, a formula handling the time derivative of rotation tensors or operators. This formula is relatively unknown and hardly applied in the Western literature on continuum theory. However, it is very effective during algebraic manipulations of kinematic quantities as will be demonstrated by several examples. In

the same context another item of interest, the complexity of which is frequently underestimated, especially in the engineering literature, is the *angular velocity vector*, which arises in two variants, called left and right angular velocity vector, and its relation to the spin (or rotational) operator. In particular, its transformation characteristics during changes of FoRs will be studied.
Because of its appearance in the aforementioned time-derivatives another quantity subjected to critical scrutiny is the *nabla operator*, which is known to assess spatial changes of field quantities. Its transformation properties are also non-trivial, especially if two ways of field description are carefully distinguished, namely the material and the spatial representation, in particular, in context with the notion of a reference configuration. For a comprehensive study the following issues will be of particular importance. First, for physical reasons, observers in different FoRs must, at a given time, study the same point in order to study changes in its vicinity. This point of observation can be a fixed or a moving space point or it can be a material particle. Second, it is required to study the action of the nabla operator in combination with a scalar field being invariant during changes of FoRs and, third, the scalar field must be described by using the same variables in different FoRs. Once an invariance of that action is established the invariance of the operator can be established. We will show that the nabla operator in the current configuration is invariant, whereas the nabla operator of the reference placement is not when changing FoRs.
These investigations will eventually lead us to time derivatives of fields and their objectivity properties. As we shall see the same principles as in the case of the frame-indifference analysis of the nabla operator apply. In the end it will be established that both the total as well as the substantial (material) time-derivative are objective during a change of FoR, in material as well as in the spatial description of fields.
Summarizing we may say that, on the one hand side, this paper emphasizes didactic aspects in context with continuum mechanics notions. On the other hand, it raises awareness for the current situation by commenting on possible die-hard conundrums and offers solutions. Moreover, it is our intention to clarify the situation, in particular, in context with the spatial *vs.* material description of fields which, most recently, has attracted much attention, in particular from the numerical community.

11.2 Frames of Reference – Fundamental Definitions

In view of the ambiguity of possible interpretations we start with a formal introduction of the physical concept called *frame of reference*. One of its aspects concerns the capability of surveying three-dimensional space. Hence:

Definition 11.1. Imagine in an arbitrary point O of three-dimensional Euclidean space, the *origin*, three rigidly connected, non-collinear pointers ("arrows"), $\boldsymbol{e}_1$, $\boldsymbol{e}_2$, and $\boldsymbol{e}_3$ of different unit lengths[1]. The set $\{O, \boldsymbol{e}_1, \boldsymbol{e}_2, \boldsymbol{e}_3\}$ is called a *frame*.

[1] Orthonormality of the pointers is not necessary but possible. Note the oblique angles and the different pointer sizes in Fig. 11.1. The additional requirement of orthonormality sometimes sim-

By suitable extension of these pointers the frame can be used to quantify the distance and the direction of every location in space with respect to O. At this point it should be realized that a frame is not just a mathematical construct like a coordinate system. It is "real" in the sense that it requires us to fabricate yardsticks and survey three different orientations in our physical world.

Definition 11.2. A *body of reference* is defined by a frame to which a set of points (in space) have been added, where a rigid body motion of all the points together with the frame is allowed[2]. The position of the points are labeled relatively to the frame by establishing the *reference coordinate system* x_1, x_2, x_3 with origin O in context with a position vector, $\boldsymbol{x}$:[3]

$$\boldsymbol{x} = x_1 \boldsymbol{e}_1 + x_2 \boldsymbol{e}_2 + x_3 \boldsymbol{e}_3 =: x_k \boldsymbol{e}_k \,, \quad -\infty < (x_1, x_2, x_3) < +\infty, \tag{11.1}$$

where the summation convention has been employed.

Definitions 11.1 and 11.2 are illustrated in Fig. 11.1 in two dimensions for graphical simplicity.
The frame and the reference coordinate system determine the reference body. They are "immutable." This is supposed to mean that once introduced they cannot be changed or this would lead to a different frame of reference. However, in order to describe quantitative characteristics of motion we must be able to measure distance *and* time. Hence a "clock" for measuring time t, where $-\infty < t < +\infty$, is needed is well:

Definition 11.3. The reference body with a "clock" is called the *Frame of Reference* (FoR).

Once again, note that a frame of reference is much more than a mathematical construct. Physics is involved due to the requirement of measuring distances in three independent directions and corresponding lengths as well as time. It is impossible to say anything about the motion of the reference body itself, because it stands for itself alone. It is possible to observe and quantify motions of other bodies with respect

---

plifies calculations, for example, if scalar products between the pointers are involved. However, unless stated otherwise, most of the relations presented in this paper do not require orthonormality to hold.

[2] This is to say that the added points are fixed w.r.t. the frame. If the frame moves the points move with it. They are firmly attached to the frame not to "an absolute space." The rigid body motion is just a term to describe that the distance between any two points is the same (as well as their position w.r.t. the frame).

[3] We note in passing that the use of the adjective "reference" can be treacherous: It is customary in many texts on continuum theory to use the notion "reference placement" or "reference configuration" in order to emphasize an initial state of matter. However, this is not meant here. In our context "reference" is w.r.t. the frame or "observer," if one so wishes, and all that needs to be kept in mind is that the coordinates x_k are fixed w.r.t. the frame. In order to say it explicitly: They cannot be time-dependent by way of construction. Moreover, the use of the word "body" in Definition 11.2 is not meant in the sense of a body consisting of matter. The points constituting the body are purely mathematical.

to the reference body. All physical qualities describing motion, such as velocity for example, are measured with respect to the frame of reference and do not have any meaning without it.

Definition 11.4. The capability of determining and of surveying three independent directions of space as well as measuring time in an FoR is called an *observer*.

There are different approaches to the change of an FoR. The difference is mainly due to the difference in the interpretation of tensor quantities. From one point of view, a second-order tensor is a linear operator mapping of one vector onto another, see, *e.g.*, Bowen and Wang (1976), Das (2007), Itskov (2013), or Malvern (1969). Traditionally an index (a.k.a. coordinate or component) notation is used in this approach. Another point-of-view is based on the so-called symbolic (or direct) tensor calculus (see Lebedev et al (2010), or Mase (1970) for both methods). In what follows we will make use of the latter according to which the definition of a second rank tensor reads:

Definition 11.5. Consider a set of vectors $\boldsymbol{a}$, $\boldsymbol{b}$, $\boldsymbol{c}$, $\boldsymbol{d}$, ..., *etc*. Create a formal sum of formal tensor products (a.k.a. dyadic products):

$$\boldsymbol{A} = \boldsymbol{a} \otimes \boldsymbol{b} + \boldsymbol{c} \otimes \boldsymbol{d} + \boldsymbol{e} \otimes \boldsymbol{f} + \dots . \tag{11.2}$$

An element $\boldsymbol{A}$ is called a *second-order tensor* if the tensor product satisfies the usual properties of a product, see Lebedev et al (2010):

$$\begin{aligned} (\lambda \boldsymbol{a}) \otimes \boldsymbol{b} &= \boldsymbol{a} \otimes (\lambda \boldsymbol{b}) = \lambda (\boldsymbol{a} \otimes \boldsymbol{b}), \\ (\boldsymbol{a} + \boldsymbol{b}) \otimes \boldsymbol{c} &= \boldsymbol{a} \otimes \boldsymbol{c} + \boldsymbol{b} \otimes \boldsymbol{c} . \end{aligned} \tag{11.3}$$

However, the tensor product is non-symmetric: If $\boldsymbol{a}$ is not proportional to $\boldsymbol{b}$ then $\boldsymbol{a} \otimes \boldsymbol{b} \neq \boldsymbol{b} \otimes \boldsymbol{a}$.

In agreement with the idea of direct tensor calculus vector and tensor concepts have no meaning without the notion of a reference frame. Therefore, a discussion of vector or tensor quantities has to start by indicating in which frame of reference they are

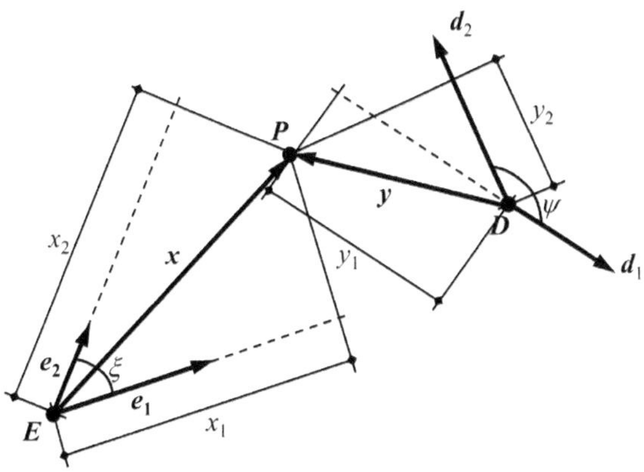

Fig. 11.1 Two FoRs surveying two-dimensional space.

determined. An algebraic operation between objects determined in different frames of reference is impossible. Of course, an observer in one FoR can examine another FoR and describe its motion with respect to its FoR in terms of vector and tensor quantities determined in its FoR. Thus, using direct tensor calculus complicates mathematical transformations regarding a change of reference frame since an additional operation of vector and tensor quantities transferring from one frame of reference to another is needed, namely the concept of *image operation*, which we shall explain shortly.

Moreover, there is also an "operator viewpoint," which allows us to consider a rotational tensor as an object outside of any FoR, see for example Muschik and Restuccia (2008), Liu (2004), or Liu (2005). In this case, from the mathematical point of view, a rotation of one FoR with respect to another is the same as a rotation of one coordinate system with respect to another coordinate system. However, a problem arises if one wants to introduce an angular velocity of the rotation since being a vector it has to be determined in a FoR. It is, therefore, no coincidence that some authors do not introduce an angular velocity at all and deal exclusively with spin tensors instead, see, *e.g.*, Truesdell and Noll (1965) or Müller (1985). In what follows we shall not make use of that "operator viewpoint" of rotation.

In order to evaluate the advantages and disadvantages of such notions it is necessary to understand the essential difference between the operation of a change between frames of reference and a change between coordinate systems. It is fair to say that quite often people do not distinguish between the concepts of frame of reference and coordinate system. But as pointed out in Cornille (1993), pg. 149:

> ... a distinction between mathematical sets of coordinates and physical frames of reference must be made. The ignorance of such distinction is the source of much confusion ...

or in Nerlich (1994), pg. 64-65:

> ... the idea of a reference frame is really quite different from that of a coordinate system. Frames differ just when they define different *spaces* (sets of *rest* points) or times (sets of simultaneous events). So the ideas of a space, a time, of rest and simultaneity, go inextricably together with that of frame. However, a mere shift of origin, or a purely spatial rotation of space coordinates results in a new coordinate system. So frames correspond at best to *classes* of coordinate systems. ...

In other words, in addition to the reference coordinate system one is free to choose any mathematical coordinate system in which the equations are specified. However, the reference coordinate system is a distinctive one since it determines the frame of reference. As an example consider a first coordinate transformation within an FoR, $x_i' = \bar{x}_i'(x_j)$, $i, j \in (1,2,3)$. On top of that we now impose a second coordinate transformation $x_i'' = \check{x}_i''(x_j') = \check{x}_i''(\bar{x}_j'(x_k)) \equiv \bar{x}_i''(x_j)$, $i, j, k \in (1,2,3)$. Note that these operations are applied in context with the spatial dependence of a physical field quantity and lead to no change of the meaning or value of that physical quantity. Indeed, a change of the coordinate system is a purely mathematical operation, where an observer "sensing" vector quality is not needed. That is why in this case there is no difference how a vector is considered: as a directed segment or as a set of three components. Even more, then we can completely exclude base vectors from our considerations and deal only with vector components.

However, if we perform a change of the FoR the situation is completely different. For a change of the FoR the existence of an observer is assumed, who can "sense" vector qualities and provide some operations with them. Two aspects should be mentioned. First, the observer can sense a directed segment (by virtue of a starting point plus a direction with a certain length in that direction), but not vector components. Second, if there are two frames of reference then all vectorial and tensorial quantities are to be determined exclusively in one of them. *A priori* the observer in the first frame of reference cannot perform any operations with quantities determined in the second frame. He has to create *images* of these quantities in his frame of reference first.[4]

The operation of "image creation" or vector "transfer" from one frame of reference to another is an informal[5] operation assuming an observer's existence. This informal operation (and only this) accounts for the difference between changes of frames of reference and changes of coordinate systems. Unfortunately in the majority of papers this operation is either silently implied or not provided at all. This definitely adds some confusion to dealing with a change of frame of reference.

Thus within the direct tensor calculus every observer may determine in his FoR a rotational tensor describing the orientation of another FoR with respect to him and use it in order to calculate the spin tensor and angular velocity corresponding to this rotational tensor. An observer in another FoR may create an image of these rotational and spin tensors and this angular velocity vector in his frame. This approach increases the amount of vectorial and tensorial quantities characterizing rotation, but at the same time every quantity is determined in its corresponding FoR, has a clear physical meaning, and all quantities are consistently related. Recall that within the operator viewpoint the same rotational tensor is used by two different observers (to be precise, the second observer uses the transposed rotational tensor). In fact, within this approach the rotational tensor performs implicitly two operations: a formal one, which is just a rotation, and an informal one where a vector is "transferred" from one frame of reference to another. In other words, reducing the number of variables leads to dubious physical meaning of the quantities, and this involves the danger of mistakes or confusion. What is more preferable: brevity or eliminating ambiguity in meaning? We believe that we should not strive for mathematical simplicity at

---

[4] One might think that by using the term "image" we follow the mathematical literature where it is synonymously used with the word "mapping," *e.g.*, in Truesdell (1991), pg. 319 for connecting second order tensors without a concrete physical meaning. However, this is not meant here. Our use of this term is physics-based: The quantities leading to the image of an external object were originally obtained by measurement in the corresponding FoR. This subtle distinction goes back to the work of Zhilin, who uses the term "отпечаток" in this context, which may be translated as print, imprint, impression, mark, or sometimes also the word "образ" meaning image, reflection, or representation, see Zhilin (2001), pp. 172 or Zhilin (2003), pp. 87.

[5] This choice of word was adapted from the Russian ("формально-математический" *vs.* "неформальнльно-физический," see, for example, Zhilin (2003), pg. 190): A *formal* operation can be expressed in purely mathematical terms. An *informal* one is physics-based and requires further rules and explanations as to how this operation can be performed measurement-wise.

the expense of clarity of meaning. Mathematical errors are rarely permitted, and if permitted, soon corrected. Dubious meanings lead to a debate, ongoing for many years.

11.3 Changing Frames of Reference

Imagine now two frames of reference that move relatively to each other in an arbitrary manner. The first one is determined by the reference frame $\{E, \boldsymbol{e}_1, \boldsymbol{e}_2, \boldsymbol{e}_3\}$, the reference coordinate system x_1, x_2, x_3, and the clock t. This frame of reference is called the *x-system*. The second one is characterized by the reference frame $\{D, \boldsymbol{d}_1, \boldsymbol{d}_2, \boldsymbol{d}_3\}$, the reference coordinate system y_1, y_2, y_3, and the (same) clock t (also compare the symbols in Fig. 11.1). This frame of reference is called the *y-system*. Note that neither the $\boldsymbol{e}_i$, $i = 1, 2, 3$ nor the $\boldsymbol{d}_j$, $j = 1, 2, 3$ are time-dependent. They cannot be, because they were chosen in a specific FoR and they are only relevant to that FoR, where they stay put.
Quantities determined in the x-system are denoted by a tilde, $\tilde{(\cdot)}$, and quantities determined in the y-system are denoted by a chevron-shaped circumflex, $\hat{(\cdot)}$. Quantities characterizing rotation of the y-system with respect to the x-system are denoted by the index d, quantities characterizing rotation of the x-system with respect to the y-system are denoted by the index e. Finally, quantities related to particle characteristics with respect to the x-system and to the y-system are denoted by indices x and y, respectively. This additional indicator besides the tilde and the hat will prove valuable when we begin the discussion of frame indifference of such objects. Moreover, note that we assume that both frames have the same clocks. In fact, *a priori* the clocks of both FoRs should be different and, consequently one would expect the times to run differently, t^x and t^y, just like the reference coordinates, x_k and y_k, are different. However, in classical physics there is no need to distinguish between times in different FoRs. Hence we shall write $t = t^x = t^y$.
The vector diagrams shown in Fig. 11.2 lay the foundation for the most general kinematic relations between two FoRs in classical physics, the so-called Euclidean transformation (see Müller (1985), pg. 4, Holzapfel (2000), pp. 180, or Hutter and Jöhnk (2004), pg. 117 for this terminology). The viewpoint of the x-system is expressed in the left inset and that of the y-system on the right. We must carefully distinguish:

$$\tilde{\boldsymbol{x}} - \tilde{\boldsymbol{x}}^D = \tilde{\boldsymbol{y}} \quad \text{or} \quad \hat{\boldsymbol{y}} - \hat{\boldsymbol{y}}^E = \hat{\boldsymbol{x}}. \tag{11.4}$$

The first equation describes the situation completely w.r.t. to the x-FoR with the base $\boldsymbol{e}_k$[6] and the second one completely w.r.t. to the y-FoR with the base $\boldsymbol{d}_k$. Following Eq. (11.1) we must write for the quantities in the first relation:

$$\tilde{\boldsymbol{x}}(t) = \tilde{x}_k(t)\boldsymbol{e}_k \,, \; \tilde{\boldsymbol{x}}^D(t) = \tilde{x}^D_k(t)\boldsymbol{e}_k \,, \; \tilde{\boldsymbol{y}}(t) = \tilde{y}_k(t)\boldsymbol{e}_k, \tag{11.5}$$

[6] For reasons of graphical simplicity the pointers in Fig. 11.2 are shown as orthonormal vectors. However, the subsequent relations hold without this constraint.

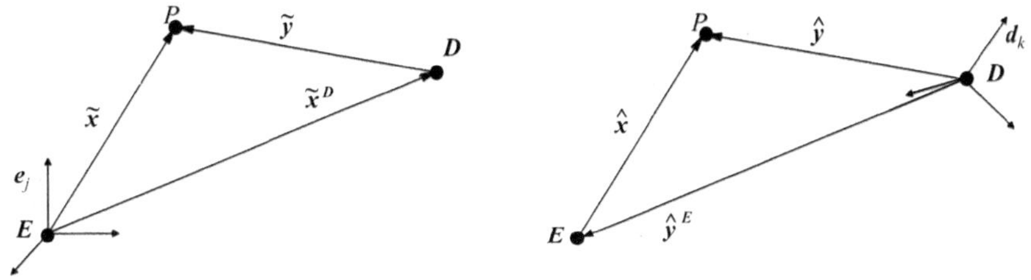

Fig. 11.2: Position vectors in two FoRs (see text).

and for the quantities in the second relation:

$$\hat{\boldsymbol{y}}(t) = \hat{y}_k(t)\boldsymbol{d}_k\,,\ \hat{\boldsymbol{y}}^E(t) = \hat{y}^E_k(t)\boldsymbol{d}_k\,,\ \hat{\boldsymbol{x}}(t) = \hat{x}_k(t)\boldsymbol{d}_k. \tag{11.6}$$

We now proceed to explain how the quantities $\tilde{\boldsymbol{x}}$ and $\hat{\boldsymbol{x}}$ as well as $\hat{\boldsymbol{y}}$ and $\tilde{\boldsymbol{y}}$ are related. We shall learn that $\hat{\boldsymbol{x}}$ is the image of $\tilde{\boldsymbol{x}}$ in the y-system and that $\tilde{\boldsymbol{y}}$ is the image of $\hat{\boldsymbol{y}}$ in the x-system, $\tilde{\boldsymbol{x}}$ and $\hat{\boldsymbol{y}}$ originally being introduced in the x and in the y-system, respectively.

11.3.1 Kinematic Quantities and Their Images

An observer belonging to the x-system watches the reference frame of the y-system, $\boldsymbol{d}_k$, and determines its orientation with respect to the reference frame of his own system at every moment of his time. In other words, he creates images of the $\boldsymbol{d}_k$'s in his x-system as vectors $\tilde{\boldsymbol{d}}_k(t)$, $k = 1, 2, 3$. These vectors have certain orientations w.r.t. the base $\boldsymbol{e}_k$ of the x-system. This is to say that they form angles, which are measurable quantities as a function of time, t and rotate w.r.t. each other.

In order to describe this rotation as general and yet as easily as possible we argue as follows. In an abstract manner of speech a time-dependent rotation tensor ${}^d\tilde{\boldsymbol{Q}}(t)$, pertinent to the x-system "turns" the base vectors $\boldsymbol{e}_k$ onto $\tilde{\boldsymbol{d}}_k(t)$:

$$\tilde{\boldsymbol{d}}_k(t) = {}^d\tilde{\boldsymbol{Q}}(t)\cdot\boldsymbol{e}_k, \quad k = 1, 2, 3. \tag{11.7}$$

Note that in this equation it can be assumed without loss of generality that at the initial time t_0 we have ${}^d\tilde{\boldsymbol{Q}}(t_0) = \tilde{\mathbf{I}}$, where $\tilde{\mathbf{I}}$ stands for the identity tensor of the x-system. In order to see this a more refined explanation is required. Assume that initially at time t_0 the oblique "pointers," $\boldsymbol{e}_k$ and $\tilde{\boldsymbol{d}}_k(t_0)$, as pertinent to the x-FoR, are situated as shown in the left inset of Fig. 11.3. Then, in a first effort a (constant) tensor $\tilde{\boldsymbol{A}}$ is applied to the pointers $\tilde{\boldsymbol{d}}_k(t_0)$, such that $\tilde{\boldsymbol{A}}\cdot\tilde{\boldsymbol{d}}_k(t_0)$ leads to the same lengths and to the same included angles as for the vectors of the initial base, $\boldsymbol{e}_k$, *i.e.*, $|\boldsymbol{e}_k| = |\tilde{\boldsymbol{A}}\cdot\tilde{\boldsymbol{d}}_k(t_0)|$, and

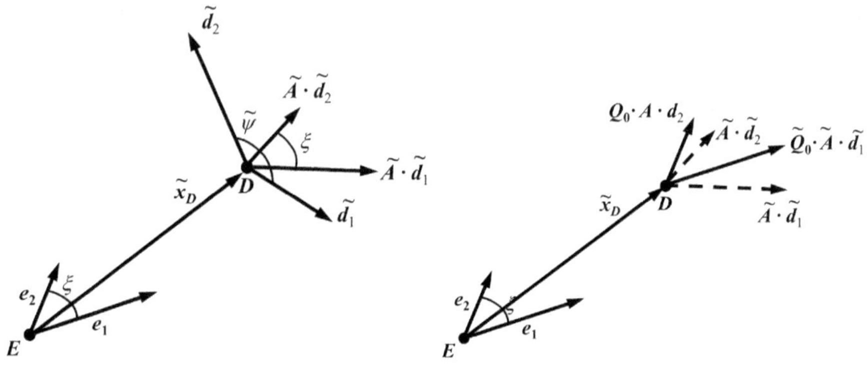

Fig. 11.3: Initial positioning of base vector images.

$$\frac{\boldsymbol{e}_k \cdot \boldsymbol{e}_l}{|\boldsymbol{e}_k||\boldsymbol{e}_l|} = \left.\frac{(\tilde{\boldsymbol{A}} \cdot \tilde{\boldsymbol{d}}_k) \cdot (\tilde{\boldsymbol{A}} \cdot \tilde{\boldsymbol{d}}_l)}{|\tilde{\boldsymbol{A}} \cdot \tilde{\boldsymbol{d}}_k||\tilde{\boldsymbol{A}} \cdot \tilde{\boldsymbol{d}}_l)|}\right|_{t=t_0} .$$

However, $\boldsymbol{e}_k$ and $\tilde{\boldsymbol{A}} \cdot \tilde{\boldsymbol{d}}_k(t_0)$ do not point in the same direction yet. In order to achieve that we apply a second operation (see the right inset in Fig. 11.3) namely a constant initial rotation (or even a combination of rotation and reflection if the FoRs have different orientations) tensor, $\tilde{\boldsymbol{Q}}_0$, such that $\boldsymbol{e}_k = \tilde{\boldsymbol{Q}}_0 \cdot \tilde{\boldsymbol{A}} \cdot \tilde{\boldsymbol{d}}_k(t_0)$. With this configuration in mind further rotations ${}^d\tilde{\boldsymbol{Q}}(t)$ can now be performed and we must obviously conclude that ${}^d\tilde{\boldsymbol{Q}}(t_0) = \tilde{\mathbf{I}}$. In what follows we shall for simplicity omit the multiplications by $\boldsymbol{A}$ and $\boldsymbol{Q}_0$, because these are constant tensors. Moreover, in what follows we will assume that $\boldsymbol{e}_k$ is an orthonormal base in order to facilitate some intermediate calculations.

Note that in contrast to the pointers $\boldsymbol{d}_j$, $j = 1, 2, 3$, their images $\tilde{\boldsymbol{d}}_k$, $k = 1, 2, 3$, are time-dependent. They must be, because from the standpoint of the x-observer the y-observer is rotating. Also note that Eq. (11.7) is a relation completely based in the x-system. In fact, it determines the operation of transfer of the base vectors of the y-system to the x-system. $\tilde{\boldsymbol{d}}_k$ is how the x-observer describes y-base vectors (that initially coincide with $\boldsymbol{e}_k$) with respect to the x-system. In other words (11.7) says how we must transfer the $\boldsymbol{d}_k$ to the x-system, *i.e.*, we introduce the image operation formally by

$$\boldsymbol{d}_k \leftrightarrow \tilde{\boldsymbol{d}}_k(t) = {}^d\tilde{\boldsymbol{Q}}(t) \cdot \boldsymbol{e}_k, \; k = 1, 2, 3. \tag{11.8}$$

The operator ${}^d\tilde{\boldsymbol{Q}}(t)$ has the usual properties of an orthogonal tensor, namely:

$${}^d\tilde{\boldsymbol{Q}} \cdot {}^d\tilde{\boldsymbol{Q}}^\top = {}^d\tilde{\boldsymbol{Q}}^\top \cdot {}^d\tilde{\boldsymbol{Q}} = \tilde{\mathbf{I}}, \; \det {}^d\tilde{\boldsymbol{Q}} = +1, \tag{11.9}$$

where the symbol $(\cdot)^\top$ refers to transposition. Moreover,we choose the plus sign in the relation for the determinant so that ${}^d\tilde{\boldsymbol{Q}}$ is a true rotation. Possible reflections are handled by the formerly introduced matrix $\tilde{\boldsymbol{Q}}_0$.

We note in passing that like any second order tensor of the x-FoR ${}^{d}\tilde{\boldsymbol{Q}}(t)$ can also be expressed component-wise as follows:

$$ {}^{d}\tilde{\boldsymbol{Q}} = {}^{d}\tilde{Q}_{ij}\boldsymbol{e}_i \otimes \boldsymbol{e}_j. \tag{11.10}$$

We shall use this relation soon in order to connect with expressions frequently found in the literature. Moreover, further below we will encounter a possibility of how to express a rotation tensor in terms of a momentary unit axis and an angle of rotation based on the so-called Rodrigues formula.

We now return to the vector relations at the beginning of Sect. 11.3. We wish to find an expression for the image of $\hat{\boldsymbol{y}}$, called $\tilde{\boldsymbol{y}}$, an operation that we denote formally by $\hat{\boldsymbol{y}} \leftrightarrow \tilde{\boldsymbol{y}}$. The image of any vector originally defined in one FoR is obtained in another FoR by keeping the components and by replacing the fixed base by the time-dependent image base vectors. Hence, in view of Eq. (11.7) we write for the image of $\hat{\boldsymbol{y}}$:

$$\hat{\boldsymbol{y}}(t) \leftrightarrow \tilde{\boldsymbol{y}}(t) = \hat{y}_k(t)\tilde{\boldsymbol{d}}_k(t). \tag{11.11}$$

Thus the relation $(11.4)_1$ can be expressed in the x-system by using Eqs. (11.7) and (11.10) as follows:

$$\begin{aligned}\tilde{\boldsymbol{x}}(t) - \tilde{\boldsymbol{x}}^D(t) = \tilde{\boldsymbol{y}}(t) = \hat{y}_k(t)\tilde{\boldsymbol{d}}_k = \hat{y}_k(t)\,{}^{d}\tilde{\boldsymbol{Q}}(t) \cdot \boldsymbol{e}_k \\ = \hat{y}_k(t)\,{}^{d}\tilde{Q}_{ml}(t)\boldsymbol{e}_m \otimes \boldsymbol{e}_l \cdot \boldsymbol{e}_k = {}^{d}\tilde{Q}_{lk}(t)\hat{y}_k(t)\boldsymbol{e}_l.\end{aligned} \tag{11.12}$$

Of course, the last step is only possible if we assume orthonormality of the base $\boldsymbol{e}_k$. Then, however, we may conclude in view of Eq. $(11.5)_3$ that:

$$\tilde{y}_l(t) = {}^{d}\tilde{Q}_{lk}(t)\hat{y}_k(t). \tag{11.13}$$

Note that the quality of this equation is different from Eq. (11.7): Components from the y-system, $\hat{y}_k$, are related to components from the x-system, $\tilde{y}_l$, by means of components in the x-system, ${}^{d}\tilde{Q}_{lk}$. Because of Eqs. (11.4) and $(11.5)_{1,2}$ we may also write:

$$\tilde{x}_k(t) = {}^{d}\tilde{Q}_{kl}(t)\hat{y}_l(t) + \tilde{x}_k^D(t). \tag{11.14}$$

Relations like this can frequently be found in continuum textbooks, see for example Müller (1985), pg. 4, or Hutter and Jöhnk (2004), pg. 117, where they are referred to as Euclidean transformations, however, frequently without making the subtle distinction of images and proper representations w.r.t. to just one FoR. In this form they hold for an orhonormal base $\boldsymbol{e}_k$.

A similar strategy holds if we view vectors originally declared in the x-FoR as images in the y-FoR. Now the observer belonging to the y-system watches the reference frame of the x-system, $\boldsymbol{e}_k$, and determines their images, *i.e.*, $\boldsymbol{e}_k \leftrightarrow \hat{\boldsymbol{e}}_k(t)$, in the y-system as vectors $\hat{\boldsymbol{e}}_k(t)$ by a rotation tensor ${}^{e}\hat{\boldsymbol{Q}}$ as follows:

$$\hat{\boldsymbol{e}}_k(t) = {}^{e}\hat{\boldsymbol{Q}} \cdot \boldsymbol{d}_k, \quad k = 1, 2, 3, \tag{11.15}$$

with

$$\boldsymbol{e}_k \leftrightarrow \hat{\boldsymbol{e}}_k(t) = {}^e\hat{\boldsymbol{Q}}(t) \cdot \boldsymbol{d}_k,\ k = 1, 2, 3, \tag{11.16}$$

and

$$ {}^e\hat{\boldsymbol{Q}} \cdot {}^e\hat{\boldsymbol{Q}}^\top = {}^e\hat{\boldsymbol{Q}}^\top \cdot {}^e\hat{\boldsymbol{Q}} = \hat{\mathbf{I}},\ \det {}^e\hat{\boldsymbol{Q}} = \pm 1, \tag{11.17}$$

where the symbol $\hat{\mathbf{I}}$ refers to the unit tensor of the y-FoR.
In coordinate form we may write for the rotation tensor ${}^e\hat{\boldsymbol{Q}}$, a quantity of the y-FoR:

$$ {}^e\hat{\boldsymbol{Q}} = {}^e\hat{Q}_{ij} \boldsymbol{d}_i \otimes \boldsymbol{d}_j. \tag{11.18}$$

In complete analogy to (11.7) the relations (11.15) determine the operation of transfer of the base vectors of the x-system to the y-system. The rotation of the x-system with respect to the y-system is described by means of the rotation tensor ${}^e\hat{\boldsymbol{Q}}(t)$ determined in the y-system. As far as the corresponding Euclidean transformation is concerned we first write for the image of $\tilde{\boldsymbol{x}}$ in the y-system:

$$\tilde{\boldsymbol{x}} \leftrightarrow \hat{\boldsymbol{x}} = \tilde{x}_k(t) \hat{\boldsymbol{e}}_k(t). \tag{11.19}$$

Thus the relation $(11.4)_2$ can be expressed in the y-system by using Eq. (11.15) and (11.18) as follows:

$$\hat{\boldsymbol{y}}(t) - \hat{\boldsymbol{y}}^E(t) = \hat{\boldsymbol{x}}(t) = {}^e\hat{Q}_{lk}(t) \tilde{x}_k(t) \boldsymbol{d}_l \tag{11.20}$$

by assuming orthonormality of the base $\boldsymbol{d}_k$ similarly as in Eq. (11.12). And now, in view of Eq. $(11.6)_3$, we conclude that:

$$\hat{x}_l(t) = {}^e\hat{Q}_{lk}(t) \tilde{x}_k(t), \tag{11.21}$$

and because of Eqs. $(11.6)_{1,2}$ we may write:

$$\hat{y}_k(t) = {}^e\hat{Q}_{kl}(t) \tilde{x}_l(t) + \hat{y}_k^E, \tag{11.22}$$

in perfect analogy to Eq. (11.14).
In order to study the course of motion time derivatives of rotation tensors will play a roll, which can easily be mastered by the so-called Poisson relations. At this point we simply present them:

$$\frac{\mathrm{d}\,{}^d\tilde{\boldsymbol{Q}}}{\mathrm{d}t} = \tilde{\boldsymbol{\omega}}^d \times {}^d\tilde{\boldsymbol{Q}}, \quad \frac{\mathrm{d}\,{}^d\tilde{\boldsymbol{Q}}}{\mathrm{d}t} = {}^d\tilde{\boldsymbol{Q}} \times \tilde{\boldsymbol{\Omega}}^d, \tag{11.23}$$

and

$$\frac{\mathrm{d}\,{}^e\hat{\boldsymbol{Q}}}{\mathrm{d}t} = \hat{\boldsymbol{\omega}}^e \times {}^e\hat{\boldsymbol{Q}}, \quad \frac{\mathrm{d}\,{}^e\hat{\boldsymbol{Q}}}{\mathrm{d}t} = {}^e\hat{\boldsymbol{Q}} \times \hat{\boldsymbol{\Omega}}^e. \tag{11.24}$$

Note that the symbols $\tilde{\boldsymbol{\omega}}^d$, $\hat{\boldsymbol{\omega}}^e$ and $\tilde{\boldsymbol{\Omega}}^d$, $\hat{\boldsymbol{\Omega}}^e$ refer to left and right angular velocity vectors of the corresponding FoRs, respectively, and that the cross product between a vector $\boldsymbol{a} = a_k \boldsymbol{e}_k$ and a tensor $\boldsymbol{A} = A_{ij} \boldsymbol{e}_i \otimes \boldsymbol{e}_j$ in the orthonormal base $\boldsymbol{e}_k$ can be evaluated by:

$$\boldsymbol{a} \times \boldsymbol{A} = \varepsilon_{lki} a_k A_{ij} \boldsymbol{e}_l \otimes \boldsymbol{e}_j\,,\ \boldsymbol{A} \times \boldsymbol{a} = \varepsilon_{ljk} A_{ij} a_k \boldsymbol{e}_i \otimes \boldsymbol{e}_l\,, \tag{11.25}$$

This holds analogously for tensors in the (orthonormal) base $\boldsymbol{d}_k$, ε_{ljk} being the totally antisymmetric Levi-Civita symbol in that base. More details can be found in the Appendix to this paper.

11.3.2 Rotation of one Reference Frame with Respect to Another

In order to interpret the various rotation tensors of the x- and of the y-systems in a rather suggestive geometrical way we start from the Rodrigues formula.[7] It reads when applied to ${}^d\tilde{\boldsymbol{Q}}$:

$$ {}^d\tilde{\boldsymbol{Q}}(t) = [1 - \cos\tilde{\psi}^d(t)]\,\tilde{\boldsymbol{m}}^d(t) \otimes \tilde{\boldsymbol{m}}^d(t) + \cos\tilde{\psi}^d(t)\tilde{\mathbf{I}} + \sin\tilde{\psi}^d(t)\,\tilde{\boldsymbol{m}}^d(t) \times \tilde{\mathbf{I}}, \tag{11.26} $$

where $\tilde{\psi}^d(t)$ is the angle of rotation of the y-system with respect to the x-system, and $\tilde{\boldsymbol{m}}^d(t)$ is a unit vector, which determines the direction of the axis of rotation in the x-system. The image of the tensor (11.26) in the y-system is obtained by keeping the components and exchanging the base vectors:

$$ {}^d\hat{\boldsymbol{Q}}(t) = \left[1 - \cos\tilde{\psi}^d(t)\right]\hat{\boldsymbol{m}}^d(\hat{t}) \otimes \hat{\boldsymbol{m}}^d(t) + \cos\tilde{\psi}^d(t)\hat{\mathbf{I}} + \sin\tilde{\psi}^d(t)\,\hat{\boldsymbol{m}}^d(t) \times \hat{\mathbf{I}}, \tag{11.27} $$

where the vector $\hat{\boldsymbol{m}}^d(t)$ determining the direction of the axis of rotation in the y-system is the image of the vector $\tilde{\boldsymbol{m}}^d(t)$, *i.e.*, again, the components are kept and the base vectors are exchanged:

$$ \tilde{\boldsymbol{m}}^d(t) \leftrightarrow \hat{\boldsymbol{m}}^d(t) = \tilde{m}^d_k(t)\hat{\boldsymbol{e}}_k(t). \tag{11.28} $$

Analogously, when starting from the rotation tensor ${}^e\hat{\boldsymbol{Q}}(t)$ in the y-system:

$$ {}^e\hat{\boldsymbol{Q}}(t) = \left[1 - \cos\hat{\psi}^e(t)\right]\hat{\boldsymbol{m}}^e(t) \otimes \hat{\boldsymbol{m}}^e(t) + \cos\hat{\psi}^e(t)\hat{\mathbf{I}} + \sin\hat{\psi}^e(t)\,\hat{\boldsymbol{m}}^e(t) \times \hat{\mathbf{I}}, \tag{11.29} $$

its image in the x-system, ${}^e\hat{\boldsymbol{Q}}(t) \leftrightarrow {}^e\tilde{\boldsymbol{Q}}(t)$, is given by:

$$ {}^e\tilde{\boldsymbol{Q}}(t) = \left[1 - \cos\hat{\psi}^e(t)\right]\tilde{\boldsymbol{m}}^e(t) \otimes \tilde{\boldsymbol{m}}^e(t) + \cos\hat{\psi}^e(t)\tilde{\mathbf{I}} + \sin\hat{\psi}^e(t)\,\tilde{\boldsymbol{m}}^e(t) \times \tilde{\mathbf{I}}, \tag{11.30} $$

where the vector $\tilde{\boldsymbol{m}}^e(t)$ determining the direction of the axis of rotation in the x-system is an image of the vector $\hat{\boldsymbol{m}}^e(t)$:

$$ \hat{\boldsymbol{m}}^e(t) \leftrightarrow \tilde{\boldsymbol{m}}^e(t) = \hat{m}^e_k(t)\tilde{\boldsymbol{d}}_k(t). \tag{11.31} $$

For physical reasons the axes of rotation of the x-system with respect to the y-system and those of the y-system with respect to the x-system must coincide in direction and in orientation. Moreover, the angles of rotation must be equal in magnitude but opposite in sign. Thus the following relations hold:

---

[7] A.k.a. Euler's theorem in the Russian literature, *cf.*, Zhilin (2001), pg. 124.

$$\tilde{\boldsymbol{m}}^e(t) = \tilde{\boldsymbol{m}}^d(t), \quad \hat{\boldsymbol{m}}^d(t) = \hat{\boldsymbol{m}}^e(t), \quad \hat{\psi}^e(t) = -\tilde{\psi}^d(t). \tag{11.32}$$

Since $\left(\tilde{\boldsymbol{m}}^d \times \tilde{\mathbf{I}}\right)^{\top} = -\tilde{\boldsymbol{m}}^d \times \tilde{\mathbf{I}}$ and $\left(\hat{\boldsymbol{m}}^e \times \hat{\mathbf{I}}\right)^{\top} = -\hat{\boldsymbol{m}}^e \times \hat{\mathbf{I}}$ we arrive once more at the relations (11.130).

The corresponding relation between the angular velocities is not so obvious. First note that the image of the angular velocity does not correspond to the image of the rotation tensor. Hence, in particular, naively postulated Poisson relations do not hold and we must state that:

$$\frac{\mathrm{d}^d\hat{\boldsymbol{Q}}}{\mathrm{d}t} \neq \hat{\boldsymbol{\omega}}^d \times {}^d\hat{\boldsymbol{Q}}, \; \frac{\mathrm{d}^d\hat{\boldsymbol{Q}}}{\mathrm{d}t} \neq {}^d\hat{\boldsymbol{Q}} \times \hat{\boldsymbol{\Omega}}^d, \; \frac{\mathrm{d}^e\tilde{\boldsymbol{Q}}}{\mathrm{d}t} \neq \tilde{\boldsymbol{\omega}}^e \times {}^e\tilde{\boldsymbol{Q}}, \; \frac{\mathrm{d}^e\tilde{\boldsymbol{Q}}}{\mathrm{d}t} \neq {}^e\tilde{\boldsymbol{Q}} \times \tilde{\boldsymbol{\Omega}}^e. \tag{11.33}$$

As an example we will now derive an "extended" Poisson relation that replaces $(11.33)_1$. We start by observing that Eq. $(11.23)_1$ in combination with Eqs. (11.26) is fulfilled by the following expression for the left angular velocity:

$$\tilde{\boldsymbol{\omega}}^d = \frac{\mathrm{d}\tilde{\psi}^d}{\mathrm{d}t}\tilde{\boldsymbol{m}}^d + \sin\tilde{\psi}^d \frac{\mathrm{d}\tilde{\boldsymbol{m}}^d}{\mathrm{d}t} + (1-\cos\tilde{\psi}^d)\tilde{\boldsymbol{m}}^d \times \frac{\mathrm{d}\tilde{\boldsymbol{m}}^d}{\mathrm{d}t}. \tag{11.34}$$

However, an analogous equation does not hold for the image vectors, because the image of the time derivative of $\tilde{\boldsymbol{m}}^d(t)$ does not equal the time derivative of the image of $\tilde{\boldsymbol{m}}^d(t)$ as can be seen from Eq. (11.28).

Let us determine the angular velocities corresponding to the rotation tensors ${}^d\tilde{\boldsymbol{Q}}$ and ${}^e\tilde{\boldsymbol{Q}}$. Hence, we start from Eq. (11.34) and obtain by using (11.15), (11.24), (11.28) and (11.34):

$$\frac{\mathrm{d}\tilde{\psi}^d}{\mathrm{d}t}\hat{\boldsymbol{m}}^d + \sin\tilde{\psi}^d \frac{\mathrm{d}\hat{\boldsymbol{m}}^d}{\mathrm{d}t} + (1-\cos\tilde{\psi}^d)\hat{\boldsymbol{m}}^d \times \frac{\mathrm{d}\hat{\boldsymbol{m}}^d}{\mathrm{d}t} = \tag{11.35}$$

$$\hat{\boldsymbol{\omega}}^d + \sin\tilde{\psi}^d\,\hat{\boldsymbol{\omega}}^e \times \hat{\boldsymbol{m}}^d + (1-\cos\tilde{\psi}^d)\hat{\boldsymbol{m}}^d \times (\hat{\boldsymbol{\omega}}^e \times \hat{\boldsymbol{m}}^d).$$

Now by employing Eq. (11.27) this may be rearranged to:

$$\frac{\mathrm{d}\tilde{\psi}^d}{\mathrm{d}t}\hat{\boldsymbol{m}}^d + \sin\tilde{\psi}^d \frac{\mathrm{d}\hat{\boldsymbol{m}}^d}{\mathrm{d}t} + (1-\cos\tilde{\psi}^d)\hat{\boldsymbol{m}}^d \times \frac{\mathrm{d}\hat{\boldsymbol{m}}^d}{\mathrm{d}t} = \hat{\boldsymbol{\omega}}^d + \hat{\boldsymbol{\omega}}^e - {}^d\hat{\boldsymbol{Q}} \cdot \hat{\boldsymbol{\omega}}^e. \tag{11.36}$$

In view of the resemblance of this result to the left angular velocity pertinent to ${}^d\tilde{\boldsymbol{Q}}$ as shown in Eq. (11.34) we must conclude that the right hand side of Eq. (11.36) is the sought left angular velocity corresponding to ${}^d\hat{\boldsymbol{Q}}$.

Similarly, by substituting the expression for the rotation tensor (11.26) into $(11.23)_2$, we obtain the expression for the right angular velocity:

$$\tilde{\boldsymbol{\Omega}}^d = \frac{\mathrm{d}\tilde{\psi}^d}{\mathrm{d}t}\tilde{\boldsymbol{m}}^d + \sin\tilde{\psi}^d \frac{\mathrm{d}\tilde{\boldsymbol{m}}^d}{\mathrm{d}t} - (1-\cos\tilde{\psi}^d)\tilde{\boldsymbol{m}}^d \times \frac{\mathrm{d}\tilde{\boldsymbol{m}}^d}{\mathrm{d}t}. \tag{11.37}$$

We obtain the expression for the right angular velocity corresponding to ${}^d\hat{\boldsymbol{Q}}$ by a derivation similar to the one given above:

$$\frac{\mathrm{d}\tilde{\psi}^d}{\mathrm{d}t}\hat{\boldsymbol{m}}^d + \sin\tilde{\psi}^d \frac{\mathrm{d}\hat{\boldsymbol{m}}^d}{\mathrm{d}t} - (1-\cos\tilde{\psi}^d)\hat{\boldsymbol{m}}^d \times \frac{\mathrm{d}\hat{\boldsymbol{m}}^d}{\mathrm{d}t} = \tag{11.38}$$

$$\hat{\boldsymbol{\Omega}}^d + {}^d\hat{\boldsymbol{Q}}^\top \cdot \hat{\boldsymbol{\omega}}^e - \hat{\boldsymbol{\omega}}^e.$$

Thus the following "extended" Poisson relations hold in context with the rotation tensor ${}^d\hat{\boldsymbol{Q}}$ (*cf.*, Eqs. (11.23) and (11.24)):

$$\frac{\mathrm{d}{}^d\hat{\boldsymbol{Q}}}{\mathrm{d}t} = (\hat{\boldsymbol{\omega}}^d + \hat{\boldsymbol{\omega}}^e - {}^d\hat{\boldsymbol{Q}} \cdot \hat{\boldsymbol{\omega}}^e) \times {}^d\hat{\boldsymbol{Q}}, \quad \frac{\mathrm{d}{}^d\hat{\boldsymbol{Q}}}{\mathrm{d}t} = {}^d\hat{\boldsymbol{Q}} \times (\hat{\boldsymbol{\Omega}}^d + {}^d\hat{\boldsymbol{Q}}^\top \cdot \hat{\boldsymbol{\omega}}^e - \hat{\boldsymbol{\omega}}^e), \tag{11.39}$$

and analogously for ${}^e\tilde{\boldsymbol{Q}}$:

$$\frac{\mathrm{d}{}^e\tilde{\boldsymbol{Q}}}{\mathrm{d}t} = (\tilde{\boldsymbol{\omega}}^e + \tilde{\boldsymbol{\omega}}^d - {}^e\tilde{\boldsymbol{Q}} \cdot \tilde{\boldsymbol{\omega}}^d) \times {}^e\tilde{\boldsymbol{Q}}, \quad \frac{\mathrm{d}{}^e\tilde{\boldsymbol{Q}}}{\mathrm{d}t} = {}^e\tilde{\boldsymbol{Q}} \times (\tilde{\boldsymbol{\Omega}}^e + {}^e\tilde{\boldsymbol{Q}}^\top \cdot \tilde{\boldsymbol{\omega}}^d - \tilde{\boldsymbol{\omega}}^d). \tag{11.40}$$

By observing Eq. (11.130) and noting that:

$$\frac{\mathrm{d}{}^d\tilde{\boldsymbol{Q}}^\top}{\mathrm{d}t} = -\tilde{\boldsymbol{\Omega}}^d(t) \times {}^d\tilde{\boldsymbol{Q}}^\top, \quad \frac{\mathrm{d}{}^e\hat{\boldsymbol{Q}}^\top}{\mathrm{d}t} = -\hat{\boldsymbol{\Omega}}^e(t) \times {}^e\hat{\boldsymbol{Q}}^\top, \tag{11.41}$$

which hold because of Eqs. $(11.23)_2$ and $(11.24)_2$, the following relations between the angular velocities can be obtained:

$$\tilde{\boldsymbol{\omega}}^e + \tilde{\boldsymbol{\omega}}^d - {}^e\tilde{\boldsymbol{Q}} \cdot \tilde{\boldsymbol{\omega}}^d = -\tilde{\boldsymbol{\Omega}}^d(t), \quad \hat{\boldsymbol{\omega}}^d + \hat{\boldsymbol{\omega}}^e - {}^d\hat{\boldsymbol{Q}} \cdot \hat{\boldsymbol{\omega}}^e = -\hat{\boldsymbol{\Omega}}^e(t). \tag{11.42}$$

From Eq. $(11.42)_1$ it follows via Eqs. $(11.119)_3$ and $(11.24)_3$ that the left angular velocities are related by the following simple equations:

$$\tilde{\boldsymbol{\omega}}^e(t) = -\tilde{\boldsymbol{\omega}}^d(t), \quad \hat{\boldsymbol{\omega}}^d(t) = -\hat{\boldsymbol{\omega}}^e(t). \tag{11.43}$$

Note that these relations express exactly what one would intuitively expect. For example: The image of the left angular velocity, $\tilde{\boldsymbol{\omega}}^e$, *i.e*, how the y-observer views the rotation of the x-observer, is equal to the angular velocity, $\tilde{\boldsymbol{\omega}}^d$, *i.e.*, how the x-observer describes the rotation of the y-observer, with the exception of orientation, *i.e.*, the sign.

The relations for the right angular velocities are slightly more involved:

$${}^e\tilde{\boldsymbol{Q}} \cdot \tilde{\boldsymbol{\Omega}}^e(t) = -{}^d\tilde{\boldsymbol{Q}}(t) \cdot \tilde{\boldsymbol{\Omega}}^d(t), \quad {}^d\hat{\boldsymbol{Q}}(t) \cdot \hat{\boldsymbol{\Omega}}^d(t) = -{}^e\hat{\boldsymbol{Q}}(t) \cdot \hat{\boldsymbol{\Omega}}^e(t). \tag{11.44}$$

Comment I. Note the difference between changes of frame of reference and changes of coordinate system. Consider two different coordinate systems in one frame of reference. The first one, a, co-moving with a rigid body and the second one, b, fixed with respect to the frame of reference. The rotation of the body (or the first coordinate system) with respect to the second one is determined by the rotation tensor ${}^a\tilde{\boldsymbol{Q}}(t) = \boldsymbol{Q}(t)$. And ${}^b\tilde{\boldsymbol{Q}}(t) = \boldsymbol{Q}^\top(t)$ is the rotation tensor of the second coordinate system with respect to the first one. Then, the angular velocities are related by:

$$\tilde{\boldsymbol{\omega}}^a =: \boldsymbol{\omega}, \quad \tilde{\boldsymbol{\Omega}}^a =: \boldsymbol{\Omega}, \quad \tilde{\boldsymbol{\omega}}^b = -\boldsymbol{\Omega}, \quad \tilde{\boldsymbol{\Omega}}^b = -\boldsymbol{\omega}. \tag{11.45}$$

because of relations such as:

$$\frac{\mathrm{d}^a\tilde{\boldsymbol{Q}}}{\mathrm{d}t} = \tilde{\boldsymbol{\omega}}^a \times {}^a\tilde{\boldsymbol{Q}} = {}^a\tilde{\boldsymbol{Q}} \times \tilde{\boldsymbol{\Omega}}^a, \quad \frac{\mathrm{d}^b\tilde{\boldsymbol{Q}}}{\mathrm{d}t} = \tilde{\boldsymbol{\omega}}^b \times {}^b\tilde{\boldsymbol{Q}} = {}^b\tilde{\boldsymbol{Q}} \times \tilde{\boldsymbol{\Omega}}^b. \tag{11.46}$$

Thus, we obtain:

$$ {}^b\tilde{\boldsymbol{Q}}(t) = {}^a\tilde{\boldsymbol{Q}}^\top(t), \quad \tilde{\boldsymbol{\omega}}^b \neq -\tilde{\boldsymbol{\omega}}^a, \quad {}^b\tilde{\boldsymbol{Q}}(t)\cdot\tilde{\boldsymbol{\Omega}}^b \neq -{}^a\tilde{\boldsymbol{Q}}\cdot\tilde{\boldsymbol{\Omega}}^a. \tag{11.47}$$

By comparing (11.47) with (11.130), (11.43), (11.44) one can see a significant difference. Obviously, the reason lies in transferring vectorial and tensorial quantities from one frame of reference to another. More specifically, the reason is that the initial rotation tensors and the angular velocities are related by Eqs. (11.23), (11.24), however, their images are related by Eqs. (11.39), (11.40).

11.3.3 Motion of FoRs with Respect to Each Other

Recall that the position of the origin, $\boldsymbol{D}$, of the y-system with respect to the origin, $\boldsymbol{E}$, of the x-system is given by $\tilde{\boldsymbol{x}}^D(t)$ pertinent to the x-system. $\hat{\boldsymbol{x}}^D(t)$ is the image of this vector in the y-system. In coordinate form we have:

$$\tilde{\boldsymbol{x}}^D(t) \leftrightarrow \hat{\boldsymbol{x}}^D(t) = \hat{x}_k^D(t)\hat{\boldsymbol{e}}_k(t), \tag{11.48}$$

where the $\hat{\boldsymbol{e}}_k(t)$ are given by (11.15). Analogously, $\hat{\boldsymbol{y}}^E(t)$ and $\tilde{\boldsymbol{y}}^E(t)$ are the position vectors of $\boldsymbol{E}$ with respect to $\boldsymbol{D}$ in the y-system and its image determined in the x-system, respectively, such that:

$$\hat{\boldsymbol{y}}^E(t) \leftrightarrow \tilde{\boldsymbol{y}}^E(t) = \tilde{y}_k^E(t)\tilde{\boldsymbol{d}}_k(t), \tag{11.49}$$

where the $\tilde{\boldsymbol{d}}_k(t)$ are given by (11.7). For geometrical reasons the following must hold:

$$\tilde{\boldsymbol{y}}^E(t) = -\tilde{\boldsymbol{x}}^D(t), \; \hat{\boldsymbol{x}}^D(t) = -\hat{\boldsymbol{y}}^E(t). \tag{11.50}$$

The velocity of the point $\boldsymbol{D}$ with respect to the x-system and the velocity of the point $\boldsymbol{E}$ with respect to the y-system are:

$$\tilde{\boldsymbol{v}}_D^x(t) = \frac{\mathrm{d}\tilde{\boldsymbol{x}}^D}{\mathrm{d}t}, \; \hat{\boldsymbol{v}}_E^y(t) = \frac{\mathrm{d}\hat{\boldsymbol{y}}^E}{\mathrm{d}t}. \tag{11.51}$$

The velocities $\tilde{\boldsymbol{v}}_D^x(t)$ and $\hat{\boldsymbol{v}}_E^y(t)$ are determined in different reference frames and cannot be compared *a priori*. However, what we can compare are suitable images, which we have to create in the first place. In coordinate form we can write:

$$\tilde{\boldsymbol{v}}_D^x(t) \leftrightarrow \hat{\boldsymbol{v}}_D^x(t) = \frac{\mathrm{d}\tilde{x}_k^D}{\mathrm{d}t}\hat{\boldsymbol{e}}_k(t), \; \hat{\boldsymbol{v}}_E^y(t) \leftrightarrow \tilde{\boldsymbol{v}}_E^y(t) = \frac{\mathrm{d}\hat{y}_k^E}{\mathrm{d}t}\tilde{\boldsymbol{d}}_k(t). \tag{11.52}$$

Note that the points $\boldsymbol{D}$ and $\boldsymbol{E}$ are immovable in their reference frames. Therefore, $\hat{\boldsymbol{v}}_D^x(t)$ is not the velocity of $\boldsymbol{D}$ with respect to the y-system, and $\tilde{\boldsymbol{v}}_E^y(t)$ is not the velocity of $\boldsymbol{E}$ with respect to the x-system, and they are both not equal to zero. By using Eqs. (11.50) and (11.51) we obtain:

$$\tilde{\boldsymbol{v}}^x_D(t) = -\frac{\mathrm{d}\tilde{\boldsymbol{y}}^E}{\mathrm{d}t}\,,\ \ \hat{\boldsymbol{v}}^y_E(t) = -\frac{\mathrm{d}\hat{\boldsymbol{x}}^D}{\mathrm{d}t}\,. \tag{11.53}$$

We now start by differentiating $\tilde{\boldsymbol{y}}^E(t)$ and $\hat{\boldsymbol{x}}^D(t)$, respectively, with respect to time. Then, by taking into account Eqs. (11.49), $(11.52)_2$, (11.7), $(11.23)_1$, (11.49) and (11.48), $(11.52)_1$, (11.15), $(11.24)_1$, (11.48), respectively, we arrive at:

$$\tilde{\boldsymbol{v}}^y_E = -\tilde{\boldsymbol{v}}^x_D + \tilde{\boldsymbol{\omega}}^d \times \tilde{\boldsymbol{x}}^D\,,\ \ \hat{\boldsymbol{v}}^x_D = -\hat{\boldsymbol{v}}^y_E + \hat{\boldsymbol{\omega}}^e \times \hat{\boldsymbol{y}}^E\,, \tag{11.54}$$

because of Eq. (11.53).

Comment II Note the difference between changes of frame of reference and changes of coordinate system in context with translational movement. Consider two moving coordinate systems in one frame of reference, a and b, respectively. Let the positions of their origins be determined by $\boldsymbol{r}^a(t)$ and $\boldsymbol{r}^b(t)$. The velocities of these points are given by $\boldsymbol{v}^a(t) = \mathrm{d}\boldsymbol{r}^a/\mathrm{d}t$ and $\boldsymbol{v}^b(t) = \mathrm{d}\boldsymbol{r}^b/\mathrm{d}t$, respectively. The velocity of point b with respect to point a is $\boldsymbol{v}^{ba}(t) = \boldsymbol{v}^b(t) - \boldsymbol{v}^a(t)$ and the velocity of point a with respect to point b is $\boldsymbol{v}^{ab}(t) = -\boldsymbol{v}^{ba}(t)$. Evidently a rotation of one coordinate system with respect to another has no influence on their relative velocities. In contrast to this, when speaking about changes between FoRs, relations like $\tilde{\boldsymbol{v}}^y_E = -\tilde{\boldsymbol{v}}^x_D$ and $\hat{\boldsymbol{v}}^x_D = -\hat{\boldsymbol{v}}^y_E$ are valid only if the frames of reference move translationally with respect to each other.

Finally in this section we turn to the Euclidean transformations shown in Eqs. (11.12) and (11.20), respectively. We differentiate both w.r.t. time, t, and obtain by observing Eqs. $(11.52)_1$, (11.11), (11.7), $(11.23)_1$ and $(11.52)_2$, (11.19), (11.15), $(11.24)_1$, respectively, the following result:

$$\tilde{\boldsymbol{v}}^x = \tilde{\boldsymbol{v}}^x_D + \tilde{\boldsymbol{v}}^y + \tilde{\boldsymbol{\omega}}^d \times \tilde{\boldsymbol{y}}\,,\ \ \hat{\boldsymbol{v}}^y = \hat{\boldsymbol{v}}^y_E + \hat{\boldsymbol{v}}^x + \hat{\boldsymbol{\omega}}^e \times \hat{\boldsymbol{x}}. \tag{11.55}$$

In this context we have used the following symbols for the so-called relative velocity:[8]

$$\tilde{\boldsymbol{v}}^x(t) \leftrightarrow \hat{\boldsymbol{v}}^x(t) = \frac{\mathrm{d}\tilde{x}_k}{\mathrm{d}t}\,\hat{\boldsymbol{e}}_k(t)\,,\ \hat{\boldsymbol{v}}^y(t) \leftrightarrow \tilde{\boldsymbol{v}}^y(t) = \frac{\mathrm{d}\hat{y}_k}{\mathrm{d}t}\,\tilde{\boldsymbol{d}}_k(t). \tag{11.56}$$

Note that Eqs. $(11.55)_1$ $(11.55)_2$ hold exclusively in the x and in the y-system, respectively. One can find similarly looking equations in textbooks, for example in Hutter and Jöhnk (2004), pp. 118/119 or in Hibbeler (2010), pp. 377. However, the subtle distinction of images and various angular velocities is not made there.

Two special case are of interest: First, let us put $\boldsymbol{P} \to \boldsymbol{D}$. Then $\tilde{\boldsymbol{v}}^x \to \tilde{\boldsymbol{v}}^x_D$, $\tilde{\boldsymbol{v}}^y \to \boldsymbol{0}$, $\tilde{\boldsymbol{y}} \to \boldsymbol{0}$ and $\hat{\boldsymbol{v}}^y \to \boldsymbol{0}$, $\hat{\boldsymbol{v}}^x \to \hat{\boldsymbol{v}}^x_D$, $\hat{\boldsymbol{x}}^x \to -\hat{\boldsymbol{y}}^E$ and Eq. $(11.55)_1$ turns into the identity $\tilde{\boldsymbol{v}}^x_D = \tilde{\boldsymbol{v}}^x_D$ whereas Eq. $(11.55)_2$ becomes $\boldsymbol{0} = \hat{\boldsymbol{v}}^y_E + \hat{\boldsymbol{v}}^x_D - \hat{\boldsymbol{\omega}}^e \times \hat{\boldsymbol{y}}^E$, which

[8] According to a suggestion on pp. 118/119 in Hutter and Jöhnk (2004) or Hibbeler (2010), pg. 87, we may call the (vector) sum of *translation* velocity, *i.e.*, $\tilde{\boldsymbol{v}}^x_D$ or $\hat{\boldsymbol{v}}^y_E$, plus *relative* velocity, *i.e.*, $\tilde{\boldsymbol{v}}^y$ or $\hat{\boldsymbol{v}}^x$, *frozen* velocity.

is nothing else but Eq. $(11.54)_2$. Now we put $\boldsymbol{P} \to \boldsymbol{E}$. Then $\tilde{\boldsymbol{v}}^x \to \boldsymbol{0}$, $\tilde{\boldsymbol{v}}^y \to \tilde{\boldsymbol{v}}^y_E$, $\tilde{\boldsymbol{y}} \to -\tilde{\boldsymbol{x}}^D$ and $\hat{\boldsymbol{v}}^y \to \hat{\boldsymbol{v}}^y_E$, $\hat{\boldsymbol{v}}^x \to \boldsymbol{0}$, $\hat{\boldsymbol{x}} \to \boldsymbol{0}$ and Eq. $(11.55)_1$ becomes $\boldsymbol{0} = \tilde{\boldsymbol{v}}^x_D + \tilde{\boldsymbol{v}}^y_E - \tilde{\boldsymbol{\omega}}^d \times \tilde{\boldsymbol{x}}^D$, which is nothing else but Eq. $(11.54)_1$, and Eq. $(11.55)_1$ turns into an identity, $\hat{\boldsymbol{v}}^y_E = \hat{\boldsymbol{v}}^y_E$.

11.4 Frame Indifference of Operators

The main objective in this section is to apply the concept of images to find out about the transformation behavior of the total and of the substantial (material) time-derivatives. These, however, contain gradients and, therefore, we shall start our discussion with the behavior of the nabla operator during a change of FoR.

11.4.1 Transformation Properties of Spatial Gradients

This section is dedicated to a discussion on the gradient of continuum theory fields. Formally speaking, this operation will be described by a vectorial quantity, the so-called nabla operator, $\boldsymbol{\nabla}$, (see, *e.g.*, Lebedev et al, 2010, , pg. 95) for the choice of nomenclature). Several issues will be dealt with here, which are unfortunately not completely separable.

First, we must agree on what exactly the gradient of a continuous field is supposed to describe and what it means in terms of mathematics and, more importantly to us, in continuum physics. Once this has been decided we have to study how the operator $\boldsymbol{\nabla}$ is expressed w.r.t. different frames, how images of these expressions are defined, and how these images relate to the original gradient operation in a particular FoR or, in other words, how the operator nabla transforms during changes between FoRs. Third, we will have to study the gradient operation from the two viewpoints of continuum theory, namely material as well as spatial description.

We start by exploring the formal mathematical point-of-view. Consider a continuous, multivariant function $f(x_1, x_2, \ldots, x_n)$. We wish to know by how much it changes when moving from $x_1, x_2, \ldots, x_n$ to $x_1 + \Delta x_1, x_2 + \Delta x_2, \ldots, x_n + \Delta x_n$. Recall that in order to find out we simply have to compute

$$
\Delta f := f(x_1 + \Delta x_1, x_2 + \Delta x_2, \ldots, x_n + \Delta x_n) - f(x_1, x_2, \ldots, x_n) =
$$

$$
\frac{f(x_1 + \Delta x_1, x_2 + \Delta x_2, \ldots, x_n + \Delta x_n) - f(x_1, x_2 + \Delta x_2, \ldots, x_n + \Delta x_n)}{\Delta x_1} \pm \cdots +
$$

$$
\frac{f(x_1, x_2, \ldots, x_n + \Delta x_n) - f(x_1, x_2, \ldots, x_n)}{\Delta x_1} \tag{11.57}
$$

and due to the assumed continuity conclude in the limit the following well-known relation:

$$\mathrm{d}f(x_1,x_2,\dots,x_n) = \frac{\partial f}{\partial x_1}\,\mathrm{d}x_1 + \frac{\partial f}{\partial x_2}\,\mathrm{d}x_2 + \dots + \frac{\partial f}{\partial x_n}\,\mathrm{d}x_n. \tag{11.58}$$

The row vector $\left(\frac{\partial}{\partial x_1}, \frac{\partial}{\partial x_2}, \dots \frac{\partial}{\partial x_n},\right)$ is known as the gradient of a scalar-valued function $f(x_1,x_2,\dots,x_n)$. Obviously in mathematics the "coordinates" of the "point," $x_1,x_2,\dots,x_n$, have only an abstract meaning, and it is possible but unnecessary to assign base vectors $\boldsymbol{e}_k$, $k = 1,\dots,n$ indicating directions and units of length to them. They are simply the variables the function f depends upon. It goes almost without saying that this gradient addresses the change of the function in a given point with the aforementioned coordinates, and, if it is intended to compare different functions depending on the same coordinates, it makes sense to compare the gradients of these functions only in the neighborhood of a point with the same coordinates.

However, moving on to continuum physics, we now limit ourselves to the case $n = 3$, *i.e.*, we think in terms of physical space. Moreover, because we will initially deal with the material description and reference configuration, we interpret x_1,x_2,x_3 as the coordinates of a frame of reference as indicated in Definition 11.1, and identify the scalar-valued function $f(x_k)$ with the reference density distribution $\tilde{\rho}_0(x_k)$ of the x-FoR (say), which may be heterogeneous:

$$\mathrm{d}\tilde{\rho}_0(x_k) = \frac{\partial \tilde{\rho}_0}{\partial x_i}\,\mathrm{d}x_i. \tag{11.59}$$

Let us now consider a continuum within the material description. Recall that in this case we imagine the continuum, $\mathscr{B}$, to consist of undestructible entities of matter, the material points, $P \in \mathscr{B}$. These material points must first be identified and then they will be traced along their passage in space and time. Typically their identification is based on using their positions in space at some (arbitrary but then fixed) time. This identification is made within an FoR and therefore we write $\tilde{\boldsymbol{X}}(x_i)$ in the x-system and $\hat{\boldsymbol{Y}}(y_i)$ in the y-system, the arguments x_i, y_i being shorthand notations for x_1,x_2,x_3 and y_1,y_2,y_3, respectively:

$$\tilde{\boldsymbol{X}}(x_i) = x_k\boldsymbol{e}_k\,,\;\; \hat{\boldsymbol{Y}}(y_i) = y_k\boldsymbol{d}_k, \tag{11.60}$$

and, analogously to the situation illustrated for $\tilde{\boldsymbol{x}}$ and $\hat{\boldsymbol{y}}$ in Fig. 11.2, $\tilde{\boldsymbol{X}}$ and $\hat{\boldsymbol{Y}}$ point to the same location in space.

In order to investigate the invariance of the gradient operator we have to investigate its effect on an invariant scalar (*e.g.*, the mass density, ρ_0) transforming under a change of FoR. We call a scalar expressed in material description invariant, or an Euclidean scalar in material description, if the values of the functions $\tilde{\rho}_0(\tilde{\boldsymbol{X}},t)$ and $\hat{\rho}_0(\hat{\boldsymbol{Y}},t)$ are the same at every moment of time under the condition that the position vectors $\tilde{\boldsymbol{X}}$ and $\hat{\boldsymbol{Y}}$ refer to the same point in space.

We may now rewrite Eq. (11.59) in the following equivalent form:

$$\mathrm{d}\tilde{\rho}_0(\tilde{\boldsymbol{X}}) = \frac{\partial \tilde{\rho}_0}{\partial \tilde{\boldsymbol{X}}}\cdot \mathrm{d}\tilde{\boldsymbol{X}} \quad \text{or} \quad \mathrm{d}\tilde{\rho}_0(\tilde{\boldsymbol{X}}^x) = \tilde{\boldsymbol{\nabla}}_0^x(\tilde{\rho}_0)\cdot \mathrm{d}\tilde{\boldsymbol{X}}, \tag{11.61}$$

where we have defined the gradient or (synonymously) the nabla operator of the reference configuration in the x-system spanned by an orthonormal base by:

$$\tilde{\nabla}_0^x := \frac{\partial}{\partial \tilde{X}} = e_k \frac{\partial}{\partial x_k}. \tag{11.62}$$

Analogous definitions and relations hold in the y-system:

$$\mathrm{d}\hat{\rho}_0(\hat{Y}) = \frac{\partial \hat{\rho}_0}{\partial \hat{X}} \cdot \mathrm{d}\hat{Y} \quad \text{or} \quad \mathrm{d}\hat{\rho}_0(\hat{Y}) = \hat{\nabla}_0^y(\hat{\rho}_0) \cdot \mathrm{d}\hat{Y}, \tag{11.63}$$

and

$$\hat{\nabla}_0^y := \frac{\partial}{\partial \hat{X}} = d_k \frac{\partial}{\partial y_k}. \tag{11.64}$$

Summarizing we conclude that the "coefficients" $\partial/\partial\tilde{x}$ and $\partial/\partial\hat{Y}$ represent the gradients of the scalar-valued functions $\tilde{\rho}_0$ and $\hat{\rho}_0$, respectively. In a formal manner of speech one may carry forward the operation of transfer of the vector $\partial\hat{\rho}_0/\partial\hat{Y}$ into the x-FoR and of the vector $\partial\tilde{\rho}_0/\partial\tilde{X}$ into the y-FoR, and then ask the question regarding the comparison of the image of the vector $\partial\hat{\rho}_0/\partial\hat{Y}$ with the vector $\partial\tilde{\rho}_0/\partial\tilde{X}$ and of the image of the vector $\partial\tilde{\rho}_0/\partial\tilde{X}$ with the vector $\partial\hat{\rho}_0/\partial\hat{Y}$, respectively. However, we shall see that from a physical point-of-view this comparison is pointless. The following arguments are presented here with the objective to show why this comparison does not make sense.

First, we note that the reference position vectors do not depend upon time but that their images do, because of Eqs. (11.15) and (11.7):

$$\hat{X}(t) = x_k \hat{e}_k(t) = x_k {}^d\hat{Q}(t) \cdot d_k, \; \tilde{Y}(t) = y_k \tilde{d}_k(t) = y_k {}^e\tilde{Q}(t) \cdot e_k. \tag{11.65}$$

This means that even if the two position vectors $\tilde{X}$ and $\tilde{Y}$ refer to the same point in physical space at time t_0 they will point to different points at other times. Correspondingly the gradients $\partial\tilde{\rho}_0/\partial\tilde{X}$ and $\partial\tilde{\rho}_0/\partial\tilde{Y}$ reflect variations of mass density in the neighborhood of different points of space. An analogous remark holds for $\hat{Y}$ and $\hat{X}$ and the gradients $\partial\hat{\rho}_0/\partial\hat{Y}$ and $\partial\hat{\rho}_0/\partial\hat{X}$.

In the same manner, even if in both FoRs both mass densities represent functions of a time-independent position vector, the mass densities $\tilde{\rho}_0(\tilde{Y})$ and $\hat{\rho}_0(\hat{X})$ correspond to different points of physical space, and to one point of physical space correspond different values of density in different FoRs. This means that a comparison of the gradient $\partial\hat{\rho}_0/\partial\hat{Y}$ with the image of the gradient $\partial\tilde{\rho}_0/\partial\tilde{X}$ in the y-system just like a comparison of the gradient $\partial\tilde{\rho}_0/\partial\tilde{X}$ with the image of the gradient $\partial\hat{\rho}_0/\partial\hat{Y}$ in the x-system is bare of any meaning.

For a solution of this conundrum it is necessary to introduce comoving position vectors in both systems, which determine at every moment of time the same point of physical space. In material description these are the position vectors $\tilde{x}(\tilde{X},t)$ and $\hat{y}(\hat{Y},t)$ in the x and y-FoRs, respectively, which determine the current configuration of the medium. Then, in the x and y-FoR, the current density of the medium is given by the functions $\tilde{\rho}_*\left(\tilde{x}(\tilde{X},t),t\right)$ and $\hat{\rho}_*\left(\hat{y}(\hat{Y},t),t\right)$, respectively. Since in this framework the functions are explicitly introduced as time-dependent, the definition of the gradient from above requires a clarification.

Let us consider an infinitesimally small change of the current mass density of the medium. In the x-FoR this change can be assessed by the increment of the function $\tilde{\rho}_* (\tilde{\boldsymbol{x}},t)$ and in the y-FoR by the increment of the function $\hat{\rho}_* (\hat{\boldsymbol{y}},t)$, respectively:

$$\mathrm{d}\tilde{\rho}_*(\tilde{\boldsymbol{x}},t) = \frac{\partial \tilde{\rho}_*}{\partial \tilde{\boldsymbol{x}}} \cdot \mathrm{d}\tilde{\boldsymbol{x}}^x + \frac{\partial \tilde{\rho}_*}{\partial t}\,\mathrm{d}t,\ \mathrm{d}\hat{\rho}_*(\hat{\boldsymbol{x}}^y,t) = \frac{\partial \hat{\rho}_*}{\partial \hat{\boldsymbol{y}}} \cdot \mathrm{d}\hat{\boldsymbol{y}} + \frac{\partial \hat{\rho}_*}{\partial t}\,\mathrm{d}t. \tag{11.66}$$

The coefficients in front of the vectors $\mathrm{d}\tilde{\boldsymbol{x}}$ and $\mathrm{d}\hat{\boldsymbol{y}}$ are gradients of the functions $\tilde{\rho}_*(\tilde{\boldsymbol{x}},t)$ and $\hat{\rho}_*(\hat{\boldsymbol{y}},t)$ in the x- and in the y-FoR, respectively. The symbol $\tilde{\boldsymbol{\nabla}}^x := \partial/\partial\tilde{\boldsymbol{x}}$ will be used for the gradient (or nabla) operator in the current configuration, as visible from the x-FoR, and $\tilde{\boldsymbol{\nabla}}^y := \partial/\partial\hat{\boldsymbol{y}}$ for the gradient (or nabla) operator in the current configuration, as visible in the y-FoR, respectively. If the vectors $\tilde{\boldsymbol{x}}(\tilde{\boldsymbol{X}},t)$ and $\hat{\boldsymbol{y}}(\hat{\boldsymbol{Y}},t)$ are represented in terms of base vector expansions,

$$\tilde{\boldsymbol{x}}(\tilde{\boldsymbol{X}},t) = \tilde{x}_k(x_i,t)\boldsymbol{e}_k\ ,\ \hat{\boldsymbol{y}}(\hat{\boldsymbol{Y}},t) = \hat{y}_k(y_i,t)\boldsymbol{d}_k, \tag{11.67}$$

then the explicit form of the gradient operators is given by the formulae

$$\tilde{\boldsymbol{\nabla}}^x := \frac{\partial}{\partial \tilde{\boldsymbol{x}}} = \boldsymbol{e}_k \frac{\partial}{\partial \tilde{x}_k}\ ,\ \tilde{\boldsymbol{\nabla}}^y := \frac{\partial}{\partial \hat{\boldsymbol{y}}} = \boldsymbol{d}_k \frac{\partial}{\partial \hat{y}_k}. \tag{11.68}$$

Likewise one may write for the current mass densities:

$$\mathrm{d}\tilde{\rho}(\tilde{\boldsymbol{X}},t) = \frac{\partial \tilde{\rho}}{\partial \tilde{\boldsymbol{X}}} \cdot \mathrm{d}\tilde{\boldsymbol{X}} + \frac{\partial \tilde{\rho}}{\partial t}\,\mathrm{d}t\ ,\ \mathrm{d}\hat{\rho}(\hat{\boldsymbol{Y}},t) = \frac{\partial \hat{\rho}}{\partial \hat{\boldsymbol{Y}}} \cdot \mathrm{d}\hat{\boldsymbol{Y}} + \frac{\partial \hat{\rho}}{\partial t}\,\mathrm{d}t. \tag{11.69}$$

The coefficients in front of the vectors $\mathrm{d}\tilde{\boldsymbol{X}}$ and $\mathrm{d}\hat{\boldsymbol{Y}}$ in Eq. (11.69) denote the gradients in the reference configuration of the current density functions $\tilde{\rho}(\tilde{\boldsymbol{X}},t)$ and $\hat{\rho}(\hat{\boldsymbol{Y}},t)$, respectively. They have been introduced in Eqs. (11.62) and (11.64) for the x and for the y-FoR before in concrete form.

The gradient operators of the current and of the reference configurations are connected among each other by the following relations:

$$\tilde{\boldsymbol{\nabla}}_0^x = \left(\frac{\partial \tilde{\boldsymbol{x}}}{\partial \tilde{\boldsymbol{X}}}\right)^{\mathsf{T}} \cdot \tilde{\boldsymbol{\nabla}}^x\ ,\ \hat{\boldsymbol{\nabla}}_0^y = \left(\frac{\partial \hat{\boldsymbol{y}}}{\partial \hat{\boldsymbol{Y}}}\right)^{\mathsf{T}} \cdot \hat{\boldsymbol{\nabla}}^y. \tag{11.70}$$

From the relations (11.62) and (11.64) it follows that

$$\frac{\partial}{\partial \tilde{\boldsymbol{x}}} = \left(\frac{\partial \tilde{\boldsymbol{x}}}{\partial \tilde{\boldsymbol{X}}}\right)^{-\mathsf{T}} \cdot \boldsymbol{e}_k \frac{\partial}{\partial x_k}\ ,\ \frac{\partial}{\partial \hat{\boldsymbol{y}}} = \left(\frac{\partial \hat{\boldsymbol{y}}}{\partial \hat{\boldsymbol{Y}}}\right)^{-\mathsf{T}} \cdot \boldsymbol{d}_k \frac{\partial}{\partial y_k}. \tag{11.71}$$

The relations (11.68) and (11.71) depict two well known representations of the gradient operators in the current configuration. Of course they can be written in other representation since the gradient operator, like any other vector, does not depend on the choice of the system coordinates of a given FoR.

When the material description for continuous media is used two gradient operators are frequently adapted, the gradient operator in the current and the gradient operator of the reference configuration. At the same time one can take into consideration a multitude of different gradient operators, which are related to each other according to (11.70). The gradient operator in a given FoR are determined by position vectors, which not necessarily point to the position of a continuum particle. These position vectors may point to the positions of speculative points, which move w.r.t. the given FoR according to a given rule. Further down we shall denote them as observational points.
We now state what is required in order to compare gradients in different FoRs:

- First, the points of observation chosen in different FoRs must reside in one and the same point of physical space.
- Second, we must not compare the gradient operators but rather the action of these operators on some invariant scalar quantity during a change of the FoR. If the gradient of this quantity is an invariant it can be concluded that the gradient operator is an invariant, too.
- Third, during the transfer operation it is necessary that the arguments of the functions match and only the base vectors are distinguished. If one succeeded to transfer and compare such a gradient then all other gradients can be compared by using a formula of the type shown in Eq. (11.70).

We will now show that the operator $\hat{\boldsymbol{\nabla}}^y$ is the image of the operator $\tilde{\boldsymbol{\nabla}}^x$. For this purpose we introduce an auxiliary z-coordinate system in the x-FoR (also see Fig. 11.4). This z-coordinate system is moving in the x-system. We use the following notations in context with this coordinate system: O is its origin moving with respect to the x-system, $\tilde{\boldsymbol{i}}_k$ are its unit base vectors rotating w.r.t. the x-system, and z_k are the coordinates w.r.t. the base $\tilde{\boldsymbol{i}}_k$. Because of the movement of the origin and of the base vector coordinate representations of vectors defined in the z-systems become time-dependent w.r.t. the x-FoR. We will see this shortly. A rigid body movement of the coordinate system with respect to the x-FoR is determined by the position vector of the origin O, *i.e.*, $\tilde{\boldsymbol{x}}^O(t)$, and the rotation tensor, ${}^z\tilde{\boldsymbol{Q}}^x(t)$:

$$\tilde{\boldsymbol{x}}^O(t) = \tilde{x}_k^O(t)\boldsymbol{e}_k\,,\ \ \tilde{\boldsymbol{i}}_k(t) = {}^z\tilde{\boldsymbol{Q}}^x(t)\cdot\boldsymbol{e}_k, \qquad k = 1,2,3, \tag{11.72}$$

where it is assumed that $\tilde{\boldsymbol{i}}_k(0) = \boldsymbol{e}_k$.
Then the current position of the observation point can be written in the following form (see Fig. 11.4):

$$\tilde{\boldsymbol{x}}(\tilde{\boldsymbol{X}},t) = \tilde{\boldsymbol{x}}^O(t) + \tilde{\boldsymbol{x}}^z\left(z_i(\tilde{\boldsymbol{X}},t),t\right). \tag{11.73}$$

By taking into account that the offset $\tilde{\boldsymbol{x}}_O(t)$ with respect to the origin of the reference frame does not depend on the observation point and by taking into account the following expression for the position vector:

$$\tilde{\boldsymbol{x}}^z(z_i,t) = z_k(t)\tilde{\boldsymbol{i}}_k(t), \tag{11.74}$$

we obtain:

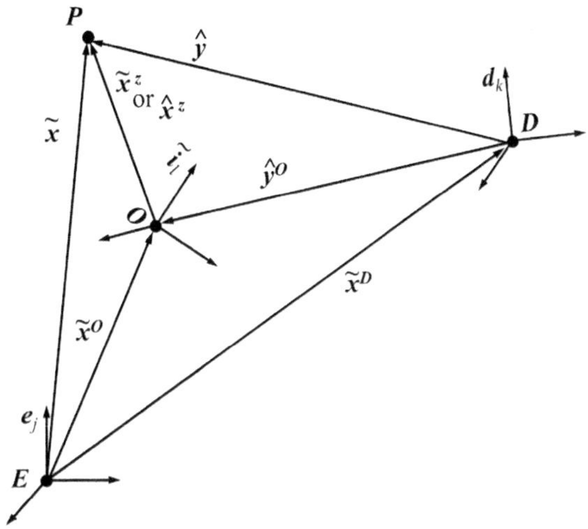

Fig. 11.4 A moving coordinate system (see text).

$$\tilde{\boldsymbol{\nabla}}^x := \frac{\partial}{\partial \tilde{\boldsymbol{x}}} = \left(\frac{\partial\,(\tilde{\boldsymbol{x}} - \tilde{\boldsymbol{x}}^O)}{\partial \tilde{\boldsymbol{x}}} \right)^{\top} \cdot \frac{\partial}{\partial\,(\tilde{\boldsymbol{x}} - \tilde{\boldsymbol{x}}^O)} = \tilde{\mathbf{I}} \cdot \frac{\partial}{\partial\,(\tilde{\boldsymbol{x}} - \tilde{\boldsymbol{x}}^O)} = \frac{\partial}{\partial \tilde{\boldsymbol{x}}^z} = \tilde{\boldsymbol{i}}_k \frac{\partial}{\partial z_k}. \tag{11.75}$$

We now introduce the image of the z-coordinate system in the y FoR. Considering the relations between the distance vectors (see Fig. 11.4)

$$\hat{\boldsymbol{y}}(\hat{\boldsymbol{Y}}, t) = \hat{\boldsymbol{y}}^O(t) + \hat{\boldsymbol{x}}^z(z_i, t), \tag{11.76}$$

and the expression

$$\hat{\boldsymbol{x}}^z(z_i, t) = z_k(t)\hat{\boldsymbol{i}}_k(t) \tag{11.77}$$

for the image of the distance vector $\tilde{\boldsymbol{x}}^z$, where $\hat{\boldsymbol{i}}_k$ are images of $\tilde{\boldsymbol{i}}_k$, we find:

$$\hat{\boldsymbol{\nabla}}^y := \frac{\partial}{\partial \hat{\boldsymbol{y}}} = \frac{\partial}{\partial\,(\hat{\boldsymbol{y}} - \hat{\boldsymbol{y}}^O)} = \frac{\partial}{\partial \hat{\boldsymbol{x}}^z} = \hat{\boldsymbol{i}}_k \frac{\partial}{\partial z_k}. \tag{11.78}$$

For a comparison between the operators $\tilde{\boldsymbol{\nabla}}^x$ and $\hat{\boldsymbol{\nabla}}^y$ and the corresponding images we consider the result of the action of these operators on the mass densities. We substitute the variables (see Fig. 11.4 for the meaning of the symbols):

$$\tilde{\rho}_*(\tilde{x}, t) = \tilde{\rho}^z_*(\tilde{x}^z, t) = \rho^z(z_i, t)\ ,\ \ \hat{\rho}_*(\hat{y}, t) = \hat{\rho}^z_*(\hat{x}^z, t) = \rho^z(z_i, t). \tag{11.79}$$

Then:

$$\tilde{\nabla}^x(\tilde{\rho}^z_*(\tilde{x}^z, t)) = \tilde{i}_k \frac{\partial \rho^z(z_i, t)}{\partial z_k},\ \hat{\nabla}^y(\hat{\rho}^z_*(\hat{x}^z, t)) = \hat{i}_k \frac{\partial \rho^z(z_i, t)}{\partial z_k}. \tag{11.80}$$

Since the vectors $\hat{i}_k(t)$ present themselves in the y-FoR as the images of the vectors $\tilde{i}_k(t)$ it follows from Eq. (11.80) that for the nabla operators in the current configuration the following relations hold:

$$\tilde{\nabla}^x \leftrightarrow \hat{\nabla}^x = \hat{\boldsymbol{i}}_k \frac{\partial}{\partial z_k} = \hat{\nabla}^y, \; \hat{\nabla}^y \leftrightarrow \tilde{\nabla}^y = \tilde{\boldsymbol{i}}_k \frac{\partial}{\partial z_k} = \tilde{\nabla}^x. \tag{11.81}$$

In order to enable a comparison of the nabla operators in the reference configurations fixed in the FoRs we make use of the relations (11.70) and (11.81):

$$\begin{aligned} \hat{\nabla}_0^x &= \left(\frac{\partial \hat{\boldsymbol{X}}}{\partial \hat{\boldsymbol{x}}}\right)^{-\top} \cdot \hat{\nabla}^x = \left(\hat{\nabla}^x \hat{\boldsymbol{X}}\right)^{-1} \cdot \hat{\nabla}^y = \\ &\left(\hat{\nabla}^x \left(\hat{\boldsymbol{x}}^O(t) + \hat{\boldsymbol{X}}^z\right)\right)^{-1} \cdot \hat{\nabla}^y = \left(\hat{\nabla}^x {}^e\hat{\boldsymbol{Q}}(t) \cdot \hat{\boldsymbol{Y}}^z\right)^{-1} \cdot \hat{\nabla}^y = \\ &\left(\hat{\nabla}^x {}^e\hat{\boldsymbol{Q}}(t) \cdot (\hat{\boldsymbol{Y}} - \hat{\boldsymbol{y}}^O(t))\right)^{-1} \cdot \hat{\nabla}^y = \left(\hat{\nabla}^y {}^e\hat{\boldsymbol{Q}}(t) \cdot \hat{\boldsymbol{Y}}\right)^{-1} \cdot \hat{\nabla}^y = \\ &\left(\left(\frac{\partial \hat{\boldsymbol{Y}}}{\partial \hat{\boldsymbol{y}}}\right)^{\top} \cdot {}^e\hat{\boldsymbol{Q}}^{\top}(t)\right)^{-1} \cdot \hat{\nabla}^y = {}^e\hat{\boldsymbol{Q}}(t) \cdot \left(\frac{\partial \hat{\boldsymbol{y}}}{\partial \hat{\boldsymbol{Y}}}\right)^{\top} \cdot \hat{\nabla}^y = {}^e\hat{\boldsymbol{Q}}(t) \cdot \hat{\nabla}_0^y. \end{aligned} \tag{11.82}$$

A few remarks are in order for the first manipulations in Eq. (11.4.1). We have $\tilde{\boldsymbol{X}}^z = Z_k \tilde{\boldsymbol{i}}_k(0), \hat{\boldsymbol{Y}}^z = Z_k \boldsymbol{i}_k\hat{(0)}$, where the vectors $\tilde{\boldsymbol{i}}_k(0)$ and $\boldsymbol{i}_k\hat{(0)}$ are z-base vectors in the x-FoR and its image in the y-FoR created at t=0, respectively. They are fixed in their FoRs. However the image of $\tilde{\boldsymbol{i}}_k(0)$ in the y-FoR depends on time $\hat{\boldsymbol{i}}_k(0) = {}^e\hat{\boldsymbol{Q}}(t) \cdot \boldsymbol{i}_k\hat{(0)}$. This means $\hat{\boldsymbol{X}}^z = Z_k {}^e\hat{\boldsymbol{Q}}(t) \cdot \boldsymbol{i}_k\hat{(0)} = {}^e\hat{\boldsymbol{Q}}(t) \cdot \hat{\boldsymbol{Y}}^z$.

Thus for the nabla operator in the reference configuration unmoving in the For the relations (11.81) do not hold. This is primarily due to that the reference configurations in material description merely serve as a parametrization of the body particles in a given FoR representing just a mathematical construction, which can have no bearing to the real configuration of a material medium and turns out differently in different FoRs.

All arguments provided for the material description can easily be transferred to the spatial one. The only formal difference in this context is that the current position vector fixate a geometrical point in space and nor a concrete material particle.

This way, if the observational points in both FoRs are motionless (which is often assumed in the spatial description), then the nabla operator determined in one FoR does not coincide with the image of the nabla operator as determined in the other FoR. The relations (11.81) are performable in spatial description only for the case if the observational points in both FoRs designate one and the same point of physical space in every moment of time, *i.e.*, if they are movable.

11.4.2 Transformation Properties of the Total and Material Time Derivatives

It is a widespread opinion that the total and the material derivatives are different names for the same concept. However, as it was shown in Ivanova et al (2016) they are, in general, different concepts having different meanings and different values. It was shown that the material derivative characterizes the rate of change a physical property of the material point that was in the observation point at the certain moment of time, while the total derivative is the rate of change of the property in the observation point. Within the material description the observation point coincides with the material particle and as a result in this particular case the value of the total and material derivatives coincide. Moreover, it is ill-fated that the adjective *material* is used in context with material description and with the material time derivative: The material time derivative is the expression for rates to be used in spatial description (see Ivanova et al, 2016, , pp. 19).

We will investigate the situation in material description first. In order to investigate the invariance of the total time derivative we have to analyze how the total derivative of an invariant scalar (*e.g.*, the mass density) transforms under a change of FoR.

At this point recall our definition of a Euclidean scalar in material description after Eq. (11.64). The total derivative in the frame of reference is the partial time derivative with the reference coordinates held constant (see Ivanova et al, 2016, pg. 17). Thus, in the x-system:

$$\frac{\mathrm{d}^x \tilde{\rho}(\tilde{\boldsymbol{X}},t)}{\mathrm{d}t} = \left.\frac{\partial \tilde{\rho}(\tilde{\boldsymbol{X}},t)}{\partial t}\right|_{x_i} = \left.\frac{\partial \tilde{\rho}(\tilde{\boldsymbol{X}},t)}{\partial t}\right|_{\tilde{\boldsymbol{X}}}, \tag{11.83}$$

and analogously in the y-system:

$$\frac{\mathrm{d}^y \hat{\rho}(\hat{\boldsymbol{Y}},t)}{\mathrm{d}t} = \left.\frac{\partial \hat{\rho}(\hat{\boldsymbol{Y}},t)}{\partial t}\right|_{y_i} = \left.\frac{\partial \hat{\rho}(\hat{\boldsymbol{Y}},t)}{\partial t}\right|_{\hat{\boldsymbol{Y}}}, \tag{11.84}$$

where x_i and y_i, $i = 1,2,3$ refer to the coordinates of the two FoRs and symbols at the bottom of vertical bars indicate quantities that are kept constant. In order to compare the total derivatives determined in different reference frames we have to create their images. But, as was pointed out in the previous section, the images $\hat{\boldsymbol{X}}$ and $\tilde{\boldsymbol{Y}}$ depend on time. They relate to the position of the same space point only at one special moment, t_0, and they correspond to different points at other moments of time. Therefore such relations as

$$\frac{\partial \tilde{\rho}\,(\tilde{\boldsymbol{X}},t)}{\partial t}\Big|_{\tilde{\boldsymbol{X}}} = \frac{\partial \hat{\rho}\,(\tilde{\boldsymbol{Y}},t)}{\partial t}\Big|_{\tilde{\boldsymbol{Y}}}, \qquad \frac{\partial \hat{\rho}\,(\hat{\boldsymbol{Y}},t)}{\partial t}\Big|_{\hat{\boldsymbol{Y}}} = \frac{\partial \tilde{\rho}\,(\hat{\boldsymbol{X}},t)}{\partial t}\Big|_{\hat{\boldsymbol{X}}}$$

or

$$\frac{\partial \tilde{\rho}\,(\tilde{\boldsymbol{X}},t)}{\partial t}\Big|_{x_i} = \frac{\partial \tilde{\rho}\,(\hat{\boldsymbol{X}},t)}{\partial t}\Big|_{\hat{\boldsymbol{X}}} \quad \text{and} \quad \frac{\partial \hat{\rho}\,(\tilde{\boldsymbol{Y}},t)}{\partial t}\Big|_{y_i} = \frac{\partial \hat{\rho}(\tilde{\boldsymbol{Y}},t)}{\partial t}\Big|_{\tilde{\boldsymbol{Y}}}$$

seem dubious. Thus, a simple representation of the total derivative as the derivative of function of the referential variables is not convenient for transferring into another frame of reference. Switching to expressions of the function in terms of spatial variables, namely $\tilde{\rho}(\tilde{\boldsymbol{X}},t) = \tilde{\rho}_*(\tilde{\boldsymbol{x}},t)$ and $\hat{\rho}(\hat{\boldsymbol{Y}},t) = \hat{\rho}_*(\hat{\boldsymbol{y}},t)$, is more suitable. Then by virtue of the chain rule the total derivative of the mass density in the x-system simply reads:

$$\frac{\mathrm{d}^x \tilde{\rho}(\tilde{\boldsymbol{X}},t)}{\mathrm{d}t} = \left.\frac{\partial \tilde{\rho}_*(\tilde{\boldsymbol{x}},t)}{\partial t}\right|_{\tilde{\boldsymbol{x}}} + \frac{\mathrm{d}^x \tilde{\boldsymbol{x}}}{\mathrm{d}t} \cdot \frac{\partial \tilde{\rho}_*(\tilde{\boldsymbol{x}},t)}{\partial \tilde{\boldsymbol{x}}}, \tag{11.85}$$

and in the y-system:

$$\frac{\mathrm{d}^y \hat{\rho}(\hat{\boldsymbol{Y}},t)}{\mathrm{d}t} = \left.\frac{\partial \hat{\rho}_*(\hat{\boldsymbol{y}},t)}{\partial t}\right|_{\hat{\boldsymbol{y}}} + \frac{\mathrm{d}^y \hat{\boldsymbol{y}}}{\mathrm{d}t} \cdot \frac{\partial \hat{\rho}_*(\hat{\boldsymbol{y}},t)}{\partial \hat{\boldsymbol{y}}}. \tag{11.86}$$

After introducing images of $\tilde{\boldsymbol{x}}$ and $\hat{\boldsymbol{y}}$, namely $\hat{\boldsymbol{x}}$ and $\tilde{\boldsymbol{y}}$, respectively, we can write:

$$\tilde{\boldsymbol{x}}(x_i,t) = \tilde{\boldsymbol{x}}^D(t) + \tilde{\boldsymbol{y}}(y_i,t), \quad \hat{\boldsymbol{y}}(y_i,t) = \hat{\boldsymbol{x}}^E(t) + \hat{\boldsymbol{x}}(x_i,t). \tag{11.87}$$

In contrast to the reference position the vectors $\tilde{\boldsymbol{x}}$ and $\tilde{\boldsymbol{y}}$ in the x-system as well as $\hat{\boldsymbol{y}}$ and $\hat{\boldsymbol{x}}$ in the y-system always determine the same point in space. But they start from different points (namely from the origins of the corresponding FoRs) moving with respect to each other. Hence we cannot be sure that

$$\left.\frac{\partial \tilde{\rho}_*(\tilde{\boldsymbol{x}},t)}{\partial t}\right|_{\tilde{\boldsymbol{x}}} = \left.\frac{\partial \hat{\rho}_*(\tilde{\boldsymbol{y}},t)}{\partial t}\right|_{\tilde{\boldsymbol{y}}} \quad \text{and} \quad \left.\frac{\partial \hat{\rho}_*(\hat{\boldsymbol{y}},t)}{\partial t}\right|_{\hat{\boldsymbol{y}}} = \left.\frac{\partial \tilde{\rho}_*(\hat{\boldsymbol{x}},t)}{\partial t}\right|_{\hat{\boldsymbol{x}}}.$$

In addition to that, (11.86) contains stand-alone velocity terms and not relative (differences) of velocities, which are known for their objective character. In fact, it follows from Eq. (11.87) that

$$\frac{\mathrm{d}^x \tilde{\boldsymbol{x}}}{\mathrm{d}t} \neq \frac{\mathrm{d}^x \tilde{\boldsymbol{y}}}{\mathrm{d}t} \quad \text{and} \quad \frac{\mathrm{d}^y \hat{\boldsymbol{y}}}{\mathrm{d}t} \neq \frac{\mathrm{d}^y \hat{\boldsymbol{x}}}{\mathrm{d}t}.$$

In order to improve the situation we will use the total derivative in a moving coordinate system.

Recall the introduction of the z-coordinate system in the previous subsection before Eq. (11.72). The velocity of the origin of the z-coordinate system, $\tilde{\boldsymbol{v}}_O^x$, and its angular velocity, $\tilde{\boldsymbol{\omega}}^{zx}$, are then given by:

$$\tilde{\boldsymbol{v}}_O^x = \frac{\mathrm{d}\tilde{\boldsymbol{x}}^O}{\mathrm{d}t}, \quad \frac{\mathrm{d}\tilde{\boldsymbol{Q}}^x}{\mathrm{d}t} = \tilde{\boldsymbol{\omega}}^{zx} \times {}^z\tilde{\boldsymbol{Q}}^x, \tag{11.88}$$

the former as in *Comment I* and the latter analogously to *Comment II*, Eq. (11.46).

The position vector determining the position of a point $\boldsymbol{P}$ with respect to the moving coordinate system is:

$$\tilde{\boldsymbol{x}}^z(z_i,t) = z_k(t)\tilde{\boldsymbol{i}}_k(t). \tag{11.89}$$

We observe the following relation, which becomes obvious in view of Fig. 11.4:

$$\tilde{\boldsymbol{x}}(\tilde{\boldsymbol{X}},t) = \tilde{\boldsymbol{x}}^O(t) + \tilde{\boldsymbol{x}}^z(z_i,t), \tag{11.90}$$

which is formally analogous to Eq. (11.87). By combining Eqs. $(11.72)_2$ and (11.89) we introduce an auxiliary position vector $\tilde{\boldsymbol{Z}}^x(z_i)$, such that:

$$\tilde{\boldsymbol{x}}^z(z_i,t) = {}^{z}\tilde{\boldsymbol{Q}}^x(t)\cdot\tilde{\boldsymbol{Z}}^x(z_i), \quad \tilde{\boldsymbol{Z}}^x(z_i) = z_k\boldsymbol{e}_k. \tag{11.91}$$

Note that the vector $\tilde{\boldsymbol{Z}}^x(z_i)$ can indirectly depend on time through the movement of the coordinates $z_i, i = 1,2,3$ in the x-FoR. In order to comprehend this suppose, in a first step, a point $\boldsymbol{P}$ is fixed w.r.t. the z-system. Then $\tilde{\boldsymbol{x}}^z(z_i,t) = z_k\tilde{\boldsymbol{i}}_k(t)$. Now introduce an auxiliary position vector in accordance with Eq. (11.91). In this case $\tilde{\boldsymbol{Z}}^x(z_i) = z_k\boldsymbol{e}_k$ and it does not depend on time. In a second step let $\boldsymbol{P}$ move w.r.t. the z-system, $\tilde{\boldsymbol{x}}^z(z_i,t) = z_k(t)\tilde{\boldsymbol{i}}_k(t)$. An auxiliary position vector is $\tilde{\boldsymbol{Z}}^x(z_i) = z_k(t)\boldsymbol{e}_k$. The idea is that a direct dependence on time $\tilde{\boldsymbol{Z}}^x(\cdot,t)$ is eliminated and $\tilde{\boldsymbol{Z}}^x$ depends on time only indirectly (through z_i). With Eq. (11.91) Eq.(11.90) takes the form:

$$\tilde{\boldsymbol{x}}(\tilde{\boldsymbol{X}},t) = \tilde{\boldsymbol{x}}^O(t) + {}^{z}\tilde{\boldsymbol{Q}}^x(t)\cdot\tilde{\boldsymbol{Z}}^x(z_i). \tag{11.92}$$

We now study the situation in the y-FoR. Here the position vectors are related by:

$$\hat{\boldsymbol{y}}(\hat{\boldsymbol{Y}},t) = \hat{\boldsymbol{y}}^O(t) + \hat{\boldsymbol{x}}^z(z_i,t), \tag{11.93}$$

where $\hat{\boldsymbol{x}}^z$ is the image of $\tilde{\boldsymbol{x}}^z$. In analogy to Eq. (11.91) we may write:

$$\hat{\boldsymbol{x}}^z(z_i,t) = {}^{z}\hat{\boldsymbol{Q}}^y(t)\cdot\hat{\boldsymbol{Z}}^y(z_i), \quad \hat{\boldsymbol{Z}}^y(z_i) = z_k\boldsymbol{d}_k, \tag{11.94}$$

where we have defined another auxiliary vector $\hat{\boldsymbol{Z}}^y(z_i)$. Note that it is not the image of $\tilde{\boldsymbol{Z}}^x(z_i)$, because this would be $\hat{\boldsymbol{Z}}^x = z_k\hat{\boldsymbol{e}}_k(t) = z_k\,{}^{e}\hat{\boldsymbol{Q}}(t)\cdot\boldsymbol{d}_k$. Moreover, the rotation tensor ${}^{z}\hat{\boldsymbol{Q}}^y(t)$ determines the orientation of the images of the base vectors of the z-system with respect to the y-system:

$$\hat{\boldsymbol{i}}_k(t) = {}^{z}\hat{\boldsymbol{Q}}^y(t)\cdot\boldsymbol{d}_k, \qquad k = 1,2,3. \tag{11.95}$$

It is a composition of two rotation tensors:

$${}^{z}\hat{\boldsymbol{Q}}^y(t) = {}^{z}\hat{\boldsymbol{Q}}^x(t)\cdot{}^{e}\hat{\boldsymbol{Q}}(t), \tag{11.96}$$

where ${}^{z}\hat{\boldsymbol{Q}}^x(t)$ is the image of the rotation tensor ${}^{z}\tilde{\boldsymbol{Q}}^x(t)$. With Eq. (11.94) we can rewrite Eq. (11.93) as:

$$\hat{\boldsymbol{y}}(\hat{\boldsymbol{Y}},t) = \hat{\boldsymbol{y}}^O(t) + {}^{z}\hat{\boldsymbol{Q}}^y(t)\cdot\hat{\boldsymbol{Z}}^y(z_i). \tag{11.97}$$

Just like $\tilde{\boldsymbol{Z}}^x$ the vector $\hat{\boldsymbol{Z}}^y$ is an indirectly time-dependent marker of a specific point, in fact of the same point.

Now we return to the mass density and change the variables, where $\tilde{\boldsymbol{x}}^z$ points the same point in space as $\tilde{\boldsymbol{x}}$ and $\hat{\boldsymbol{y}}$, just like $\tilde{\boldsymbol{Z}}^x$ and $\hat{\boldsymbol{Z}}^y$ characterize that same point:

$$\tilde{\rho}_*(\tilde{\boldsymbol{x}},t) = \tilde{\rho}^z_*(\tilde{\boldsymbol{x}}^z,t) = \tilde{\rho}^z(\tilde{\boldsymbol{Z}}^x,t), \; \hat{\rho}_*(\hat{\boldsymbol{y}},t) = \hat{\rho}^z_*(\hat{\boldsymbol{x}}^z,t) = \hat{\rho}^z(\hat{\boldsymbol{Z}}^y,t). \tag{11.98}$$

In context with Eq. (11.98) and in view of Eqs. (11.90)/(11.91) and (11.93)/(11.94), respectively, two comments are in order. First, note that the position vectors $\tilde{\boldsymbol{x}}^z(z_i,t)$ and $\hat{\boldsymbol{x}}^z(z_i,t)$ connect the same two points in space. Second, recall once more that $\tilde{\boldsymbol{Z}}^x$ and $\hat{\boldsymbol{Z}}^y$ are implicitly time-dependent. Thus, by using Eq. (11.98) we rewrite the total time derivative in the x-FoR, Eq. (11.83), as follows:

$$\frac{\mathrm{d}^x\tilde{\rho}_*(\tilde{\boldsymbol{x}},t)}{\mathrm{d}t} = \frac{\mathrm{d}^x\tilde{\rho}^z(\tilde{\boldsymbol{Z}}^x,t)}{\mathrm{d}t} = \left.\frac{\partial\tilde{\rho}^z(\tilde{\boldsymbol{Z}}^x,t)}{\partial t}\right|_{\tilde{\boldsymbol{Z}}^x} + \frac{\mathrm{d}^x\tilde{\boldsymbol{Z}}^x}{\mathrm{d}t}\cdot\frac{\partial\tilde{\rho}^z(\tilde{\boldsymbol{Z}}^x,t)}{\partial\tilde{\boldsymbol{Z}}^x}. \tag{11.99}$$

After taking the total time derivative of Eq. (11.92) and by using Eqs. (11.88), (11.91) we obtain:

$$\frac{\mathrm{d}^x\tilde{\boldsymbol{x}}(\tilde{\boldsymbol{X}},t)}{\mathrm{d}t} = \tilde{\boldsymbol{v}}^x_O(t) + \tilde{\boldsymbol{\omega}}^{zx}(t)\times\tilde{\boldsymbol{x}}^z(z_i,t) + {}^z\tilde{\boldsymbol{Q}}^x(t)\cdot\frac{\mathrm{d}^x\tilde{\boldsymbol{Z}}^x(z_i)}{\mathrm{d}t}. \tag{11.100}$$

The first two terms on the right hand side of Eq. (11.100) we denote by $\tilde{\boldsymbol{v}}^{zx}$:

$$\tilde{\boldsymbol{v}}^{zx}(z_i,t) := \tilde{\boldsymbol{v}}^x_O(t) + \tilde{\boldsymbol{\omega}}^{zx}(t)\times\tilde{\boldsymbol{x}}^z(z_i,t). \tag{11.101}$$

This is what is called by Hutter and Jöhnk (2004), pg. 119 the "frozen velocity" or *Führungsgeschwindigkeit* (also see Hauger et al, 1993, , pp. 241) in German. It is obtained by momentarily freezing the particle (at the point whose position vector is $\tilde{\boldsymbol{x}}^z$) with the moving z-system. Since $\tilde{\boldsymbol{x}}$ determines the current position of a material particle in the x-system, we can write:

$$\frac{\mathrm{d}^x\tilde{\boldsymbol{x}}}{\mathrm{d}t} = \tilde{\boldsymbol{v}}^x \quad\Rightarrow\quad \frac{\mathrm{d}^x\tilde{\boldsymbol{Z}}^x}{\mathrm{d}t} = \left(\tilde{\boldsymbol{v}}^x - \tilde{\boldsymbol{v}}^{zx}\right)\cdot{}^z\tilde{\boldsymbol{Q}}^x, \tag{11.102}$$

where $\tilde{\boldsymbol{v}}^x$ is the velocity of the particle in the x-system. After substituting Eq. (11.102) into Eq. (11.99) we get:

$$\frac{\mathrm{d}^x\tilde{\rho}}{\mathrm{d}t} = \left.\frac{\partial\tilde{\rho}^z}{\partial t}\right|_{\tilde{\boldsymbol{Z}}^x} + \left(\tilde{\boldsymbol{v}}^x - \tilde{\boldsymbol{v}}^{zx}\right)\cdot{}^z\tilde{\boldsymbol{Q}}^x\cdot\frac{\partial\tilde{\rho}^z}{\partial\tilde{\boldsymbol{Z}}^x}, \tag{11.103}$$

and by using Eq. (11.91) we have:

$$\frac{\mathrm{d}^x\tilde{\rho}}{\mathrm{d}t} = \left.\frac{\partial\tilde{\rho}^z}{\partial t}\right|_{z_i} + \left(\tilde{\boldsymbol{v}}^x - \tilde{\boldsymbol{v}}^{zx}\right)\cdot\frac{\partial\tilde{\rho}^z}{\partial\tilde{\boldsymbol{x}}^z}. \tag{11.104}$$

The image of Eq. (11.104) in the y-FoR is:

$$\frac{\hat{\mathrm{d}}^x\hat{\rho}}{\mathrm{d}t} = \left.\frac{\partial\hat{\rho}^z}{\partial t}\right|_{z_i} + \left(\hat{\boldsymbol{v}}^x - \hat{\boldsymbol{v}}^{zx}\right)\cdot\frac{\partial\hat{\rho}^z}{\partial\hat{\boldsymbol{x}}^z}. \tag{11.105}$$

This is the total derivative determined in the x-FoR and transferred to the y-FoR. On the other hand, the total time derivative in the y-FoR (11.84) can be written by using Eq. (11.98):

$$\frac{\mathrm{d}^y \hat{\rho}(\hat{\boldsymbol{Y}},t)}{\mathrm{d}t} = \frac{\mathrm{d}^y \hat{\rho}^z(\hat{\boldsymbol{Z}}^y,t)}{\mathrm{d}t} = \left.\frac{\partial \hat{\rho}^z(\hat{\boldsymbol{Z}}^y,t)}{\partial t}\right|_{\hat{\boldsymbol{Z}}^y} + \frac{\mathrm{d}^y \hat{\boldsymbol{Z}}^y}{\mathrm{d}t} \cdot \frac{\partial \hat{\rho}^z(\hat{\boldsymbol{Z}}^y,t)}{\partial \hat{\boldsymbol{Z}}^y}. \tag{11.106}$$

Since:

$$\frac{\mathrm{d}^y \hat{\boldsymbol{y}}(\hat{\boldsymbol{Y}},t)}{\mathrm{d}t} = \hat{\boldsymbol{v}}^y = \hat{\boldsymbol{v}}^{zy}(z_i,t) + {}^z\hat{\boldsymbol{Q}}^y(t) \cdot \frac{\mathrm{d}^y \hat{\boldsymbol{Z}}^y(z_i)}{\mathrm{d}t}, \tag{11.107}$$

with a definition analogous to Eq. (11.101):

$$\hat{\boldsymbol{v}}^{zy} := \hat{\boldsymbol{v}}^y_O(t) + \hat{\boldsymbol{\omega}}^{zy}(t) \times \hat{\boldsymbol{x}}^z(z_i,t) \tag{11.108}$$

we have:

$$\frac{\mathrm{d}^y \hat{\rho}}{\mathrm{d}t} = \left.\frac{\partial \hat{\rho}^z}{\partial t}\right|_{\hat{\boldsymbol{Z}}^y} + (\hat{\boldsymbol{v}}^y - \hat{\boldsymbol{v}}^{zy}) \cdot {}^z\hat{\boldsymbol{Q}}^y \cdot \frac{\partial \hat{\rho}^z}{\partial \hat{\boldsymbol{Z}}^y} \tag{11.109}$$

By using Eq. (11.94) we may write Eq. (11.109) as:

$$\frac{\mathrm{d}^y \hat{\rho}}{\mathrm{d}t} = \left.\frac{\partial \hat{\rho}^z}{\partial t}\right|_{z_i} + (\hat{\boldsymbol{v}}^y - \hat{\boldsymbol{v}}^{zy}) \cdot \frac{\partial \hat{\rho}^z}{\partial \hat{\boldsymbol{x}}^z}. \tag{11.110}$$

The image of the total derivative determined in the y-FoR is:

$$\frac{\tilde{\mathrm{d}}^y \tilde{\rho}}{\mathrm{d}t} = \left.\frac{\partial \tilde{\rho}^z}{\partial t}\right|_{z_i} + (\tilde{\boldsymbol{v}}^y - \tilde{\boldsymbol{v}}^{zy}) \cdot \frac{\partial \tilde{\rho}^z}{\partial \tilde{\boldsymbol{x}}^z}. \tag{11.111}$$

From the point of view of the y-FoR an absolute velocity is a velocity of the object measured w.r.t. the y-FOR, a relative velocity is a velocity of the object measured w.r.t. x-FoR, a frozen velocity is a velocity of the point of the x-FoR where the object is located. In Eq. (11.110) we have two velocities. First, the velocity of the point. It is determined by

$$\hat{\boldsymbol{v}}^y = \hat{\boldsymbol{v}}^y_E + \frac{\mathrm{d}\hat{\boldsymbol{x}}}{\mathrm{d}t} = \hat{\boldsymbol{v}}^y_E + \hat{\boldsymbol{\omega}}^e \times \hat{\boldsymbol{x}} + \hat{\boldsymbol{v}}^x = \hat{\boldsymbol{v}}^z + \hat{\boldsymbol{v}}^x. \tag{11.112}$$

Second, there is a velocity of a point of the z-coordinate system where a material point is located. Thus a relative velocity of this point is the image of the frozen velocity in (11.101). And the frozen velocity in this case is $\hat{v}^z$ (as in the first case), because the locations of objects (in the x-FoR) are the same.

In here $\hat{\boldsymbol{v}}^z$ is a frozen velocity defined by:

$$\hat{\boldsymbol{v}}^z := \hat{\boldsymbol{v}}^y_E + \hat{\boldsymbol{\omega}}^e \times \hat{\boldsymbol{x}}. \tag{11.113}$$

Note that here the frozen velocity it is the velocity of the point in the x-FoR (where the point $\boldsymbol{P}$ is located) w.r.t. the y-FoR. In (11.101) the frozen velocity is the velocity of the point of z-coordinate system (where the point $\boldsymbol{P}$ is located) w.r.t. the x-FoR.

In order to determine the velocity of $\hat{v}^{zy}$ in the y-FoR we have to add to the velocity $\hat{\boldsymbol{v}}^{zx}$ the velocity of that very point of the x-FoR where the point $\boldsymbol{P}$ is located. Thus:

$$\hat{\boldsymbol{v}}^{zy}(z_i,t) = \hat{\boldsymbol{v}}^{z}(z_i,t) + \hat{\boldsymbol{v}}^{zx}(z_i,t). \tag{11.114}$$

From Eqs. (11.112) and (11.114) it follows that:

$$\hat{\boldsymbol{v}}^{y} - \hat{\boldsymbol{v}}^{zy} = \hat{\boldsymbol{v}}^{x} - \hat{\boldsymbol{v}}^{zx}. \tag{11.115}$$

Taking this into account it shows that the expressions (11.110) and (11.105) coincide. Thus, the total derivative in the y-FoR coincides with the image of the total derivative determined in the x-FoR. Analogously, it can be shown that the total derivative in the x-FoR coincides with the image of the total derivative determined in the y-FoR.

Now let us consider the material time derivative. An expression for the operator of the material derivative in spatial description with a moving observation point was suggested in Altenbach et al (2003) and in Zhilin (2012). It contains the total time derivative and the nabla operator:

$$\frac{\delta}{\delta t} = \frac{\mathrm{d}}{\mathrm{d}t} + \left(\boldsymbol{v}\big(\boldsymbol{r}(t),t\big) - \frac{\mathrm{d}\boldsymbol{r}(t)}{\mathrm{d}t}\right)\cdot\boldsymbol{\nabla}. \tag{11.116}$$

Here $\boldsymbol{v}$ is the velocity of the particle located in the observation point at the moment t, and $\mathrm{d}\boldsymbol{r}(t)/\mathrm{d}t$ is the velocity of the observation point. This expression contains the total derivative, the gradient operator (nabla is the differentiation w.r.t. the position vector of the observation point, $\partial/\partial\boldsymbol{x}$), and a speed difference. The images of the total derivative, of the difference in velocities, and of the nabla operator determined in one frame of reference coincide with the corresponding quantities determined in another frame of reference. Therefore, the material derivative as well as the total derivative are invariant operators.

11.5 Conclusions and Outlook

Two main intentions were pursued in this paper. First, based on a thorough introduction of the notion Frame of Reference (FoR), the usefulness and necessity of creating images of tensorial quantities was emphasized. Tensorial quantities can be compared only if they are based in the same FoR. In order to enable a comparison of physical quantities existing in various FoRs, images must first be created, since a direct comparison is meaningless.

Second, based on this idea frame indifference properties of the current nabla operator were shown. This property could then be used to prove frame indifference of time derivative operators during a change of FoR as well, because the latter contain nabla operators. In this context three issues arise: The observation point considered in context with all operators must be the same for all FoRs at all times; the action of all operators in combination with (and not in absence of) an invariant scalar must be investigated, and, finally, the arguments of the scalar field must be the same for all FoRs in order to compare the action of the operators from various FoRs.

Following this strategy it was shown that the nabla operator in the current configuration is invariant, that the nabla operator in the reference configuration is not, and that the aforementioned time derivatives in material and in spatial description are invariant.
The idea of image creation could potentially be helpful when studying the principle of objectivity, a.k.a. principle of material frame indifference, which is used for reducing the amount of possible forms of constitutive equations. However, the details of such an investigation are left to future work.

Acknowledgements Support of this work by a grant from the Russian Foundation for Basic Research (16-01-00815) is gratefully acknowledged.

9.A Appendix

The following passages were primarily written for the benefit of readers who do not want to get distracted from the main line of thought of this paper – which is the proof of frame indifference of nabla and time derivative operators. Yet they might want to obtain some additional information on other kinematic quantities and relations, such as the spin tensors, the so-called Poisson relations, and angular velocities, all within the framework of the image concept.

9.A.1 Rotational Tensors and Angular Velocity Vectors

Differentiation of Eq. (11.12) will lead to velocities. In what follows we shall pay special attention to velocities associated with rotation. For this purpose a few preparatory considerations are in order. We start by introducing the left and right spin tensors, identifiable by upper indices l and r, respectively:

$$\sigma^{\mathrm{l}} := \frac{\mathrm{d}^d\tilde{\boldsymbol{Q}}}{\mathrm{d}t} \cdot {}^d\tilde{\boldsymbol{Q}}^{\top}, \quad \sigma^{\mathrm{r}} := {}^d\tilde{\boldsymbol{Q}}^{\top} \cdot \frac{\mathrm{d}^d\tilde{\boldsymbol{Q}}}{\mathrm{d}t}. \tag{11.117}$$

Because of the orthogonality relations shown in Eq. (11.9) it follows that:

$$\begin{aligned} \frac{\mathrm{d}^d\tilde{\boldsymbol{Q}}}{\mathrm{d}t} \cdot {}^d\tilde{\boldsymbol{Q}}^{\top} + {}^d\tilde{\boldsymbol{Q}} \cdot \frac{\mathrm{d}^d\tilde{\boldsymbol{Q}}^{\top}}{\mathrm{d}t} &= \sigma^{\mathrm{l}} + (\sigma^{\mathrm{l}})^{\top} = \tilde{\mathbf{0}}, \\ \frac{\mathrm{d}^d\tilde{\boldsymbol{Q}}^{\top}}{\mathrm{d}t} \cdot {}^d\tilde{\boldsymbol{Q}} + {}^d\tilde{\boldsymbol{Q}}^{\top} \cdot \frac{\mathrm{d}^d\tilde{\boldsymbol{Q}}}{\mathrm{d}t} &= \sigma^{\mathrm{r}} + (\sigma^{\mathrm{r}})^{\top} = \tilde{\mathbf{0}}. \end{aligned} \tag{11.118}$$

Thus the tensors σ^{l} and σ^{r} are antisymmetric and corresponding conjugate vectors $\tilde{\boldsymbol{\omega}}^d$ and $\tilde{\boldsymbol{\Omega}}^d$ exist, such that:

$$\sigma^{\mathrm{l}} = \tilde{\boldsymbol{\omega}}^d \times \tilde{\mathbf{I}}, \quad \sigma^{\mathrm{r}} = \tilde{\mathbf{I}} \times \tilde{\boldsymbol{\Omega}}^d, \quad \tilde{\boldsymbol{\omega}}^d = {}^d\tilde{\boldsymbol{Q}} \cdot \tilde{\boldsymbol{\Omega}}^d, \tag{11.119}$$

where cross products between a second order tensor $\boldsymbol{A} = \boldsymbol{a} \otimes \boldsymbol{b}$ and a vector $\boldsymbol{x}$ are defined by the following formulae:

$$\boldsymbol{A} \times \boldsymbol{x} = \boldsymbol{a} \otimes (\boldsymbol{b} \times \boldsymbol{x}), \quad \boldsymbol{x} \times \boldsymbol{A} = (\boldsymbol{x} \times \boldsymbol{a}) \otimes \boldsymbol{b}. \tag{11.120}$$

After dot-multiplication of Eq. $(11.119)_1$ by ${}^d\tilde{\boldsymbol{Q}}$ from the right and of $(11.119)_2$ from the left we get the following equations relating the left angular velocity $\tilde{\boldsymbol{\omega}}^d$ and the right angular velocity $\tilde{\boldsymbol{\Omega}}^d$ through the rotational tensor ${}^d\tilde{\boldsymbol{Q}}$ – the so-called Poisson relations:[9]

$$\frac{\mathrm{d}{}^d\tilde{\boldsymbol{Q}}}{\mathrm{d}t} = \tilde{\boldsymbol{\omega}}^d \times {}^d\tilde{\boldsymbol{Q}}, \quad \frac{\mathrm{d}{}^d\tilde{\boldsymbol{Q}}}{\mathrm{d}t} = {}^d\tilde{\boldsymbol{Q}} \times \tilde{\boldsymbol{\Omega}}^d. \tag{11.121}$$

If the left and right angular velocities are known, Eqs. (11.23) may serve as differential equations for ${}^d\tilde{\boldsymbol{Q}}$.
At this point note that many textbooks look at the situation from a different angle, see for example Truesdell (1972), pg. 48 (footnote), Truesdell and Toupin (1960), pg. 437 and pp. 439, Eringen (1980), pg. 85, Hutter and Jöhnk (2004), pg. 119. Hence, before we proceed to discuss the connection note the following mathematical preliminaries. We introduce the Gibbsian cross, which is nothing else but a dyad replacement operator, $\times$, in a second order tensor $\boldsymbol{A}_\times = (A_{ij}\boldsymbol{e}_i \otimes \boldsymbol{e}_j)_\times$ as follows:

$$\boldsymbol{A}_\times = (A_{ij}\boldsymbol{e}_i \otimes \boldsymbol{e}_j)_\times := A_{ij}\boldsymbol{e}_i \times \boldsymbol{e}_j, \tag{11.122}$$

a notation that is frequently used in Russian textbooks (see, *e.g.*, Zhilin (2003), pg. 48). Because in an orthonormal base we have $\boldsymbol{e}_i \times \boldsymbol{e}_j = \varepsilon_{kij}\boldsymbol{e}_k$, ε_{kij} being the totally antisymmetric Levi-Cevita symbol, it can be converted into the following expression based on the Levi-Cevita tensor, $\boldsymbol{\varepsilon} = \varepsilon_{kij}\boldsymbol{e}_k \otimes \boldsymbol{e}_i \otimes \boldsymbol{e}_j$:

$$\boldsymbol{A}_\times = \boldsymbol{\varepsilon} \bullet \boldsymbol{A}, \tag{11.123}$$

where the outer double dot product is defined by $(\boldsymbol{a} \otimes \boldsymbol{b}) \bullet (\boldsymbol{c} \otimes \boldsymbol{d}) = (\boldsymbol{a} \cdot \boldsymbol{c})(\boldsymbol{b} \cdot \boldsymbol{d})$ (see Lebedev et al, 2010, , pg. 40).
Once more we now perform a calculation for simplicity in an orthonormal frame $\boldsymbol{e}_k$. Then $\tilde{\mathbf{I}} = \boldsymbol{e}_k \otimes \boldsymbol{e}_k$, and if we now apply the Gibbsian cross operator to Eq. $(11.119)_1$ we obtain:

$$\sigma^1_\times = (\tilde{\boldsymbol{\omega}}^d \times \boldsymbol{e}_k \otimes \boldsymbol{e}_k)_\times = (\tilde{\boldsymbol{\omega}}^d \times \boldsymbol{e}_k) \times \boldsymbol{e}_k = -2\tilde{\boldsymbol{\omega}}^d. \tag{11.124}$$

However, by observing Eq. (11.123) this is usually written as:

$$\tilde{\boldsymbol{\omega}}^d = -\frac{1}{2}\boldsymbol{\varepsilon} \bullet \sigma^1, \tag{11.125}$$

and considered as the definition of the angular velocity vector (see for example Eringen (1980), pg. 85 or Hutter and Jöhnk (2004), pg. 119). Typically no reference is made to the fact that this relation concerns the left angular velocity vector and that there is a right equivalent as well:

[9] Sometimes also referred to as the "strapdown equations" in the American aero-astro literature, see Stevens et al (2016), pg. 45.

$$\sigma^{\mathrm{r}}_{\times} = -2\tilde{\boldsymbol{\Omega}}^{d}, \; \tilde{\boldsymbol{\Omega}}^{d} = -\frac{1}{2}\boldsymbol{\varepsilon} \bullet \sigma^{\mathrm{r}}. \tag{11.126}$$

These subtle facts are simply ignored, because the structure of the equations stays the same. We may conclude that what is known as Poisson relations in the Russian literature is nothing else but Eqs. (11.125) or $(11.126)_2$ in disguise. Nevertheless, the connection between Eqs. (11.124) and (11.125) will prove valuable if it comes to the transformation properties of vectors and tensors during a change of FoR.

So far the situation w.r.t. the x-FoR. The viewpoint of the y-FoR is analogous: The left angular velocity, $\hat{\boldsymbol{\omega}}^{e}(t)$, and the right angular velocity, $\hat{\boldsymbol{\Omega}}^{e}(t)$, are connected by the rotational tensor ${}^{e}\hat{\boldsymbol{Q}}(t)$:

$$\frac{\mathrm{d}\,{}^{e}\hat{\boldsymbol{Q}}}{\mathrm{d}t} = \hat{\boldsymbol{\omega}}^{e} \times {}^{e}\hat{\boldsymbol{Q}}, \quad \frac{\mathrm{d}\,{}^{e}\hat{\boldsymbol{Q}}}{\mathrm{d}t} = {}^{e}\hat{\boldsymbol{Q}} \times \hat{\boldsymbol{\Omega}}^{e}, \quad \hat{\boldsymbol{\omega}}^{e} = {}^{e}\hat{\boldsymbol{Q}} \cdot \hat{\boldsymbol{\Omega}}^{e}. \tag{11.127}$$

Consistently we should now ask as to how the rotational tensors ${}^{d}\tilde{\boldsymbol{Q}}(t)$ and ${}^{e}\hat{\boldsymbol{Q}}(t)$ (and the corresponding left and right angular velocities, $\tilde{\boldsymbol{\omega}}^{d}$, $\hat{\boldsymbol{\omega}}^{e}$ and $\tilde{\boldsymbol{\Omega}}^{d}$, $\hat{\boldsymbol{\Omega}}^{e}$, respectively) are related. However, this question is inadequately posed, because the tensors ${}^{d}\tilde{\boldsymbol{Q}}(t)$ and ${}^{e}\hat{\boldsymbol{Q}}(t)$ are determined in different frames of reference (*cf.*, Eqs. (11.10) and (11.18)). First an imaging operation has to be performed. In order to transfer a vector or a tensor from the y-system to the x-system and *vice versa* one has to replace $\boldsymbol{d}_k \leftrightarrow \tilde{\boldsymbol{d}}_k(t)$ and $\boldsymbol{e}_k \leftrightarrow \hat{\boldsymbol{e}}_k(t)$. Hence, for imaging ${}^{d}\tilde{\boldsymbol{Q}}(t) \leftrightarrow {}^{d}\hat{\boldsymbol{Q}}(t)$ and ${}^{e}\hat{\boldsymbol{Q}}(t) \leftrightarrow {}^{e}\tilde{\boldsymbol{Q}}(t)$ the best way to start is with the following relations, which hold for orthonormal frames:

$${}^{d}\tilde{\boldsymbol{Q}} = \tilde{\boldsymbol{d}}_k(t) \otimes \boldsymbol{e}_k, \; {}^{e}\hat{\boldsymbol{Q}} = \hat{\boldsymbol{e}}_k(t) \otimes \boldsymbol{d}_k, \tag{11.128}$$

If we now first argue intuitively and assume that the image operation $\tilde{(\cdot)}$ on $\hat{\boldsymbol{e}}_k$ annihilates the previous operation $\hat{(\cdot)}$ to obtain $\boldsymbol{e}_k$ again (and similarly for the $\boldsymbol{d}_k$-vectors) we find:

$${}^{e}\tilde{\boldsymbol{Q}} = \boldsymbol{e}_k \otimes \tilde{\boldsymbol{d}}_k(t), \quad {}^{d}\hat{\boldsymbol{Q}} = \boldsymbol{d}_k \otimes \hat{\boldsymbol{e}}_k(t). \tag{11.129}$$

Then the rotational tensors should be related by the following two expressions, which, in hindsight, seem rather obvious:

$${}^{e}\tilde{\boldsymbol{Q}}(t) = {}^{d}\tilde{\boldsymbol{Q}}^{\mathsf{T}}(t), \quad {}^{d}\hat{\boldsymbol{Q}}(t) = {}^{e}\hat{\boldsymbol{Q}}^{\mathsf{T}}(t). \tag{11.130}$$

In the main text we shall present a less abstract proof that, however, is also based on informal but seemingly obvious arguments.

These relations come in handy for rewriting Eqs. (11.13) and (11.21) as follows:

$$\tilde{y}_l(t) = {}^{e}\tilde{Q}_{kl}(t)\hat{y}_k(t), \; \hat{x}_l(t) = {}^{d}\hat{Q}_{kl}(t)\tilde{x}_k(t) \tag{11.131}$$

both valid in orthonormal bases.

At this point reference can again be made to textbooks, such as Müller (1985), pg. 4, or Hutter and Jöhnk (2004), pg. 117, which show equations like that but without making all the subtle distinctions.

References

Altenbach H, Naumenko K, Zhilin PA (2003) A micro-polar theory for binary media with application to phase-transitional flow of fiber suspensions. Continuum Mechanics and Thermodynamics 15(6):539–570

Bampi F, Morro A (1980) Objectivity and objective time derivatives in continuum physics. Foundations of Physics 10(11-12):905–920

Bertram A, Svendsen B (2001) On material objectivity and reduced constitutive equations. Archives of Mechanics 53:6535–675

Bowen R, Wang C (1976) Introduction to Vectors and Tensors Volume 1: Linear and Multilinear Algebra. Springer

Bressan A (1972) On the principles of material indifference and local equivalence. Meccanica 7:3–12

Cornille P (1993) Inhomogeneous waves and Maxwell's equations, Chapter 4 in: Essays on the formal aspects of electromagnetic theory. World Scientific, Singapore

Das A (2007) Tensors. The mathematics of relativity theory and continuum mechanics. Springer, New York

Edelen D, McLennan J (1973) Material indifference: A principle or a convenience. Int J Engng Sci 11:381–393

Eringen C (1980) Mechanics of Continua. Robert E. Krieger Publishing Company, Huntington, New York

Frewer M (2009) More clarity on the concept of material frame-indifference in classical continuum mechanics. Acta Mechanica 202(1-4):213–246

Hauger W, Schnell W, Gross D (1993) Technische Mechanik: Band 3: Kinetik, 4. Auflage. Springer Verlag, Berlin, Heidelberg

Hibbeler R (2010) Engineering Mechanics Dynamics, 12th Edition. Pearson Prentice Hall, Upper Saddle River, New Jersey 07458

Holzapfel EA (2000) Nonlinear Solid Mechanics: A Continuum Approach for Engineering. John Wiley & Sons, Ltd., Chicester, Weinheim, New York, Brisbane, Singapore, Toronto

Hutter K, Jöhnk K (2004) Continuum Methods of Physical Modeling. Continuum Mechanics, Dimensional Analysis, Turbulence. Springer, Springer-Verlag Berlin Heidelberg

Itskov M (2013) Tensor algebra and tensor analysis for engineers. With applications to continuum mechanics. Springer, Springer-Verlag Berlin Heidelberg

Ivanova E, Vilchevskaya E, Müller WH (2016) Time derivatives in material and spatial description - what are the differences and why do they concern us? In: Naumenko K, Aßmus M (eds) Advanced Methods of Mechanics for Materials and Structures, Springer, pp 3–28

Lebedev L, Cloud M, Eremeev V (2010) Tensor analysis with applications in mechanics. World Scientific, New Jersey

Liu I (2004) On Euclidean objectivity and the principle of material frame-indifference. Continuum Mechanics and Thermodynamics 16(1-2):177–183

Liu I (2005) Further remarks on Euclidean objectivity and the principle of material frame-indifference. Continuum Mechanics and Thermodynamics 17(2):125–133

Malvern E (1969) Introduction to the Mechanics of a Continuous Medium. Prentice-Hall, Inc., Englewood Cliffs, New Jersey

Mase G (1970) Theory and Problems of Continuum Mechanics. McGraw-Hill Book Company, New York

Massoudi M (2002) On the importance of material frame-indifference and lift forces in multiphase fows. Chemical Engineering Science 57:3687 – 3701

Matolcsi T (1986) On material frame-indifference. Arch Rat Mech Anal 91(2):99–118

Müller I (1985) Thermodynamics. Pitman

Muschik W, Restuccia L (2008) Systematic remarks on objectivity and frame-indifference, liquid crystal theory as an example. Arch Appl Mech 78:837–854

Nerlich G (1994) What spacetime explains: Metaphysical essays on space and time. Cambridge University Press

Noll W (1958) A mathematical theory of the mechanical behavior of continuous media. Arch Rat Mech Anal 2:197–226

Rivlin R (2006) Some thoughts on frame indifference. Mathematics and Mechanics of Solids 11:113–122

Ryskin G (1985) Misconception which led to the "material frame-indifference" controversy. Phys Rev A 32(2):1239–1240

Stevens BL, Lewis FL, Johnson EN (2016) Aircraft Control and Simulation: Dynamics, Controls Design, and Autonomous Systems, Third Edition. John Wiley & Sons, Inc.

Svendsen B, Bertram A (2001) On frame-indifference and form-invariance in constitutive theory. Acta Mechanica 132:195–207

Truesdell C (1972) A First Course in Rational Continuum Mechanics. The John's Hopkins University, Baltimore, Maryland

Truesdell C, Noll W (1965) The Non-Linear Field Theories of Mechanics. Springer, Heidelberg

Truesdell C, Toupin RA (1960) The Classical Field Theories. Springer, Heidelberg

Truesdell CA (1991) A First Course in Rational Continuum Mechanics. Pure and applied mathematics, vol. 71, Academic Press

Zhilin PA (2001) Vectors and Tensors of Second Rank in Three-Dimensional Space (in Russ.). Nestor, St. Petersburg

Zhilin PA (2003) Theoretical Mechanics. The Fundamental Laws of Mechanics (in Russ.). St. Petersburg Polytechnic University publisher, St. Petersburg

Zhilin PA (2012) Rational Continuum Mechanics (in Russ.). St. Petersburg Polytechnic University publisher, St. Petersburg

Chapter 12
On Electronically Restoring an Imperfect Vibratory Gyroscope to an Ideal State

Stephan V. Joubert, Michael Y. Shatalov & Hiltje Spoelstra

Dedicated to Charlotta E. Coetzee 1955 - 2012.

Abstract With regard to G.H. Bryan's publication in 1890, we call the following Bryan's law (or Bryan's effect): "The vibration pattern of a revolving cylinder or bell revolves at a rate proportional to the inertial rotation rate of the cylinder or bell". Bryan's factor is the proportionality constant that can be theoretically calculated for an ideal vibratory gyroscope (VG). If a perfectly symmetric VG is not ideal, that is, if imperfections and damping are present, then the precession rate (pattern rotation rate) depends on a number of factors. Indeed it depends on the rotation rate of the vehicle it is attached to, mass-stiffness and symmetry imperfections as well as any anisotropic damping (linear or nonlinear) that may be present in the VG. Assuming perfect axissymmetry for the VG, we show how to negate the effects of manufacturing mass-stiffness imperfections as well as the effects of any type of tangentially anisotropic damping that might occur. We achieve this by showing exactly how to symmetrically arrange an electronic array about the symmetry axis. This array consists of curved capacitors under a mixture of a constant (fixed) charge and a small meander charge. We show exactly how the fixed voltage on the capacitor should be adjusted in order to eliminate the frequency split caused by the mass-stiffness imperfection. Furthermore, we show how the meander voltages of the capacitors should be adjusted in order to maintain principal vibration, eliminate quadrature vibration and restore spurious pattern drift in the VG so that it obeys Bryan's law, restoring the precession rate to the ideal rate so that Bryan's factor can be used for calibration purpose. Equations of motion are derived in the form of averaged ODEs that provide us insight into VG behaviour.

Key words: Vibratory gyroscope · Capacitor array · Mass-stiffness imperfections · Anisotropic nonlinear damping · Nonlinear prestress

---

Stephan V. Joubert · Michael Y. Shatalov · Hiltje Spoelstra
Department of Mathematics and Statistics, Tshwane Univesity of Technology, South Africa,
e-mail: JoubertSV@tut.ac.za, ShatalovM@tut.ac.za, SpoelstraH@tut.ac.za

H. Altenbach et al. (eds.), *Mechanics for Materials and Technologies*,
Advanced Structured Materials 46, DOI 10.1007/978-3-319-56050-2_12

12.1 Introduction

Engineers exploit Bryan's law (Bryan, 1890) for calibrating a navigational vibratory gyroscope (VG). However, when a VG is manufactured, inevitable manufacturing imperfections will be present if the manufacturing process is not stringent and/or the materials used are not chemically pure. Because a VG operates using Bryan's effect and Bryan's factor (the proportionality constant mentioned in the law), we will call the vibration dynamics of a VG an *ideal state of vibration* if Bryan's effect is the same as that of a VG where manufacturing imperfections have been eliminated or controlled and no anisotropic damping is present or has been controlled.
It is folklore that a frequency split occurs affecting Bryan's law in a VG if

- it is not perfectly axissymmetric about its axis of rotation (see e.g. Wang et al, 2015, 2016, for a list of citations) and/or
- mass imperfections are present (see e.g. Shatalov et al, 2011; Ma and Su, 2015, for a list of citations) and/or
- stiffness imperfections are present (see e.g. Joubert et al, 2011, 2015).

With the sophisticated computer-assisted turning machinery available nowadays we will assume that the VG *is manufactured to a fine tolerance* so that *axissymmetry is established initially* and we consequently *eliminate this source* of frequency splitting *immediately* from our considerations.
Isotropic damping does not affect Bryan's law in a VG as discussed in Joubert et al (2013), while the presence of any of the following three "anisotropies" in the VG does:

- *mass imperfections*,
- *stiffness imperfections*,
- any form of *anisotropic damping*.

The work presented here extends the work presented in Shatalov and Coetzee (2011) by

1. using a theory that embraces a not necessarily thin shelled VG;
2. introducing anisotropic nonlinear damping into the equations of motion;
3. extending the averaging process to include even order anisotropic damping;
4. including numerical experiments to visualise and explain VG behaviour and equations of motion;
5. clarifying symbol usage;
6. simplifying all equations;
7. showing exactly how the controlling electrodes should be manipulated; and
8. correcting a "concluding statement".

A number of contemporary engineers are contemplating the manufacture of VGs that *are not fused quartz hemispherical vibratory gyroscopes* (HVGs). They use geometries that are easier to manufacture (such as bell, cylindrical, cone and paraboloid shapes) as well as materials such as metals that are readily machined, but that have a lower Q-factor than fused quartz. Recently, Ma and Su (2015), Liu et al

(2013), and Liu et al (2016) manufactured what is claimed to be an *impact-resistant*, bell-shaped VG *using Ni43CrTi*. Hence the sensitivity of these VGs is lower than a fused quartz HVG. Consequently, we aim to make our ideas about increasing sensitivity accessible to these manufacturing engineers who do not necessarily have the time or inclination to interpret a highly technical, succinct paper such as Shatalov and Coetzee (2011).

While the exact nature of the frequency split caused by mass-stiffness imperfections is discussed in Joubert et al (2015) Eq. (10) and for mass-imperfections in Ma and Su (2015) Eq. (8), engineers have measured frequency split and gyroscopic effects in a VG using a ring model with mass imperfections simulated by FEM (see e.g. Ma and Su, 2015; Liu et al, 2016). Unfortunately, even using an ideal model for the VG, the standard FEM generates a frequency split. Hence this type of numerical analysis is flawed unless the mesh is extremely fine. Using their superb equipment, Ma and Su (2015) measured the resulting frequency split (caused by the glued-on mass of the piezoelectric electrodes used for "driving" and "sensing") and they suggested a mass-balancing scheme that they verified experimentally.

Before engineers attempt to build apparatus based on our electrode array, an *optimal scheme* (such as that discussed in Shatalov et al, 2015) for a mechanical mass-balancing scheme such as that suggested by Ma and Su (2015) (and some of the citations therein) must be applied so that any remaining frequency split is small. Indeed, if the frequency split is large, then the capacitor voltages per small capacitor gap length that may be necessary to eliminate the split might exceed the *dielectric strength* (of air, say, with $E_{\max} = 3\,\mathrm{MV\,.\,m^{-1}}$ (Reitz and Milford, 1969)) and the dielectric will break down, rendering the capacitor array useless.

Another motivation for using our electrode array is the following: Even if (after optimal mass-balancing) the frequency split is small enough to neglect, unfortunately, with time, mass-stiffness and damping anisotropy distributions could change slightly in the VG (fatigue effects). This could negate the mass-balancing and significant frequency split and/or vibration pattern drift could reappear. This is undesirable in a working VG that would then have to be disassembled in order to rebalance it, if at all possible.

We do not attempt to verify our theory by means of experiment, but tentatively propose a theory that we trust will enable engineers with well-equipped laboratories to design the electrode array described here and verify (and modify if necessary) the theory via repeatable experiments.

With regard to vibratory gyroscopes and electronic operation, it is worth consulting the pioneering works of Zhuravlev and Klimov (1988) and Zhuravlev and Linch (1995) as well as the contemporary work of Zhuravlev (2015).

Section 12.2 discusses notation and lists some useful definite integrals. Nonlinear prestress, kinetic and potential energy are contemplated in Sect. 12.3, while Sect. 12.4 focusses on a generalised Rayleigh dissipation function, leading to nonlinear tangentially anisotropic damping of any order. An electrode array, which is introduced in Sect. 12.5, controls both frequency split (Sect. 12.6) and nonlinear anisotropic damping via parametric excitation (Sect. 12.7). Exact equations of motion (ODEs), that include nonlinear anisotropic damping of any order, are obtained in Sect. 12.8 and a numerical experiment is conducted using these ODEs in Sect. 12.9. Averaged versions of the ODEs of motion are obtained in Sect. 12.10, while some graphical comparisons are made and a quantitative analysis of the averaged ODE yields some insight into VG behaviour in Sects 12.11 and 12.12. Conclusions are drawn in Sect. 12.13.

12.2 Notation

Studies have been made of VGs for various slowly rotating structures that include thin-shelled cylinders, rings and discs as well as composite bodies consisting possibly of layers of discs, rings, spheres, cylinders, cones, paraboloids and bells (see e.g. Liu et al, 2013, 2016; Joubert et al, 2009; Shatalov et al, 2012; Loveday and Rogers, 1998; Shatalov et al, 2009; Joubert et al, 2014a,b; Ma et al, 2015).
While we consider a polar coordinate system $Or\varphi$, we might have considered other curvilinear systems $Or\varphi k$ with k an "axial" variable (axis of symmetry variable), r a "radial" variable and φ a "tangential" variable, see for instance Ma et al (2015) for a paraboloid system and Liu et al (2016) for a bell-shaped system.
Indeed, in the sequel we will be working with an annular disc vibratory gyroscope (DVG) with inner radius p and outer radius q using the polar coordinate system (see Fig. 12.1), because all of the characteristics of a VG may be illustrated in a technically easy manner when compared to the calculation details of a more complicated VG structure.
By "*smallness*" of the inertial (angular) rate of rotation about the axis of symmetry $\varepsilon\Omega$, we mean that this rate is *substantially smaller than the lowest eigenvalue* ω_0 of the vibrating system. In practice, if $\varepsilon\Omega$ is less than 10% of the eigenvalue for the mode of vibration being studied, then results are satisfactory. Consequently $O(\varepsilon^2)$ terms such as centripetal forces (proportional to $\varepsilon^2\Omega^2$) are neglected and hence, together with the *nonlinear* form of tangential strain, the standard equations of the linear theory of elasticity are used to determine the total potential energy E_{p} of the particles in the object under consideration. Furthermore, both the total kinetic energy E_{k} of the particles in the object under consideration (see e.g. Shatalov et al,

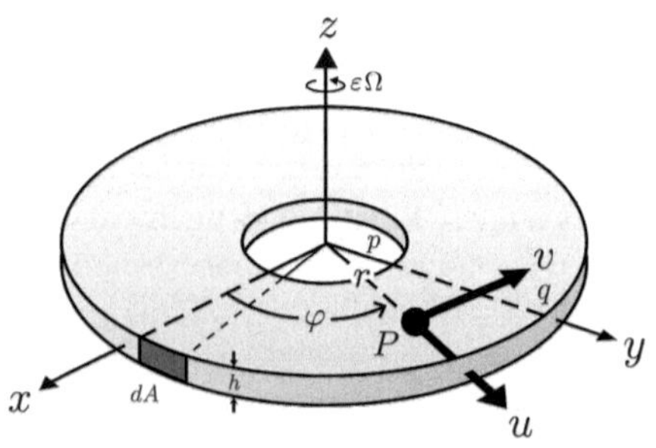

Fig. 12.1 An annular disc showing polar coordinates $Or\varphi$; its thickness h; inner radius p; outer radius q; infinitesimal capcitor area dA; radial and tangential displacements u and v of point P; and slow rotation rate $\varepsilon\Omega$ about the symmetry axis z

2011; Joubert et al, 2011) as well as the total electrical energy E_e of the controlling capacitor array are determined and consequently the Lagrangian $L = E_k - E_p + E_e$ for the system of vibrating particles is formulated (up to $O(\varepsilon^2)$).

Consider a particle P in the VG under consideration. Assume that u is "radial", v is "tangential" and w is "axial" displacement with

$$u = U\left[C(t)\cos m\varphi + S(t)\sin m\varphi\right], \tag{12.1}$$

$$v = V\left[C(t)\sin m\varphi - S(t)\cos m\varphi\right], \tag{12.2}$$

$$w = W\left[C(t)\cos m\varphi + S(t)\sin m\varphi\right], \tag{12.3}$$

where m is the *circumferential wave number* U, V and W are *eigenfunctions* of one or two variables appropriate to the coordinate system (see e.g. Fig. 4. and Eqs. (5) of Ma et al (2015) for a non-rotating paraboloid system or Shatalov et al (2009) Eqs. (9) for a layered spherical system). These eigenfunctions have the *dimensions of length,* while the functions $C(t)$ and $S(t)$ are *dimensionless* functions of time t. If the structure is a ring or annulus, then we take $w = 0$ and assume that all unbalanced forces in the axial z-direction are zero. It is shown in Joubert et al (2014b) that in a DVG the eigenfunctions $U = U(r)$ and $V = V(r)$ for a stationary DVG remain invariant under slow rotation and are hence readily calculated numerically (Joubert et al, 2014a) or (not so readily) analytically in terms of Bessel and Neumann functions. Ultimately, the equations of gyroscopic motion for each system under similar constraints *will be similar*, see e.g. Joubert et al (2009) Eqs. (24) and (25)] where a layered DVG was studied, Joubert et al (2011) Eqs. (19) where Bryan's effect for a planet with various layers was studied and Liu et al (2016) Eqs. (20) where a bell-shaped VG was studied).

For completeness we reproduce the definite integrals $I_0, I_1,$ I_2 and I_3 given in Eqs. (14), (15), (16) and (17) of Joubert et al (2014a), which occur in the calculation of the kinetic and potential energy (see Eqs. (12.9) and (12.14) below) for a DVG. We note that *in general* they might be sums of integrals for various layers of substance in another symmetrically shaped VG. For the annular disc in Fig. 12.1 they are

$$I_0 = \frac{1}{2}\rho_0 h \int_p^q \left(U^2 + V^2\right) r\mathrm{d}r,\ I_1 = \rho_0 h \int_p^q UVr\mathrm{d}r,\ I_3 = \rho_0 h \int_p^q \left(U^2 - V^2\right) r\mathrm{d}r, \tag{12.4}$$

$$I_2 = \frac{\mathbb{E}h}{4\left(1-\mu^2\right)} \int_p^q \frac{1}{r} \left\{ \begin{array}{c} 2rU'\left(\mu\left(mV+U\right)+rU'\right)+ \\ \left(1-\mu\right)\left(mU+V-rV'\right)^2+ \\ 2(mV+U)\left(mV+U+\mu rU'\right) \end{array} \right\} \mathrm{d}r, \tag{12.5}$$

where $\mathbb{E}$ is Young's modulus of elasticity, μ is Poisson's ratio (see e.g. Redwood, 1960), ρ_0 is average mass density and h is the height of the disc.

In this paper, for a DVG, our assumptions reveal two more integrals during the calculation of the potential energy that are readily determined, namely

$$I_4 = \mathbb{E}h \int_p^q \left[(1+m^2)(U^2+V^2) + 4mUV \right] r\mathrm{d}r, \tag{12.6}$$

$$I_5 = \mathbb{E}h \int_p^q \left((1-m^2) \right) (U^2 - V^2)\, r\mathrm{d}r. \tag{12.7}$$

It was demonstrated in Shatalov et al (2009) (for the appropriate definite integrals of the spherical VG studied there), that Bryan's factor η and the *ideal* VG natural angular rate or eigenvalue $\omega_0 = 2\pi f$ (where f is the natural frequency or eigenfrequency of vibration) are given respectively by

$$\eta = \frac{I_1}{I_0} \quad \& \quad \omega_0^2 = \frac{I_2}{I_0}. \tag{12.8}$$

12.3 Kinetic Energy, Prestress and Potential Energy

Assuming that mass imperfections occur tangentially in the annular disc (see e.g. Eq. (21) of Joubert et al (2014a)), the kinetic energy of all of the vibrating particles in the disc is given by

$$E_\mathrm{k} \approx \frac{h}{2} \int_0^{2\pi} \int_p^q \rho(\varphi) \left[\left(\dot{u}^2 + \dot{v}^2 \right) + 2\varepsilon\Omega \left(u\dot{v} - \dot{u}v \right) + 2\varepsilon\Omega \dot{v} r \right] r\mathrm{d}r\mathrm{d}\varphi, \tag{12.9}$$

where the density may be chosen as the sum of the zeroth and $2m^{\mathrm{th}}$ harmonics:

$$\rho(\varphi) := \rho_0 \left[1 + 4\,\varepsilon \frac{I_0}{I_3} \left(\rho_c \cos 2m\varphi + \rho_s S \sin 2m\varphi \right) \right]. \tag{12.10}$$

Here ρ_0 is the average density of the disc and ρ_c and ρ_s are dimensionless constants that we may readily associate with the coefficients of the $2m^{\mathrm{th}}$ harmonics. According to Eq. (29) of Joubert et al (2014a), this yields

$$E_\mathrm{k} \approx \pi \left[(\dot{C}^2 + \dot{S}^2) I_0 + \varepsilon \left[2\Omega (\dot{C}S - C\dot{S}) I_1 + \left(\rho_c (\dot{C}^2 - \dot{S}^2) + 2\rho_s \dot{C}\dot{S} \right) I_0 \right] \right]. \tag{12.11}$$

We assume that during the manufacturing process, some small prestress is introduced in the tangential direction. Using an argument similar to that discussed for density variations in Joubert et al (2014a), we assume that tangential prestress has a Fourier series that is essentially the sum of the zeroth and $2m^{\mathrm{th}}$ harmonics, yielding tangential stress

$$\sigma_\varphi := \frac{\mathbb{E}}{1-\mu^2} (\varepsilon_\varphi + \mu \varepsilon_r) + 8\,\varepsilon\, \mathbb{E}\, I_3 \left(\frac{\sigma_0}{I_4} + \frac{\sigma_c}{I_5} \cos 2m\varphi + \frac{\sigma_s}{I_5} \sin 2m\varphi \right), \tag{12.12}$$

where ε_φ is tangential strain, ε_r is radial strain and μ is Poisson's ratio. Here σ_0, σ_c and σ_s are dimensionless constants that we may readily associate with the coefficients of the zeroth and $2m^{\text{th}}$ harmonics (we have chosen the form of the stress and density coefficients for the purposes of later simplification). In order to analyse the effects of this prestress, we need to introduce the *nonlinear* form of tangential strain:

$$\varepsilon_\varphi := \frac{1}{r}\left(\frac{\partial v}{\partial \varphi} + u\right) + \frac{1}{2r^2}\left[\left(\frac{\partial v}{\partial \varphi} + u\right)^2 + \left(\frac{\partial u}{\partial \varphi} - v\right)^2\right]. \tag{12.13}$$

The second-order strain term is necessary to yield nonzero potential energy due to prestress. After a long, technical calculation (we used MATHEMATICA to do the book-keeping) we determined that the potential energy E_p of the particles in the annular disc is given by

$$\begin{aligned} E_\mathrm{p} &= \frac{h}{2}\int_0^{2\pi}\int_p^q \left\{\sigma_r\varepsilon_r + \sigma_\varphi\varepsilon_\varphi + \sigma_{r\varphi}\varepsilon_{r\varphi}\right\} r\mathrm{d}r\mathrm{d}\varphi \\ &= \pi I_2\left\{\left(C^2+S^2\right)(1+\varepsilon\sigma_0) + \varepsilon\left[\left(C^2-S^2\right)\sigma_c + 2CS\sigma_s\right]\right\} \end{aligned} \tag{12.14}$$

(see Redwood, 1960; Joubert et al, 2014b, Section 4 for the usual definitions of linear stresses $\sigma_r, \sigma_{r\varphi}$ and linear strains $\varepsilon_r, \varepsilon_{r\varphi}$ etc.).

12.4 Tangentially Anisotropic Damping

In Joubert et al (2017) it was demonstrated that the **generalised** Rayleigh dissipation function R for tangentially anisotropic nonlinear damping yields

$$\begin{aligned} \frac{\partial R}{\partial \dot{C}} \approx 2\pi I_0 \sum_{n=1}^{\infty} \Bigg\{ &\varepsilon\left(\sqrt{(\dot{C}^2+\dot{S}^2)}\right)^{n-1} \frac{\left(\delta_{0,n} + \tilde{\varepsilon}\mathbb{C}\sqrt[n+1]{\delta_{0,n}^{n-1}\delta_{c,n}^2}\right)\dot{C} + \tilde{\varepsilon}\mathbb{S}\sqrt[n+1]{\delta_{0,n}^{n-1}\delta_{s,n}^2}\dot{S}}{\tilde{\omega}^{n-1}} \\ &+ \varepsilon\tilde{\varepsilon}\left(\sqrt{(\dot{C}^2+\dot{S}^2)}\right)^{n-3} \frac{(n-1)\left(\mathbb{C}\sqrt[n+1]{\delta_{0,n}^{n-1}\delta_{s,n}^2}(\dot{C}^2-\dot{S}^2) + 2\mathbb{S}\sqrt[n+1]{\delta_{0,n}^{n-1}\delta_{s,n}^2}\dot{C}\dot{S}\right)\dot{C}}{2\tilde{\omega}^{n-1}} \Bigg\}, \end{aligned} \tag{12.15}$$

$$\begin{aligned} \frac{\partial R}{\partial \dot{S}} \approx 2\pi I_0 \sum_{n=1}^{\infty} \Bigg\{ &\varepsilon\left(\sqrt{(\dot{C}^2+\dot{S}^2)}\right)^{n-1} \frac{\left(\delta_{0,n} - \tilde{\varepsilon}\mathbb{C}\sqrt[n+1]{\delta_{0,n}^{n-1}\delta_{c,n}^2}\right)\dot{S} + \tilde{\varepsilon}\mathbb{S}\sqrt[n+1]{\delta_{0,n}^{n-1}\delta_{s,n}^2}\dot{C}}{\tilde{\omega}^{n-1}} \\ &+ \varepsilon\tilde{\varepsilon}\left(\sqrt{(\dot{C}^2+\dot{S}^2)}\right)^{3-n} \frac{(n-1)\left(\mathbb{C}\sqrt[n+1]{\delta_{0,n}^{n-1}\delta_{s,n}^2}(\dot{C}^2-\dot{S}^2) + 2\mathbb{S}\sqrt[n+1]{\delta_{0,n}^{n-1}\delta_{s,n}^2}\dot{C}\dot{S}\right)\dot{S}}{2\tilde{\omega}^{n-1}} \Bigg\} \end{aligned} \tag{12.16}$$

where $\delta_{0,n} > 0$ is the "isotropic damping" coefficient that is proportional to the zeroth harmonic of a Fourier series, $\delta_{c,n} > 0$ (respectively $\delta_{s,n} > 0$) is the "anisotropic damping coefficient" proportional to the cosine (respectively sine) $2m^{\text{th}}$ harmonic of the Fourier series, $\mathbb{C}$ (respectively $\mathbb{S}$) is the sign of the cosine (respectively sine)

$2m^{\text{th}}$ harmonic of the Fourier series and $\tilde{\omega}$ is a fixed angular frequency chosen close to the vibration frequency of the mode of vibration m under consideration. The small quantity ε reminds us that damping is light, while the small quantity $\tilde{\varepsilon}$ reminds us that the anisotropic part of the damping is small and we may neglect $O(\tilde{\varepsilon}^2)$.

12.5 Electrical Energy

Consider the square wave with amplitude $A = 1$, period π, on-off ratio 1 and its approximate Fourier series

$$\mathrm{Sq}(x) = \begin{cases} 1 \text{ if } 0 \leq x \leq \frac{\pi}{2} \\ 0 \text{ if } \frac{\pi}{2} < x \leq \pi \end{cases} \approx \frac{1}{2} + \frac{2}{\pi} \sin 2x. \tag{12.17}$$

Now consider a driving function for the VG given in terms of a *driving angular rate* $\hat{\omega}$, namely the *meander function*

$$M(t) = \frac{1}{2} + \frac{2}{\pi} \sin 2\hat{\omega} t. \tag{12.18}$$

We proceed to control the three anisotropies by using an array of circular-surfaced electronic plates arranged symmetrically about the vibrating annular disc. For instance, for $m \geq 2$, to control the mth mode of vibration we place $8m$ circular-plate electrodes around the disc, evenly spaced. Each plate is placed in such a way that its surface is "parallel" to the cylindrical surface of the disc and the small gap between the plate and disc is d. These circular plates, together with the cylindrical surface of the disc, *approximate a parallel plate capacitor* array. We make the following assumptions about the array for the mth mode of vibration:

1. The polar axis runs from the centre of the disc through the centre of the first electrode and the angle subtended at the centre of the disc by the arc length of each symmetrically arranged circular plate is $2\Delta\varphi$. For the sake of brevity we call $2\Delta\varphi$ the "angular length". The plates are numbered (from the first plate) 1 to $8m$ in the direction of increasing φ.
2. Small potential differences $\sqrt{\varepsilon}V_1(t), \sqrt{\varepsilon}V_2(t), \sqrt{\varepsilon}V_3(t)$ and $\sqrt{\varepsilon}V_4(t)$ are maintained between each plate and the disc for capacitors numbered one to four respectively with

$$V_i(t) = W_i + \hat{\varepsilon}\Delta W_i M(t); \quad W_i \geq 0 \quad \& \quad \hat{\varepsilon}\Delta W_i \geq 0 \quad i = 1,2,3,4, \tag{12.19}$$

where the small parameter $\tilde{\varepsilon} > 0$ is introduced into the meander term $\hat{\varepsilon}\Delta W_i M(t)$ in order to remind us that W_i is at least two orders larger than $\hat{\varepsilon}\Delta W_i$, that is

$$\frac{\hat{\varepsilon}\Delta W_i}{W_i} < 1\%. \tag{12.20}$$

We may consequently make the assumption that

$$W_i^2 + W_i \hat{\varepsilon} \Delta W_i \approx W_i^2. \tag{12.21}$$

The meander term will introduce *parametric excitation (resonance)* into the equations of motion. We will not neglect terms of $O(\varepsilon\hat{\varepsilon})$ because the meander part of each $V_i(t)$ plays a significant role in maintaining vibration in the presence of damping. However, where necessary we will neglect $O(\hat{\varepsilon}^2)$ to simplify matters. In order to calculate electrical capacitance, we need to work with the square of the potential difference. Hence, neglecting $O(\hat{\varepsilon}^2)$, using Eqs. (12.18) and (12.21) we have

$$\begin{aligned} V_i^2(\varphi,t) &= W_i^2 + 2W_i\hat{\varepsilon}\Delta W_i M(t) + \hat{\varepsilon}^2 \Delta W_i^2 M^2(t) \\ &= U_i^2 + \frac{4}{\pi}\Delta U_i^2 \sin 2\hat{\omega}t, \end{aligned} \tag{12.22}$$

where

$$U_i^2 = W_i^2; \quad \Delta U_i^2 = W_i\hat{\varepsilon}\Delta W; \quad \frac{\Delta U_i^2}{U_i^2} = \frac{\hat{\varepsilon}\Delta W_i}{W_i} < 1\%. \tag{12.23}$$

3. The other potential differences around the disc are $\frac{\pi}{m}$ periodic in the sense that capacitor number five has potential difference $\sqrt{\varepsilon}V_1$, capacitor number six has potential difference $\sqrt{\varepsilon}V_2, \cdots$ and capacitor number $8m$ has potential difference $\sqrt{\varepsilon}V_4$.
4. The plates are large enough to ensure that the electric field is almost constant and are far enough away from each other so that mutual inductance is negligible.

Figure 12.2 illustrates the array for the $m = 3$ mode. Now consider a small surface area $\mathrm{d}A = hq\mathrm{d}\varphi$ on the cylindrical surface of the disc as depicted in Fig. 12.1. If there is part of a plate covering $\mathrm{d}A$, then this *"infinitesimal parallel plate capacitor"* has infinitesimal capacitance (see Reitz and Milford, 1969, Eq. (6-30))

$$\mathrm{d}C = \frac{\varepsilon_0}{d-u_q}\mathrm{d}A = \frac{\varepsilon_0 hq}{d-u_q}\mathrm{d}\varphi; \quad \varepsilon_0 \approx 8.854 \times 10^{-12}\mathrm{F.m^{-1}} \tag{12.24}$$

where ε_0 is the electromagnetic permittivity of air (vacuum), d is the gap between the nonvibrating disc and the plate and $u_q = u(q,\varphi,t)$ is the radial displacement of a vibrating particle at the edge of the DVG where $r = q$ (see Eq. (12.1)). If there is no part of a plate covering $\mathrm{d}A$, then $\mathrm{d}C = 0$. Because $u_q << d$, the total electrical potential is (see Reitz and Milford, 1969, Eq. (6-29))

$$E_\mathrm{e} = \varepsilon\frac{1}{2}\int_C V^2(\varphi,t)\mathrm{d}C = \varepsilon\frac{\varepsilon_0 hq}{2d}\int_0^{2\pi} V^2(\varphi,t)\left[1 + \frac{u_q}{d} + \frac{u_q^2}{d^2}\right]\mathrm{d}\varphi. \tag{12.25}$$

Because of the periodicity involved with the square of the potentials, for the function $V^2(\varphi,t)$ we may determine a Fourier series

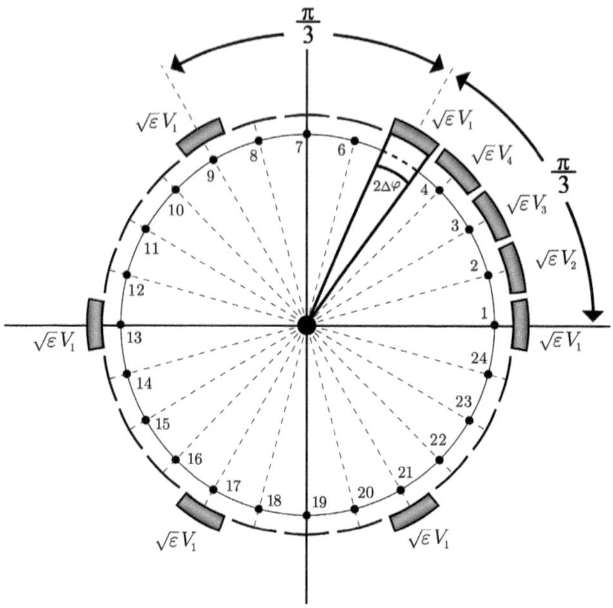

Fig. 12.2: The electrode array for the $m = 3$ vibration mode showing the orientation of the small potential differences $\sqrt{\varepsilon}V_i$, $i = 1,2,3,4$

$$V^2(\alpha,t) = \frac{a_0}{2} + \sum_{n=0}^{\infty} a_n \cos n\varphi + b_n \sin n\varphi. \tag{12.26}$$

This function is depicted for the $m = 3$ mode of vibration in Fig. 12.3 and we used MATHEMATICA to effortlessly provide the coefficients of the Fourier series yielding coefficients as depicted in Fig. 12.4. From Eq. (12.1), the "TrigReduce" command in MATHEMATICA yields

$$\frac{u_q^2}{d^2} = \frac{u^2(q,\varphi,t)}{d^2} = \frac{U^2(q)}{d^2}\left[\frac{C^2+S^2}{2} + \frac{C^2-S^2}{2}\cos 2m\varphi + CS\sin 2m\varphi\right]. \tag{12.27}$$

Consequently, because of the *orthogonality* of the sine and cosine functions, when we substitute the Fourier series for $V^2(\varphi,t)$ given by Eq. (12.26) into Eq. (12.25), only the zeroth and the $2m^{\text{th}}$ harmonics are salient. Furthermore, because the mth harmonics are zero (see Fig. 12.4), we may also neglect $\frac{u_q}{d}$.

Hence using the tables of Fourier coefficients given in Fig. 12.4, the electrical potential energy of the array is

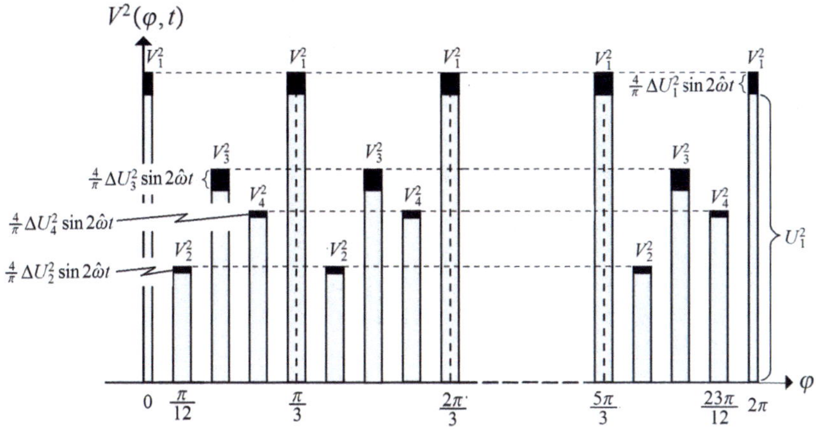

Fig. 12.3: A depiction of the profile of the square of the potential difference function $V^2(\varphi,t)$ for the capacitor array surrounding the disc vibratory gyroscope where $V_i^2(\varphi,t) = U_i^2 + \frac{4}{\pi}\Delta U_i^2 \sin 2\hat{\omega}t$; $i = 1,2,3,4$ and the $m = 3$ vibration mode

$$E_{\mathrm{e}} = \varepsilon\frac{\varepsilon_0 hq}{2d}\int_0^{2\pi}\left\{\frac{2m\Delta\varphi\left(V_1^2+V_2^2+V_3^2+V_4^2\right)}{\pi} + \frac{2\left(V_1^2-V_3^2\right)\sin(2m\Delta\varphi)}{\pi}\cos 2m\varphi + \frac{2\left(V_2^2-V_4^2\right)\sin(2m\Delta\varphi)}{\pi}\sin 2m\varphi\right\}\left\{1+\frac{u_q^2}{d^2}\right\}d\varphi. \tag{12.28}$$

In[3]:= $a_{k\_} :=$

$$\frac{1}{\pi}\left(\sum_{n=0}^{3}\int_{n\frac{\pi}{12}-\Delta\varphi}^{n\frac{\pi}{12}+\Delta\varphi} V_{n+1}^2 \,\mathrm{Cos}[k\,\varphi]\, d\varphi + \sum_{n=4}^{7}\int_{n\frac{\pi}{12}-\Delta\varphi}^{n\frac{\pi}{12}+\Delta\varphi} V_{n-3}^2 \,\mathrm{Cos}[k\,\varphi]\, d\varphi + \sum_{n=8}^{11}\int_{n\frac{\pi}{12}-\Delta\varphi}^{n\frac{\pi}{12}+\Delta\varphi} V_{n-7}^2 \,\mathrm{Cos}[k\,\varphi]\, d\varphi + \sum_{n=12}^{15}\int_{n\frac{\pi}{12}-\Delta\varphi}^{n\frac{\pi}{12}+\Delta\varphi} V_{n-11}^2 \,\mathrm{Cos}[k\,\varphi]\, d\varphi + \sum_{n=16}^{19}\int_{n\frac{\pi}{12}-\Delta\varphi}^{n\frac{\pi}{12}+\Delta\varphi} V_{n-15}^2 \,\mathrm{Cos}[k\,\varphi]\, d\varphi + \sum_{n=20}^{23}\int_{n\frac{\pi}{12}-\Delta\varphi}^{n\frac{\pi}{12}+\Delta\varphi} V_{n-19}^2 \,\mathrm{Cos}[k\,\varphi]\, d\varphi\right)$$

In[5]:= `Table[FullSimplify[a_k], {k, 0, 8}]`

Out[5]= $\left\{\frac{12\,\Delta\varphi\,\left(V_1^2+V_2^2+V_3^2+V_4^2\right)}{\pi}, 0, 0, 0, 0, 0, \frac{2\,\mathrm{Sin}[6\,\Delta\varphi]\,\left(V_1^2-V_3^2\right)}{\pi}, 0, 0\right\}$

In[6]:= `Table[FullSimplify[b_k], {k, 1, 10}]`

Out[6]= $\left\{0, 0, 0, 0, 0, \frac{2\,\mathrm{Sin}[6\,\Delta\varphi]\,\left(V_2^2-V_4^2\right)}{\pi}, 0, 0, 0, 0\right\}$

Fig. 12.4: The Mathematica code for determining the Fourier coefficients of the function $V^2(\varphi,t)$ depicted in Fig. 12.3 for the $m = 3$ vibration mode

Keeping Eq. (12.22) in mind, we use MATHEMATICA to do the "book-keeping" during the calculation of Eq. (12.28) to yield

$$E_e = \pi\varepsilon\left\{k_1\left(C^2+S^2\right)+k_2(C^2-S^2)+2k_3CS+k_0\right\}, \tag{12.29}$$

where

$$\begin{aligned}
k_1 &= \frac{mhq\Delta\varphi\varepsilon_0 U^2(q)}{\pi d^3}\left(V_1^2+V_2^2+V_3^2+V_4^2\right),\\
k_2 &= \frac{hq\varepsilon_0\sin(2m\Delta\varphi)U^2(q)}{2\pi d^3}\left(V_1^2-V_3^2\right),\\
k_3 &= \frac{hq\varepsilon_0\sin(2m\Delta\varphi)U^2(q)}{2\pi d^3}\left(V_2^2-V_4^2\right),\\
k_0 &= \frac{2mhq\Delta\varphi\varepsilon_0}{\pi d}\left(V_1^2+V_2^2+V_3^2+V_4^2\right).
\end{aligned} \tag{12.30}$$

12.6 Eliminating Frequency Split

We include the electrical potential energy into the Lagrangian L as follows

$$L = E_k - E_p + E_e. \tag{12.31}$$

Hence using Eqs. (12.11), (12.14) and (12.29), we obtain

$$\begin{aligned}
L &= \pi I_0(\dot{C}^2+\dot{S}^2)\\
&+ \varepsilon\pi\left[2I_1\Omega(\dot{C}S-C\dot{S})+I_0\rho_c(\dot{C}^2-\dot{S}^2)+2I_0\rho_s\dot{C}\dot{S}\right]\\
&- \pi I_2\left\{\left(C^2+S^2\right)(1+\varepsilon\sigma_0)+\varepsilon\left[\left(C^2-S^2\right)\sigma_c+2CS\sigma_s\right]\right\}\\
&+ \varepsilon\pi\left[k_0+k_1\left(C^2+S^2\right)+k_2(C^2-S^2)+2k_3CS\right].
\end{aligned} \tag{12.32}$$

With Eqs. (12.15) and (12.16) in mind, the Euler-Lagrange equations of motion

$$\begin{aligned}
\frac{\mathrm{d}}{\mathrm{d}t}\frac{\partial L}{\partial \dot{C}} - \frac{\partial L}{\partial C} &= -\frac{\partial R}{\partial \dot{C}},\\
\frac{\mathrm{d}}{\mathrm{d}t}\frac{\partial L}{\partial \dot{S}} - \frac{\partial L}{\partial S} &= -\frac{\partial R}{\partial \dot{S}},
\end{aligned} \tag{12.33}$$

yield the following matrix equation of motion for the VG:

$$\begin{aligned}
&\begin{pmatrix} 1+\varepsilon\rho_c & \varepsilon\rho_s \\ \varepsilon\rho_s & 1-\varepsilon\rho_s \end{pmatrix}\begin{pmatrix}\ddot{C}\\ \ddot{S}\end{pmatrix} + 2\eta\varepsilon\Omega\begin{pmatrix}0 & 1\\ -1 & 0\end{pmatrix}\begin{pmatrix}\dot{C}\\ \dot{S}\end{pmatrix}\\
&+\frac{1}{I_0}\begin{pmatrix} I_2\left[1+\varepsilon(\sigma_0+\sigma_c)\right] - \varepsilon(k_1+k_2) & \varepsilon(I_2\sigma_s-k_3) \\ \varepsilon(I_2\sigma_s-k_3) & I_2\left[1+\varepsilon(\sigma_0-\sigma_c)\right] - \varepsilon(k_1-k_2)\end{pmatrix}\begin{pmatrix}C\\ S\end{pmatrix}\\
&= -\varepsilon D
\end{aligned} \tag{12.34}$$

where the matrix D representing damping is

$$D = \begin{pmatrix} \dfrac{\partial R}{\partial \dot{C}} \\ \dfrac{\partial R}{\partial \dot{S}} \end{pmatrix}. \tag{12.35}$$

Multiplying Eq. (12.34) by the inverse of the leading coefficient matrix, neglecting $O(\varepsilon^2)$ and identifying the *mass-stiffness coefficients*

$$\mu_c = \rho_c + \sigma_c; \quad \mu_s = \rho_s + \sigma_s \tag{12.36}$$

we obtain

$$\begin{pmatrix} \ddot{C} \\ \ddot{S} \end{pmatrix} + \frac{1}{I_0} \begin{pmatrix} I_2(1+\varepsilon\sigma_0) - \varepsilon k_1 \\ -\varepsilon(k_2 + I_2\mu_c) & -\varepsilon(k_3 + I_2\mu_s) \\ -\varepsilon(k_3 + I_2\mu_s) & I_2(1+\varepsilon\sigma_0) - \varepsilon k_1 \\ & +\varepsilon(k_2 + I_2\mu_c) \end{pmatrix} \begin{pmatrix} C \\ S \end{pmatrix}$$
$$= 2\eta\varepsilon\Omega \begin{pmatrix} 0 & -1 \\ 1 & 0 \end{pmatrix} \begin{pmatrix} \dot{C} \\ \dot{S} \end{pmatrix} - \varepsilon D. \tag{12.37}$$

If all of the electrodes (capacitors) are switched off, then the $k_i = 0$ for $i = 1,2,3,4$ and the matrix equation of motion will be

$$\begin{pmatrix} \ddot{C} \\ \ddot{S} \end{pmatrix} + \omega_0 \begin{pmatrix} 1+\varepsilon\sigma_0 - \varepsilon\mu_c & -\varepsilon\mu_s \\ -\varepsilon\mu_s & 1+\varepsilon\sigma_0 + \varepsilon\mu_c \end{pmatrix} \begin{pmatrix} C \\ S \end{pmatrix}$$
$$= 2\eta\varepsilon\Omega \begin{pmatrix} 0 & -1 \\ 1 & 0 \end{pmatrix} \begin{pmatrix} \dot{C} \\ \dot{S} \end{pmatrix} - \varepsilon D. \tag{12.38}$$

with ω_0 the ideal (or natural) angular frequency given by Eq. (12.8).

The eigenvalues of the matrix

$$\omega_0^2 \begin{pmatrix} 1+\varepsilon\sigma_0 - \varepsilon\mu_c & -\varepsilon\mu_s \\ -\varepsilon\mu_s & 1+\varepsilon\sigma_0 + \varepsilon\mu_c \end{pmatrix} \tag{12.39}$$

show that there exists a frequency split with frequency of beats up to $O(\varepsilon^2)$

$$f_{\text{beats}} = \frac{\varepsilon\omega_0\sqrt{\mu_c^2 + \mu_s^2}}{2\pi} \tag{12.40}$$

caused by the mass-stiffness anisotropy. Notice that the *constant prestress* term σ_0 does not play a role in the frequency split.

This frequency split can be controlled by the capacitor array. Indeed, keeping Eq. (12.22) in mind, from Eq. (12.30) each k_i can be written

$$k_i = K_i + \Delta K_i \sin 2\hat{\omega}t; i = 1,2,3 \tag{12.41}$$

where

$$\begin{aligned} K_1 &\propto \left(U_1^2+U_2^2+U_3^2+U_4^2\right), \\ \Delta K_1 &\propto \left(\Delta U_1^2+\Delta U_2^2+\Delta U_3^2+\Delta U_4^2\right), \\ K_2 &\propto \left(U_1^2-U_3^2\right),\ \Delta K_2 \propto \left(\Delta U_1^2-\Delta U_3^2\right), \\ K_3 &\propto \left(U_2^2-U_4^2\right), \Delta K_3 \propto \left(\Delta U_2^2-\Delta U_4^2\right). \end{aligned} \tag{12.42}$$

Consequently, with the capacitors switched on, Eq. (12.37) yields

$$\begin{pmatrix} \ddot{C} \\ \ddot{S} \end{pmatrix} + \frac{1}{I_0} \begin{pmatrix} \begin{matrix} I_2(1+\varepsilon\sigma_0)-\varepsilon K_1- \\ \varepsilon\left(K_2+I_2\mu_c\right) \end{matrix} & -\varepsilon\left(K_3+I_2\mu_s\right) \\ -\varepsilon\left(K_3+I_2\mu_s\right) & \begin{matrix} I_2(1+\varepsilon\sigma_0)-\varepsilon K_1+ \\ \varepsilon\left(K_2+I_2\mu_c\right) \end{matrix} \end{pmatrix} \begin{pmatrix} C \\ S \end{pmatrix} \\ = \varepsilon F(C,S,\dot{C},\dot{S}) \tag{12.43}$$

where on the right hand side of Eq. (12.43)

$$F(C,S,\dot{C},\dot{S}) = \frac{\sin 2\hat{\omega}t}{I_0} \begin{pmatrix} \Delta K_1+\Delta K_2 & \Delta K_3 \\ \Delta K_3 & \Delta K_1-\Delta K_2 \end{pmatrix} \begin{pmatrix} C \\ S \end{pmatrix} + \\ 2\eta\Omega \begin{pmatrix} 0 & -1 \\ 1 & 0 \end{pmatrix} \begin{pmatrix} \dot{C} \\ \dot{S} \end{pmatrix} - D. \tag{12.44}$$

Notice that in Term (12.43) it is possible to achieve

$$K_2 = -\mu_c I_2 \quad \& \quad K_3 = -\mu_s I_2 \tag{12.45}$$

because we may manipulate capacitors, changing the size and sign of K_2 and K_3 according to Expressions (12.42).

Manipulating capacitors as indicated, the equations of motion simplify to

$$\begin{pmatrix} \ddot{C} \\ \ddot{S} \end{pmatrix} + \omega^2 \begin{pmatrix} C \\ S \end{pmatrix} = \varepsilon F \tag{12.46}$$

where the new *electronically induced angular frequency* of vibration is

$$\omega = \sqrt{\frac{I_2(1+\varepsilon\sigma_0)-\varepsilon K_1}{I_0}} = \sqrt{\omega_0^2(1+\varepsilon\sigma_0)-\varepsilon\frac{K_1}{I_0}}, \tag{12.47}$$

keeping Eq. (12.8) in mind. Note that the new angular frequency ω is

1. *slightly different* from the *ideal* angular frequency ω_0;
2. *affected slightly by constant prestress* σ_0; *and*
3. *affected slightly* by *all of the nonmeander potential differences* because

$$K_1 \propto U_1^2+U_2^2+U_3^2+U_4^2. \tag{12.48}$$

If it exists, it is not possible to eliminate constant prestress σ_0 once the VG has been manufactured; nor is it possible to eliminate K_1 once the capacitors have been adjusted to eliminate the frequency split.

12.7 Parametric Excitation

There are instruments capable of measuring the frequency split (see e.g. Figure 14 of the experimental system in Wang et al (2016)) and we may thus simplify our equations by assuming the following:

1. the *new angular frequency* ω *has been achieved* by suitable adjustment of capacitors as suggested by Eqs. (12.42) and (12.45);
2. the excitation electrodes have been adjusted so that $\Delta U_1 = \Delta U_3$ and $\Delta U_2 = \Delta U_4$. This yields
$$\Delta K_2 = \Delta K_3 = 0, \tag{12.49}$$
according to Expressions (12.42);
3. the meander signal (see Eqs. (12.18) and (12.22)) is adjusted so that the driving frequency $2\hat{\omega}$ is set to the new angular frequency 2ω; and
4. the angular rate $\tilde{\omega}$ in the Fourier series for the damping coefficient is set to the new angular frequency ω (see Eqs. (12.15) and (12.16)).

Hence Eq. (12.44) now simplifies to

$$F = \frac{\sin 2\omega t}{I_0}\begin{pmatrix}\Delta K_1 & 0\\ 0 & \Delta K_1\end{pmatrix}\begin{pmatrix}C\\ S\end{pmatrix} + 2\eta\Omega\begin{pmatrix}0 & -1\\ 1 & 0\end{pmatrix}\begin{pmatrix}\dot{C}\\ \dot{S}\end{pmatrix} - D. \tag{12.50}$$

We now rearrange the matrix equation Eq. (12.46) into a system of two ODEs:

$$\ddot{C} + \omega^2 C = \varepsilon f_1(\dot{C}, \dot{S}, C, S)) \tag{12.51}$$

$$\ddot{S} + \omega^2 S = \varepsilon f_2(\dot{C}, \dot{S}, C, S). \tag{12.52}$$

where, noting Eqs. (12.35) and (12.50), we obtain

$$f_1(\dot{C}, \dot{S}, C, S)) = \frac{\sin 2\omega t}{I_0}\Delta K_1 C - 2\eta\varepsilon\Omega\dot{S} - \frac{1}{2\pi I_0 \varepsilon}\frac{\partial R}{\partial \dot{C}}, \tag{12.53}$$

and

$$f_2(\dot{C}, \dot{S}, C, S) = \frac{\sin 2\omega t}{I_0}\Delta K_1 S + 2\eta\varepsilon\Omega\dot{C} - \frac{1}{2\pi I_0 \varepsilon}\frac{\partial R}{\partial \dot{S}} \tag{12.54}$$

We could conduct a numerical experiment using Eqs. (12.51) and (12.52), comparing the graphs of C and S, but little will be achieved because these are “fast variables” depending on the fast independent variable ω. In order to conduct a numerical analysis we need to transform to “slow variables”, because, as observed by Friedland

and Hutton (1978) just after Eq. (2.10), the rapidly varying quantities C and S are "difficult to relate to" the inertial angular rate $\varepsilon\Omega$, while the behaviour of the slowly varying quantities "may be expected to be more indicative of" the inertial angular rate.

12.8 Principal and Quadrature Vibration

We introduce a change from "fast" to "slow" variables

$$\left(C(t),S(t),\dot{C}(t),\dot{S}(t)\right) \longrightarrow (P(t),Q(t),\Theta(t),\psi(t)) \tag{12.55}$$

using the transformation

$$C\cos m\varphi + S\sin m\varphi = P\cos[m(\varphi-\Theta)]\cos\gamma + Q\sin[m(\varphi-\Theta)]\sin\gamma \tag{12.56}$$

where P is the amplitude of the *principal vibration*, Q is the amplitude of the *quadrature vibration*, $m\Theta$ is the *precession angle* (the rotation angle of the vibrating pattern) and ψ is a *phase angle* with fast variable

$$\gamma = \omega t + \psi. \tag{12.57}$$

With $P > Q \geq 0$, as explained in detail in Shatalov et al (2011), Eqs. (36) to (48), the equations of motion of the VG are

$$\dot{P} = -\frac{\varepsilon}{\omega}[f_1\cos m\Theta + f_2\sin m\Theta]\sin\gamma, \tag{12.58a}$$

$$\dot{Q} = -\frac{\varepsilon}{\omega}[f_1\sin m\Theta - f_2\cos m\Theta]\cos\gamma, \tag{12.58b}$$

$$\begin{aligned} m\dot{\Theta} = {} & \frac{\varepsilon}{\omega(P^2-Q^2)}\{f_1[P\sin m\Theta\sin\gamma + Q\cos m\Theta\cos\gamma] \\ & - f_2[P\cos m\Theta\sin\gamma - Q\sin m\Theta\cos\gamma]\}, \end{aligned} \tag{12.58c}$$

$$\begin{aligned} \dot{\psi} = {} & -\frac{\varepsilon}{\omega(P^2-Q^2)}\{f_1[P\cos m\Theta\cos\gamma + Q\sin m\Theta\sin\gamma] \\ & + f_2[P\sin m\Theta\cos\gamma - Q\cos m\Theta\sin\gamma]\}. \end{aligned} \tag{12.58d}$$

12.9 Numerical Experiment

For various orders and/or combinations of orders of anisotropic damping, numerical experiments can be conducted using Eqs. (12.58). First we need to determine reasonable estimates for quantities in Eqs. (12.53) and (12.54). In order to determine a good estimate for $\frac{\Delta K_1}{I_0}$, we use the "ideal" aluminium disc described in Joubert et al (2014b) with inner radius 1.45 m and outer radius 1.5 m, which at the $m = 2$ vibration mode, has an angular frequency of approximately $30\pi\,\mathrm{rad\,.\,s^{-1}}$ and Bryan's factor $\eta = -\frac{8}{10}$ (see Eqs. (12.8)). Because the eigenfunctions are only determined up to a multiplicative constant, their magnitude is never determined. Hence the size of I_0 (see Eqs. (12.4)) is never determined, although the ratio $\sqrt{\frac{I_2}{I_0}} \approx 30\pi\,\mathrm{rad\,.\,s^{-1}}$ is. However, we can estimate a suitable size for $\frac{\Delta K_1}{I_0}$ as follows: Consider that, neglecting prestress σ_0, Eq. (12.47) yields two facts. **Firstly**

$$\varepsilon\frac{K_1}{I_0} \approx \omega_0^2 - \omega^2, \tag{12.59}$$

and **secondly** that if $\omega = 30\pi\,\mathrm{rad\,.\,s^{-1}}$, then it is not much smaller than ω_0, say, $\omega_0 = 31\pi\,\mathrm{rad\,.\,s^{-1}}$. Hence it is reasonable to assume that with $\varepsilon = \frac{1}{100}$

$$\frac{K_1}{I_0} \approx 6000\pi^2\,\mathrm{Hz}^2. \tag{12.60}$$

Consequently, using Eqs. (12.41) and (12.23),

$$\frac{\Delta K_1}{I_0} \approx 60\pi^2\,\mathrm{Hz}^2. \tag{12.61}$$

We use anisotropic *quadratic* damping ($\delta_{0,n\neq 2} = \delta_{c,n\neq 2} = \delta_{s,n\neq 2} = 0$) in Eqs. (12.53) and (12.54); and we consider the following values that have been chosen in order to *highlight the influence* of the *driving electrodes* and the *anisotropic damping*:

$$\begin{gathered} m = 2, \varepsilon = \frac{1}{100}, \tilde{\varepsilon} = \frac{1}{100}, \eta = -\frac{8}{10}, \Omega = \pi\,\mathrm{rad\,.\,s^{-1}}, \\ \omega = 30\pi\,\mathrm{rad\,.\,s^{-1}}, \mathbb{C} = 1, \mathbb{S} = 1, \delta_{0,2} = M \times 600\,\mathrm{Hz}, \delta_{c,2} = M \times 3000\,\mathrm{Hz}, \\ \delta_{c,2} = M \times 7500\,\mathrm{Hz}, \frac{\Delta K_1}{\mathbb{I}_0} = L \times 60\pi^2\,\mathrm{Hz}^2, \\ P(0) = \frac{1}{2}, Q(0) = \frac{1}{20}, \Theta(0) = 0\,\mathrm{rad}, \psi(0) = 0\,\mathrm{rad}, 0 \le t \le 2000\,\mathrm{s} \end{gathered} \tag{12.62}$$

We used M and L to vary the size of the damping and driving electrodes respectively. Using similar values as given in (12.62) for nonlinear damping of any order, we choose $M = 1$ and $L = 1$ to switch the meander (driving) electrodes on and $L = 0$ to switch them off. However, for anisotropic linear damping ($\delta_{0,n\neq 1} = \delta_{c,n\neq 1} = \delta_{s,n\neq 1} = 0$), we had to make $M = \frac{1}{100}$ and $L = \frac{191}{100}$ in order to see parametric resonance. With $L = \frac{190}{100}$, P eventually vanishes and with $L = \frac{192}{100}$, P appears to grow without bound. In the presence of anisotropic quadratic damping, using Eqs. (12.58) with values from (12.62), the NDSolve routine of MATHEMATICA produced Fig. 12.5, which reveals the following: If the driving electrodes are switched on ($L = 1$), then the principal vibration amplitude P is initially damped down from 0.5 to about 0.0065 during a *transient state*, but at the *steady-state* (about $t > T_{ss} = 300\,\mathrm{s}$), *it oscillates at an average nonzero level of about* $P = 0.00615$, with $0.0060 \le P \le 0.0063$. Clearly we are close to a *state of parametric resonance*

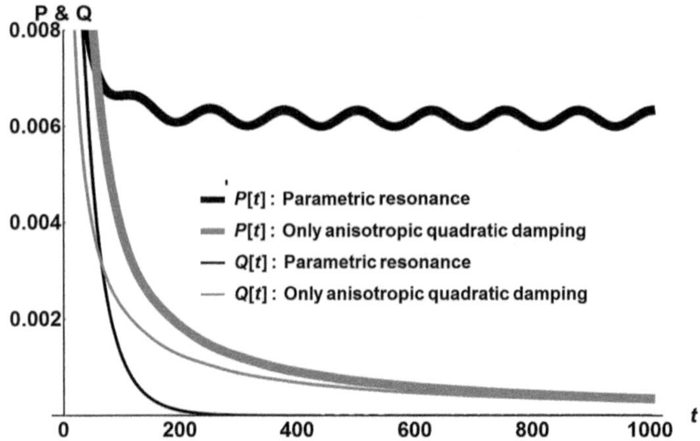

Fig. 12.5: The "exact" numerical solution showing the behaviour of the amplitudes of the principal P and quadrature Q vibrations for the $m = 2$ mode of vibration and anisotropic quadratic damping $(\delta_{0,n\neq 2} = \delta_{c,n\neq 2} = \delta_{s,n\neq 2} = 0)$

here. Another desirable feature is also exhibited here, namely the *amplitude* of the quadrature vibration Q *is quite rapidly damped down* from 0.05 *to become negligible* at the steady-state. However, if the driving electrodes are switched off $(L = 0)$, then *both* the principal amplitude P and the quadrature amplitude Q are initially damped down with what appears to be *different damping rates* (as it should be with anisotropic damping). However, as time evolves, neither Q nor P appears to vanish completely, but seems to maintain a small but nonzero value.

With the electrodes switched on and suitably adjusted, Fig. 12.6 shows that during the transient state, the precession angle $m\Theta$ does not increase "ideally". However, during the steady-state $(t > T_{ss} = 300\,\text{s})$, the precession angle appears to become linear with the same slope $(\eta\varepsilon\Omega)$ as the ideal precession angle $\eta\varepsilon\Omega t$. Thus $m\Theta[t]$ increases at a rate *proportional to the inertial angular rate* (Bryan (1890)) and so the VG is behaving like an "ideal" VG that utilises Bryan's factor for calibration purposes.

We note that swapping the signs to $\mathbb{C} = -1, \mathbb{S} = -1$ in (12.62) does not significantly change P and Q from what is presented in Fig. 12.5. However, the precession angle $m\Theta$ and the phase angle ψ show significant change, as can be seen in Fig. 12.7 for the phase angle ψ. This is a phenomenon that appears to be more general than merely swapping the signs $\mathbb{C}$ and $\mathbb{S}$. Indeed, with any choice of signs $\mathbb{C}$ and $\mathbb{S}$, numerical experiments indicate that *no matter what value we assign to the initial value* $\psi(0)$, the phase angle *appears to obey*

$$\lim_{t\to T>T_{ss}} \psi(t) = \tfrac{2k+1}{2}\pi \tag{12.63}$$

for some integer k (not necessarily 0 or -1), as can be seen in Fig. 12.7.

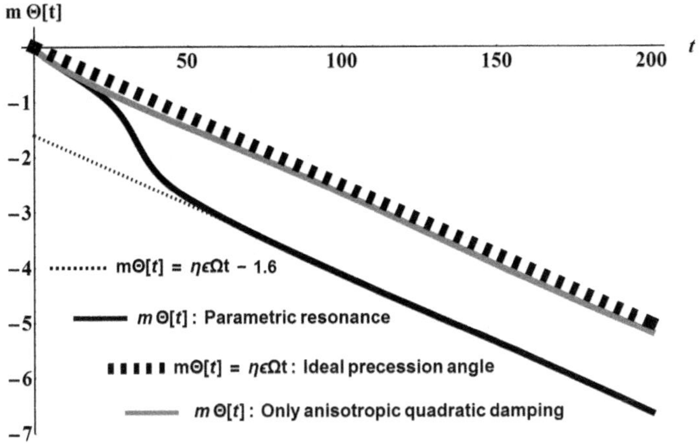

Fig. 12.6: The "exact" numerical solution showing the behaviour of the precession angle $m\Theta$

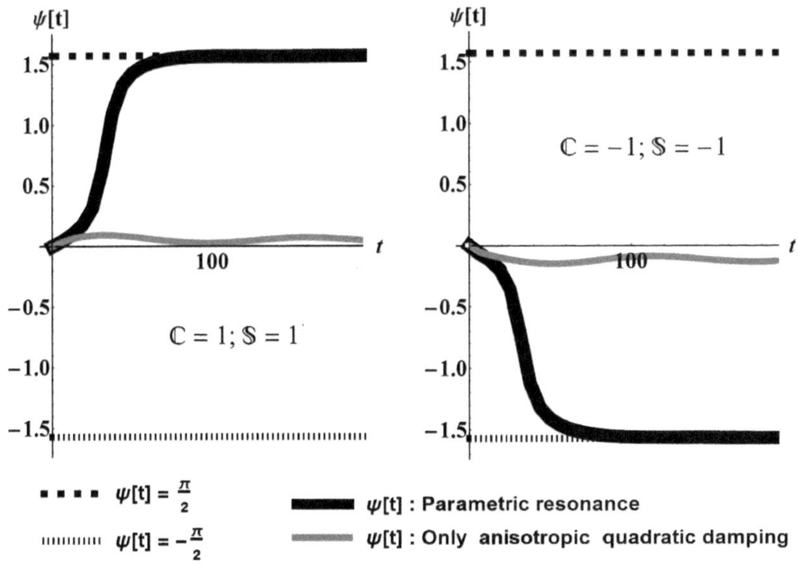

Fig. 12.7: Swapping the signs of C and S of the $2m^{\text{th}}$ harmonics of the anisotropic quadratic damping coefficient significantly change the phase angle ψ (as well as the precession angle $m\Theta$ (not shown)) demonstrating that anisotropic damping plays a significant role in both quantities

12.10 Averaging

While it is easy enough to analyse the behaviour of ODE (12.58) qualitatively (graphically), it is not immediately evident how to predict behaviour by examining the ODE (12.58). In order to better understand the system of ODE (12.58) we consider using the averaging operator

$$\langle - \rangle = \frac{1}{2\pi} \int_0^{2\pi} (-)\, d\gamma, \tag{12.64}$$

as discussed in Joubert et al (2017), which we attempt to apply to the ODEs. Hopefully these averaged ODEs will give us insight into how the VG behaves if it is subject to nonlinear anisotropic damping and parametric excitation.

Linear (order one) anisotropic damping has been dealt with in Joubert et al (2011) and Shatalov and Coetzee (2011), while the averaging process with *odd-order anisotropic damping* is routine. For *anisotropic even-order damping,* with say $n = 2k$, when one applies the transformation Eq. (12.56) to the Eqs. (12.53) and (12.54), one encounters the expression

$$\left(\sqrt{\dot{C}^2 + \dot{S}^2}\right)^{2k-1} = \left(\omega\sqrt{P^2 \sin^2\gamma + Q^2\cos^2\gamma}\right)^{2k-1}. \tag{12.65}$$

Consequently it is **not a routine exercise** by pen and paper (or by using MATHEMATICA) to calculate the average of the ODE (12.58).

In Joubert et al (2017), using values similar to those given in (12.62), for various vibration modes $m \leq 10$ and (combinations of) orders of anisotropic damping $n \leq 10$, it was demonstrated numerically that the root of the mean square (RMS) value of P and Q satisfies

$$\sqrt{\tfrac{P^2+Q^2}{2}} \approx \sqrt{\left(P^2 \sin^2\gamma + Q^2\cos^2\gamma\right)}. \tag{12.66}$$

This approximation appears to be *valid* over the range $0 \leq t \leq 2000$ for *any initial values* $0 < Q(0) < P(0)$.

In the absence of electronic control using the meander function, it was demonstrated in Joubert et al (2017) that a good approximation is obtained by neglecting the second line $O(\varepsilon\tilde{\varepsilon})$ terms in Eqs. (12.15) and (12.16) when calculating $\dot{P}$ and $\dot{Q}$ and $m\dot{\Theta}$. We therefore

1. use Eq. (12.56) to transform Eqs. (12.53) and (12.54), neglecting some $O(\varepsilon\tilde{\varepsilon})$ terms in Eqs. (12.15) and (12.16) when calculating $\dot{P}$ and $\dot{Q}$ and $m\dot{\Theta}$;
2. make the replacement suggested by the RMS (Eq. (12.66)) in the transformed equations; and
3. use the averaging operator Eq. (12.64) and MATHEMATICA to determine the "averaged" versions of ODE (12.58) as follows:

$$\dot{P} = -\varepsilon \sum_{n=1}^{\infty} P\left(P^2+Q^2\right)^{\frac{n-1}{2}} \frac{\delta_{0,n}+\tilde{\varepsilon}\left[\mathbb{C}\,{}^{n+1}\!\sqrt{\delta_{0,n}^{n-1}\delta_{c,n}^2}\cos 2m\Theta+\mathbb{S}\,{}^{n+1}\!\sqrt{\delta_{0,n}^{n-1}\delta_{s,n}^2}\sin 2m\Theta\right]}{\sqrt{2^{n+1}}} - \varepsilon\frac{\Delta K_1}{4\mathbb{I}_0\omega}P\cos 2\psi, \tag{12.67}$$

$$\dot{Q} = -\varepsilon \sum_{n=1}^{\infty} Q\left(P^2+Q^2\right)^{\frac{n-1}{2}} \frac{\delta_{0,n}-\tilde{\varepsilon}\left[\mathbb{C}\,{}^{n+1}\!\sqrt{\delta_{0,n}^{n-1}\delta_{c,n}^2}\cos 2m\Theta+\mathbb{S}\,{}^{n+1}\!\sqrt{\delta_{0,n}^{n-1}\delta_{s,n}^2}\sin 2m\Theta\right]}{\sqrt{2^{n+1}}} + \varepsilon\frac{\Delta K_1}{4\mathbb{I}_0\omega}Q\cos 2\psi, \tag{12.68}$$

$$m\dot{\Theta} = \eta\varepsilon\Omega + \varepsilon\tilde{\varepsilon}\sum_{n=1}^{\infty}\frac{\left(\sqrt{P^2+Q^2}\right)^{n+1}}{\left(P^2-Q^2\right)}\frac{\left[\mathbb{C}\,{}^{n+1}\!\sqrt{\delta_{0,n}^{n-1}\delta_{c,n}^2}\sin 2m\Theta-\mathbb{S}\,{}^{n+1}\!\sqrt{\delta_{0,n}^{n-1}\delta_{s,n}^2}\cos 2m\Theta\right]}{\sqrt{2^{n+1}}} - \varepsilon\frac{\Delta K_1}{\mathbb{I}_0\omega}\frac{PQ}{\left(P^2-Q^2\right)}\cos\psi\sin\psi, \tag{12.69}$$

$$\dot{\psi} = -\varepsilon\tilde{\varepsilon}\sum_{n=1}^{\infty}\frac{PQ\left(\sqrt{P^2+Q^2}\right)^{n-1}}{\left(P^2-Q^2\right)}\frac{(n+3)\left[\mathbb{C}\,{}^{n+1}\!\sqrt{\delta_{0,n}^{n-1}\delta_{c,n}^2}\sin 2m\Theta-\mathbb{S}\,{}^{n+1}\!\sqrt{\delta_{0,n}^{n-1}\delta_{s,n}^2}\cos 2m\Theta\right]}{\sqrt{2^{n+3}}} + \varepsilon\frac{\Delta K_1}{4\mathbb{I}_0\omega}\frac{P^2+Q^2}{\left(P^2-Q^2\right)}\sin 2\psi. \tag{12.70}$$

12.11 Graphical Comparisons and Quantitative Analysis of the Exact and Averaged ODE

We will talk of the "exact solution" when we mean a suitable numerical solution produced by using Eqs. (12.53) and (12.54)) in ODEs (12.58). Using the values in (12.62), the solutions to the averaged ODEs (12.67) to (12.70) show the same global behaviour as the exact solutions, as may be observed in Fig. (12.8). Consequently we assume that the equations of motion of the VG are well represented by the averaged ODEs (12.67) to (12.70), and hence we assume that they may be used to conduct some quantitative analyses of VG behaviour.

Firstly, observe Eq. (12.68): The first term in Eq. (12.68) is an anisotropic damping term with a sinusoidally varying damping coefficient - we will call such terms "*anisotropic variable damping*". The second (meander) term in Eq. (12.68) represents either linear (viscous) damping (or growth) with a variable coefficient that becomes a constant value as ψ becomes constant - we will call such terms "*meander terms*" and refer to them appropriately as "*meander damping*" or "*meander growth*". Numerical evidence suggests that ψ rapidly approaches $\frac{2k+1}{2}\pi$ for some integer k and so $\cos 2\psi < 0$. Hence we rapidly have a meander damping term present. This combination of anisotropic damping and meander linear damping rapidly causes Q to vanish for all practical purposes (see Fig. 3 of Joubert et al (2013) for an example of a combination of isotropic linear and quadratic damping). Consequently, for all

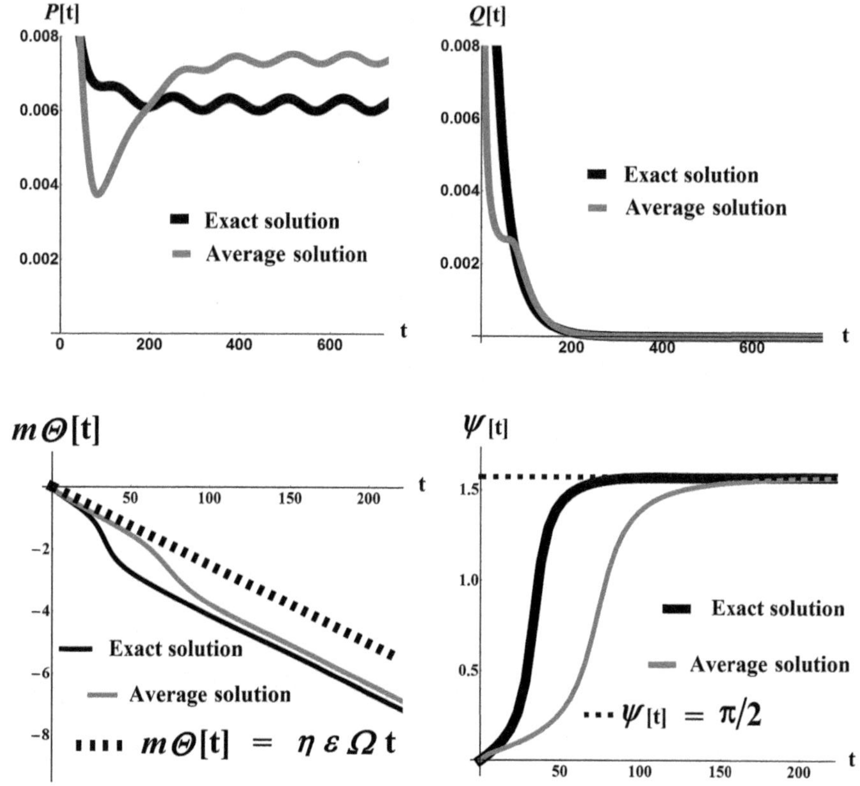

Fig. 12.8: The exact and averaged solutions for all variables P, Q, $m\Theta$ and ψ show the same global trends

practical purposes,

$$Q \approx 0; 0 < t \leq T_{\mathrm{ss}} \quad \text{and} \quad \lim_{t \to T > T_{\mathrm{ss}}} Q = 0. \tag{12.71}$$

Secondly, observe Eq. (12.70):

1. Numerical experiments all indicate that the phase angle eventually achieves a constant value, that is

$$\lim_{t \to T > T_{\mathrm{ss}}} \dot{\psi} = 0. \tag{12.72}$$

   Consequently, taking limits in Eq. (12.70) we have

$$0 = \frac{\varepsilon \Delta K_1}{4\mathbb{I}_0 \omega} \cos 2 \left(\lim_{t \to T > T_{\mathrm{ss}}} \psi \right) \tag{12.73}$$

   and hence

$$\lim_{t \to T > T_{ss}} \psi = n\frac{\pi}{2} \tag{12.74}$$

for some integer n. Numerical evidence indicates that $n = 2k+1$ for some integer k.

2. In Shatalov and Coetzee (2011) in the last paragraph before the "Conclusions" section, it is stated that "For the realization of the control $\psi \to 0$, it is possible to use a reference phase generator which generates reference excitation signals so that $\psi \approx 0$." This appears to be an impossible realisation when electronic meander is present.

Penultimately, observe Eq. (12.67). Notice that the anisotropic variable damping coefficient in Eq. (12.67) is different to the first term in Eq. (12.68), as one would expect for anisotropic damping. Initially the anisotropic variable damping term, the first term in Eq. (12.67), might combine with the meander term in Eq. (12.67) (during a possible damping phase), causing rapid damping of P. However, the above argument shows that $\cos 2\psi$ becomes negative rapidly and the second term in Eq. (12.67) therefore represents a meander growth term, becoming linear when $\cos 2\psi = -1$. Consequently, with correctly adjusted excitation electrodes, the linear meander **growth** term (the second term in Eq. (12.67)), **will balance** the anisotropic **damping** term (the first term in Eq. (12.67)). Indeed, at the steady-state, P appears to be held in a constant nonzero small amplitude sinusoidal oscillatory state that appears to be *parametric resonance*.

Ultimately, observe Eq. (12.69).

1. Even if the meander voltage is $\Delta K_1 = 0\,\mathrm{V}$ and the inertial rotation rate $\varepsilon\Omega = 0\,\mathrm{rad.s^{-1}}$, if anisotropic damping is present, then the pattern will drift. Pattern drift under such circumstances *indicates that anisotropic damping is indeed present.*
2. With regard to the second term in Eq. (12.69), because of the product $\varepsilon\tilde{\varepsilon}$, this could be either a small growth or damping term, depending on the signs of $\mathbb{C}$ and $\mathbb{S}$. During the transient state, the second term in Eq. (12.69) is dominated by the nonmeander value of P and so has a substantial influence on the growth or shrinkage of $m\Theta$ at this stage. The size of this growth or shrinkage can be estimated by numerical integration of the averaged solution. In MATHEMATICA this solution is stored as an interpolating function that is readily integrated. For instance, using Eq. (12.62),

$$C_{\mathrm{damp}} = \varepsilon\tilde{\varepsilon}\sqrt[3]{\delta_{o,2}} \int_0^{t>T_{ss}} \frac{\left(P^2+Q^2\right)^{\frac{3}{2}}}{4\left(P^2-Q^2\right)} \left(\mathbb{S}\sqrt[3]{\delta_{s,2}^2}\cos 2m\Theta - \mathbb{C}\sqrt[3]{\delta_{c,2}^2}\sin 2m\Theta\right) \mathrm{d}t \approx -0.2 \tag{12.75}$$

$$C_{\mathrm{meander}} = -\frac{\varepsilon\Delta K_1}{4\mathbb{I}_0\omega} \int_0^{t>T_{ss}} \frac{PQ}{\left(P^2-Q^2\right)} \cos\psi \sin\psi dt \approx -1.4 \tag{12.76}$$

Consequently for $t > T_{ss}$, the precession angle is linear

$$m\Theta = \eta\varepsilon\Omega t + C_{\text{damp}} + C_{\text{meander}} = \eta\varepsilon\Omega t - 1.6 \tag{12.77}$$

(see Fig. 12.6). Consequently, at the steady-state, the VG obeys Bryan's law, behaving like a perfect VG with its precession angle growing at a rate proportional to the inertial angular rate

$$m\dot{\Theta} = \eta\varepsilon\Omega. \tag{12.78}$$

12.12 Isotropic Damping and the Meander Electrodes

Assume that the ever-present damping is a combination of any order of *isotropic* damping. With $\tilde{\varepsilon} = 0$, we have suitably adjusted averaged ODEs (12.67) to (12.70) and numerically investigated this type of damping using (12.62). There is *a steady-state* $t > T_{\text{ss}}$ at which

1. the principal vibration amplitude P will be sustained at a constant nonzero level;
2. quadrature vibration amplitude Q will vanish;
3. the precession angular rate $m\dot{\Theta}$ *is not affected by the isotropic damping* (as demonstrated in Joubert et al (2013)), although the meander electrodes induce $m\Theta = \eta\varepsilon\Omega t + C_{\text{meander}}$ where C_{meander} is obtained from an integral similar to that in Eq. (12.76) (using (12.62), quadratic isotropic damping yields $C_{\text{meander}} \approx 1.5$) - the VG thus obeys Bryan's law Eq. (12.78);
4. the rate of change of the phase angle $\dot{\psi} = 0$ and, apparently, $\psi = \frac{2k+1}{2}\pi$ for some integer k.

12.13 Conclusion

We have shown how to negate the effects of manufacturing mass-stiffness imperfections as well as the effects of any type of tangentially anisotropic or isotropic damping that might occur in a perfectly axissymmetric VG. We have achieved this by showing exactly how to symmetrically arrange an electronic array about the symmetry axis. This array consists of curved capacitors under a mixture of a constant charge and a small meander charge. We have shown exactly how the voltage of the fixed charges should be adjusted in order to eliminate the frequency split caused by the mass-stiffness imperfection. Furthermore, we have shown how the meander voltages of the capacitors may be adjusted in order to maintain principal vibration, eliminate quadrature vibration and restore spurious pattern drift in the VG so that it obeys Bryan's law. Equations of motion were derived in the form of averaged ODEs that gave us insight into VG behaviour.

Acknowledgements This material is based upon work supported financially by the Tshwane University of Technology (TUT) and the National Research Foundation (NRF) of South Africa (NRF grant reference number IFR160211157784). Any opinions, findings and conclusions or recommendations expressed in this material are those of the authors, and TUT and the NRF therefore do not accept any liability in regard thereto.

References

Bryan GH (1890) On the beats in the vibrations of a revolving cylinder or bell. Proc Camb Philos Soc Math Phys Sci 7:101–114

Friedland B, Hutton M (1978) Theory and error analysis of vibrating member gyroscope. IEEE Transactions on Automatic Control AC-23(4):545 – 556

Joubert SV, Shatalov MY, Fay TH (2009) Rotating structures and Bryan's effect. Am J Phys 77(6):520–525

Joubert SV, Shatalov MY, Coetzee CE (2011) Analysing manufacturing imperfections in a spherical vibratory gyroscope. In: 4th IEEE International Workshop on Advances in Sensors and Interfaces (IWASI), pp 165–170

Joubert SV, Shatalov MY, Manzhirov AV (2013) Bryan's effect and isotropic nonlinear damping. J Sound Vib 332(23):6169–6176

Joubert SV, Shatalov MY, Coetzee CE (2014a) Using Fourier series to analyse mass imperfections in vibratory gyroscopes. J Symb Comput 61-62:116–127

Joubert SV, Shatalov MY, Fay TH (2014b) On numerically solving an eigenvalue problem arising in a resonator gyroscope. Appl Math and Comput 246:561–571

Joubert SV, Shatalov MY, Spoelstra H (2015) Modelling and controlling imperfections in vibratory gyroscopes. In: Crocker MJ, Pawelczyk M, Pedrielli F, Carletti E, Luzzi S (eds) International conference on sound and vibration (ICSV22), Florence, Italy, International Institute of Acoustics and Vibration, pp 1–8

Joubert SV, Shatalov MY, Sedebo GT (2017) Anisotropic nonlinear damping and vibratory gyrsoscopes. Submitted

Liu N, Su Z, Li Q, Fu M, Liu H, Fan J (2013) Characterization of the bell-shaped vibratory angular rate gyro. Sensors 13:10,123–10,150

Liu N, Su Z, Li Q (2016) Design and experiment of a novel bell-shaped vibratory gyro. Sensors and Actuators A: Physical 238:37–50

Loveday PW, Rogers CA (1998) Free vibrations of elastically supported thin cylinders including gyroscopic effects. J Sound Vib 217(3):547–562

Ma X, Su Z (2015) Analysis and compensation of mass imperfection effects on 3-d sensitive structure of bell-shaped vibratory gyro. Sensors and Actuators A: Physical 224:14–23

Ma X, Su Z, Liu H, Liu N (2015) Research on paraboloid-shaped resonator of novel solid vibratory gyroscope. Journal of Vibroengineering 17(1):82–91

Redwood M (1960) Mechanical Waveguides. Pergamon Press, Oxford

Reitz JR, Milford FJ (1969) Foundations of Electromagnetic Theory. Addison–Wesely, Reading, MA

Shatalov MY, Coetzee CE (2011) Dynamics of rotating and vibrating thin hemispherical shell with mass and damping imperfections and parametrically driven by discrete electrodes. Gyroscopy and Navigation 2:27–33

Shatalov MY, Joubert SV, Coetzee CE, Fedotov I (2009) Free vibration of rotating hollow spheres containing acoustic media. J Sound Vib 322(4-5):1038–1047

Shatalov MY, Joubert SV, Coetzee CE (2011) The influence of mass imperfections on the evolution of standing waves in slowly rotating spherical bodies. J Sound Vib 330:127–135

Shatalov MY, Joubert SV, Coetzee CE (2012) The vibration of a layered rotating planet and Bryan's effect. In: Moustafa A (ed) Advances in Geotechnical Earthquake Engineering - Soil Liquefaction and Seismic Safety of Dams and Monuments, InTech, Rijeka, pp 405–424

Shatalov MY, Joubert SV, Spoelstra H (2015) Optimisation of vibratory gyroscopes mass-balancing procedures. In: Crocker MJ, Pawelczyk M, Pedrielli F, Carletti E, Luzzi S (eds)

International conference on sound and vibration (ICSV22), Florence, Italy, International Institute of Acoustics and Vibration, pp 1–8

Wang Y, Wang S, Zhu D (2015) Dual-mode frequency splitting elimination of ring periodic structures via feature shifting. J Mech Eng Sci 203–210:1989–1996

Wang Z, Wu Y, Xi X, Zhang Y, Wu X (2016) Analysis on node position of imperfect resonators for cylindrical shell gyroscopes. Sensors 16(8):1206

Zhuravlev VF, Klimov DM (1988) Applied methods in the theory of oscillations (in Russian). Nauka, Moscow

Zhuravlev VP (2015) Hemispherical resonator gyro with m data electrodes and n control electrodes. Mech Solids 50(4):375–378

Zhuravlev VP, Linch DD (1995) Electric model of a hemispherical resonator gyro. Mech Solids (Engl Transl) 30(5):10 – 21

Chapter 13
Shock Wave Rise Time and the Viscosity of Liquids and Solids

Gennady I. Kanel, Andrey S. Savinykh, Gennady V. Garkushin, Alexander V.Pavlenko & Sergey V. Razorenov

Abstract With the goal to verify universality of the well-known Swegle-Grady fourth-power relation between the shock pressure and the rate of compression in plastic shock waves, we performed direct comparison of the results of measurements of shock waves in titanium and in glycerol. The universality has not been confirmed, but has been explained in terms of the different natures of the viscosities of liquids and crystalline solids.

Key words: Shock waves · Viscosity · Titanium · Glycerol · High strain rates

13.1 Introduction

Although shock waves are often considered to be discontinuities of infinitesimally small thickness, the thickness of shock waves of moderate strength in solids and liquids may be quite resolvable. The thickness, or the rise time, of shock wave is controlled by dissipative processes, mainly by the matter viscosity (Zel'dovich and Raizer, 1967). This means that, from measurements of the rise time, we can obtain unique information about the viscosity of condensed matter at high pressures and very high strain rates. Swegle and Grady (1985) observed for solids that the maximum strain rate in the plastic wave and the peak shock stress, or the stress jump, from the state at the Hugoniot elastic limit (HEL) to the peak shock stress, are related by the fourth-power dependence. It is believed that the fourth-power low is satisfied for very different solids: for crystalline and amorphous, ductile and brittle, for metals and nonmetals. Such invariance of the matter viscosity at ultrahigh strain

Gennady I. Kanel · Andrey S. Savinykh · Gennady V. Garkushin · Alexander V. Pavlenko · Sergey V. Razorenov
Joint Institute for High Temperatures of Russian Academy of Sciences, Izhorskaya st. 13 bldg 2, 125412 Moscow, Russia,
e-mail: kanel@ficp.ac.ru

H. Altenbach et al. (eds.), *Mechanics for Materials and Technologies*,
Advanced Structured Materials 46, DOI 10.1007/978-3-319-56050-2_13

rates looks intriguing and has caused a many theoretical speculations, although the universal nature of the fourth-power behaviour of steady shock waves actually was not completely proved. In this paper we compare our recent results of measurements of steady shock waves in a solid metal (titanium) and in liquid glycerol, and we aim at discussing them from the position of expected universality of the Swegle-Grady relationship and its nature.

13.2 Experiments and Their Results

In the experiments, we created plane shock waves in plane samples by impacts of flyer plates. The flyer plates were launched with maximum velocities in the range of 300 m/s to 900 m/s using a gas gun or explosive facilities. In the experiments, parameters of shock waves were recorded by means of the monitoring of the velocity histories of the sample free rear surface or the interface between the sample and a transparent window. The measurements were undertaken with the VISAR velocimeter (Barker and Hollenbach, 1972) at around 1 ns of the temporal resolution. Figures 13.1 and 13.2 present results of experiments with commercial-grade titanium (Kanel et al, 2016) and chemically-pure glycerol (Savinykh et al, 2017). In the case of titanium, the wave-form consists of an elastic precursor wave followed by the plastic shock wave. Parameters behind the elastic precursor front correspond to the Hugoniot Elastic Limit (HEL) and practically do not vary with an increase in the impact velocity. The rise time of the plastic shock wave rapidly decreases with an increase in the impact velocity. In the case of glycerol, we recorded only one shock wave with no signature of an elastic-plastic response. The rise time of the shock wave is much less than the rise time for the titanium, and it rapidly decreases

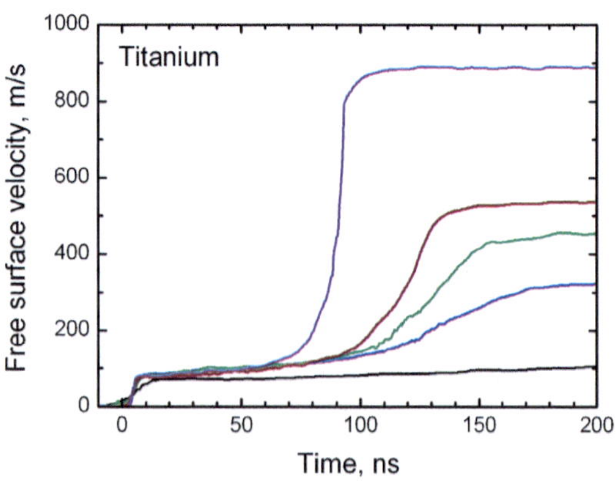

Fig. 13.1 The free surface velocity histories of titanium sample plates of 4 mm in thickness at different peak stress (Kanel et al, 2016)

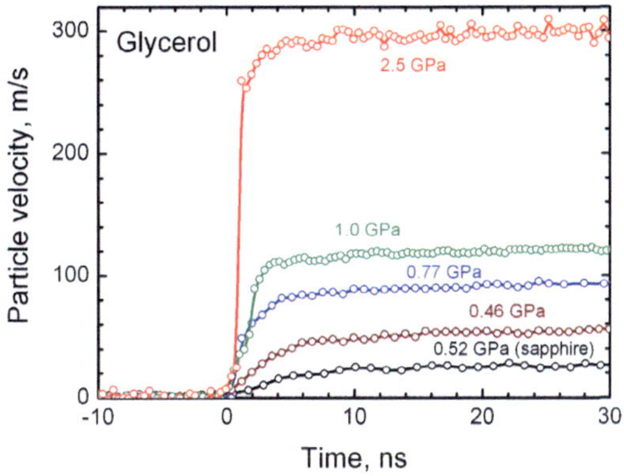

Fig. 13.2 The velocity histories of inter-faces between a layer of glycerol and a lithium fluoride (LiF) or sapphire window at various peak pressures (noted at the waveforms) (Savinykh et al, 2017). The time interval between points of digital recording is 0.4 ns, or 0.8 ns, or 1.2 ns.

with increasing peak shock pressure and approaches the limit of temporal resolution at about 2.5 GPa of the shock pressure.

The relationship between the strain rate in shock wave $\dot{\varepsilon}_x = -\rho_0 \mathrm{d}V/\mathrm{d}t - (1/U_s)\mathrm{d}u/\mathrm{d}x$ and the matter viscosity η can be found in Zel'dovich and Raizer (1967). It is:

$$-\frac{4}{3}\eta\frac{\mathrm{d}u}{\mathrm{d}x} = \rho_0 U_s^2\left(1-\frac{V}{V_0}\right) - p \tag{13.1}$$

The right side of the relationship (13.1) is the difference between the compressive stress σ_x along the Rayleigh line $\sigma_x = \rho_0 U_s^2(1 - V/V_0)$ (σ_x is assumed positive under compression) and the pressure p at the same compression V/V_0. Here, U_s is shock wave speed, V is specific volume, ρ is matter density, u is particle velocity. The difference between the normal stress and the pressure is the deviator component of the stress which, under uniaxial strain condition, is equal to 4/3 of maximum shear stress $\tau : \sigma_x - p = (4/3)\tau$. Thus, for estimation of the viscosity we have

$$\eta = \tau/\dot{\varepsilon}_x. \tag{13.2}$$

Usually, only maximum strain rate and maximum shear stress in the middle of shock wave are used in the relationship (13.2). It can be shown that the maximum shear stress within a shock wave is approximately proportional to the squared value of the peak compressive stress. In this case, the squared dependence of the maximum strain rate in the shock wave upon the peak shock pressure (or stress) means a constant coefficient of viscosity, a higher exponent value means the viscosity decrease, and a smaller exponent means increasing viscosity with increasing peak pressure.

Figures 13.3 and 13.4 show results of treatment of experimental data in terms of maximum strain rates as functions of the peak shock stress. While the peak shock stress (or shock pressure) is varied by almost an order of magnitude, the relationship between the strain rate and the peak stress in both cases may be represented by power functions:

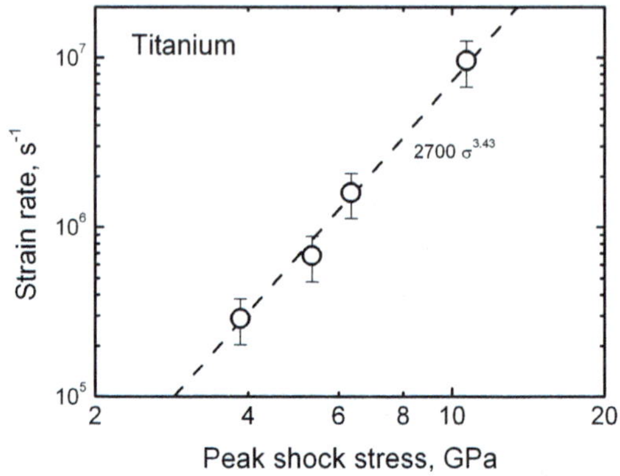

Fig. 13.3 Maximum rate of compression in plastic shock wave in titanium as a function of the peak stress

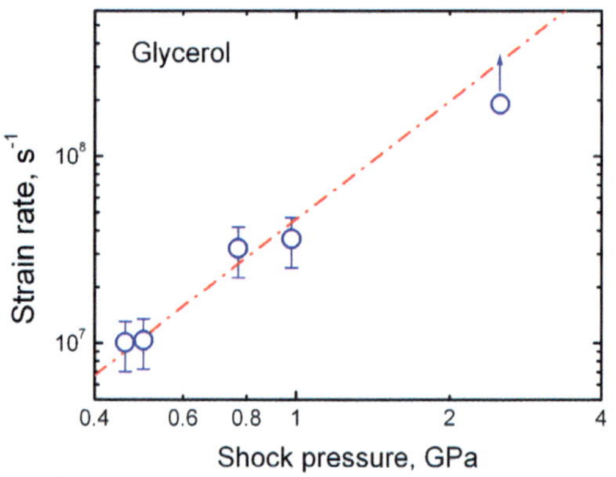

Fig. 13.4 The compression rate in shock wave in glycerol as a function of shock pressure

$$\dot{\varepsilon}_x = 2.7 \cdot 10^3 (p_1/p_0)^{3.43} \qquad \text{fortitanium, and}$$

$$\dot{\varepsilon}_x = 4.6 \cdot 10^7 (p_1/p_0)^{2.1} \qquad \text{forglycerol.}$$

The 3.4 of the exponent value for titanium, in general, is close to the expected value of 4 in the Swegle-Grady relationship. However, experiments with glycerol show a much smaller value of the exponent – 2.1. In this regard, one should note the different nature of the viscosity of liquids and solids. The viscosity of liquids is a result of transportation of momentum by chaotically moving molecules with a relatively small contribution of the additional forces between them. It depends upon the temperature and the pressure, but should not depend upon the strain.

The viscosity of solids in the plastic deformation regime is controlled by the dynamics of dislocations. The plastic strain rate, in terms of the theory of dislocation dynamics, is described by the Orowan equation (Gilman, 1969) as a function of the average dislocation velocity $\bar{v}$ and the mobile dislocation density N_m:

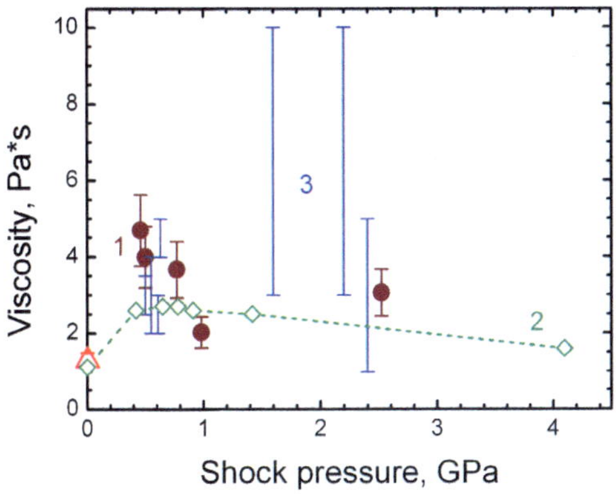

Fig. 13.5 Viscosity of shock-compressed glycerol measured in three different ways: by means of measurements of rise times of shock waves – 1 (Savinykh et al, 2017), by means of measurements of electrical conductivity – 2 (Dremin et al, 1980), and by means of measurements of acceleration of metal cylinders by shocked liquids – 3 (Al'tshuler et al, 1986)

$$\dot{\gamma} = N_m b \bar{v} \tag{13.3}$$

At moderate stresses, the dislocation velocity is controlled by the phonon drag forces and is practically proportional to the shear stress. The mobile dislocation density increases with plastic deformation due to processes of multi-plication and heterogeneous nucleation, and decreases due to their immobilisation and annihilation.

Figure 13.5 shows the results of measurements of the viscosity of glycerol as a function of shock pressure. The diagram presents three sets of data. In the first series, the viscosity was estimated from measurements of electrical conductivity of a dilute solution of CsI in glycerol (Dremin et al, 1980). These measurements obviously correspond to low strain rates. In the second type of experiments, an acceleration of metal cylinders (actually wires) by shocked liquid was monitored (Al'tshuler et al, 1986). The strain rate was not identified in these experiments, but it certainly should be rather high. Finally, in the third series the viscosity was evaluated from the rise time of the shock wave. All of the data are within one order of magnitude and the data from the rise time measurements at the strain rate of 107 – 108 s^{-1} are very close to the low-rate data from electrical measurements. The initial growth of the viscosity is the pressure effect; whereas, the contribution of the temperature increase becomes larger at higher intensities of shock waves. In general, the viscosity does not vary much within the shock pressure range of 0.5 GPa to 2.5 GPa, which explains the square dependence of the strain rate upon the shock pressure.

Figure 13.6 shows results of estimations of plastic strain rate in a shock wave and immediately behind the elastic precursor front. Both dependencies are strongly nonlinear and certainly show that plastic strain rate in the middle of shock wave (i.e. after some small deformation) is, by order of magnitude, larger than the initial plastic strain rate at the same stress in the elastic precursor wave. Such a difference indicates an intense multiplication of dislocations. In the considered stress range, the multiplication is directly associated with plastic strain, and a larger strain in a plastic shock wave leads to a larger density of mobile dislocations. Dunn and Grady

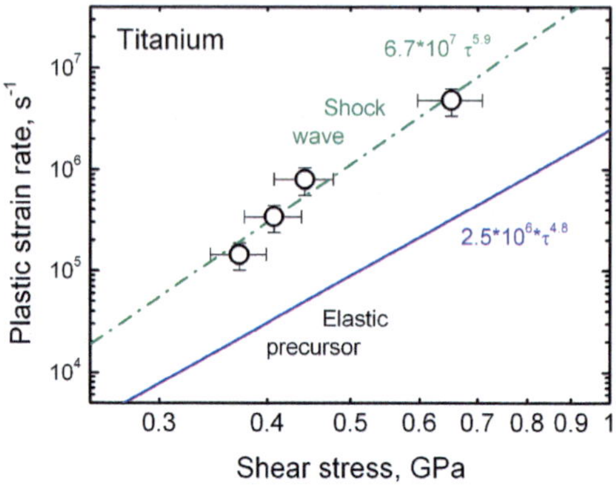

Fig. 13.6 The dependencies of plastic strain rate upon the shear stress in the elastic precursor wave and in plastic shock wave in titanium (Kanel et al, 2016)

(1986) demonstrate that, in contrast to the unique determination of stress-dependent relaxation, a strain dependent viscosity (as a result, for example, of multiplication of mobile dislocations) may also satisfy the empirical Swegle-Grady relation, but to the best of our knowledge, this result has not attracted enough attention, to date.

13.3 Conclusions

Comparison of shock wave rise times in solid metal and liquid glycerol do not confirm the universality of the well-known Swegle-Grady fourth-power relation. The difference in the strain rate dependencies upon the shock pressure is caused by the different nature of the dissipative properties of liquids and crystalline solids. The squared dependence recorded for glycerol is the evidence of its approximately constant viscosity, as a result of the competition between the effects of pressure and temperature. The power dependence for titanium, with the exponent close to four, is an example of evidence of the intense multiplication of dislocation on earlier stages of the shock compression. At moderate stresses, the multiplication occurs during plastic deformation, the value of which is proportional to shock pressure. Thus, we can speak of the universality of the low multiplication rather than of the universality of the fourth-power relation between the shock pressure and the compression rate in the shock wave.

Acknowledgements This work was supported by the Russian Science Founda-tion (project no. 14-50-00124).

References

Al'tshuler LV, Doronin GS, Kim GK (1986) Viscosity of shock-compressed fluids. Journal of Applied Mechanics and Technical Physics 27(6):887–894

Barker LM, Hollenbach RE (1972) Laser interferometer for measuring high velocities of any reflecting surface. J Appl Phys 43:4669

Dremin AN, Kuznetsov DI, Shunin VM, Yakushev VV (1980) Estimation of the viscosity of glycerol using a width of weak shock wave. Zhurnal Fizicheskoi Khimii (Russian Journal of Physical Chemistry A: Focus on Chemistry) 54(1):135–139

Dunn JE, Grady DE (1986) Strain rate dependence in steady, plastic shock waves. In: Gupta YM (ed) Shock Waves in Condensed Matter, Plenum Press, New York, pp 359–364

Gilman JJ (1969) Micromechanics of Flow in Solids. McGraw-Hill, New York

Kanel GI, Razorenov SV, Garkushin GV, Pavlenko AV, Malyugina SN (2016) Change of the kinetics of shock-wave deformation and fracture of vt1-0 titanium as a result of annealing. Physics of the Solid State 58(6):1191–1198

Savinykh GIKAS, Garkushin GV, Razorenov SV (2017) Estimation of the viscosity of glycerol using a width of weak shock wave. High Temperatures 55(3):in print

Swegle JW, Grady DE (1985) Shock viscosity and the prediction of shock wave rise times. J Appl Phys 58(2):692

Zel'dovich YB, Raizer YP (1967) Physics of Shock Waves and High-Temperature Hydrodynamic Phenomena. Academic Press, New York

Chapter 14
Lowest Vibration Modes of Strongly Inhomogeneous Elastic Structures

Julius Kaplunov, Danila A. Prikazchikov & Olga Sergushova

Abstract Lowest vibration modes of strongly inhomogeneous layered elastic structures are studied. A perturbation procedure is developed for evaluating the almost rigid body motions of stiffer components. The procedure is adapted for the antiplane motion of a two-layered cylindrical body of arbitrary cross section. The asymptotic formulae for lowest natural frequencies are derived. Illustrative examples are presented, including comparison with the exact solution of the eigenvalue problem for a two-layered circular cylinder.

Key words: Vibration · Contrast · Perturbation · Rigid body motion

14.1 Introduction

Dynamics of inhomogeneous elastic structures composed of materials with high-contrast mechanical and geometrical properties is an important engineering problem, having numerous advanced industrial applications, see e.g. Milton (2002); Martin et al (2012). This field is intensively studied in structural mechanics, see Horgan and Chan (1999); Elishakoff (2004); Elishakoff and Yost (2010), including a number of contributions dealing with sandwich plates, see Berdichevsky (2010); Ryazantseva and Antonov (2012); Chapman (2013); Altenbach et al (2015). High contrast is also investigated from the prospective of homogenization of periodic media, see Cherednichenko et al (2006); Smyshlyaev (2009); Cherdantsev and Cherednichenko (2012); Kaplunov and Nobili (2016). A similarity of long wave dynamic behaviour of periodic and thin structures is reported in Craster et al (2014).

Julius Kaplunov · Danila A. Prikazchikov · Olga Sergushova
School of Computing and Mathematics, Keele University, Keele, Staffordshire, ST5 5BG, UK,
e-mail: `j.kaplunov@keele.ac.uk`, `d.prikazchikov@keele.ac.uk`
`o.sergushova@keele.ac.uk`

H. Altenbach et al. (eds.), *Mechanics for Materials and Technologies*,
Advanced Structured Materials 46, DOI 10.1007/978-3-319-56050-2_14

In this paper we generalise the consideration in Kaplunov et al (2016) analysing low-frequency vibrations of strongly inhomogeneous rods. For the sake of simplicity, the proposed perturbation approach is specified for antiplane motion of a two-layered strongly inhomogeneous cylinder of arbitrary cross section. The focus is on the vibration modes associated with the so-called "‘global"’ low-frequency regime, see Kaplunov et al (2016), assuming quasi-static behaviour of both stiffer and softer structure components.

If the imposed boundary conditions exclude overall rigid body motion, then, in line with the previous considerations, the almost rigid body motion of the stiffer component still occurs as a solution of the Neumann boundary value problem for the Laplace equation. Then, the displacement of the softer component is given at leading order by a plane harmonic function satisfying Dirichlet boundary condition on the outer contour and on the interface. At next order, we arrive at the sought for estimate for the lowest natural frequency following from the solvability of the boundary value problem for the stiffer component. The derived formula contains a line integral of the normal derivative of the aforementioned plane harmonic function along the interface.

To illustrate the efficiency of the developed scheme, we present two examples. The first one considers a two-layered circular cylinder with a softer outer layer fixed along its contour. This problem allows a comparison with the exact solution of the antiplane eigenvalue problem confirming the obtained results for the low-frequency vibration mode. The second example addresses a hollow cylinder of a square cross section with a circular annular inclusion. Although the latter problem does not allow a simple analytical solution, explicit asymptotic formulae are presented.

14.2 Antiplane Shear Motion

Consider a two-component cylindrical body of an arbitrary cross section with a homogeneous or hollow (annular) inclusion. In case of a hollow two-layered cylinder the inner and outer domains, D_1 and D_2, respectively, are located between non-intersecting closed boundaries Γ_0, Γ_1 and Γ_2 with the origin of the Cartesian system $Oxyz$ set inside the inner hole, see Fig. 14.1. We consider both traction-free and fixed boundaries of the inner and outer domains assuming conventional continuity conditions along the interface.

We study antiplane shear motion with the out-of-plane displacement independent of z. Then, in absence of body forces, the governing Helmholtz equations are given by

$$\Delta u_i + \frac{\omega^2}{c_i^2} u_i = 0, \qquad i = 1,2, \tag{14.1}$$

where Δ is a 2D Laplace operator in variables x and y, u_i are the displacements, $c_i = \sqrt{\mu_i/\rho_i}$ are the associated shear wave speeds, μ_i are the Lamé shear moduli,

ρ_i are the volume mass densities, and ω is the vibration frequency. Here and below indices 1 and 2 correspond to the inner and outer domains, respectively.

We introduce the scaled frequencies

$$\lambda_i = \frac{\omega l_i}{c_i}, \qquad i = 1,2,$$

where l_i are typical length scales, along with the contrast parameters

$$\mu = \frac{\mu_1}{\mu_2}, \qquad l = \frac{l_1}{l_2}, \qquad \rho = \frac{\rho_1}{\rho_2},$$

and dimensionless variables

$$X_i = \frac{x}{l_i}, \qquad Y_i = \frac{y}{l_i}, \qquad i = 1,2.$$

Then, the equations of motion (14.1) take the form

$$\Delta u_i + \lambda_i^2 u_i = 0, \qquad i = 1,2, \tag{14.2}$$

where Δ is now the Laplace operator in the variables X_i, Y_i.

Below we consider the cases of the homogeneous Dirichlet

$$u_i|_{\Gamma_j} = 0, \tag{14.3}$$

or Neumann

$$\left.\frac{\partial u_i}{\partial n}\right|_{\Gamma_j} = 0, \tag{14.4}$$

boundary conditions on the inner and outer contours, where $i = 1,2$, and n is the unit normal to the appropriate curve Γ_j, $j = 0,2$.

The continuity conditions along the interface Γ_1 are formulated as

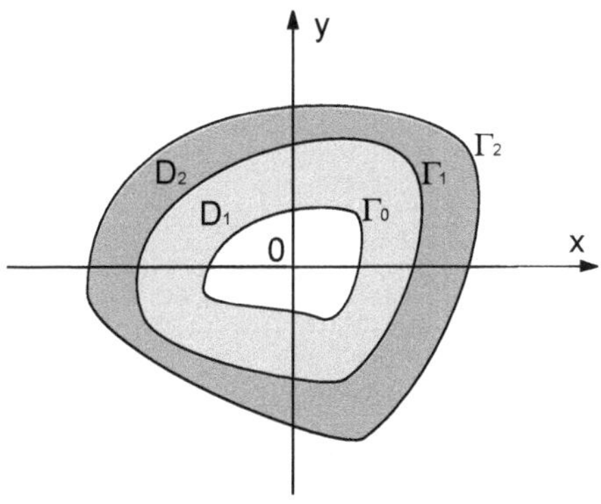

Fig. 14.1 A two-layered cylindrical body

$$(u_1 - u_2)|_{\Gamma_1} = 0, \qquad \left(\frac{\mu}{l}\frac{\partial u_1}{\partial n} - \frac{\partial u_2}{\partial n}\right)\bigg|_{\Gamma_1} = 0. \tag{14.5}$$

Let us implement a perturbation technique, focusing on the low-frequency regime for which $\lambda_1^2 \sim \lambda_2^2 \sim \varepsilon$, where $\varepsilon \ll 1$ is a small parameter associated with contrast material properties. In what follows we assume, for the sake of definiteness, that $l \sim 1$ and

$$\mu \sim \rho \sim \varepsilon \qquad \text{or} \qquad \mu \sim \rho \sim \varepsilon^{-1}. \tag{14.6}$$

Hence, we consider two contrast cases, namely $\varepsilon = \dfrac{l}{\mu} \ll 1$ $\left(\varepsilon \sim \mu^{-1}\right)$ corresponding to a stiffer inner component, or $\varepsilon = \dfrac{\mu}{l} \ll 1$ $(\varepsilon \sim \mu)$ associated with a stiffer outer domain.

The displacements and frequencies are now expanded as asymptotic series

$$\begin{aligned} u_i &= u_i^{(0)} + \varepsilon u_i^{(1)} + \varepsilon^2 u_i^{(2)} + \ldots, \\ \lambda_i^2 &= \varepsilon(\lambda_{i0}^2 + \varepsilon\lambda_{i1}^2 + \varepsilon^2\lambda_{i2}^2 + \ldots). \end{aligned} \tag{14.7}$$

14.2.1 Stiffer Outer Domain

First, we draw attention to a body with a stiffer outer component D_2 having fixed inner and free outer contours, when $\varepsilon = \dfrac{\mu}{l} \ll 1$

$$u_1|_{\Gamma_0} = \frac{\partial u_2}{\partial n}\bigg|_{\Gamma_2} = 0. \tag{14.8}$$

On substituting (14.7) into the dimensionless equation of motion (14.2) at $i = 2$ for the outer domain, and conditions (14.5) and (14.8), at leading order we obtain the Laplace equation

$$\Delta u_2^{(0)} = 0, \tag{14.9}$$

subject to the Neumann boundary conditions

$$\frac{\partial u_2^{(0)}}{\partial n}\bigg|_{\Gamma_1} = \frac{\partial u_2^{(0)}}{\partial n}\bigg|_{\Gamma_2} = 0. \tag{14.10}$$

This implies

$$u_2^{(0)} = C = \text{const}, \tag{14.11}$$

corresponding to the rigid body motion of the stiffer part, similarly to 1D consideration in Kaplunov et al (2016), see also Kudaibergenov et al (2016).

Then, the leading order problem for the inner domain is given by

$$\Delta u_1^{(0)} = 0, \tag{14.12}$$

subject to the Dirichlet boundary conditions

$$u_1^{(0)}|_{\Gamma_0} = 0, \qquad u_1^{(0)}|_{\Gamma_1} = C. \tag{14.13}$$

Thus, we arrive at

$$u_1^{(0)} = CH_1(X_1, Y_1), \tag{14.14}$$

where $H_1(X_1, Y_1)$ is a harmonic function satisfying the conditions

$$H_1(X_1, Y_1)|_{\Gamma_0} = 0, \qquad H_1(X_1, Y_1)|_{\Gamma_1} = 1. \tag{14.15}$$

Treatment of the stiffer domain D_2 at next order provides an estimate for the sought for frequency. The boundary value problem can be written as

$$\Delta u_2^{(1)} = -\lambda_{20}^2 u_2^{(0)}, \tag{14.16}$$

subject to

$$\left.\frac{\partial u_2^{(1)}}{\partial n}\right|_{\Gamma_1} = C \left.\frac{\partial H_1}{\partial n}\right|_{\Gamma_1}, \tag{14.17}$$

and

$$\left.\frac{\partial u_2^{(1)}}{\partial n}\right|_{\Gamma_2} = 0. \tag{14.18}$$

In view of (14.11), integration of (14.16) over the outer domain D_2 gives

$$\iint\limits_{D_2} \Delta u_2^{(1)} dX_2 dY_2 = -\lambda_{20}^2 C \iint\limits_{D_2} dX_2 dY_2. \tag{14.19}$$

Then, on employing the first Green's identity Strauss (2007), we have

$$\oint\limits_{\Gamma_2} \frac{\partial u_2^{(1)}}{\partial n} ds - \oint\limits_{\Gamma_1} \frac{\partial u_2^{(1)}}{\partial n} ds = -\lambda_{20}^2 C A_2, \tag{14.20}$$

where A_2 is the area of the inner domain D_2.
Due to (14.18), the first term in the left hand side of (14.20) vanishes, hence, on making use of (14.17) the leading order estimate for frequency is given by

$$\lambda_2^2 \sim \frac{\varepsilon}{A_2} \oint\limits_{\Gamma_1} \frac{\partial H_1}{\partial n} ds. \tag{14.21}$$

It should be noted that the eigenfrequencies of antiplane motion of a two-layered cylinder of arbitrary cross section correspond to the cut-off frequencies of thin rods, see e.g. Le (1999).

It may be readily verified that the global low frequency regime $\lambda_1^2 \sim \lambda_2^2 \sim \varepsilon$ is not possible for other types of boundary conditions on Γ_0 and Γ_2 in case of a stiffer outer domain.

14.2.2 Stiffer Inner Domain

Let us now discuss the case of a stiffer inner domain D_1, with the small parameter $\varepsilon = \dfrac{l}{\mu} \ll 1$. Following the same procedure as in the previous subsection, it may be shown that the global low frequency regime ($\lambda_1^2 \sim \lambda_2^2 \sim \varepsilon$) occurs only for fixed inner and free outer contours, i.e.

$$\left.\frac{\partial u_1}{\partial n}\right|_{\Gamma_0} = u_2|_{\Gamma_2} = 0. \tag{14.22}$$

Similarly to the previous subsection, the leading order displacements are found as

$$u_1^{(0)} = C, \qquad u_2^{(0)} = CH_2(X_2, Y_2), \tag{14.23}$$

where $H_2(X_2, Y_2)$ is a plane harmonic function, satisfying the boundary conditions

$$H_2(X_2, Y_2)|_{\Gamma_1} = 1, \qquad H_2(X_2, Y_2)|_{\Gamma_2} = 0. \tag{14.24}$$

At next order, we have for the stiffer domain

$$\Delta u_1^{(1)} = -\lambda_{10}^2 u_1^{(0)}, \tag{14.25}$$

subject to

$$\left.\frac{\partial u_1^{(1)}}{\partial n}\right|_{\Gamma_1} = C \left.\frac{\partial H_2}{\partial n}\right|_{\Gamma_1}, \tag{14.26}$$

and

$$\left.\frac{\partial u_1^{(1)}}{\partial n}\right|_{\Gamma_0} = 0. \tag{14.27}$$

On integrating (14.25) over the inner domain D_1 and employing the first Green's identity along with conditions (14.26) and (14.27), we deduce for the lowest natural frequency

$$\lambda_1^2 \sim -\frac{\varepsilon}{A_1} \oint\limits_{\Gamma_1} \frac{\partial H_2}{\partial n}\, ds. \tag{14.28}$$

where A_1 is the area of D_1.

14.3 Model Examples

Consider now two examples illustrating the methodology described in the previous section.

14.3.1 Two-Layered Circular Cylinder

First, let us study antiplane motion of a two-layered solid circular cylindrical elastic body with stiffer inner and softer outer components. Suppose the boundaries Γ_1 and Γ_2 are cylinders of radii l_1 and $l_1 + l_2$, respectively, see Fig. 14.2. This type of geometry allows an exact solution, presented in the Appendix. As follows from the previous section, in order to have the global low-frequency regime ($\lambda_1^2 \sim \lambda_2^2 \sim \varepsilon$), we consider a fixed outer contour.

In terms of dimensionless polar coordinates $R_i = r/l_i, \quad i = 1,2$, the domains D_1 and D_2 are given by

$$D_1: \quad 0 \le R_1 \le 1, \qquad \text{and} \qquad D_2: \quad l \le R_2 \le l+1.$$

The equations of axisymmetric motion are then written as

$$\frac{d^2u_i}{dR_i^2} + \frac{1}{R_i}\frac{du_i}{dR_i} + \lambda_i^2 u_i = 0. \tag{14.29}$$

For the studied two-layered cylinder the conditions (14.5) and (14.8) become

$$u_1\Big|_{R_1=1} = u_2\Big|_{R_2=l}, \qquad \frac{du_1}{dR_1}\Big|_{R_1=1} = \varepsilon\frac{du_2}{dR_2}\Big|_{R_2=l}, \qquad u_2\Big|_{R_2=l+1} = 0, \tag{14.30}$$

where, as previously, $\varepsilon = \dfrac{l}{\mu}$.

On expanding the displacements and frequencies as (14.7) and applying the procedure described in the previous section, we have for the leading order eigensolution

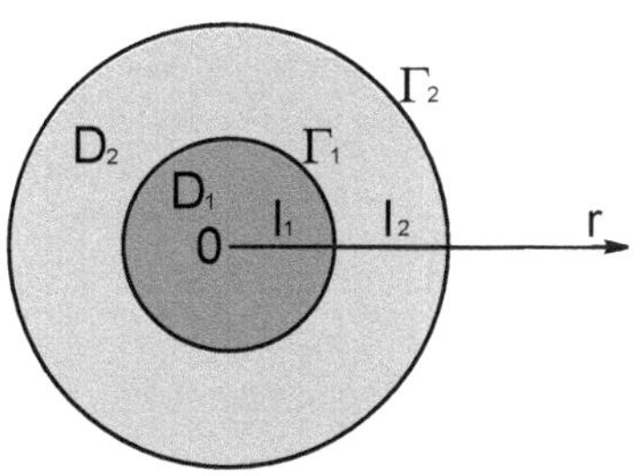

Fig. 14.2 A two-layered cylinder

$$u = \begin{cases} 1, & \text{in } D_1; \\ H_2(R_2), & \text{in } D_2, \end{cases} \tag{14.31}$$

where it is assumed that $C = 1$. Here H_2 is a harmonic function, satisfying

$$\frac{d^2 H_2}{dR_2^2} + \frac{1}{R_2}\frac{dH_2}{dR_2} = 0, \tag{14.32}$$

along with the conditions (14.24), which take the form

$$H_2|_{R_2=l+1} = 0, \qquad H_2|_{R_2=l} = 1, \tag{14.33}$$

thus

$$H(R_2) = \frac{\ln R_2 - \ln(l+1)}{\ln l - \ln(l+1)}. \tag{14.34}$$

The illustrations of the displacement profile (14.31) are shown in Fig. 14.3, containing the overall axisymmetric plot along with its axial cross section. It is worth noting that the displacement variation of the softer component is no longer polynomial, as was observed previously in 1D case, see Fig. 9 in Kaplunov et al (2016).
The leading order estimate for frequency follows from (14.28), resulting in

$$\lambda_1^2 \sim \frac{2}{\mu \ln\left(1 + l^{-1}\right)}, \tag{14.35}$$

On employing the relation $c\lambda_1 = l\lambda_2$, we also deduce

$$\lambda_2^2 \sim \frac{2}{\rho\, l^2 \ln(1 + l^{-1})}. \tag{14.36}$$

Figure 14.4 illustrates numerically the exact solution of the frequency equation (14.46), see Appendix, showing the eigenfrequencies expressed in terms of the parameters λ_1 and λ_2, together with the leading order approximation (14.35). It is clear from Fig. 14.4 that for the chosen values $\mu = \rho = 100$ satisfying the condition

a) b)

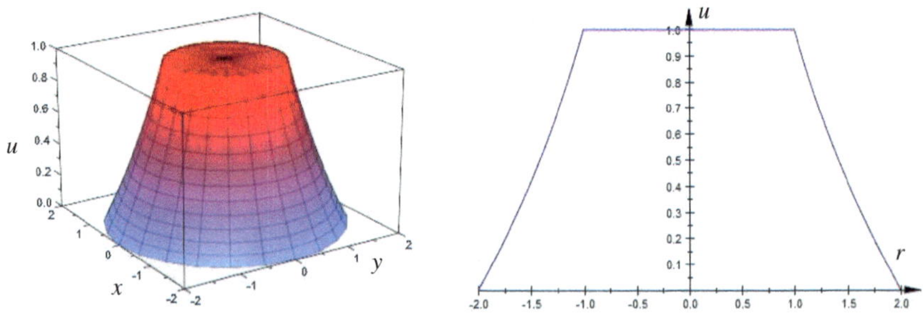

Fig. 14.3: Displacement profile (14.31) for $l_1 = l_2 = 1$. a) Displacement profile, b) axial cross section

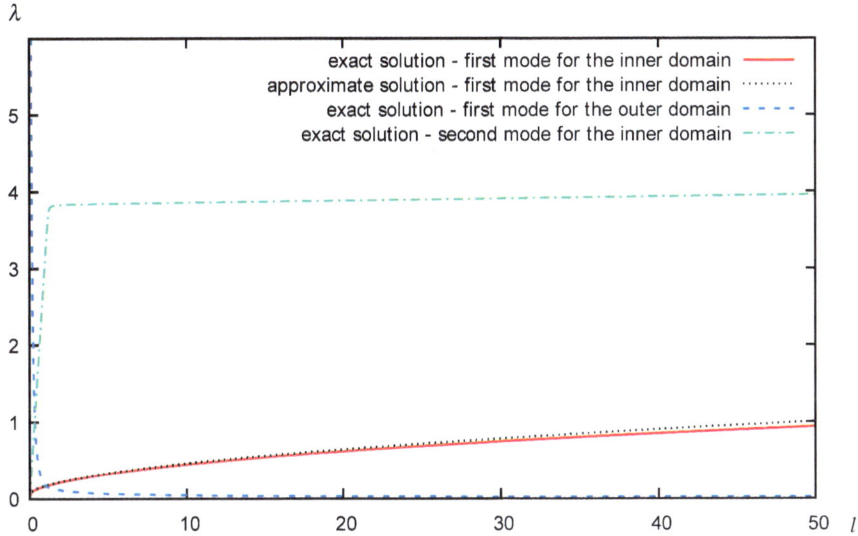

Fig. 14.4: Frequency vs. relative thickness l: $\mu = 100$, $\rho = 100$.

$\mu \sim \rho \sim \varepsilon^{-1}$, see (14.6), $\lambda_1 \ll 1$ and $\lambda_2 \ll 1$ oer a range of relative thickness l. A good agreement is observed between the exact solution shown by a solid line and the asymptotic formula (14.35) depicted by dotted line. It is also visible that the deviation between the first and second eigenfrequencies is substantial, replicating the behaviour in the 1D problem Kaplunov et al (2016).

14.3.2 Square Cylinder with a Circular Annular Inclusion

As the next example, we consider antiplane motion of a cylinder of a square cross section with a circular inclusion having an inner hole of radius l_0, see Fig. 5. The softer inner annular domain D_1 is specified in polar coordinates by $l_0 \leq r \leq l_0 + l_1$, whereas the stiffer outer domain D_2 is located between the circumference Γ_1 of radius $l_0 + l_1$ and the square Γ_2 with side length of $2(l_0 + l_1 + l_2)$, see Fig. 14.5. In contrast to the previous example, this problem does not allow a straightforward analytical treatment leading to a closed-form solution.

The equations of motion are then taken in the form (14.2) and (14.29) for the outer and inner domains, respectively. On adapting the general scheme of Sect. 14.2, consider the fixed inner and free outer face contours

$$u_1\big|_{R_1=L_1} = \frac{\partial u_2}{\partial n}\bigg|_{\Gamma_2} = 0, \tag{14.37}$$

with the continuity conditions

$$u_1\Big|_{R_1=L_1+1} = u_2\Big|_{R_2=L_2}, \qquad \varepsilon \frac{du_1}{dR_1}\Big|_{R_1=L_1+1} = \frac{du_2}{dR_2}\Big|_{R_2=L_2}, \tag{14.38}$$

along the interface Γ_1. In the above

$$L_0 = \frac{l_0}{l_1}, \qquad L_2 = \frac{l_0+l_1}{l_2}, \tag{14.39}$$

and $\varepsilon = \dfrac{\mu}{l}$ is the small parameter.

As above, we deduce the leading order displacements in the form

$$u = \begin{cases} H_1(R_1), & \text{in } D_1; \\ 1, & \text{in } D_2, \end{cases} \tag{14.40}$$

where the harmonic function

$$H_1(R_1) = \frac{\ln R_1 - \ln L_1}{\ln(1+L_1^{-1})}. \tag{14.41}$$

is the solution of (14.32) subject to

$$H_1|_{R_1=L_1} = 0, \qquad H_1|_{R_1=L_1+1} = 1. \tag{14.42}$$

The eigensolution (14.40) is shown in Fig. 14.6, containing both the displacement profile and its crossection at $y=0$, calculated at $l_0 = l_1 = l_2 = 1$.

Finally, on employing (14.21), the leading order estimate for the eigenfrequency is found as

$$\lambda_2^2 = \frac{2\pi\varepsilon l}{\ln(L_1^{-1}+1)(4(L_2+1)^2 - \pi L_2^2)}. \tag{14.43}$$

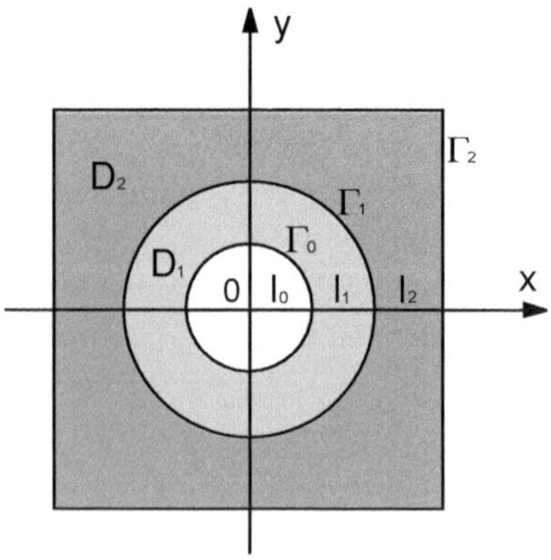

Fig. 14.5 A square cylinder with a circular annular inclusion.

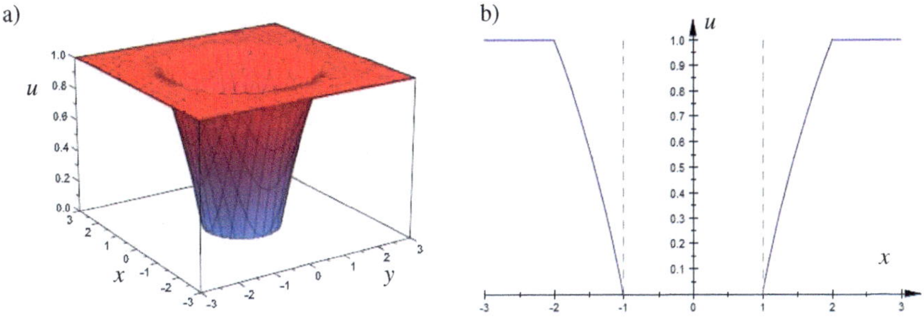

Fig. 14.6: Displacement profile (14.40) and its cross section at $y = 0$ for $l_0 = l_1 = l_2 = 1$. a) Displacement profile, b) cross section of the profile at $y = 0$

14.4 Concluding Remarks

The obtained results develop further the methodology of "‘almost rigid body motion"’, see Kaplunov et al (2016, 2015), extending it to antiplane low-frequency motion of two-layered cylindrical bodies of arbitrary cross section. The example of a hollow cylinder of a square cross section with a circular annular inclusion clearly demonstrates the power of the perturbation technique well beyond the scope of exact analysis. It is also worth noting that the derived asymptotic formulae are also valid for the lowest cut-off frequencies of high-contrast layered plates and shells, see Ryazantseva and Antonov (2012).

The proposed perturbation approach may be readily adapted for more general 2D and 3D eigenvalue problems for multi-layered strongly inhomogeneous structures. Higher order asymptotic expansions for natural frequencies may be also derived. However, this would usually assume numerical solution of the Laplace equation subject to the Dirichlet on Neumann boundary conditions. Nevertheless, such hybrid procedure still has obvious advantages against straightforward numerical treatment.

12.A Appendix

Let us present the exact solution for vibrations of a two-layered circular cylinder considered in Subsect. 14.3.1, see Fig. 14.2.

The solution of (14.29) is given by

$$\begin{aligned} u_1 &= AJ_0(\lambda_1 R_1), \\ u_2 &= BJ_0(\lambda_2 R_2) + CY_0(\lambda_2 R_2). \end{aligned} \tag{14.44}$$

On substituting (14.44) into the conditions (14.30) and employing basic properties of Bessel functions, we arrive at a set of linear algebraic equations given by

$$\begin{aligned} AJ_0(\lambda_1) - BJ_0(\lambda_2 l) - CY_0(\lambda_2 l) &= 0,\\ A\mu J_1(\lambda_1) - c(BJ_1(\lambda_2 l) + CY_1(\lambda_2 l)) &= 0,\\ BJ_0(\lambda_2(1+l)) + CY_0(\lambda_2(1+l)) &= 0. \end{aligned} \tag{14.45}$$

The solvability of the latter implies the frequency equation

$$\mu J_1(\lambda_1)F_0(\lambda_2 l) - cJ_0(\lambda_1)F_1(\lambda_2 l) = 0, \tag{14.46}$$

where

$$F_j(x) = J_j(x)Y_0(\lambda_2(l+1)) - Y_j(x)J_0(\lambda_2(l+1)), \qquad j = 0, 1.$$

Then, on using (14.45) and setting $A = 1$, the displacement profile is found as

$$u_1 = J_0(\lambda_1 R_1), \qquad u_2 = J_0(\lambda_1)\frac{F_0(\lambda_2 R_2)}{F_0(\lambda_2 l)}. \tag{14.47}$$

At the global low-frequency approximation $\lambda_1 \ll 1$, and $\lambda_2 \ll 1$ the expressions (14.46) and (14.47) become ($l \sim 1$, $\mu \gg 1$, $\rho \gg 1$)

$$u_1 \sim 1, \qquad u_2 \sim \frac{\ln R_2 - \ln(l+1)}{\ln l - \ln(l+1)}, \tag{14.48}$$

and

$$\lambda_1^2 \sim \frac{2}{\mu \ln(1+l^{-1})}, \qquad \lambda_2^2 \sim \frac{2}{\rho\, l^2 \ln(1+l^{-1})}, \tag{14.49}$$

which coincide with the results (14.31), (14.35), and (14.36) of the perturbation approach.

References

Altenbach H, Eremeyev VA, Naumenko K (2015) On the use of the first order shear deformation plate theory for the analysis of three-layer plates with thin soft core layer. Zeitschrift für Angewandte Mathematik und Mechanik 95:1004–1011

Berdichevsky VL (2010) An asymptotic theory of sandwich plates. International Journal of Engineering Science 48(3):383–404

Chapman CJ (2013) An asymptotic decoupling method for waves in layered media. Proceedings of the Royal Society of London A: Mathematical, Physical and Engineering Sciences 469(2153):20120,659

Cherdantsev M, Cherednichenko KD (2012) Two-scale Γ-convergence of integral functionals and its application to homogenisation of nonlinear high-contrast periodic composites. Archive for Rational Mechanics and Analysis 204:445–478

Cherednichenko K, Smyshlyaev VP, Zhikov V (2006) Non-local homogenized limits for composite media with highly anisotropic periodic fibres. Proceedings of the Royal Society of Edinburgh: Section A Mathematics 136(01):87–114

Craster RV, Joseph LM, Kaplunov J (2014) Long-wave asymptotic theories: the connection between functionally graded waveguides and periodic media. WaveMotion 51(4):581–588

Elishakoff I (2004) Eigenvalues of inhomogeneous structures: unusual closed-form solutions. CRC Press

Elishakoff I, Yost J (2010) Design for a specified natural frequency of elastically constrained axially graded bars. Acta Mechanica Sinica 26(2):313–316

Horgan C, Chan A (1999) Vibration of inhomogeneous strings, rods and membranes. Journal of Sound and Vibration 225(3):503–513

Kaplunov J, Nobili A (2016) Multi-parametric analysis of strongly inhomogeneous periodic waveguides with internal cutoff frequencies. Mathematical Methods in the Applied Sciences

Kaplunov J, Shestakova A, Aleynikov I, Hopkins B, Talonov A (2015) Low-frequency perturbations of rigid body motions of a viscoelastic inhomogeneous bar. Mechanics of Time-Dependent Materials 19(2):135–151

Kaplunov J, Prikazchikov D, Sergushova O (2016) Multi-parametric analysis of the lowest natural frequencies of strongly inhomogeneous elastic rods. Journal of Sound and Vibration 366:264–276

Kudaibergenov A, Nobili A, Prikazchikova LA (2016) On low-frequency vibrations of a composite string with contrast properties for energy scavenging fabric devices. Journal of Mechanics of Materials and Structures 11(3):231–243

Le KC (1999) Vibrations of Shells and Rods. Springer Berlin

Martin TP, Layman CN, Moore KM, Orris GJ (2012) Elastic shells with high-contrast material properties as acoustic metamaterial components. Physical Review B 85(16):161,103

Milton GW (2002) The Theory of Composites. Cambridge University Press

Ryazantseva MY, Antonov FK (2012) Harmonic running waves in sandwich plates. International Journal of Engineering Science 59:184–192

Smyshlyaev VP (2009) Propagation and localization of elastic waves in highly anisotropic periodic composites via two-scale homogenization. Mechanics of Materials 41(4):434–447

Strauss WA (2007) Partial Differential Equations: An Introduction, 2nd edn. Wiley, New York

Chapter 15
Geometrical Inverse Thermoelastic Problem for Multiple Inhomogeneities

Alexander V. Kaptsov & Efim I. Shifrin

Abstract Geometrical inverse thermoelastic problem is considered. It is assumed that a finite number of inhomogeneities (cavities, cracks, inclusions) are located inside a linearly elastic, mechanically and thermally isotropic 3D body. It is assumed also that mechanical and steady-state thermal loads are applied to the external boundary of the body in a single experiment. As a result of the experiment the displacements and the applied mechanical loads are measured on the external boundary. A method for identification of the number and locations of the inhomogeneities, which are considered as point defects, is developed. It is important to stress that the identification method is based only on the knowledge of the displacements and the applied mechanical loads, if any, on the external boundary of the body. The knowledge of the temperature field on the external boundary of the body and the conditions of heat exchange between the matrix and inclusions are not assumed. Numerical examples illustrating the efficiency of the developed method are considered.

Key words: Geometric inverse problem · Thermoelasticity · Multiple defects · Reciprocity gap functional

15.1 Introduction

Geometrical inverse elastostatic problems are of great importance for the nondestructive testing of the materials and structures. That is why such problems were considered in a number of publications. Most publications addressed the problem of identification of one or a maximum of two defects in 2D and 3D elastic solids, see,

Alexander V. Kaptsov · Efim I. Shifrin
Ishlinsky Institute for Problems in Mechanics RAS, Vernadskiy Ave, 101 bldg 1, 119526 Moscow, Russia,
e-mail: kaptsov@ipmnet.ru, shifrin@ipmnet.ru

H. Altenbach et al. (eds.), *Mechanics for Materials and Technologies*,
Advanced Structured Materials 46, DOI 10.1007/978-3-319-56050-2_15

for example, Steinhorst and Sändig (2012); Khodadad and Dashti-Ardakani (2009); Karageorghis et al (2012); Ammari et al (2002); Engelhardt et al (2006); Shifrin and Shushpannikov (2011, 2013b). At the same time, the methods of identification of multiple defects are developed in a much lesser extent. Several approaches for solving inverse elastostatic problems by means of boundary data in case of multiple defects has been suggested in the recent publications by Dehghan Manshadi et al (2014); Andrieux (2015); Shifrin and Shushpannikov (2013a, 2015). Dehghan Manshadi et al (2014) extended the linear sampling method, previously used for solving inverse scattering problems, for the solution of the inverse elastostatic problems. The paper deals with only 2D inverse problems. Generally, the linear sampling method enables to reconstruct multiple defects, but it requires the data, which can be obtained from a large number of experiments. Such data are usually not available. Andrieux (2015) proposed to use some modification of the reciprocity gap method for identifying multiple cracks and point sources. The approach was illustrated with an example, in which the problem of identifying multiple point sources for the 2D Laplace equation was considered. Shifrin and Shushpannikov (2013a, 2015) developed a method for the identification of small, multiple defects in both 3D isotropic and anisotropic bodies using boundary data. The method is also based on the use of the reciprocity gap functional (RGF). The purpose of the paper to extend the method to solve the 3D inverse problems of thermoelasticity.

To date, methods for solving geometric, inverse, thermoelastic problems are not developed enough. It should be noted the papers of Ben Ameur et al (2007); Karageorghis et al (2014) where the problems of identification of one and two cavities in 2D thermoelastic bodies were solved by the level set method and the method of fundamental solutions, respectively.

The paper is organized as follows. The direct problem is formulated in Sect. 15.2. The reciprocity gap functional is defined in Sect. 15.3. Statement of the inverse problem and the method of its solving are presented in Sect. 15.4. Numerical examples illustrating efficiency of the proposed method are considered in Sect. 15.5. Conclusions are presented in Sect. 15.6.

15.2 Mathematical Formulation of the Direct Problem

Let $V \subset R^3$ be a bounded simply connected domain with a boundary ∂V. $G_k \subset V$, $k = 1, 2, \cdots, n$ are simply connected subdomains. We suppose that $\bar{G}_i \cap \bar{G}_j = \emptyset$, $i \neq j$, $\bar{G} = \bigcup\limits_{k=1}^{n} \bar{G}_k \subset V$, where $\bar{G}_k$ is a closure of the subdomain G_k. Let us suppose that a linearly elastic, mechanically and thermally isotropic body occupies the domain $\Omega = V \backslash \bar{G}$. Introduce some notation for the thermal and mechanical characteristics of the body, which occupies the domain Ω. The thermal characteristics are as follows: α_M is the coefficient of thermal expansion, κ_M is the heat conductivity, h_M is the convective heat transfer coefficient. The elastic properties are: μ_M is the shear modulus and ν_M is the Poisson ratio. The defects G_k can be cavities or inclusions (rigid or linear elastic). If G_k is a cavity we suppose that its boundary ∂G_k is free of mechanical loads. If G_k is an inclusion, it is supposed complete mechanical bonding between the matrix and inclusion. Below we consider the case of linear elastic inclusions because the cavities and rigid inclusions can be considered as particular cases when the shear modulus tends to zero and infinity, respectively. The thermal and elastic characteristics of the inclusion G_k we denote by the index Ik: α_{Ik}, κ_{Ik}, μ_{Ik}, ν_{Ik}. The temperature, displacements, strains and stresses in the body Ω we will mark with the superscript d : T^d is the temperature, $\mathbf{u}^d = \left(u_1^d, u_2^d, u_3^d\right)$ is the displacements vector, e_{ij}^d is the strain tensor and σ_{ij}^d is the stress tensor. The similar values in the inclusion G_k we will mark with the superscript Ik : T^{Ik}, $\mathbf{u}^{Ik} = \left(u_1^{Ik}, u_2^{Ik}, u_3^{Ik}\right)$, e_{ij}^{Ik}, σ_{ij}^{Ik}. Here and below, we suppose that some Cartesian coordinates $Ox_1x_2x_3$ are introduced and all values are considered in these coordinates. The steady-state heat equations have the form:

$$\Delta T^d(x) = 0, \quad x \in \Omega \tag{15.1}$$

$$\Delta T^{Ik}(x) = 0, \quad x \in G_k \tag{15.2}$$

Here Δ is the Laplace operator.

We consider the following thermal boundary conditions: given temperature T_o over $\partial V_0 \subset \partial V$, given heat-flux q in the direction normal to the part of the boundary $\partial V_1 \subset \partial V$ and prescribed ambient temperature T_∞ over the convective part of the boundary $\partial V_2 \subset \partial V$. We assume that $\partial V_i \cap \partial V_j = \emptyset$ for $i \neq j$ and $\partial \bar{V}_0 \cup \partial \bar{V}_1 \cup \partial \bar{V}_2 = \partial V$. Here $\partial \bar{V}_i$ is the closure of ∂V_i. Thus, the boundary conditions on the external boundary of the body are as follows:

$$\begin{gathered} T^d(x') = T_0(x'), \quad x' \in \partial V_0 \\ \kappa_M \frac{\partial T^d(x')}{\partial x_i} n_i(x') = q(x'), \quad x' \in \partial V_1 \\ \kappa_M \frac{\partial T^d(x')}{\partial x_i} n_i(x') = h_M\left(T_\infty - T^d(x')\right), \quad x' \in \partial V_2 \end{gathered} \tag{15.3}$$

where $\mathbf{n} = (n_1, n_2, n_3)$ is a unit outward normal to the boundary ∂V and convention of summation for repeated indices is used. Conditions of heat transfer between the matrix and inclusions are of the form

$$T^d(x^*) = T^{Ik}(x^*), \quad \kappa_M \frac{\partial T^d(x^*)}{\partial x_i} N_i(x^*) = \kappa_{Ik} \frac{\partial T^{Ik}(x^*)}{\partial x_i} N_i(x^*), \quad x^* \in \partial G_k^1 \tag{15.4}$$

$$\kappa_M \frac{\partial T^d(\tilde{x})}{\partial x_i} N_i(\tilde{x}) = \kappa_{Ik} \frac{\partial T^{Ik}(\tilde{x})}{\partial x_i} N_i(\tilde{x}) = h_{Mk}\left(T^{Ik}(\tilde{x}) - T^d(\tilde{x})\right), \quad \tilde{x} \in \partial G_k^2 \tag{15.5}$$

Here $\mathbf{N} = (N_1, N_2, N_3)$ is a unit normal to ∂G_k, external to the inclusion G_k, h_{Mk} is a coefficient of heat transfer between the matrix Ω and inclusion G_k, $\partial G_k^1 \cap \partial G_k^2 = \emptyset$, $\partial \bar{G}_k^1 \cup \partial \bar{G}_k^2 = \partial G_k$, ∂G_k^1 is a portion of the inclusions boundary ∂G_k, where an ideal thermal contact between the matrix and inclusion is realized, ∂G_k^2 is a portion of the boundary, where non-ideal thermal contact is realized. The elastic part of the system of thermoelastic equations is as follows.

$$\begin{aligned} & e_{ij}^d(x) = \frac{1}{2}\left(u_{i,j}^d(x) + u_{j,i}^d(x)\right), \quad (i,j=1,2,3), \quad x \in \Omega \\ & \sigma_{ij}^d = \lambda_M \theta^d \delta_{ij} + 2\mu_M e_{ij}^d - (3\lambda_M + 2\mu_M)\,\alpha_M T^d \delta_{ij} \\ & \theta^d = \sum_{k=1}^{3} e_{kk}^d, \quad \lambda_M = \frac{2\mu_M \nu_M}{1-2\nu_M} \\ & \sigma_{ij,j}^d = 0 \end{aligned} \tag{15.6}$$

Here δ_{ij} is the Kronecker delta.
The elastic equations inside the domains G_k have the same form as Eq. (15.6), where subscript M is substituted with the subscript Ik and the superscript d is substituted with the superscript Ik. The conditions on the external boundary of the body have the form:

$$\mathbf{u}^d(x') = \mathbf{u}^0(x'), \quad x' \in \partial V_D \tag{15.7}$$

$$\sigma_{ij}^d(x')\,n_j(x') = t_i^0(x'), \quad x' \in \partial V_N \tag{15.8}$$

where $\partial V_D \cap \partial V_N = \emptyset$, $\partial \bar{V}_D \cup \partial \bar{V}_N = \partial V$.
The conditions of complete mechanical bonding between the matrix and inclusion G_k have the following form:

$$\mathbf{u}^{Ik}(x^*) = \mathbf{u}^d(x^*), \quad \sigma_{ij}^{Ik}(x^*)N_j(x^*) = \sigma_{ij}^d(x^*)N_j(x^*), \quad x^* \in \partial G_k \tag{15.9}$$

Equations (15.1)–(15.9) form a system of equations of steady-state thermoelasticity, see Nowacki (1986).

15.3 Reciprocity Principle and Reciprocity Gap Functional

Let $\tilde{T}(x), \tilde{\mathbf{u}}(x), \tilde{e}_{ij}(x), \tilde{\sigma}_{ij}(x)$ be a solution of the equations (15.1) and (15.6) in the domain Ω. According to reciprocity principle the following equality is valid (see Nowacki, 1986):

$$\int\limits_{\partial\Omega} \left(\sigma_{ij}^d n_j \tilde{u}_i - \tilde{\sigma}_{ij} n_j u_i^d\right) dS + \alpha_M (3\lambda_M + 2\mu_M) \int\limits_{\Omega} \left(T^d \tilde{\theta} - \tilde{T}\theta^d\right) dx = 0 \tag{15.10}$$

where $\tilde{\theta} = \sum_{k=1}^{3} \tilde{e}_{kk}$ and the thermoelastic field, that is marked by superscript d, was defined above. In a particular case, when $\tilde{T} = 0$ and $\tilde{\theta} = 0$, Eq. (15.10) is reduced to the following equality

$$\int_{\partial\Omega} \left(\sigma_{ij}^{d} n_j \tilde{u}_i - \tilde{\sigma}_{ij} n_j u_i^{d}\right) dS = 0 \tag{15.11}$$

Taking into account, that $\partial\Omega = \partial V \cup \partial G_1 \cup \cdots \cup \partial G_n$ and the unit normal to ∂G_k, external to the domain G_k, was denoted by $\mathbf{N} = (N_1, N_2, N_3)$, Eq. (15.11) can be rewritten as follows

$$\int_{\partial V} \left(\sigma_{ij}^{d} n_j \tilde{u}_i - \tilde{\sigma}_{ij} n_j u_i^{d}\right) dS = \sum_{k=1}^{n} \int_{\partial G_k} \left(\sigma_{ij}^{d} N_j \tilde{u}_i - \tilde{\sigma}_{ij} N_j u_i^{d}\right) dS \tag{15.12}$$

Let us consider an elastic field in the domain V, satisfying the following equations:

$$\begin{aligned} &e_{ij}^{r}(x) = \tfrac{1}{2}\left(u_{i,j}^{r}(x) + u_{j,i}^{r}(x)\right), \quad (i = 1,2,3;\ j = 1,2,3), \quad x \in V \\ &\sigma_{ij}^{r} = 2\mu_M e_{ij}^{r}, \quad \theta^{r} = \sum_{k=1}^{3} e_{kk}^{r} = 0 \\ &\sigma_{ij,j}^{r} = 0 \end{aligned} \tag{15.13}$$

Elastic fields, satisfying Eq. (15.13), we will call regular elastic fields and mark with a superscript r. Now we can define the reciprocity gap functional (RGF), depending on the given thermoelastic field with the superscript d and regular elastostatic field with a superscript r

$$RG(d,r) = \int_{\partial V} \left(\sigma_{ij}^{d} n_j u_i^{r} - \sigma_{ij}^{r} n_j u_i^{d}\right) dS \tag{15.14}$$

From the Eqs. (15.12)–(15.14) one has

$$RG(d,r) = \sum_{k=1}^{n} \int_{\partial G_k} \left(\sigma_{ij}^{d} N_j u_i^{r} - \sigma_{ij}^{r} N_j u_i^{d}\right) dS \tag{15.15}$$

Since the thermoelastic field with a superscript d and elastostatic field with a superscript r are determined inside the domains G_k, the surface integrals in Eq. (15.15) can be transformed into integrals over the domains G_k. It follows from the Eqs. (15.9), (15.13) and (15.15)

$$\begin{aligned} RG(d,r) &= \sum_{k=1}^{n} \int_{\partial G_k} \left(\sigma_{ij}^{Ik} N_j u_i^{r} - \sigma_{ij}^{r} N_j u_i^{Ik}\right) dS \\ &= \sum_{k=1}^{n} \int_{G_k} \left(\sigma_{ij}^{Ik} e_{ij}^{r} - \sigma_{ij}^{r} e_{ij}^{Ik}\right) dx \\ &= \sum_{k=1}^{n} 2\left(\mu_{Ik} - \mu_M\right) \int_{G_k} e_{ij}^{Ik} e_{ij}^{r} dx \end{aligned} \tag{15.16}$$

15.4 Statement of the Inverse Problem and a Method of its Solving

We assume that typical sizes of the defects have the same order. Denote the typical size l. Assume also that the distances between the defects exceed some value L. We assume that the defects are small in the following sense

$$l \ll L \tag{15.17}$$

We suppose that in a single experiment the mechanical loads $\sigma_{ij}^{d} n_j = t_i^d$ and displacements $\mathbf{u}^d$ are measured on the whole external boundary. The problem is to detect and identify the defects G_k using available data.

In this paper we will not solve the problem in its entirety, that is, we do not propose a method for determining the sizes and shapes of defects. We develop a method for determining the number of defects, considered as points, and their locations.

It is interesting to note that we solve the problem using only the values $\sigma_{ij}^{d} n_j$ and $\mathbf{u}^d$ on the external boundary of the body without any information about thermal conditions on the external boundary and conditions of heat transfer between the matrix and inclusions. In particular, we solve the problem in case, when the external boundary is loaded only by thermal loads. In this case, the defect identification is performed using the displacements on the external boundary, which is free of mechanical loads.

To solve the problem we use the RGF defined in Eq. (15.14). Because the values $\sigma_{ij}^{d} n_j$ and u_i^d are known, the values $RG(d,r)$ can be calculated for all regular elastic fields r. It follows from Eq. (15.16), that if there are no defects in the body, then $RG(d,r) = 0$ for all regular elastic fields. Otherwise there are such regular elastic fields r, that $RG(d,r) \neq 0$ and the values $RG(d,r)$ give information about the defects. So, the problem will be solved if we express the positions of the defects G_k by means of the values $RG(d,r)$. Using approach proposed by Shifrin and Shushpannikov (2013a, 2015), we determine projections of the defects on an arbitrary plane. Consider, for example, projections of the defects on the plane x_1x_2. To determine projections of the defects on the plane x_1x_2 we will use the regular elastic fields corresponding to plane strain states. According to Muskhelishvili (1977) the displacements $\mathbf{u} = (u_1, u_2, 0)$, strains e_{ij} and stresses σ_{ij} of a plane strain field can be expressed by means of two holomorphic functions $\varphi(z)$ and $\psi(z)$, $z = x_1 + ix_2$

$$\mu_M (u_1 + iu_2) = \kappa\varphi(z) - z\overline{\varphi'(z)} - \overline{\psi(z)}, \quad \kappa = 3 - 4\nu_M \tag{15.18}$$

$$\begin{aligned} &\sigma_{11} + \sigma_{22} = 4\mathrm{Re}\left[\Phi(z)\right] \\ &\sigma_{22} - \sigma_{11} + 2i\sigma_{12} = 2\left[\bar{z}\Phi'(z) + \Psi(z)\right] \\ &\sigma_{13} = \sigma_{23} = 0, \quad \sigma_{33} = \nu_M(\sigma_{11} + \sigma_{22}) \end{aligned} \tag{15.19}$$

$$
\begin{aligned}
e_{11}+e_{22} &= \frac{2\left(1-2\nu_M\right)}{\mu_M}\operatorname{Re}\left[\Phi\left(z\right)\right] \\
e_{22}-e_{11}+2ie_{12} &= \frac{1}{\mu_M}\left[\bar{z}\Phi'\left(z\right)+\Psi\left(z\right)\right] \\
e_{13}=e_{23}=e_{33} &= 0
\end{aligned} \tag{15.20}
$$

where $\Phi\left(z\right)=\varphi'\left(z\right),\ \ \Psi\left(z\right)=\psi'\left(z\right)$, the overbar denotes the complex conjugation. Let us consider an elastic field corresponding to a plane strain state determined by the following holomorphic functions:

$$
\Phi\left(z\right)=0,\ \ \Psi\left(z\right)=-2\mu_M H\left(z\right) \tag{15.21}
$$

It follows from the Eqs. (15.20) and (15.21), that the elastic field satisfies the equation $\theta=e_{11}+e_{22}+e_{33}=0$ and according to the definition given above (see Eq. (15.13)), is a regular.

It follows from the Eqs. (15.16), (15.20) and (15.21)

$$
RG\left(d,r\right)=\sum_{k=1}^{n}2\left(\mu_{Ik}-\mu_M\right)\int\limits_{G_k}\left[\left(e_{11}^{Ik}-e_{22}^{Ik}\right)\operatorname{Re}H-2e_{12}^{Ik}\operatorname{Im}H\right]dx \tag{15.22}
$$

Let us mark with a superscript ρ a regular elastic field corresponding to a plane strain state determined by the holomorphic functions:

$$
\Phi\left(z\right)=0,\ \ \Psi\left(z\right)=-2\mu_M iH\left(z\right) \tag{15.23}
$$

From the Eqs. (15.16), (15.20) and (15.23) one has

$$
RG\left(d,\rho\right)=\sum_{k=1}^{n}2\left(\mu_{Ik}-\mu_M\right)\int\limits_{G_k}\left[-\left(e_{11}^{Ik}-e_{22}^{Ik}\right)\operatorname{Im}H-2e_{12}^{Ik}\operatorname{Re}H\right]dx \tag{15.24}
$$

From the Eqs. (15.22) and (15.24) it follows

$$
RG\left(d,r\right)-iRG\left(d,\rho\right)=\sum_{k=1}^{n}2\left(\mu_{Ik}-\mu_M\right)\int\limits_{G_k}\left[\left(e_{11}^{Ik}-e_{22}^{Ik}\right)+2ie_{12}^{Ik}\right]H\,dx \tag{15.25}
$$

Let us take $H\left(z\right)=H_m\left(z\right)=w^m\left(z\right),\ m=0,1,2,\cdots$. Denote r_m and ρ_m the regular elastic fields corresponding to the holomorphic function $w^m\left(z\right)$. Using Eq. (15.25) we obtain a system of equations.

$$
RG\left(d,r_m\right)-iRG\left(d,\rho_m\right)=\sum_{k=1}^{n}2\left(\mu_{Ik}-\mu_M\right)\int\limits_{G_k}\left[\left(e_{11}^{Ik}-e_{22}^{Ik}\right)+2ie_{12}^{Ik}\right]w^m\left(z\right)dx \tag{15.26}
$$

Now we use the assumption that the domains G_k are small and approximate the integrals in Eq. (15.26) by the principal term of the asymptotic expansion, provided that $l/L\rightarrow 0$.

$$RG(d,r_m) - iRG(d,\rho_m) =$$

$$= \sum_{k=1}^{n} 2(\mu_{Ik} - \mu_M)\left[\left(e_{11}^{Ik}\left(x^k\right) - e_{22}^{Ik}\left(x^k\right)\right) + 2ie_{12}^{Ik}\left(x^k\right)\right] w^m(z_k)\,|G_k| \tag{15.27}$$

where $x^k = \left(x_1^k, x_2^k, x_3^k\right) \in G_k$, $z_k = x_1^k + ix_2^k$, $|G_k|$ is the volume of the domain G_k. The system of Eq. (15.27) has the following form:

$$\sum_{k=1}^{n} A_k w_k^m = b_m, \quad m = 0,1,2,\cdots \tag{15.28}$$

where

$$A_k = 2(\mu_{Ik} - \mu_M)\left[\left(e_{11}^{Ik}\left(x^k\right) - e_{22}^{Ik}\left(x^k\right)\right) + 2ie_{12}^{Ik}\left(x^k\right)\right]|G_k|$$

$$b_m = RG(d,r_m) - iRG(d,\rho_m),\, w_k = w(z_k)$$

The equations having the form of Eq. (15.28), are encountered in a number of problems. Among them can be noted signal analysis (see, for example, Peter and Plonka, 2013), the problem of reconstruction of a plane polygonal domain from moments (Golub et al, 1999), identification of simple poles of a meromorphic function (see Badia and Ha-Duong, 2000; Kang and Lee, 2004), geometric inverse elastostatic problem (Shifrin and Shushpannikov, 2013a, 2015). The methods of solving the system of equations (15.28) are well developed. Below we use one of the possible approaches for solving the system (15.28). Let us suppose that the number of defects is equal N. Construct the Hankel matrices

$$\mathbf{H}_0 = \begin{pmatrix} b_0 & b_1 & \cdots & b_{N-1} \\ b_1 & b_2 & \cdots & b_N \\ \vdots & \vdots & \ddots & \vdots \\ b_{N-1} & b_{N-2} & \cdots & b_{2N-2} \end{pmatrix}, \quad \mathbf{H}_1 = \begin{pmatrix} b_1 & b_2 & \cdots & b_N \\ b_2 & b_3 & \cdots & b_{N+1} \\ \vdots & \vdots & \ddots & \vdots \\ b_N & b_{N+1} & \cdots & b_{2N-1} \end{pmatrix} \tag{15.29}$$

As it was shown by Golub et al (1999), the values w_k, $k = 1,2,\cdots,N$ are the eigenvalues of the generalized eigenvalue problem

$$\mathbf{H}_1\mathbf{a} = w\mathbf{H}_0\mathbf{a} \tag{15.30}$$

Here **a** is the eigenvector.

After determination of the values w_k we solve a system of linear algebraic equations (15.28) with respect to A_k, where $m = 0,1,2,\cdots,N-1$ and n is substituted with N. Inverting the function $w(z)$ we find the locations of defects z_k. This procedure we repeat for a sequence of values N. As a result we obtain a sequence of solutions corresponding to different values N. Among the obtained values w_k there are some roots corresponding to the defects projections and some spurious roots. To choose the correct solutions we use the following criteria:

1. The solutions z_k located outside the projection of the body are spurious and should be excluded.

2. Let $A_{\max} = \max_k |A_k|$. Solution of the Eq. (15.28) w_p and A_p for which the condition $|A_p|/A_{\max} << 1$ is fulfilled is considered as spurious and is excluded.
3. The correct solutions, starting with a certain number, should be stable with respect to the prescribed number of defects N.

Let us note that after excluding the spurious roots the number of remaining roots can exceed the number of projections of real defects because several roots can correspond to one defect, see for example Hanke and Rundell (2011). For formalization of the criteria 2 and 3 and determination of the number of defects projections on an arbitrary plane we apply the following procedure. Denote the projection of the domain V on the plane x_1x_2 by V_{12}. Divide the area V_{12} in the J subsets $V_{12} = \bigcup_{j=1}^{J} Q_j$, $meas(Q_i \cap Q_j) = 0,\ i \neq j,\ diam(Q_j) \leq \varepsilon(J)$. Here $meas(B)$ is the measure of the set B, $diam(Q_j)$ is a diameter of the set Q_j and the function $\varepsilon(J)$ satisfies the condition $\lim_{J\to\infty} \varepsilon(J) = 0$. Introduce a set of pairwise disjoint events H_k, $k = 1, 2, \cdots$. The event H_N denotes that exactly N inhomogeneities are embedded in the body V. Let $A_{1N}, w_{1N}, A_{2N}, w_{2N}, \cdots, A_{NN}, w_{NN}$ be the solution to the Eq. (15.28) corresponding to the assumption H_N. Let $z_{1N}, z_{2N}, \cdots, z_{NN}$ be the locations of point defects corresponding to the values $w_{1N}, w_{2N}, \cdots, w_{NN}$. Assume that $z_{kN} \in V_{12}$, $k = 1, \cdots, N_1$ and $z_{jN} \notin V_{12}$, $j = N_1 + 1, \cdots, N$, $N_1 \leq N$. We will interpret the values z_{kN}, $k = 1, \cdots, N_1$ as the realization of a complex-valued random variable. The coefficients A_{kN} we will interpret as the conditional probabilities

$$P_{jN} = P(z = z_j/H_N) = \frac{|A_{jN}|}{\sum_{k=1}^{N_1} |A_{kN}|}, \quad j = 1, \cdots, N_1 \tag{15.31}$$

Using the values P_{jN} it is possible to calculate the following conditional probability

$$P(z \in Q_j/H_N) = \sum_{z_{iN} \in Q_j} P_{iN} \tag{15.32}$$

Assume that the number of inclusions can not exceed the value $N_{\max}$. In this case $H_1, \cdots, H_{N_{\max}}$ are the mutually exclusive events whose probabilities sum is equal unity. Thus, we can assume that

$$P(H_N) = \frac{1}{N_{\max}} \tag{15.33}$$

From the theorem of total probability it follows

$$P(z \in Q_j) = \sum_{N=1}^{N_{\max}} P(z \in Q_j/H_N) P(H_N) = \frac{1}{N_{\max}} P(z \in Q_j/H_N) \tag{15.34}$$

Using Eqs. (15.32) and (15.34) we can calculate $P(z \in Q_j)$ for all subsets Q_j. Consider $P(z \in Q_j)$ as a function defined on the subsets Q_j of the body projection V_{12}.

The number of the projections of the inclusions corresponds to the number of clusters, where the function is considerably separated from zero.

Remark 15.1. We presented an algorithm for determining the number of projections of the inclusions and their locations in 3D bodies. It is possible to apply the same algorithm for solving 2D geometrical inverse problems. In this case, we will directly determine the number of inclusions and their locations.

15.5 Numerical Procedure and Numerical Examples

Thus, the proposed method for solving of the thermoelastic inverse problem by means of the overdetermined boundary data $\sigma_{ij}^d n_j$ and u_i^d is realized by the following algorithm.

1. First, we take some holomorphic function $w(z)$ and construct holomorphic functions $H_m(z) = w^m(z)$. Then we construct regular elastic fields with superscripts r_m and ρ_m according to the Eqs. (15.18), (15.19), (15.21), where $H(z) = H_m(z) = w^m(z)$.
2. On the second step, we calculate the values of the reciprocity gap functional $RG(d, r_m)$ and $RG(d, \rho_m)$ by means of the Eq. (15.14).
3. After that we construct the system of equations (15.28) for some number N using the values $RG(d, r_m)$ and $RG(d, \rho_m)$. The system of equations is solved by the considered above method and the values w_k and A_k are determined. The projections z_k are determined by means of the values w_k. This procedure is repeated for a sequence of values N.
4. The solutions, that are considered as spurious according to the criterion 1, are excluded.
5. The function $P(z \in Q_j)$ is constructed and the number of the projections of defects and their locations are determined.

In order to emphasize, that the defects can be identified by the data of the experiments, in which only the thermal loads are applied to the boundary of the body, below we consider examples, in which the boundary is free of mechanical loads ($\sigma_{ij}^d n_j = 0$). In all examples the thermal loading has the form:

$$\begin{gathered} T^d(x') = T_1, \quad x' \in \Gamma_1 \\ T^d(x') = T_2, \quad x' \in \Gamma_2 \\ \kappa_M \frac{\partial T^d(x')}{\partial x_i} n_i(x') = 0, \quad x' \in \Gamma_3 \end{gathered} \tag{15.35}$$

where $\Gamma_i \subset \partial V, \ \Gamma_i \cap \Gamma_j = \emptyset, \ i \neq j, \ \bigcup_{i=1}^{3} \bar{\Gamma}_i = \partial V.$

In the examples, the defects G_k are cavities and their boundaries ∂G_k are considered insulated

$$\kappa_M \frac{\partial T^d(x^*)}{\partial x_i} N_i(x^*) = 0, \quad x^* \in \partial G_k \tag{15.36}$$

The initial data for the solution of the inverse problems, that are displacements u_i^d on the boundary of the body ∂V, were obtained by solving the direct problems by finite elements method. To improve the accuracy of the results, we calculate the values $RG(d, r_m)$ and $RG(d, \rho_m)$ are not by means of the Eq. (15.14), but by the following method. Let us denote by a superscript $d0$ the thermoelastic field in the body V without any defects, in case when the body is subjected to the same conditions on the external boundary, as in the considered inverse problem. This thermoelastic field can be obtained by solving the direct problem. Because the thermoelastic field with the superscript $d0$ has no singularities inside the domain V, the following equalities are valid:

$$RG(d0, r_m) = \int_{\partial V} \left(\sigma_{ij}^{d0} n_j u_i^{r_m} - \sigma_{ij}^{r_m} n_j u_i^{d0} \right) dS = 0 \tag{15.37}$$

$$RG(d0, \rho_m) = \int_{\partial V} \left(\sigma_{ij}^{d0} n_j u_i^{\rho_m} - \sigma_{ij}^{\rho_m} n_j u_i^{d0} \right) dS = 0 \tag{15.38}$$

Using Eqs. (15.14), (15.37), (15.38) and the condition $\sigma_{ij}^{d} n_j = \sigma_{ij}^{d0} n_j$ on the boundary, one has

$$RG(d, r_m) = RG(d, r_m) - RG(d0, r_m) = -\int_{\partial V} \sigma_{ij}^{r_m} n_j \Delta u_i^d dS \tag{15.39}$$

$$RG(d, \rho_m) = RG(d, \rho_m) - RG(d0, \rho_m) = -\int_{\partial V} \sigma_{ij}^{\rho_m} n_j \Delta u_i^d dS \tag{15.40}$$

Here $\Delta u_i^d = u_i^d - u_i^{d0}$. The use of the formulas (15.39) and (15.40) instead of the Eq. (15.14) significantly improves the accuracy of the results.

To illustrate the efficiency of the proposed method, we have considered several two-dimensional problems. In all considered examples the elastic body V is a square $\{x: \ |x_i| \le 10\,\mathrm{m}, \ i = 1,2\}, w(z) = z/10$. The elastic properties of the matrix are as follows: the Young modulus $E_M = 2\mu_M(1 + \nu_M) = 68.5\,\mathrm{GPa}$, $\nu_M = 0.36$. The thermal characteristics are as follows: $\alpha_M = 19.5 \times 10^{-6}\,\mathrm{K}^{-1}$, $\kappa_M = 207\,\mathrm{W \cdot m^{-1} \cdot K^{-1}}$. These constants correspond to the aluminum.

Example 1. First, we consider the case of one cavity. The cavity has a shape of a square, which occupies the region $G_1 = \{(x_1, x_2) \in [3,5] \times [2,4]\}$, see Fig. 15.1. The thermal boundary conditions according to the notation introduced in the Eq. (15.35) are the following:

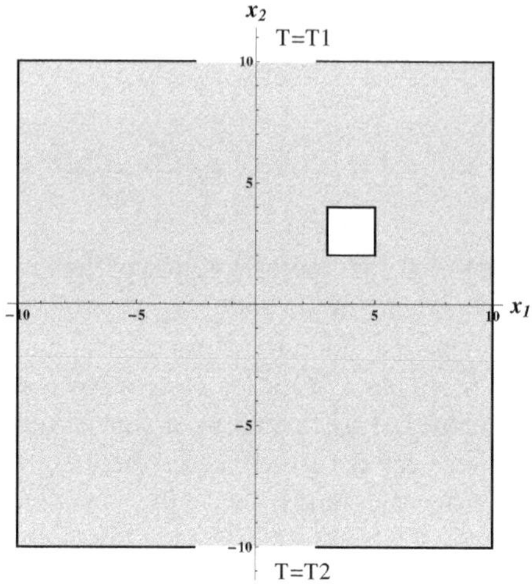

Fig. 15.1 A square with a single cavity

$$
\begin{aligned}
&\Gamma_1 = [-2.5, 2.5] \times 10, \; T_1(x') = 20^\circ\mathrm{C}, \; x' \in \Gamma_1 \\
&\Gamma_2 = [-2.5, 2.5] \times (-10), \; T_2(x') = 10^\circ\mathrm{C}, \; x' \in \Gamma_2 \\
&\Gamma_3 = \partial V \backslash (\Gamma_1 \cup \Gamma_2)
\end{aligned} \tag{15.41}
$$

Parts of the boundary of the body, where the temperature is given by the boundary conditions are indicated in Fig. 15.1.

The surface defined by the function $P(Q_j)$ is shown in Fig. 15.2. Formally, the result may depend on two parameters the value $N_{\max}$ and the decomposition of the body into the set of subsets Q_j. In Fig. 15.2, the results are presented for the case when the area V is divided into $J = 100^2$ subsets Q_j. The subsets Q_j are the squares

$$
\begin{aligned}
&Q_j = Q_{lk} = \left\{ \begin{array}{c} (x_1, x_2): \; -10 + 0.2(l-1) \le x_1 \le -10 + 0.2 \cdot l, \\ -10 + 0.2(k-1) \le x_2 \le -10 + 0.2 \cdot k \end{array} \right\} \\
&l = 1, \cdots, 100; \; k = 1, \cdots, 100
\end{aligned} \tag{15.42}
$$

The results are presented for two values of $N_{\max}$ ($N_{\max} = 20$ on the left figure and $N_{\max} = 40$ on the right figure). It follows from Fig. 2 that there is only one inhomogeneity and its position is approximately determined. The results are almost independent on the value $N_{\max}$. We also fulfilled calculations for the case of the partition of the body into $J = 200^2$ squares Q_j, whose sides twice lesser than the sides of the squares Q_j considered above. The results are presented in Fig. 15.3. The Figs 15.2 and 15.3 show that such partition does not have any noticeable influence on the results.

Because, the results presented in Figs. 15.2 and 15.3 do not depend significantly on the values J and $N_{\max}$, in the examples considered below, the results are presented for $J = 100^2$ and $N_{\max} = 20$.

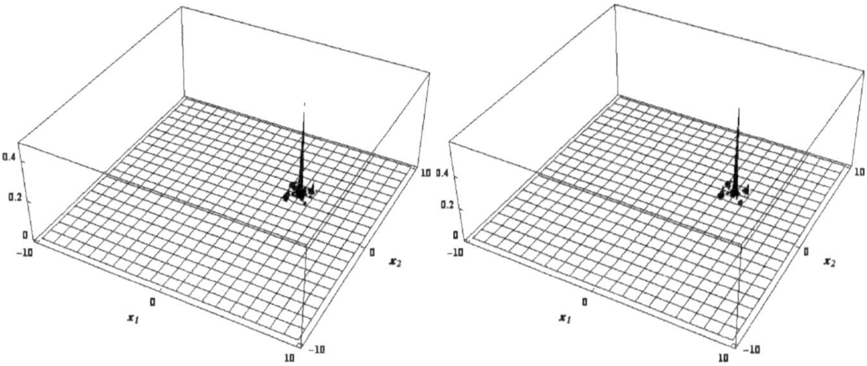

Fig. 15.2: Example 1. Results for $J = 100^2$, $N_{\max} = 20$ (left figure), $N_{\max} = 40$ (right figure)

Fig. 15.3 Example 1. Results for $J = 200^2$, $N_{\max} = 20$

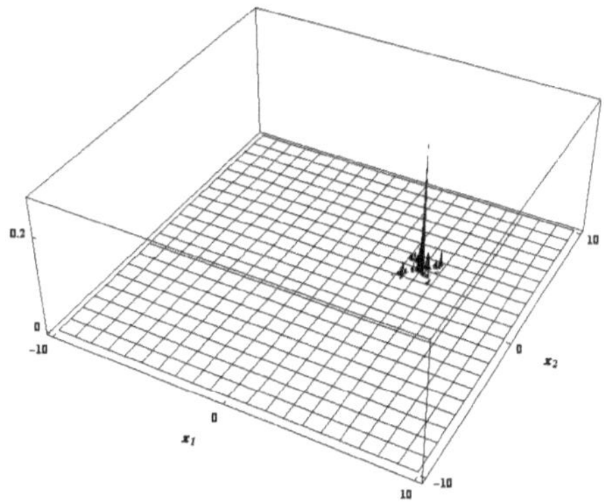

Example 2. In this example we consider a body with two cavities subjected to the same boundary conditions as in example 1. The cavities occupy the following positions: the first cavity G_1 is the same as in example 1; the cavity G_2 occupies the region $G_2 = \{[-2.5, -3.5] \times [5.5, 6.5]\}$. The body with two cavities is presented in Fig. 15.4. The results of identification are presented in Fig. 15.5.

It is possible to see from Fig. 15.5, that both cavities are detected by the proposed method. Besides the cavities, we can see one small additional defect. Such solution is obtained due to the event H_1. When we assume, that there is only one inhomogeneity, we obtain some point z_{11}, which does not belong to areas G_1 and G_2. Because the conditional probability $P(z = z_{11}/H_1) = 1$, the contribution of the point in the probability distribution is significant. The contribution of the point decreases with the growth of the value $N_{\max}$. We also exclude this solution if we assume, according to the results, presented in Fig. 15.5, that H_1 is an im-

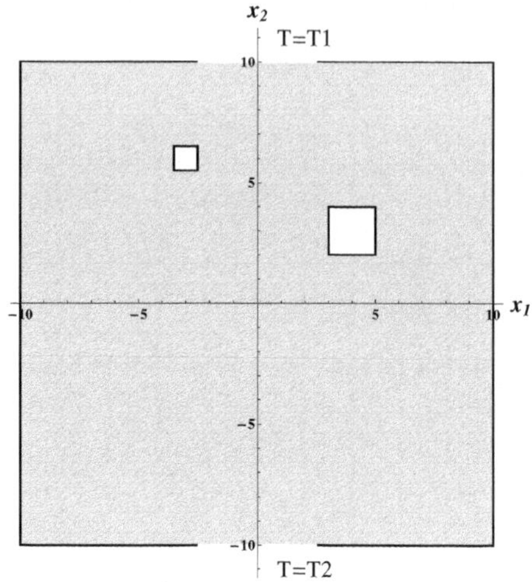

Fig. 15.4 A square with two cavities

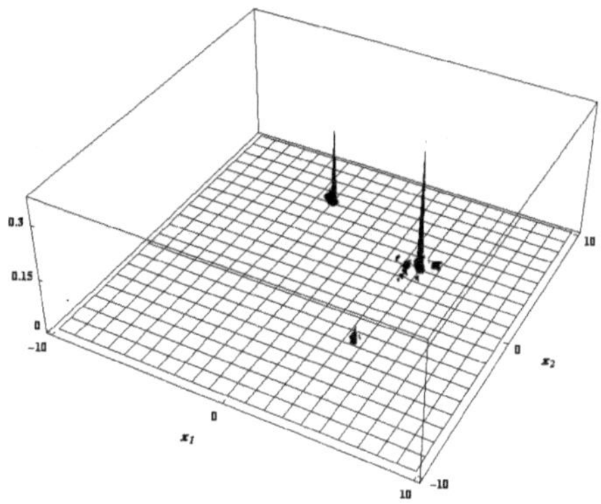

Fig. 15.5 Example 2. Results for $J = 100^2$, $N_{\max} = 20$

possible event $(P(H_1) = 0)$ and $H_2, H_3, \cdots, H_{N_{\max}}$ is a complete group of events $(P(H_k) = 1/(N_{\max} - 1),\ k = 2, \cdots, N_{\max})$. The results, obtained under this assumption, are presented in Fig. 15.6. Our assumption is confirmed in Fig. 15.6, where we can see only two clusters, corresponding to the inhomogeneities.

Example 3. In this example, we consider the case of three cavities. The cavities G_1 and G_2 are the same as in example 2. The third cavity G_3 has a rectangular shape, $G_3 = \{[-2,0] \times [-3.5, -4.5]\}$. The body with three cavities is presented in Fig. 15.7.

The obtained results are presented in Fig. 15.8. In accordance with the Fig. 15.8, we detect three real cavities and additionally we obtain three spurious solutions. Similar to example 2, the spurious solutions are appeared due to the assumption that events H_1 and H_2 are possible. In the Fig. 15.9 we present the results under assumptions $P(H_1) = 0$ (the left figure) and $P(H_1 \cup H_2) = 0$ (the right figure). In case $P(H_1) = 0$ we consider the complete group of events $H_2, H_3, \cdots, H_{N_{\max}}$ and $P(H_k) = 1/(N_{\max} - 1)$, $k = 2, \cdots, N_{\max}$. When we assume, that $P(H_1 \cup H_2) = 0$, we

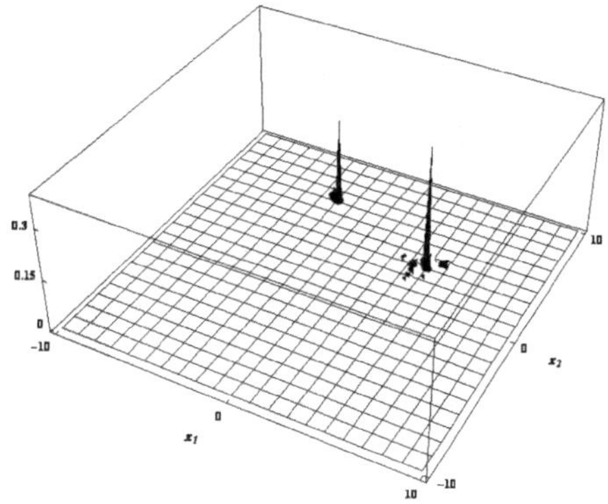

Fig. 15.6 Example 2. Results, obtained under assumption $P(H_1) = 0$

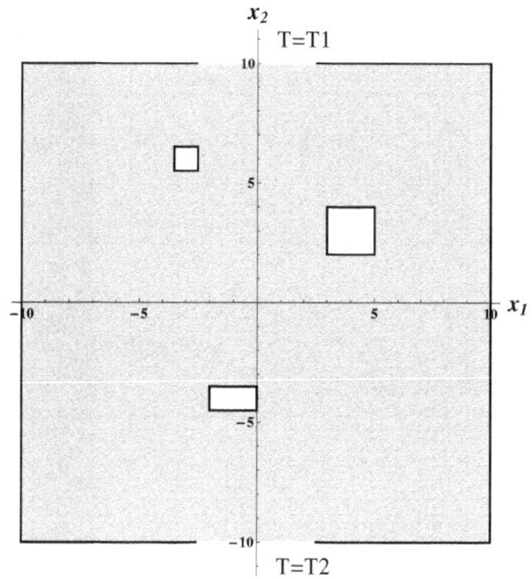

Fig. 15.7 A square with three cavities

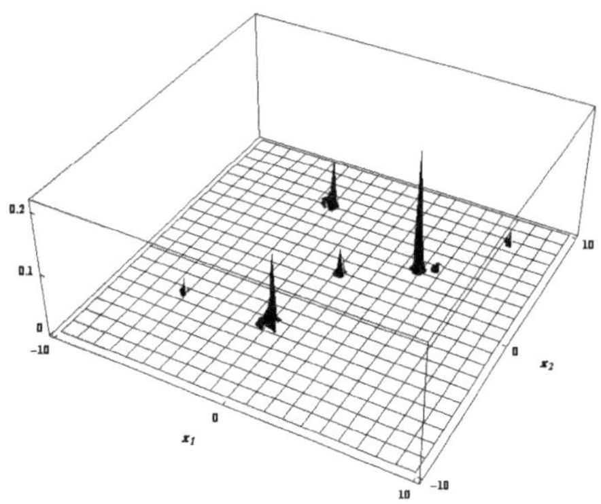

Fig. 15.8 Example 3. Results for $J = 100^2$, $N_{\max} = 20$

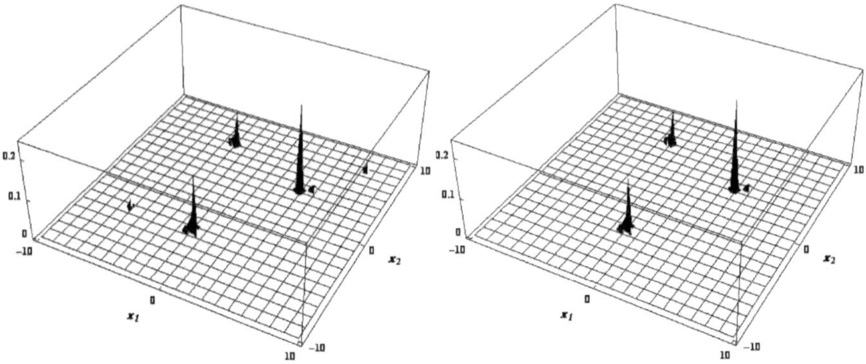

Fig. 15.9: Example 3. Results, obtained under assumptions $P(H_1) = 0$ and $P(H_1 \cup H_2) = 0$

consider $H_3, \cdots, H_{N_{\max}}$ as a complete group of events and $P(H_k) = 1/(N_{\max} - 2)$ for $k = 3, \cdots, N_{\max}$. It is possible to see, that the spurious solutions are excluded successively.

15.6 Conclusions

A method for identification of a finite number of small, well-separated inhomogeneities (cracks, cavities, inclusions) in 3D and 2D thermoelasticity is developed. The method uses the knowledge of the mechanical loads and the displacements on the boundary of the body, obtained in a single experiment, and enables to identify the number and positions of the inhomogeneities, that are considered as the point defects. The knowledge of the thermal load and the conditions of heat exchange between the matrix and the inclusions are not assumed in the method. Considered numerical examples show the efficiency of the proposed method.

Acknowledgements The support of RFBR grant 16-01-00149 is gratefully acknowledged.

References

Ammari H, Kang H, Nakamura G, Tanuma K (2002) Complete asymptotic expansions of solutions of the system of elastostatics in the presence of an inclusion of small diameter and detection of an inclusion. Journal of elasticity and the physical science of solids 67(2):97–129

Andrieux S (2015) The reciprocity likelihood maximization: a variational approach of the reciprocity gap method. Journal of Mechanics of Materials and Structures 10(3):219–237

Badia AE, Ha-Duong T (2000) An inverse source problem in potential analysis. Inverse Problems 16(3):651–663

Ben Ameur H, Burger M, Hackl B (2007) Cavity identification in linear elasticity and thermoelasticity. Mathematical Methods in the Applied Sciences 30(6):625–647

Dehghan Manshadi SH, Khaji N, Rahimian M (2014) Cavity/inclusion detection in plane linear elastic bodies using linear sampling method. Journal of Nondestructive Evaluation 33(1):93–103

Engelhardt M, Schanz M, Stavroulakis GE, Antes H (2006) Defect identification in 3-d elastostatics using a genetic algorithm. Optimization and Engineering 7(1):63–79

Golub GH, Milanfar P, Varah J (1999) A stable numerical method for inverting shape from moments. SIAM Journal on Scientific Computing 21(4):1222–1243

Hanke M, Rundell W (2011) On rational approximation methods for inverse source problems. Inverse Problems and Imaging 5(1):185–202

Kang H, Lee H (2004) Identification of simple poles via boundary measurements and an application of eit. Inverse Problems 20(6):1853–1863

Karageorghis A, Lesnic D, Marin L (2012) The method of fundamental solutions for the detection of rigid inclusions and cavities in plane linear elastic bodies. Computers & Structures 106-107:176–188

Karageorghis A, Lesnic D, Marin L (2014) A moving pseudo-boundary {MFS} for void detection in two-dimensional thermoelasticity. International Journal of Mechanical Sciences 88:276 – 288

Khodadad M, Dashti-Ardakani M (2009) Determination of the location, size and mechanical properties of an elastic inclusion using surface measurements. Inverse Problems in Science and Engineering 17:591 – 604

Muskhelishvili NI (1977) Some basic problems of the mathematical theory of elasticity. Noordhoff, Leyden

Nowacki W (1986) Thermo-elasticity, 2nd edn. Pergamon, Warsaw

Peter T, Plonka G (2013) A generalized prony method for reconstruction of sparse sums of eigenfunctions of linear operators. Inverse Problems 29(2):025,001

Shifrin E, Shushpannikov P (2011) Identification of an ellipsoidal defect in an elastic solid using boundary measurements. Int J Solids Struct 48(7-8):1154–1163

Shifrin E, Shushpannikov P (2013a) Identification of small well-separated defects in an isotropic elastic body using boundary measurements. Int J Solids Struct 50(22-23):3707–3716

Shifrin E, Shushpannikov P (2013b) Reconstruction of an ellipsoidal defect in anisotropic elastic solid, using results of one static test. Inverse Problems in Science and Engineering 21:781–800

Shifrin E, Shushpannikov P (2015) Identification of finitely many small defects in an anisotropic linearly elastic body from a single static test. Mechanics of Solids 50:421–431

Steinhorst P, Sändig AM (2012) Reciprocity principle for the detection of planar cracks in anisotropic elastic material. Inverse Problems 28(8):085,010

Chapter 16
Indentation of the Regular System of Punches into the Foundation with Routh Coating

Kirill E. Kazakov & Svetlana P. Kurdina

Abstract We study plane contact problem for ageing viscoelastic bases with regular finite systems of rigid punches. The case in which the punch and the coating surfaces are conformal (mutually repeating) is under consideration. Such problems can arise, for example, when the punch immerses into a solidifying coating before its complete solidification, e.g. into some sort of glue or new concrete. The shape of thin coating as well as the shapes of the punches bases can be described by rapidly changing functions. These problems arises when designing of machines and mechanisms, when dealing with erection of structure complexes on bases and foundations, when determining the admissible tilt angle of objects erected in close proximity to each other, or when reinforcing engineering or civil engineering structures, when designing and assembling structures that are in close vicinity to each other, etc. We obtain the system of basic integral equations with the system of additional conditions, convert it into one operator equation with two additional vector conditions, and construct its analytical solution by using a generalized projection method.

Key words: Contact problem · Conformal contact · Mixed integral equations · Generalized projection method

Kirill E. Kazakov
Ishlinsky Institute for Problems in Mechanics RAS, Vernadskiy Ave, 101 bldg 1, 119526 Moscow & Bauman Moscow State Technical University, Baumanskaya Str. 5 Bldg 1, 105005 Moscow, Russia,
e-mail: `kazakov-ke@yandex.ru`

Svetlana P. Kurdina
Bauman Moscow State Technical University, Baumanskaya Str. 5 Bldg 1, 105005 Moscow, Russia,
e-mail: `svetlana-ka@yandex.ru`

H. Altenbach et al. (eds.), *Mechanics for Materials and Technologies*,
Advanced Structured Materials 46, DOI 10.1007/978-3-319-56050-2_16

16.1 Statement of the Problem

Viscoelastic layer of an arbitrary thickness H lies on the rigid foundation. There is perfect contact or smooth contact between layer and foundation. Second thin viscoelastic layer (coating) of thickness $h(x)$ lies on lower layer (see Fig. 16.1). There is also perfect contact or smooth contact between layers. We denote the moments of their production by τ_1 and τ_2, respectively. We assume that the coating rigidity is less than the rigidity of the lower layer or they are of the same order of magnitude. We also assume that the coating thickness described by periodic function with period Δa, i.e. $h(x) = h(x + \Delta a)$.

At time τ_0, the forces $P_i(t)$ with eccentricities $e_i(t)$ starts to indent smooth rigid punches into the surface of such a foundation ($i = \overline{1,n}$, n is the number of the punches). We assume that the system of punches is regular, i.e. the distances between neighbor punches are the same and punch lengths are equal. The lengths of lines of contact area are $(b_i - a_i) = \bar{a} \gg h(x)$ for all x, where a_i and b_i are left and right coordinates of ith punch. Moreover the period Δa is equal to $(a_{i+1} - a_i)$ for all $i = \overline{1,n-1}$. We consider the special case when the coating shape (the shape of the surface of the layer packet) coincides with the punch base shape, i.e. the contact interaction is conformal. The simple case of conformal contact is the contact of a punch with a plane base and a plane part of a solid (including basements with coating of constant thickness).

Solution for a problem of conformal contact between layered foundation and one punch described in Kazakov and Manzhirov (2008); Manzhirov and Kazakov (2016).

The vertical displacement of the upper boundary of the foundation under the action of the normal forces $q_i(x,t)$ can be written as (see, for example, Alexandrov and Mkhitaryan, 1983; Aleksandrov and Manzhirov, 1987; Arutyunyan and Manzhirov, 1999; Alexandrov and Chebakov, 2007)

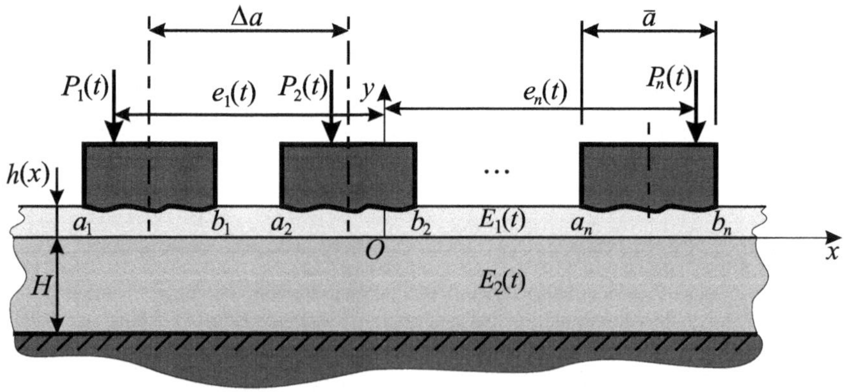

Fig. 16.1: Contact interaction of the regular system of punches and a two-layer basement

$$v_i(x,t) = k_\nu h(x)\left[\frac{q_i(x,t)}{E(t-\tau_1)} + \int_{\tau_0}^{t} K_1(t-\tau_1,\tau-\tau_1)\frac{q_i(x,\tau)}{E(\tau-\tau_1)}d\tau\right]$$
$$+\frac{2(1-\nu_2^2)}{\pi}\left[\sum_{j=1}^{n}\int_{a_i}^{b_i} k_{\rm pl}\left(\frac{x-\xi}{H}\right)\frac{q_j(\xi,t)}{E_2(t-\tau_2)}d\xi\right.$$
$$\left.+\int_{\tau_0}^{t} K_2(t-\tau_2,\tau-\tau_2)\sum_{j=1}^{n}\int_{a_i}^{b_i} k_{\rm pl}\left(\frac{x-\xi}{H}\right)\frac{q_j(\xi,\tau)}{E_2(\tau-\tau_2)}d\xi\, d\tau\right], \quad i=\overline{1,n},$$
$$K_k(t,\tau) = E_k(\tau)\frac{\partial}{\partial\tau}\left[\frac{1}{E_k(\tau)} + C_k(t,\tau)\right], \quad k=1,2, \quad x\in[a_i,b_i], \quad t\geq\tau_0,$$

where $E_1(t)$ and $E_2(t)$ are elastic moduli of the coating and the lower layer, k_ν is a coefficient depending on the contact conditions between coating and lower layer ($k_\nu = (1+\nu_1)(1-2\nu_1)/(1-\nu_1)$ in the case of an perfect contact, $k_\nu = 1-\nu_1^2$ in the case of a smooth coating-layer contact), ν_1 is the Poisson ratio of the coating; $C_k(t,\tau)$ is the tensile creep function, $k_{\rm pl}[(x-\xi)/H]$ is known kernel of the plane contact problem, which has the form (Arutyunyan and Manzhirov, 1999; Alexandrov and Chebakov, 2007)

$$k_{\rm pl}(s) = \int_0^\infty \frac{L(u)}{u}\cos(su)du,$$

and, in the case of a smooth contact between the lower layer and the rigid base,

$$L(u) = \frac{\cosh 2u - 1}{\sinh 2u + 2u},$$

and in the case of a perfect contact,

$$L(u) = \frac{2\varkappa\sinh 2u - 4u}{2\varkappa\cosh 2u + 4u^2 + 1 + \varkappa^2}, \qquad \varkappa = 3-4\nu_2.$$

By equating the vertical displacements of the upper boundary of the coating with the displacements of the rigid punches and taking into account the fact that the contact interaction is conformal, we obtain the system of integral equations of our problem in the form

$$k_\nu h(x)\left[\frac{q_i(x,t)}{E(t-\tau_1)} + \int_{\tau_0}^{t} K_1(t-\tau_1,\tau-\tau_1)\frac{q_i(x,\tau)}{E(\tau-\tau_1)}d\tau\right]$$
$$+\frac{2(1-\nu_2^2)}{\pi}\left[\sum_{j=1}^{n}\int_{a_i}^{b_i} k_{\rm pl}\left(\frac{x-\xi}{H}\right)\frac{q_j(\xi,t)}{E_2(t-\tau_2)}d\xi\right. \tag{16.1}$$
$$\left.+\int_{\tau_0}^{t} K_2(t-\tau_2,\tau-\tau_2)\sum_{j=1}^{n}\int_{a_i}^{b_i} k_{\rm pl}\left(\frac{x-\xi}{H}\right)\frac{q_j(\xi,\tau)}{E_2(\tau-\tau_2)}d\xi\, d\tau\right]$$
$$= \delta_i(t) + \alpha_i(t)\left(x - \frac{a_i+b_i}{2}\right), \quad i=\overline{1,n}, \quad x\in[a_i,b_i], \quad t\geq\tau_0,$$

where $\delta_i(t)$ and $\alpha_i(t)$ are the punches settlements and its tilt angles ($i = \overline{1,n}$). The additional condition for this problem are ($i = \overline{1,n}$, $t \geq \tau_0$)

$$P_i(t) = \int_{a_i}^{b_i} q_i(\xi,t)d\xi,$$
$$M_i(t) \equiv P_i(t)e_i(t) = \int_{a_i}^{b_i} q_i(\xi,t)\left(\xi - \frac{a_i+b_i}{2}\right)d\xi. \tag{16.2}$$

There exist 15 versions of mathematical statements for the contact problem for a system of punches in the plane case. It is easy to show that there is one of 4 types of conditions on each punch: the force and the moment of the load application are given, the tilt angle of the punch and the force of the load application are given, the punch settlement and the moment of the load application are given, the settlement and tilt angle of the punch are given. So: 1) there is 4 versions of mathematical statements if all conditions are equal ($C_4^1 = 4$); 2) there is 6 versions of mathematical statements if exists 2 groups of punches with different conditions ($C_4^2 = 6$); 3) there is 4 versions of mathematical statements if exists 3 groups of punches with different conditions ($C_4^3 = 4$); 4) there is 1 version of mathematical statement if exists 4 groups of punches with different conditions ($C_4^4 = 1$).
We will describe most common version of mathematical statement if exists 4 groups of punches with different conditions. All another version are particular cases of this version.

16.2 Dimensionless Form and Operator Representation

Let us make the change of variables in (16.1) and (16.2) by the formulas ($x \in [a_i,b_i]$, $\xi \in [a_j,b_j]$)

$$x^* = \frac{2(x-\eta_i)}{\bar{a}}, \quad \xi^* = \frac{2(\xi-\eta_j)}{\bar{a}}, \quad t^* = \frac{t}{\tau_0}, \quad \tau_1^* = \frac{\tau_1}{\tau_0}, \quad \tau_2^* = \frac{\tau_2}{\tau_0},$$
$$\lambda = \frac{2H}{\bar{a}}, \quad \delta^{i*}(t^*) = \frac{2\delta_i(t)}{\bar{a}}, \quad \alpha^{i*}(t^*) = \alpha_i(t), \quad c^*(t^*) = \frac{E_2(t-\tau_2)}{E_1(t-\tau_1)},$$
$$m^*(x^*) \equiv m^i(x^*) = \frac{k_\nu}{2(1-\nu_2^2)}\frac{2h(x)}{\bar{a}}, \quad q^{i*}(x^*,t^*) = \frac{2(1-\nu_2^2)q_i(x,t)}{E_2(t-\tau_2)},$$
$$P^{i*}(t^*) = \frac{4P_i(t)(1-\nu_2^2)}{E_2(t-\tau_2)\bar{a}}, \quad M^{i*}(t^*) = \frac{8e_i(t)P_i(t)(1-\nu_2^2)}{E_2(t-\tau_2)\bar{a}},$$
$$\mathbf{F}^{ij*}f(x^*) = \int_{-1}^{1} k^{ij}(x^*,\xi^*)f(\xi^*)d\xi^*, \quad \mathbf{L}_k^* f(t^*) = \int_{1}^{t^*} K^{k*}(t^*,\tau^*)f(\tau^*)d\tau^*,$$

$$k^{ij}(x^*,\xi^*) = \frac{1}{\pi}k_{\rm pl}\left(\frac{x-\xi}{H}\right) = \frac{1}{\pi}k_{\rm pl}\left(\frac{x^*-\xi^*+\eta^{i*}-\eta^{j*}}{\lambda}\right), \quad \eta^{i*} = \frac{2\eta_i}{\bar{a}},$$
$$K^{1*}(t^*,\tau^*) = \frac{E_1(t-\tau_1)}{E_1(\tau-\tau_1)}\frac{E_2(\tau-\tau_1)}{E_2(t-\tau_1)}K_1(t-\tau_1,\tau-\tau_1)\tau_0,$$
$$K^{2*}(t^*,\tau^*) = K(t-\tau_2,\tau-\tau_2)\tau_0, \quad i,j=1,2,\ldots,n.$$

Then we obtain a system of mixed integral equation and additional conditions in the dimensionless form

$$c(t)m(x)(\mathbf{I}-\mathbf{L}_1)q^i(x,t) + (\mathbf{I}-\mathbf{L}_2)\sum_{j=1}^{n}\mathbf{F}^{ij}q^j(x,t) = \delta^i(t)+\alpha^i(t)x,$$
$$\int_{-1}^{1} q^i(\xi,t)\,d\xi = P^i(t), \quad \int_{-1}^{1} q^i(\xi,t)\xi\,d\xi = M^i(t), \quad i=\overline{1,n},\ x\in[-1,1],\ t\geq 1. \tag{16.3}$$

We can represent this system with additional conditions (16.3) as one operator equation with two additional vector conditions

$$c(t)m(x)(\mathbf{I}-\mathbf{L}_1)\mathbf{q}(x,t) + (\mathbf{I}-\mathbf{L}_2)\mathbf{G}\mathbf{q}(x,t) = \delta(t)+\alpha(t)x,$$
$$\int_{-1}^{1} \mathbf{q}(\xi,t)\,d\xi = \mathbf{P}(t), \quad \int_{-1}^{1} \mathbf{q}(\xi,t)\xi\,d\xi = \mathbf{M}(t), \quad x\in[-1,1], \quad t\geq 1, \tag{16.4}$$

Here

$$\mathbf{q}(x,t) = q^i(x,t)\mathbf{i}^i, \quad \mathbf{P}(t) = P^i(t)\mathbf{i}^i, \quad \mathbf{M}(t) = M^i(t)\mathbf{i}^i,$$
$$\delta(t) = \delta^i(t)\mathbf{i}^i, \quad \alpha(t) = \alpha^i(t)\mathbf{i}^i, \quad \mathbf{k}(x,\xi) = k^{ij}(x,\xi)\mathbf{i}^i\mathbf{i}^j,$$
$$\mathbf{G}\mathbf{f}(x) = \int_{-1}^{1}\mathbf{k}(x,\xi)\cdot\mathbf{f}(\xi)\,d\xi. \tag{16.5}$$

16.3 Transformation of Main Equation and Special Basis

Let us multiply the operator equation (16.4) by $1/\sqrt{m(x)}$

$$c(t)\sqrt{m(x)}(\mathbf{I}-\mathbf{L}_1)\mathbf{q}(x,t) + \frac{1}{\sqrt{m(x)}}(\mathbf{I}-\mathbf{L}_2)\mathbf{G}\mathbf{q}(x,t) = \frac{\delta(t)+\alpha(t)x}{\sqrt{m(x)}}$$

By introducing notations

$$\mathbf{Q}(x,t)=\sqrt{m(x)}\mathbf{q}(x,t),\quad \mathbf{K}(x,\xi)=\frac{\mathbf{k}(x,\xi)}{\sqrt{m(x)m(\xi)}},$$
$$\mathbf{Ff}(x)=\int_{-1}^{1}\mathbf{K}(x,\xi)\cdot\mathbf{f}(\xi)\,d\xi, \tag{16.6}$$

we obtain ($x\in[-1,1], t\geq 1$)

$$c(t)(\mathbf{I}-\mathbf{L}_1)\mathbf{Q}(x,t)+(\mathbf{I}-\mathbf{L}_2)\mathbf{FQ}(x,t)=\frac{\delta(t)+\alpha(t)x}{\sqrt{m(x)}}, \tag{16.7}$$

$$\int_{-1}^{1}\frac{\mathbf{Q}(\xi,t)}{\sqrt{m(\xi)}}d\xi=\mathbf{P}(t),\quad \int_{-1}^{1}\frac{\mathbf{Q}(\xi,t)}{\sqrt{m(\xi)}}\xi\,d\xi=\mathbf{M}(t). \tag{16.8}$$

Thus, we will find solution of operator equation (16.7) with additional condition (16.8) in the class of vector–functions continuous in time t in Hilbert space $L_2([-1,1],V)$. The special functional basis $\{\mathbf{p}_k^i(x)\}_{i=\overline{1,n},k=0,1,2,\ldots}$ of this space should contain factor $1/\sqrt{m(x)}$. A system of functions satisfying the above conditions can be constructed by the formulas (Szegö, 1959):

$$\mathbf{p}_k^i(x)=\frac{\mathbf{p}_k^{i*}(x)}{\sqrt{m(x)}},\quad \mathbf{p}_k^{i*}(x)=p_k^*(x)\mathbf{i}^i,\quad d_{-1}=1,\quad J_k=\int_{-1}^{1}\frac{\xi^k\,d\xi}{m(\xi)},$$
$$d_k=\begin{vmatrix} J_0 & \cdots & J_k\\ \vdots & \ddots & \vdots\\ J_k & \cdots & J_{2k}\end{vmatrix},\quad p_k^*(x)=\frac{1}{\sqrt{d_{k-1}d_k}}\begin{vmatrix} J_0 & J_1 & \cdots & J_k\\ \vdots & & \ddots & \vdots\\ J_{k-1} & J_k & \cdots & J_{2k-1}\\ 1 & x & \cdots & x^k\end{vmatrix}, \tag{16.9}$$
$$i=\overline{1,n},\quad k=0,1,2,\ldots,\quad x\in[-1,1].$$

We can use this basis in the part of versions of statement of the problem (in particular, we can use it in the case when we know all forces and moments). But in some cases we should use another basis. It can be constructed by the formulas:

$$\hat{\mathbf{p}}_k^i(x)=\frac{\hat{\mathbf{p}}_k^{i*}(x)}{\sqrt{m(x)}},\quad \hat{\mathbf{p}}_k^{i*}(x)=\hat{p}_k^*(x)\mathbf{i}^i,\quad \hat{p}_m^*(x)=p_m^*(x),$$
$$\hat{p}_0^*(x)=\frac{x}{\sqrt{J_2}}=\frac{J_1}{\sqrt{J_0J_2}}p_0^*(x)+\sqrt{\frac{J_0J_2-J_1^2}{J_0J_2}}p_1^*(x),$$
$$\hat{p}_1^*(x)=\frac{J_2-J_1x}{\sqrt{J_2(J_0J_2-J_1^2)}}=\sqrt{\frac{J_0J_2-J_1^2}{J_0J_2}}p_0^*(x)-\frac{J_1}{\sqrt{J_0J_2}}p_1^*(x), \tag{16.10}$$
$$i=\overline{1,n},\quad k=0,1,2,\ldots,\quad m=2,3,4,\ldots,\quad x\in[-1,1],$$

where $p_k^*(x)$, J_0, J_1, J_2 determined by (16.9). We should use this basis in the case when we know all tilt angles and applied forces. And if we want to solve mixed problem (i.e., for example, on one part of punches we know forces and moments and on another part of punches we know tilt angles and applied forces) we should construct the basis by the formulas ($x \in [-1,1]$, $k = 0,1,2,\ldots$, $i = \overline{1,n}$, $i_1 = \overline{1,n_1}$, $i_2 = \overline{n_1+1,n}$)

$$\tilde{\mathbf{p}}_k^i(x) = \frac{\tilde{\mathbf{p}}_k^{i*}(x)}{\sqrt{m(x)}}, \quad \tilde{\mathbf{p}}_k^{i_1*}(x) = \hat{p}_k^*(x)\mathbf{i}^{i_1}, \quad \tilde{\mathbf{p}}_k^{i_2*}(x) = p_k^*(x)\mathbf{i}^{i_2}, \tag{16.11}$$

where $p_k^*(x)$ and $\hat{p}_k^*(x)$ determining by (16.9) and (16.10), respectively.

16.4 Solving the Problem

Consider the case when the system of punches consist of 4 groups. On first group forces moments and punch settlements are given ($i_1 = \overline{1,n_1}$), on second group forces and moments are given ($i_2 = \overline{n_1+1,n_2}$), on third group punch settlements and tilt angles are given ($i_3 = \overline{n_2+1,n_3}$), and on fourth group forces and tilt angles are given ($i_4 = \overline{n_3+1,n}$). It is necessary to find contact pressures $q^i(x,t)$, forces $P^{i_1}(t)$, $P^{i_3}(t)$, moments $M^{i_3}(t)$, $M^{i_4}(t)$, settlements $\delta^{i_2}(t)$, $\delta^{i_3}(t)$, and tilt angles $\alpha^{i_1}(t)$, $\alpha^{i_2}(t)$ ($i = \overline{1,n}$, $i_1 = \overline{1,n_1}$, $i_2 = \overline{n_1+1,n_2}$, $i_3 = \overline{n_2+1,n_3}$, $i_4 = \overline{n_3+1,n}$). Hereinafter, it will be the summation over repeated upper indices i and j from 1 to n, over index i_1 from 1 to n_1, over index i_2 form n_1+1 to n_2, over index i_3 from n_2+1 to n_3, over index i_4 form n_3+1 to n if the left side of the formula is independent of the index.

For this problem we will use the basis (16.11) for the Hilbert space $L_2([-1,1],V)$. The Hilbert space $L_2([-1,1],V)$ can be represented as the direct sum of two orthogonal subspaces $L_2([-1,1],V) = L_2^{(0)}([-1,1],V) \oplus L_2^{(1)}([-1,1],V)$. Functions

$$\{\tilde{\mathbf{p}}_0^{i_1}(x), \tilde{\mathbf{p}}_0^{i_2}(x), \tilde{\mathbf{p}}_0^{i_4}(x), \tilde{\mathbf{p}}_1^{i_2}(x)\}_{i_1=\overline{1,n_1};i_2=\overline{n_1+1,n_2};i_4=\overline{n_3+1,n}}$$

are the basis in $L_2^{(0)}([-1,1],V)$ and functions

$$\{\tilde{\mathbf{p}}_0^{i_3}(x), \tilde{\mathbf{p}}_1^{i_1}(x), \tilde{\mathbf{p}}_1^{i_3}(x), \tilde{\mathbf{p}}_1^{i_4}(x), \tilde{\mathbf{p}}_k^{i}(x)\}_{\substack{i_1=\overline{1,n_1};i_3=\overline{n_2+1,n_3};\\ i_3=\overline{n_3+1,n};i=\overline{1,n};k=2,3,4,\ldots}}$$

are the basis in $L_2^{(1)}([-1,1],V)$. For the integrand and the right-hand side of (16.7)

$$\mathbf{Q}(x,t) = \mathbf{Q}_0(x,t) + \mathbf{Q}_1(x,t), \qquad \frac{\delta(t)+\alpha(t)x}{\sqrt{m(x)}} = \Delta_0(x,t) + \Delta_1(x,t), \tag{16.12}$$

where $\mathbf{Q}_0(x,t)$, $\Delta_0(x,t)$ and $\mathbf{Q}_1(x,t)$, $\Delta_1(x,t)$ are functions continuous in time t and ranging in $L_2^{(0)}([-1,1],V)$ and $L_2^{(1)}([-1,1],V)$, respectively:

$$\mathbf{Q}_0(x,t) = z_0^{i_1}(t)\tilde{\mathbf{p}}_0^{i_1}(x) + z_0^{i_2}(t)\tilde{\mathbf{p}}_0^{i_2}(x) + z_0^{i_4}(t)\tilde{\mathbf{p}}_0^{i_4}(x) + z_1^{i_2}(t)\tilde{\mathbf{p}}_1^{i_2}(x),$$
$$\Delta_0(x,t) = \left[\sqrt{J_2}\alpha^{i_1}(t) + \frac{J_1}{\sqrt{J_2}}\delta^{i_1}(t)\right]\tilde{\mathbf{p}}_0^{i_1}(x) + \left[\sqrt{J_0}\delta^{i_2}(t) + \frac{J_1}{\sqrt{J_0}}\alpha^{i_2}(t)\right]\tilde{\mathbf{p}}_0^{i_2}(x)$$
$$+ \left[\sqrt{J_0}\delta^{i_4}(t) + \frac{J_1}{\sqrt{J_0}}\alpha^{i_4}(t)\right]\tilde{\mathbf{p}}_0^{i_4}(x) + \left[\sqrt{\frac{J_0 J_2 - J_1^2}{J_0}}\alpha^{i_2}(t)\right]\tilde{\mathbf{p}}_1^{i_2}(x). \tag{16.13}$$

The representation for $\mathbf{Q}(x,t)$ contains the known first term $\mathbf{Q}_0(x,t)$, which is determined by the additional conditions (16.8)

$$z_0^{i_1}(t) = \frac{M^{i_1}(t)}{\sqrt{J_2}}, \quad z_0^{i_2}(t) = \frac{P^{i_2}(t)}{\sqrt{J_0}}, \quad z_0^{i_4}(t) = \frac{P^{i_4}(t)}{\sqrt{J_0}},$$
$$z_1^{i_2}(t) = \frac{J_0 M^{i_2}(t) - J_1 P^{i_2}(t)}{\sqrt{J_0(J_0 J_2 - J_1^2)}}, \quad i_1 = \overline{1,n_1}, \quad i_2 = \overline{n_1+1,n_2}, \quad i_4 = \overline{n_3+1,n}, \tag{16.14}$$

and the term $\mathbf{Q}_1(x,t)$ is to be found. Conversely, for the right-hand side, one should find $\Delta_0(x,t)$, while $\Delta_1(x,t)$ is known. These peculiarities permit one to class the resulting problem as a projection problem. The method for solution of generalized projection problems described in Manzhirov (2016).

Following Manzhirov (2016), we can introduce the orthogonal projection operator, mapping the space $L_2([-1,1],V)$ into $L_2^{(0)}([-1,1],V)$:

$$\mathbf{P}_0\mathbf{f}(x) = \int_{-1}^{1} \mathbf{f}(\xi)\cdot\tilde{\mathbf{p}}_0^{i_1}(\xi)\,d\xi\,\tilde{\mathbf{p}}_0^{i_1}(x) + \int_{-1}^{1} \mathbf{f}(\xi)\cdot\tilde{\mathbf{p}}_0^{i_2}(\xi)\,d\xi\,\tilde{\mathbf{p}}_0^{i_2}(x)$$
$$+ \int_{-1}^{1} \mathbf{f}(\xi)\cdot\tilde{\mathbf{p}}_0^{i_4}(\xi)\,d\xi\,\tilde{\mathbf{p}}_0^{i_4}(x) + \int_{-1}^{1} \mathbf{f}(\xi)\cdot\tilde{\mathbf{p}}_1^{i_2}(\xi)\,d\xi\,\tilde{\mathbf{p}}_1^{i_2}(x).$$

The orthoprojector $\mathbf{P}_1 = \mathbf{I} - \mathbf{P}_0$ maps the space $L_2([-1,1],V)$ into $L_2^{(1)}([-1,1],V)$.

We apply the orthogonal projection operator $\mathbf{P}_1$ to equation (16.7). As a result, we obtain the equation for determining $\mathbf{Q}_1(x,t)$ with a known right-hand side

$$c(t)(\mathbf{I}-\mathbf{L}_1)\mathbf{Q}_1(x,t) + (\mathbf{I}-\mathbf{L}_2)\mathbf{P}_1\mathbf{F}\mathbf{Q}_1(x,t) = -(\mathbf{I}-\mathbf{L}_2)\mathbf{P}_1\mathbf{F}\mathbf{Q}_0(x,t) + \Delta_1(x,t). \tag{16.15}$$

It is necessary to construct its solution in the form of a series in the eigenfunctions of the operator $\mathbf{P}_1\mathbf{F}$, which, as one can show, is a compact strongly positive self-adjoint operator from $L_2^{(1)}([-1,1],V)$ into $L_2^{(1)}([-1,1],V)$. The system of eigenfunctions of such an operator is a basis in the space $L_2^{(1)}([-1,1],V)$. The spectral problem for the operator $\mathbf{P}_1\mathbf{F}$ can be written in the form

$$\mathbf{P}_1\mathbf{F}\varphi_k(x) = \gamma_k\varphi_k(x), \tag{16.16}$$

where γ_k are eigenvalues of the operator $\mathbf{P}_1\mathbf{F}$ and $\varphi_k(x)$ are its eigenfunctions ($k = 2,3,4,\ldots$) which can be represented as

$$\varphi_k(x) = \psi_{k0}^{i_3}\tilde{\mathbf{p}}_0^{i_3}(x) + \psi_{k1}^{i_1}\tilde{\mathbf{p}}_1^{i_1}(x) + \psi_{k1}^{i_3}\tilde{\mathbf{p}}_1^{i_3}(x) + \psi_{k1}^{i_4}\tilde{\mathbf{p}}_1^{i_4}(x) + \sum_{m=2}^{\infty}\psi_{km}^{i}\tilde{\mathbf{p}}_m^{i}(x). \quad (16.17)$$

Spectral problem (16.16) lead us to solve the system of linear algebraic equations about coefficients ψ_{km}^i:

$$\begin{aligned}
&\tilde{K}_{00}^{i_3 j_3}\psi_{k0}^{j_3} + \tilde{K}_{01}^{i_3 j_1}\psi_{k1}^{j_1} + \tilde{K}_{01}^{i_3 j_3}\psi_{k1}^{j_3} + \tilde{K}_{01}^{i_3 j_4}\psi_{k1}^{j_4} + \sum_{m=2}^{\infty}\tilde{K}_{0m}^{i_3 j}\psi_{km}^{j} = \gamma_k\psi_{k0}^{i_3},\\
&\tilde{K}_{10}^{i_1 j_3}\psi_{k0}^{j_3} + \tilde{K}_{11}^{i_1 j_1}\psi_{k1}^{j_1} + \tilde{K}_{11}^{i_1 j_3}\psi_{k1}^{j_3} + \tilde{K}_{11}^{i_1 j_4}\psi_{k1}^{j_4} + \sum_{m=2}^{\infty}\tilde{K}_{0m}^{i_1 j}\psi_{km}^{j} = \gamma_k\psi_{k1}^{i_1},\\
&\tilde{K}_{10}^{i_3 j_3}\psi_{k0}^{j_3} + \tilde{K}_{11}^{i_3 j_1}\psi_{k1}^{j_1} + \tilde{K}_{11}^{i_3 j_3}\psi_{k1}^{j_3} + \tilde{K}_{11}^{i_3 j_4}\psi_{k1}^{j_4} + \sum_{m=2}^{\infty}\tilde{K}_{0m}^{i_3 j}\psi_{km}^{j} = \gamma_k\psi_{k1}^{i_3},\\
&\tilde{K}_{10}^{i_4 j_3}\psi_{k0}^{j_3} + \tilde{K}_{11}^{i_4 j_1}\psi_{k1}^{j_1} + \tilde{K}_{11}^{i_4 j_3}\psi_{k1}^{j_3} + \tilde{K}_{11}^{i_4 j_4}\psi_{k1}^{j_4} + \sum_{m=2}^{\infty}\tilde{K}_{0m}^{i_4 j}\psi_{km}^{j} = \gamma_k\psi_{k1}^{i_4},\\
&i_1 = \overline{1,n_1},\quad i_3 = \overline{n_2+1,n_3},\quad i_4 = \overline{n_3+1,n},\quad m_2 = 2,3,4,\ldots,
\end{aligned} \quad (16.18)$$

where coefficients $\tilde{K}_{ml}^{ij}$ ($m,l = 0,1,2,\ldots,\ i,j = \overline{1,n}$) can be calculating by the formulas

$$\tilde{K}_{ml}^{ij} = \int_{-1}^{1}\int_{-1}^{1}\tilde{\mathbf{p}}_m^i(x)\cdot\mathbf{K}(x,\xi)\tilde{\mathbf{p}}_l^j(\xi)\,dx\,d\xi.$$

We expand integrand $\mathbf{Q}_1(x,t)$ and righthand side of the equation (16.15) with respect to the new basis functions $\varphi_k(x)$ ($k = 2,3,4,\ldots$) in $L_2^{(1)}([-1,1],V)$, i.e.

$$\begin{aligned}
&\mathbf{Q}_1(x,t) = \sum_{k=2}^{\infty} z_k(t)\varphi_k(x),\\
&-(\mathbf{I}-\mathbf{L}_2)\mathbf{P}_1\mathbf{F}\mathbf{Q}_0(x,t) + \Delta_1(x,t)\\
&\quad = \sum_{k=2}^{\infty}\{-(\mathbf{I}-\mathbf{L}_2)[K_{0k}^{i_1}z_0^{i_1}(t) + K_{0k}^{i_2}z_0^{i_2}(t) + K_{0k}^{i_4}z_0^{i_4}(t) + K_{1k}^{i_2}z_1^{i_2}(t)]\\
&\qquad + \hat{\delta}_{1k}^{i_1}\delta^{i_1}(t) + \delta_k^{i_3}\delta^{i_3}(t) + \alpha_k^{i_3}\alpha^{i_3}(t) + \alpha_{1k}^{i_4}\alpha^{i_4}(t)\}\varphi_k(x),
\end{aligned} \quad (16.19)$$

where coefficients $K_{0k}^{i_1}$, $K_{0k}^{i_2}$, $K_{0k}^{i_4}$, $K_{1k}^{i_2}$, $\hat{\delta}_{1k}^{i_1}$, $\delta_k^{i_3}$, $\alpha_k^{i_3}$, $\alpha_{1k}^{i_4}$ are

$$K_{0k}^{i_1} = \tilde{K}_{00}^{i_1 j_3}\psi_{k0}^{j_3} + \tilde{K}_{01}^{i_1 j_1}\psi_{k1}^{j_1} + \tilde{K}_{01}^{i_1 j_3}\psi_{k1}^{j_3} + \tilde{K}_{01}^{i_1 j_4}\psi_{k1}^{j_4} + \sum_{m=2}^{\infty}\tilde{K}_{0m}^{i_1 j}\psi_{km}^{j},$$

$$K_{0k}^{i_2} = \tilde{K}_{00}^{i_2 j_3}\psi_{k0}^{j_3} + \tilde{K}_{01}^{i_2 j_1}\psi_{k1}^{j_1} + \tilde{K}_{01}^{i_2 j_3}\psi_{k1}^{j_3} + \tilde{K}_{01}^{i_2 j_4}\psi_{k1}^{j_4} + \sum_{m=2}^{\infty}\tilde{K}_{0m}^{i_2 j}\psi_{km}^{j},$$

$$K_{0k}^{i_4} = \tilde{K}_{00}^{i_4 j_3}\psi_{k0}^{j_3} + \tilde{K}_{01}^{i_4 j_1}\psi_{k1}^{j_1} + \tilde{K}_{01}^{i_4 j_3}\psi_{k1}^{j_3} + \tilde{K}_{01}^{i_4 j_4}\psi_{k1}^{j_4} + \sum_{m=2}^{\infty}\tilde{K}_{0m}^{i_4 j}\psi_{km}^{j},$$

$$K_{1k}^{i_2} = \tilde{K}_{10}^{i_2 j_3}\psi_{k0}^{j_3} + \tilde{K}_{11}^{i_2 j_1}\psi_{k1}^{j_1} + \tilde{K}_{11}^{i_2 j_3}\psi_{k1}^{j_3} + \tilde{K}_{11}^{i_2 j_4}\psi_{k1}^{j_4} + \sum_{m=2}^{\infty}\tilde{K}_{0m}^{i_2 j}\psi_{km}^{j},$$

$$\hat{\delta}_{1k}^{i_1} = \sqrt{\frac{J_0 J_2 - J_1^2}{J_2}}\,\psi_{k1}^{i_1}, \quad \delta_k^{i_3} = \sqrt{J_0}\,\psi_{k0}^{i_3},$$

$$\alpha_k^{i_3} = \frac{J_1}{\sqrt{J_0}}\psi_{k0}^{i_3} + \sqrt{\frac{J_0 J_2 - J_1^2}{J_0}}\,\psi_{k1}^{i_3}, \quad \alpha_{1k}^{i_4} = \sqrt{\frac{J_0 J_2 - J_1^2}{J_0}}\,\psi_{k1}^{i_4},$$

$$i_1 = \overline{1,n_1}, \quad i_2 = \overline{n_1+1,n_2}, \quad i_3 = \overline{n_2+1,n_3}, \quad i_4 = \overline{n_3+1,n}, \quad k = 2,3,4,\ldots \tag{16.20}$$

Substitute representations (16.19) into (16.15), use formula (16.16), and see that the unknown expansion functions $z_k(t)$ $(k = 2,3,4,\ldots)$ can be determined by the formula:

$$z_k(t) = (\mathbf{I} + \mathbf{W}_k)\big(\{\hat{\delta}_{1k}^{i_1}\delta^{i_1}(t) + \delta_k^{i_3}\delta^{i_3}(t) + \alpha_k^{i_3}\alpha^{i_3}(t) + \alpha_{1k}^{i_4}\alpha^{i_4}(t) -$$
$$- (\mathbf{I} - \mathbf{L}_2)[K_{0k}^{i_1} z_0^{i_1}(t) + K_{0k}^{i_2} z_0^{i_2}(t) + K_{0k}^{i_4} z_0^{i_4}(t) + K_{1k}^{i_2} z_1^{i_2}(t)]\}/[c(t) + \gamma_k]\big),$$
$$\mathbf{W}_k f(t) = \int_1^t R_k(t,\tau) f(\tau)\,d\tau, \tag{16.21}$$

where $R_k(t,\tau)$ is the resolvent of the kernel $[c(t)K_1(t,\tau) + \gamma_k K_2(t,\tau)]/[c(t) + \gamma_k]$ $(k = 2,3,4,\ldots)$.

Final formulas for contact pressure $\mathbf{q}(x,t)$ under the punch take the form (it follows from the formulas (16.6), (16.7), (16.11), (16.13), (16.17), and (16.19)):

$$\mathbf{q}(x,t) = \frac{1}{m(x)}\Big\{z_0^{i_1}(t)\tilde{\mathbf{p}}_0^{i_1 *}(x) + z_0^{i_2}(t)\tilde{\mathbf{p}}_0^{i_2 *}(x) + z_0^{i_4}(t)\tilde{\mathbf{p}}_0^{i_4 *}(x) + z_1^{i_2}(t)\tilde{\mathbf{p}}_1^{i_2 *}(x)$$
$$+ \sum_{k=2}^{\infty} z_k(t)\Big[\psi_{k0}^{i_3}\tilde{\mathbf{p}}_0^{i_3 *}(x) + \psi_{k1}^{i_1}\tilde{\mathbf{p}}_1^{i_1 *}(x) + \psi_{k1}^{i_3}\tilde{\mathbf{p}}_1^{i_3 *}(x) + \psi_{k1}^{i_4}\tilde{\mathbf{p}}_1^{i_4 *}(x) + \sum_{m=2}^{\infty}\psi_{km}^{i}\tilde{\mathbf{p}}_m^{i*}(x)\Big]\Big\}, \tag{16.22}$$

where functions $z_0^{i_1}(t), z_0^{i_2}(t), z_0^{i_4}(t), z_1^{i_2}(t), z_k(t)$ determined from (16.14) and (16.21), functions $\tilde{\mathbf{p}}_k^{i*}(x)$ calculated by formulas (16.11), and coefficients $\psi_{k0}^{i_3}$, $\psi_{k1}^{i_1}$, $\psi_{k1}^{i_3}$, $\psi_{k1}^{i_4}$, and ψ_{km}^{i} can be determined from the solution of (16.18).

In the resulting contact stress formulas (16.22), we have managed to find closed-form expressions for the rapidly changing functions describing the coating form, i.e., find the fine structure of the solution. This permits obtaining effective solutions of problems of contact interaction for coated foundations in the case, where coating thickness and the punch base shape is described by rapidly changing functions.

Then we introduce orthoprojector from $L_2([-1,1],V)$ to $L_2^{(0)}([-1,1],V)$ and obtain the system of algebraic equations for determining settlements $\delta(t)$ and tilt angles $\alpha(t)$.

Determining the contact pressure under the punches, we can find unknown forces $P^{i_1}(t)$, $P^{i_3}(t)$, moments $M^{i_3}(t)$, $M^{i_4}(t)$ ($i_1 = \overline{1,n_1}$, $i_3 = \overline{n_2+1,n_3}$, $i_4 = \overline{n_3+1,n}$). To this end we should substitute $\mathbf{q}(x,t)$ from (16.22) into conditions (16.22):

$$
\begin{aligned}
&P^{i_1}(t) = z_0^{i_1}(t)\frac{J_1}{\sqrt{J_2}} + \sum_{k=1}^{\infty} z_k(t)\hat{\delta}_{1k}^{i_1}, \quad P^{i_3}(t) = \sum_{k=1}^{\infty} z_k(t)\delta_k^{i_3}, \\
&M^{i_3}(t) = \sum_{k=1}^{\infty} z_k(t)\alpha_k^{i_3}, \quad M^{i_4}(t) = \sum_{k=1}^{\infty} z_k(t)\alpha_{1k}^{i_4},
\end{aligned}
\tag{16.23}
$$

where coefficients $\hat{\delta}_{1k}^{i_1}$, $\delta_k^{i_3}$, $\alpha_k^{i_3}$, $\alpha_{1k}^{i_4}$ calculated by (16.20), functions $z_0^{i_1}(t)$ and $z_k(t)$ determined from (16.14) and (16.21), and integrals J_1 and J_2 can be found from (16.9).

We also can find settlements $\delta^{i_2}(t)$, $\delta^{i_3}(t)$ and tilt angles $\alpha^{i_1}(t)$, $\alpha^{i_2}(t)$ ($i_1 = \overline{1,n_1}$, $i_2 = \overline{n_1+1,n_2}$, $i_3 = \overline{n_2+1,n_3}$). To this end, we must apply the operator $\mathbf{P}_0$ to Eq. (16.7):

$$
\begin{aligned}
&c(t)(\mathbf{I}-\mathbf{L}_1)\mathbf{Q}_0(x,t) + (\mathbf{I}-\mathbf{L}_2)\mathbf{P}_0\mathbf{F}\mathbf{Q}(x,t) \\
&\quad = \left[\sqrt{J_2}\alpha^{i_1}(t) + \frac{J_1}{\sqrt{J_2}}\delta^{i_1}(t)\right]\tilde{\mathbf{p}}_0^{i_1}(x) + \left[\sqrt{J_0}\delta^{i_2}(t) + \frac{J_1}{\sqrt{J_0}}\alpha^{i_2}(t)\right]\tilde{\mathbf{p}}_0^{i_2}(x) \\
&\quad + \left[\sqrt{J_0}\delta^{i_4}(t) + \frac{J_1}{\sqrt{J_0}}\alpha^{i_4}(t)\right]\tilde{\mathbf{p}}_0^{i_4}(x) + \left[\sqrt{\frac{J_0J_2 - J_1^2}{J_0}}\alpha^{i_2}(t)\right]\tilde{\mathbf{p}}_1^{i_2}(x).
\end{aligned}
\tag{16.24}
$$

So, having defined $\mathbf{Q}(x,t)$, we can find the change rule of unknown settlements and tilt angles

$$
\begin{aligned}
\alpha^{i_1}(t) &= \frac{1}{\sqrt{J_2}}\Bigg\{c(t)(\mathbf{I}-\mathbf{L}_1)z_0^{i_1}(t) + (\mathbf{I}-\mathbf{L}_2)\Bigg[\tilde{K}_{00}^{i_1j_1}z_0^{j_1}(t) + \tilde{K}_{00}^{i_1j_2}z_0^{j_2}(t) \\
&\quad + \tilde{K}_{00}^{i_1j_4}z_0^{j_4}(t) + \tilde{K}_{01}^{i_1j_2}z_1^{j_2}(t) + \sum_{k=2}^{\infty} K_{0k}^{i_1}z_k(t)\Bigg]\Bigg\} - \frac{J_1}{J_2}\delta^{i_1}(t), \\
\delta^{i_2}(t) &= \frac{1}{\sqrt{J_0}}\Bigg\{c(t)(\mathbf{I}-\mathbf{L}_1)z_0^{i_2}(t) + (\mathbf{I}-\mathbf{L}_2)\Bigg[\tilde{K}_{00}^{i_2j_1}z_0^{j_1}(t) + \tilde{K}_{00}^{i_2j_2}z_0^{j_2}(t) \\
&\quad + \tilde{K}_{00}^{i_2j_4}z_0^{j_4}(t) + \tilde{K}_{01}^{i_2j_2}z_1^{j_2}(t) + \sum_{k=2}^{\infty} K_{0k}^{i_2}z_k(t)\Bigg]\Bigg\} - \frac{J_1}{J_0}\alpha^{i_2}(t),
\end{aligned}
$$

$$\alpha^{i_2}(t) = \sqrt{\frac{J_0}{J_0 J_2 - J_1^2}} \left\{ c(t)(\mathbf{I} - \mathbf{L}_1) z_1^{i_2}(t) + (\mathbf{I} - \mathbf{L}_2) \left[\tilde{K}_{10}^{i_2 j_1} z_0^{j_1}(t) + \right.\right.$$

$$\left.\left. + \tilde{K}_{10}^{i_2 j_2} z_0^{j_2}(t) + \tilde{K}_{10}^{i_2 j_4} z_0^{j_4}(t) + \tilde{K}_{11}^{i_2 j_2} z_1^{j_2}(t) + \sum_{k=2}^{\infty} K_{1k}^{i_2} z_k(t) \right] \right\},$$

$$\delta^{i_4}(t) = \frac{1}{\sqrt{J_0}} \left\{ c(t)(\mathbf{I} - \mathbf{L}_1) z_0^{i_4}(t) + (\mathbf{I} - \mathbf{L}_2) \left[\tilde{K}_{00}^{i_4 j_1} z_0^{j_1}(t) + \tilde{K}_{00}^{i_4 j_2} z_0^{j_2}(t) \right.\right.$$

$$\left.\left. + \tilde{K}_{00}^{i_4 j_4} z_0^{j_4}(t) + \tilde{K}_{01}^{i_4 j_2} z_1^{j_2}(t) + \sum_{k=2}^{\infty} K_{0k}^{i_4} z_k(t) \right] \right\} - \frac{J_1}{J_0} \alpha^{i_4}(t).$$

16.5 Main Results and Conclusions

In the present paper, we introduce the notion of conformal contact interaction or a conformal contact of bodies, which is a generalization of the interaction between bodies and plane surface. We pose and solve plane problems of conformal contact between viscoelastic aging basements with coatings and system of rigid punches. The solution of the problem is obtained analytically, and, in the expressions for the contact stresses, the shape function of the foundation is distinguished explicitly, which allows one to perform computations for actual shapes of the coating surface, which are described by rapidly changing functions. It is allow us to solve complete problems.

Acknowledgements Authors thanks A.V. Manzhirov for statement of the problem, useful discussions, and valuable advice. The reported study was funded by Russian Foundation for Basic Research and Moscow city Government (Project No. 15-31-70002 mol_a_mos).

References

Aleksandrov VM, Manzhirov AV (1987) Two-dimensional integral equations in applied mechanics of deformable solids. Journal of Applied Mechanics and Technical Physics 28(5):781–786

Alexandrov VM, Chebakov MI (2007) Introduction into Mechanics of Contact Interaction (in Russ.). Izd-vo “TsVVR”, Rostov-on-Don

Alexandrov VM, Mkhitaryan SM (1983) Contact Problems for Bodies with Thin Coatings and Interlayers (in Russ.). Nauka, Moscow

Arutyunyan NK, Manzhirov AV (1999) Contact Problems in the Theory of Creep (in Russ.). Izd-vo NAN RA, Erevan

Kazakov KE, Manzhirov AV (2008) Conformal contact between layered foundations and punches. Mechanics of Solids 43(3):512–524

Manzhirov AV (2016) A mixed integral equation of mechanics and a generalized projection method of its solution. Dokl Phys 61(10):489–493

Manzhirov AV, Kazakov KE (2016) Contact problem for a foundation with a rough coatings. Lecture Notes in Engineering and Computer Science 2224(1):877–882

Szegö G (1959) Orthogonal Polynomials. Amer. Math. Soc, Providence

Chapter 17
Physical Modeling of Rock Deformation and Fracture in the Vicinity of Well for Deep Horizons

Dmitry M. Klimov, Vladimir I. Karev & Yury F. Kovalenko

Abstract The results of experiments on identification the strain characteristics of rocks from the deep horizons (6 km or more) under the real stress conditions arising in the vicinity of wells and perforations at drilling and operation are presented. The experiments were conducted on the unique experimental facility - Triaxial Independent Loading Test System (TILTS) of IPMech RAS. Specimens for tests were cut from the core of the exploratory wells of the Astrakhan gas condensate field. Physical modeling of the pressure reduction process in deep wells with different bottom geometry is performed. The studies have shown a fundamental possibility of increasing the permeability of rocks from deep horizons by stress condition control in the vicinity of a well.

Key words: Deep reservoirs · Rock · Well · Test facility · Deformations · Stresses · Strength · Elastic moduli

17.1 Introduction

A characteristic feature of rocks is a significant dependence of their stress-strain properties on the stress condition type and stresses value. Therefore, to solve the geomechanical problems related to the creation of new technologies for the deep-seated deposits development, the wellbores stability providing at great depths and many others, it is important to know the strain characteristics of rocks under the actual stresses in the vicinity of deep wells. Such a possibility is provided by the Triaxial Independent Loading Test System (TILTS) created in IPMech RAS (Karev and Kovalenko, 2012).

Dmitry M. Klimov · Vladimir I. Karev · Yury F. Kovalenko
Ishlinsky Institute for Problems in Mechanics RAS, Vernadskiy Ave, 101 bldg 1, 119526 Moscow, Russia,
e-mail: wikarev@ipmnet.ru, perfolinkgeo@yandex.ru

H. Altenbach et al. (eds.), *Mechanics for Materials and Technologies*,
Advanced Structured Materials 46, DOI 10.1007/978-3-319-56050-2_17

17.2 Experimental Facility and Loading Programs for Specimens

TILTS is a unique test facility that allows to study the strain, strength and filtration characteristics of rocks by testing cubic specimens of size 40 or 50 mm. The TILTS is equipped with automatic permeability measurement system, which allows studying the characteristics of rocks permeability versus magnitude and type of effective stresses.

The facility allowed to perform the experiments on determination of strain, strength and filtration characteristics for the core materials of four lithotypes (argillite, aleurolite, quartz sandstone, and dolomite) taken from depths of more than 6 km of the exploratory wells Pravoberezhnaya-1 and the Devonskaya-2, drilled at the Astrakhan gas condensate field of Public Joint Stock Company Gazprom. The 7 cubic specimens with the edge 40 mm are made from this core materials.

Prior to the TILTS testing, the velocities of longitudinal elastic waves propagation along three axes of all specimens are measured by using of special equipment. Velocities measured for sandstone and dolomite appear to be approximately the same in all directions. So, for one of the sandstone specimens (depth 6058 m) velocity along the vertical axis of the core, i.e. perpendicular to the bedding plane, is 5 km/s and velocities in the plane are 5.2 and 5.3 km/s. At the same time the velocities for the argillite and aleurolite specimens in the bedding plane are measured twice larger than one along the core axis. The velocity for the argillite specimen No. ARG 2-2 (depth 6556 m) in horizontal plane is 4.4 km/s in the bedding plane, and one along the core axis is 1.85 km/s. So sandstone and dolomite specimens have elastic properties close to isotropic and by contrast the argillite and aleurolite specimens have layered structure and have transversely isotropic properties.

The tests are performed on loading programs, simulating two types of stress-strain state: the generalized shear and the generalized tension. The generalized shear is a superposition of uniform hydrostatic compression and stress state, which occurs when one principal stress grows, other decreases, and third is held constant, while the average pressure also remains constant (see Fig. 17.1). The Lode parameter (Love, 1927) corresponding to this stress state equals to 0 as for pure plane shear. The generalized tension is a superposition of uniform hydrostatic compression and

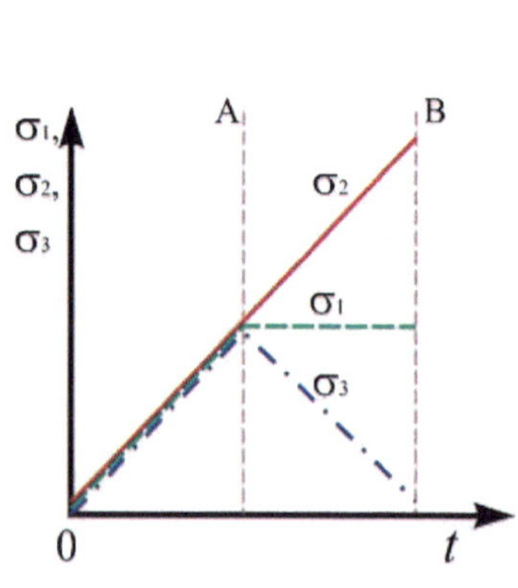

Fig. 17.1 Borehole modelling (*well*)

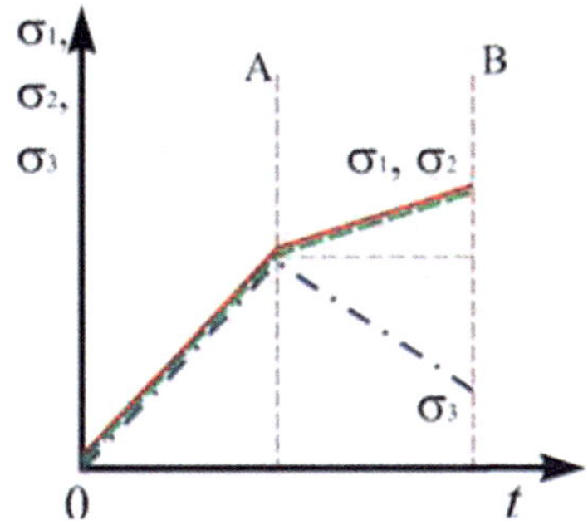

Fig. 17.2 Perforation modelling (*sphere*)

stress state, which occurs when two principal stresses grow, and third one decreases, while the average pressure remains constant (Fig. 17.2). The Lode parameter for this case is equal to +1, as under uniaxial tension.

The values $\sigma_1, \sigma_2, \sigma_3$ (in Fig. 17.1 and 17.2) correspond to the stresses applied on the axes of TILTS. The monotonically increasing stress σ_2 is the loading parameter. The specific choice of these two testing programs is due to the fact that they correspond to the real stress conditions in the reservoir for two basic bottom-hole designs (Klimov et al, 2003): generalized shear (Fig. 17.1) is realized in the vicinity of an open borehole or the perforation when the pressure in the well decreases (i.e. pressure drawdown increases) and generalized tension (Fig. 17.2) occurs in the vicinity of the spherical tip of the perforation when the pressure in the well decreases.

Effective stresses (stresses acting on the soil framework) in the vicinity of a well (Fig. 17.3) as the Lame's problem solution (Love, 1927; Khristianovich and Zheltov, 1955) are

$$\sigma_r = -(q + p_c)(R_c/r)^2 + q, \tag{17.1}$$

$$\sigma_\varphi = (q + p_c)(R_c/r)^2 + q, \tag{17.2}$$

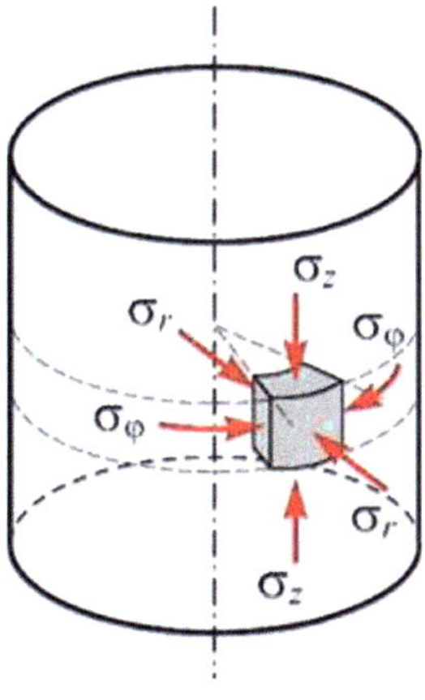

Fig. 17.3 Stresses in the vicinity of well

$$\sigma_z = q, \tag{17.3}$$

where $q = -\gamma h$ is rock pressure at the chosen depth, γ is average rock unit weight, h is formation depth, p_c is bottom-hole pressure ($p_c > 0$), R_c is bottom-hole radius, r is distance from the borehole axis, (compressing stress values are considered negative). When $r = R_c$, it can be found from (17.1) - (17.3)

$$\sigma_r = -p_c, \sigma_\varphi = 2q + p_c, \sigma_z = q \tag{17.4}$$

The expressions for effective stresses in the vicinity of sphere (Fig. 17.4) are respectively:

$$\sigma_r = -(q + p_c)(R_c/r)^2 + q, \tag{17.5}$$

$$\sigma_{\theta,\varphi} = \frac{1}{2}(q + p_c)(R_c/r)^2 + q, \tag{17.6}$$

On the perforation wall (when $r = R_c$)

$$\sigma_r = -p_c, \qquad \sigma_{\theta,\varphi} = \frac{3}{2}q + \frac{1}{2}p_c \tag{17.7}$$

Here $\sigma_r, \sigma_{\theta,\varphi}$ are stress components in radial and two circumferential directions.

During the both loading programs the specimens are first driven to the conditions of uniform overall compression, in which the rocks are supposed to be found at the depth of core capture before the well is drilled out (Fig. 17.1, 17.2, section OA, at the point A $\sigma_i = q$). Then the specimens behavior is studied under the change of stresses in the vicinity of the borehole when the pressure at the well bottom changes (Figs. 17.1, 17.2, section AB).

The correspondence of stresses applied to the faces of specimen to stress components acting in the vicinity of the borehole during the loading program *well* are

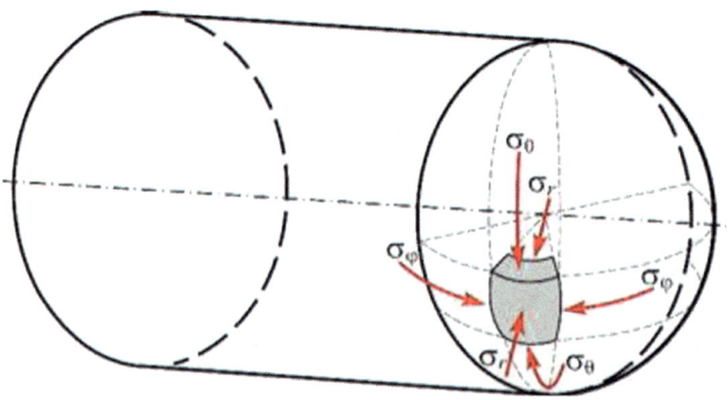

Fig. 17.4: Stresses in the vicinity of spherical hole

$\sigma_1 - \sigma_z, \sigma_2 - \sigma_\varphi, \sigma_r - \sigma_r$. During the loading program *sphere* the correspondence is $\sigma_1 - \sigma_\varphi, \sigma_2 - \sigma_\theta, \sigma_3 - \sigma_r$.

17.3 Rock Specimens Test Results

The test results for two specimens cut from the same piece of core from the Devonskaya-2 well are presented in Figs. 17.5- 17.8. The specimen is a fine-grained sandstone, depth of coring is 6058 m, which corresponds to the magnitude of the rock pressure of about 140 MPa, if $\gamma = 2.3 \cdot 10^3$ kg/m^3. The initial permeability is equal to zero.

The conditions that occur in the vicinity of uncased well, when the pressure on its bottom decreases, are simulated during the test of the specimen No. P 2-1 (see Fig. 17.5, 17.6). The conditions that occur in the vicinity of the perforation tip (hemisphere), when the bottom pressure decreases, are simulated during the test of the specimen No. P 2-2 (see Fig. 17.7, 17.8).

The illustrations show that the specimen No. P 2-1 is deformed almost elastically up to the maximum possible decrease of pressure on the bottom hole, whereas the specimen No. P 2-2 moved to the stage of inelastic deformation at the strain value corresponding to the well pressure of approximately 150 MPa and has been heavily deformed and soon destroyed.

The rock specimen No. P 2-2 after the test is shown in Fig. 17.9, where it can be easily seen that the specimen disintegrated by macroscopic fracture settled in the plane perpendicular to the direction of the lowest compressing stress. At the same

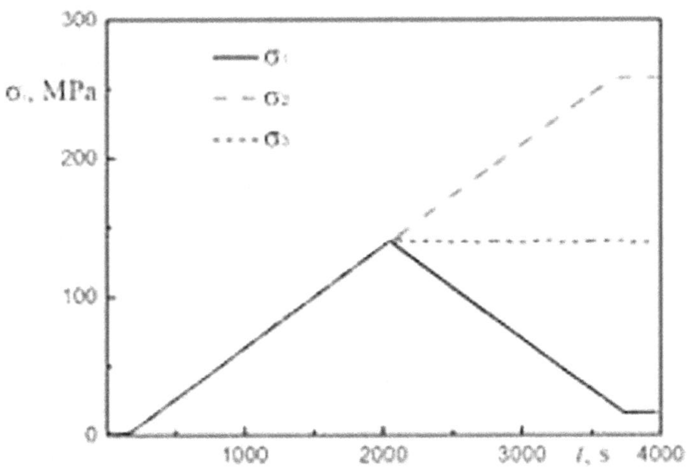

Fig. 17.5: Loading program for the specimen No. P 2-1 (*well* program)

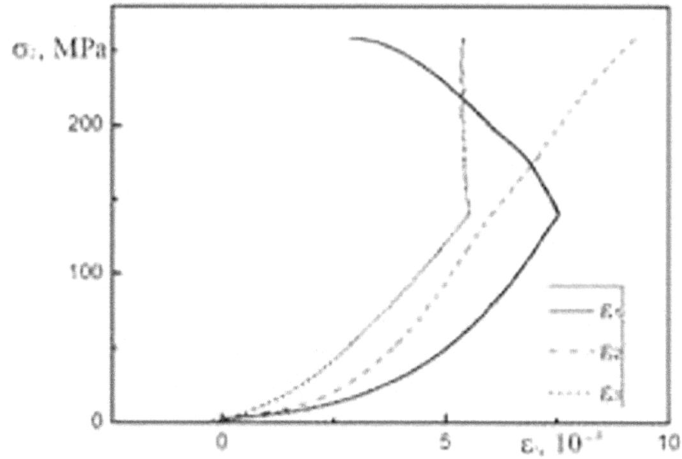

Fig. 17.6: Stress-strain curves for the specimen No. P 2-1 (*well* program)

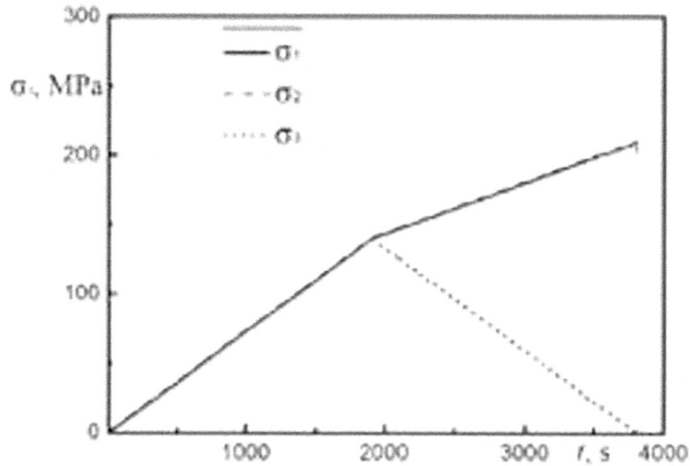

Fig. 17.7: Loading program for the specimen No. P 2-1 (*sphere* program)

time the permeability of the specimen which initially is absent in this direction has abruptly increased on the disintegration stage of loading test (Fig. 17.10).

A similar permeability pattern is observed in the test of argillite specimen No. A 2-2, but the character of fracture is distinct. The specimen is fractured in the planes perpendicular to the direction of the lowest compressing stress, but with more than one macro cracks developed in the bedding planes. Permeability of the specimen along the bedding planes direction is also increased significantly (Fig. 17.11).

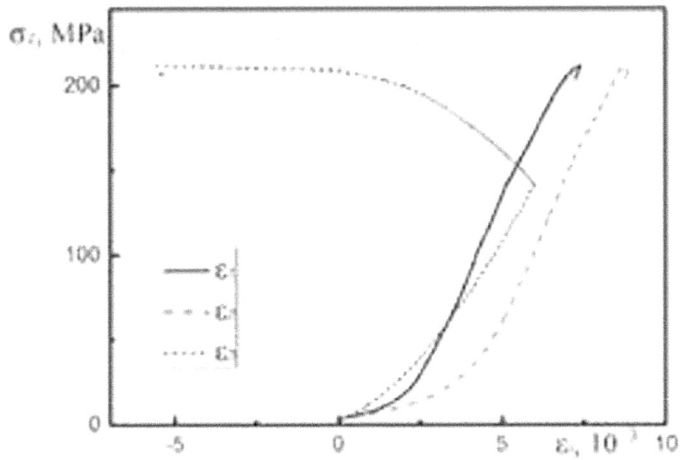

Fig. 17.8: Stress-strain curves for the specimen No. P 2-1 (*sphere* program)

Fig. 17.9 Specimen No. P 2-2 after the test

Elastic moduli are calculated using the stress-strain curves obtained for the tested rocks such as Young's moduli in the bedding planes and Poisson's ratios in the perpendicular direction. The data obtained are given in Table. 17.1.

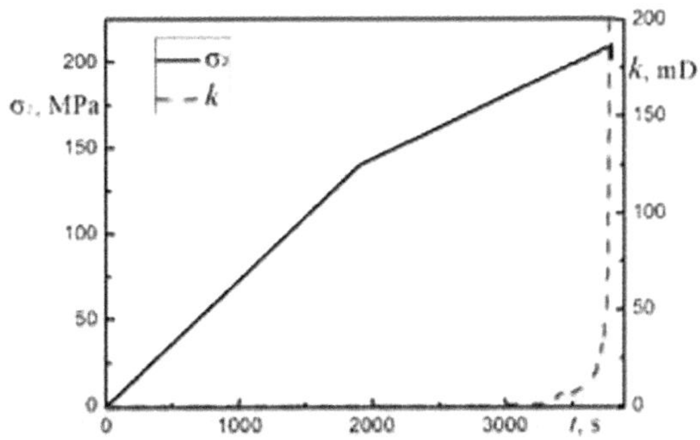

Fig. 17.10 Permeability change during the test for the specimen No. P 2-2

Fig. 17.11 Specimen No. A 2-2 after the test

Table 17.1: Elastic moduli of tested rocks

Specimen No.	Well	Depth m	Lithotype	Young's modulus 10^4 MPa	Poisson's ratio
A 1-1	Pravoberezhnaya-1	6247	argillite	3.1	0.13
AL-1	Pravoberezhnaya-1	6570	aleurolite	2.75	0.17
A 2-1	Devonskaya-2	6556	argillite	1.9	0.12
A 2-2	Devonskaya-2	6556	argillite	2.3	0.14
P 2-1	Devonskaya-2	6058	quartz sandstone	2.8	0.12
P 2-2	Devonskaya-2	6058	quartz sandstone	3.7	0.12
D 2-1	Devonskaya-2	5751	dolomite	4.1	0.12

17.4 Conclusion

The complex loading test of the rocks from depths exceeding 6 km are carried out for the first time. The tests have shown that various rocks of different lithological composition are characterized by high strength and low permeability.

The basic result of the performed research is the fundamental possibility for the rocks lying at great depths and affected by extremely high compressive stresses to significantly increase their permeability which initially is low. The purpose can be achieved in result of fracturing and disintegration of the rock in the reservoir by reducing the well pressure to a certain value, and choice the geometry of the bottom hole.

The obtained results are of great importance for solving problems connected to creating effective technologies for the development of the deeper deposits, as well as to reducing risks during drilling and operation of wells.

Acknowledgements The work is executed at financial support of the RAS Presidium program for fundamental research I.4P *Deposits of strategic resources in Russia: innovative approaches for their prediction, evaluation and production. Oil from the deep horizons of sedimentary basins - the source of replenishment of resource base of hydrocarbons: theoretical and applied aspects.*

References

Karev VI, Kovalenko YF (2012) Triaxial loading system as a tool for solving geotechnical problems of oil and gas production. In: Takahashi M (ed) True Triaxial Testing of Rocks, CRC Press, Leiden, pp 301–310

Khristianovich SA, Zheltov YP (1955) On hydraulic break of a petroliferous layer. Izv USSR Academy of Sciences OTN (5):3–41

Klimov DM, Kovalenko YF, Karev VI (2003) Practical realization of geoloosening method to increase injectivity of injection wells. Tekhnologii TEK (4):59–64

Love AEH (1927) A treatise on the mathematical theory of elasticity. Cambridge University Press, Cambridge

Chapter 18
Full Axially Symmetric Contact of a Rigid Punch with a Rough Elastic Half-Space

Ilya I. Kudish

Abstract An axially symmetric contact problem for a rigid punch indented in an elastic half-space with rough surface is considered. Friction in the contact is neglected. It is shown that there exists a normal load applied to the punch which would make the contact singly connected if the roughness distribution is a four times differentiable function. Some generalizations of the problem are considered. A relationship for the estimate of a normal load sufficient for realization of a full contact is provided.

Key words: Contact of rough elastic solids · Full contact · Roughness · Smoothness · Load sufficient for full contact

18.1 Introduction

Seals find various applications in engineering practice. The goal of seal application is to separate one chamber of a machine from another. One of the main problems with seal application is lubricant/fluid leakage associated with small gaps in seal contacts. These gaps are mainly caused by surface roughness. Therefore, one of the major goals is to minimize this leakage or eliminate it completely. However, it is practically impossible to eliminate surface roughness while it is usually possible to reduce it. The question remains: Is it possible to get a full contact for the given roughness distribution? In Ciavarella et al (2000), it is assumed that roughness is distributed according to the Weierstrass distribution which is not a smooth distribution everywhere and it is shown that a full contact of such rough surfaces is impossible. In Kudish et al (2012, 2014), on an example of a plane problem it is shown that if the surface roughness is not sufficiently smooth it is impossible to get the conditions of

Ilya I. Kudish
Department of Mathematics, Kettering University, Flint, MI, 48504, USA,
e-mail: ikudish@kettering.edu

H. Altenbach et al. (eds.), *Mechanics for Materials and Technologies*,
Advanced Structured Materials 46, DOI 10.1007/978-3-319-56050-2_18

a full contact for any applied load. On the other hand, optical and electronic force microscopy surface pictures obtained for a series of increased resolutions show that it is reasonable to assume that roughness distributions are usually sufficiently smooth (i.e. possess sufficient number of continuous derivatives) Kudish et al (2012, 2014). Also, in Kudish et al (2012, 2014) it is shown that it is sufficient for the distribution of surface roughness to be sufficiently smooth (twice continuously differentiable) for the contact of rough surfaces to be full for sufficiently high applied load.

In this paper we will examine an axially symmetric analog of plane problems studied in Kudish et al (2012, 2014). Specifically, it will be shown that for a four times differentiable surface roughness it is possible to find a normal load which would provide a full axially symmetric contact. A relationship for the estimate of such a load is provided. Some generalizations will be discussed.

18.2 Problem Formulation

Let us consider an axially symmetric rigid punch with a flat bottom of radius R indented in an elastic half-space with a rough surface described by the distribution $z=\lambda\varphi(r)$, where λ is some non-negative dimensionless constant and $r=\sqrt{x^2+y^2}$ is the radial variable in the plane of the contact. The value of λ characterizes the level of surface roughness. The punch is indented by a normal force P. The half-space is made of a homogeneous elastic material with the Young's modulus E and Poisson's ratio ν. If the contact occupies a circular region of radius R then the problem can be reduced to the following equations (Kudish and Covitch, 2010)

$$\lambda\varphi(r)+\frac{8}{\pi E'}\int\limits_0^R \frac{\rho}{\rho+r}K(\frac{2\sqrt{\rho r}}{\rho+r})p(\rho)d\rho=\delta,\ \int\limits_0^R \rho p(\rho)d\rho=\frac{P}{2\pi}, \tag{18.1}$$

where E' is the effective modulus of the half-space, $E'=\frac{E}{1-\nu^2}$, $p(r)$ is the contact pressure distribution, δ is the punch vertical displacement, $K(\cdot)$ is the full elliptic integral of the second kind (Abramowitz and Stegun, 1964). In dimensionless variables

$$r'=\frac{r}{R},\ \rho'=\frac{\rho}{R},\ p'(r')=\frac{p(r)}{p_0},\ \varphi'(r')=\frac{2\pi E'R}{3P}\varphi(r),\ \delta'=\frac{2\pi E'R}{3P}\delta,$$
$$p_0=\frac{3P}{2\pi R^2}, \tag{18.2}$$

problem (18.1) can be rewritten as follows (primes are omitted)

$$\lambda\varphi(r)+\frac{8}{\pi}\int\limits_0^1 \frac{\rho}{\rho+r}K(\frac{2\sqrt{\rho r}}{\rho+r})p(\rho)d\rho=\delta,\ \int\limits_0^1 \rho p(\rho)d\rho=\frac{1}{3}, \tag{18.3}$$

Assuming that we have a full contact of the bottom of the punch with the rough surface of the half-space the exact solution of problem (18.3) can be represented in the classic form (Kudish and Covitch, 2010)

$$p(r)=\frac{1}{2\pi\sqrt{1-r^2}}\{\delta_1-\lambda\int\limits_0^1\frac{\varphi'(\rho)d\rho}{\sqrt{1-\rho^2}}\}+\frac{\lambda}{2\pi}\int\limits_r^1\frac{1}{\sqrt{s^2-r^2}}\{s\int\limits_0^s\frac{\varphi'(\rho)d\rho}{\sqrt{s^2-\rho^2}}\}'ds,$$
$$\delta_1=\delta-\lambda\varphi(0)=\frac{2\pi}{3}+\lambda\int\limits_0^1\sqrt{1-\rho^2}\varphi'(\rho)d\rho. \tag{18.4}$$

Obviously, function $\varphi(r)$ can be chosen with the precision of a constant. Therefore, let us choose this function in such a way that $\varphi(1)=0$. If function $\varphi(r)$ is continuously differentiable then due to the axial symmetry $\varphi'(0)=0$. We will also assume that $\varphi(0)=0$.

To be able to determine the structure of the contact pressure $p(r)$ in a contact let us assume that the roughness distribution function $\varphi(r)$ can be represented by a Fourier-Bessel series (Tolstov, 1960). Let us assume that $0<\mu_1<\mu_2<\ldots<\mu_n<\ldots$ are positive roots of the Bessel function $J_0(x)$. Then according to Tolstov (1960) functions $J_0(\mu_1 x)$, $J_0(\mu_2 x)$, $\ldots$ $J_0(\mu_n x)$, $\ldots$ are orthogonal on the interval $[0,1]$ and any continuous function $\varphi(r)$ can be represented by a convergent on $[0,1]$ Fourier-Bessel series

$$\varphi(r)=\sum\limits_{k=1}^{\infty}c_kJ_0(\mu_k r),\ c_k=\frac{2}{J_1^2(\mu_k)}\int\limits_0^1 r\varphi(r)J_0(\mu_k r)dr, \tag{18.5}$$

where c_k are called the coefficients of the Fourier-Bessel series. Moreover, for sufficiently large k we have (Tolstov, 1960)

$$\mid c_k\mid\leq\frac{C}{k^{1+\varepsilon}},\ \mid c_kJ_0(\mu_k r)\mid\leq\frac{A}{k^{1+\varepsilon}}, \tag{18.6}$$

and the series from (18.5) converges to $\varphi(r)$ absolutely and uniformly (ε is a positive constant). Moreover, the Fourier-Bessel series for $\varphi(r)$ from (18.5) converges uniformly and can be differentiated term by term on the entire interval $[0,1]$ if (Tolstov, 1960)

$$c_k=O\left(\frac{1}{\mu_k^{2+\varepsilon}}\right),\ \mu_k\to\infty, \tag{18.7}$$

where ε is a positive constant. Here $\mu_k=O(k)$ for $k\to\infty$ (Tolstov, 1960). The same condition on coefficients c_k guaranties the convergence of the series involved in the expression for the pressure $p(r)$ (see below). Therefore, assuming that term by term differentiation and integration of a series are legitimate we obtain

$$\int_0^1 \frac{\varphi'(\rho)d\rho}{\sqrt{1-\rho^2}} = \sum_{k=1}^{\infty} c_k \int_0^1 \frac{J_0'(\mu_k\rho)d\rho}{\sqrt{1-\rho^2}}, \ \int_0^s \frac{\varphi'(\rho)d\rho}{\sqrt{s^2-\rho^2}} = \sum_{k=1}^{\infty} c_k \int_0^s \frac{J_0'(\mu_k\rho)d\rho}{\sqrt{s^2-\rho^2}}. \quad (18.8)$$

Taking into account the fact that $J_0'(\mu_k\rho) = -\mu_k J_1(\mu_k\rho)$ (Gradshtein and Ryzhik, 1971) we have

$$\int_0^s \frac{J_0'(\mu_k\rho)d\rho}{\sqrt{s^2-\rho^2}} = -\mu_k \int_0^s \frac{J_1(\mu_k\rho)d\rho}{\sqrt{s^2-\rho^2}} = -\mu_k \int_0^1 \frac{J_1(\mu_k sz)dz}{\sqrt{1-z^2}} = -\mu_k \frac{\pi}{2} J_{1/2}^2(\frac{\mu_k s}{2})$$
$$= -\frac{2}{s}\sin^2(\frac{\mu_k s}{2}), \ \{s\int_0^s \frac{\varphi'(\rho)d\rho}{\sqrt{s^2-\rho^2}}\}' = -\sum_{k=1}^{\infty} c_k\mu_k \sin(\mu_k s). \quad (18.9)$$

Using integration by parts we obtain

$$\int_0^1 \sqrt{1-\rho^2} J_0'(\mu_k\rho)d\rho = \frac{\sin\mu_k}{\mu_k} - 1,$$
$$\int_0^1 \sqrt{1-\rho^2}\varphi'(\rho)d\rho = \sum_{k=1}^{\infty} c_k[\frac{\sin\mu_k}{\mu_k} - 1]. \quad (18.10)$$

Therefore, the expression for $p(r)$ from (18.4) takes the form

$$p(r) = \frac{1}{2\pi\sqrt{1-r^2}}\{\frac{2\pi}{3} - \lambda\sum_{k=1}^{\infty} c_k\psi_k(\mu_k,r)\},$$
$$\psi(\mu,r) = 1 - \frac{\sin\mu}{\mu} + \mu\sqrt{1-r^2}\int_r^1 \frac{\sin(\mu s)ds}{\sqrt{s^2-r^2}}, \quad (18.11)$$

and we can make a formal conclusion that the contact would be full if the contact pressure $p(r)$ from (18.11) is positive for every $0 \le r \le 1$. For that to be true it is sufficient that

$$\lambda\sum_{k=1}^{\infty} | c_k || \psi_k(\mu_k,r) | < \frac{2\pi}{3}. \quad (18.12)$$

It is easy to make the following estimate using the expression for the integral from the expression for $\psi(\mu,r)$ from (18.11) and the fact that

$$| I(\mu,r) | = | \int_r^1 \frac{\sin(\mu s)ds}{\sqrt{s^2-r^2}} | \le \int_r^1 \frac{|\sin(\mu s)|ds}{\sqrt{s^2-r^2}} \le \mu \int_r^1 \frac{sds}{\sqrt{s^2-r^2}} = \mu\sqrt{1-r^2}. \quad (18.13)$$

However, let us find more precise details of the behavior of the integral $I(\mu,r)$ in (18.11) which would guarantee that inequality (18.12) is satisfied but, at the same time, would limit to the minimum the requirements on the roughness function $\varphi(r)$.

It is more convenient to replace $\sin(\mu s)$ by $\exp(i\mu s)$ in $I(\mu,r)$, where i is the imaginary unit. Let us make a substitution $y=\mu(s-r)$ in the integral $\widetilde{I}(\mu,r)=\int\limits_r^1 \frac{\exp(i\mu s)ds}{\sqrt{s^2-r^2}}$ for $\delta_0 \leq r \leq 1-\delta_0$, $0<\delta_0<1$, and $\mu \gg 1$. That leads to the following uniform asymptotic expression (Gradshtein and Ryzhik, 1971)

$$\widetilde{I}(\mu,r)=\int\limits_r^1 \frac{\exp(i\mu s)ds}{\sqrt{s^2-r^2}}=\frac{1}{\sqrt{\mu}}\int\limits_0^{\mu(1-r)} \frac{\exp(i\mu r+iy)dy}{\sqrt{y(2r+\frac{y}{\mu})}}$$
$$=\frac{\exp(i\mu r)}{\sqrt{2\mu r}}\int\limits_0^{\infty}\frac{e^{iy}}{\sqrt{y}}dy+o(\frac{1}{\sqrt{\mu}})=\frac{\exp(i\mu r)}{\sqrt{2\mu r}}\sqrt{\frac{\pi}{2}}(1+i)+o(\frac{1}{\sqrt{\mu}}),\ \mu \gg 1. \tag{18.14}$$

Therefore, for $\delta_0 \leq r \leq 1-\delta_0$, $0<\delta_0<1$, and $\mu \gg 1$ we have a uniform asymptotic estimate for the integral $I(\mu,r)$ from (18.11)

$$I(\mu,r)=\frac{1}{2}\sqrt{\frac{\pi}{\mu r}}[\cos(\mu r)+\sin(\mu r)]+o(\frac{1}{\sqrt{\mu}})\ldots,\ \mu \gg 1, \tag{18.15}$$

and from (18.11) we get the uniform asymptotic representation of the function

$$\psi(\mu,r)=1-\frac{\sin\mu}{\mu}+\frac{1}{2}\sqrt{\pi\mu\frac{1-r^2}{r}}[\cos(\mu r)+\sin(\mu r)]+o(\sqrt{\mu}),$$
$$\delta_0 \leq r \leq 1-\delta_0,\ 0<\delta_0<1,\ \mu \gg 1. \tag{18.16}$$

In a similar manner it can be shown that if $r=1+O(\frac{1}{\mu})$ for $\mu \gg 1$ then $I(\mu,r)=O(1)$ for $\mu \gg 1$ and if $r=O(\frac{1}{\mu})$ then $I(\mu,r)=\frac{\pi}{2}+O(\frac{1}{\mu})$ for $\mu \gg 1$. Thus, $c_k\psi_k(\mu_k,r)=O(\mu_k c_k)$ for $r=O(\frac{1}{\mu_k})$ and $\mu_k \gg 1$ while $c_k\psi_k(\mu_k,r)=O(\sqrt{\mu_k}c_k)$ for $\delta_0 \leq r \leq 1$, $0<\delta_0<1$, and $\mu_k \gg 1$. Therefore, if $\mu_k c_k=O(\frac{1}{\mu_k^{1+\varepsilon}})$ (where ε is a positive constant) then the series for $p(r)$ in (18.11) (and in (18.12)) converge and make all operations made to obtain them permissible. That makes the requirement (18.7) on the behavior of coefficients c_k sufficient for the whole analysis to be correct.

In other words, from (18.12) and (18.15) we conclude that for pressure $p(r)$ to be positive in the whole contact it is sufficient that estimate (18.7) takes place for coefficients c_k. Under these conditions the inequality in (18.12) can be satisfied for sufficiently small value of λ and/or sufficiently high applied load P.

The requirement (18.7) on coefficients c_k is related to the level of smoothness (differentiability) of the roughness distribution function $\varphi(r)$. There is a readily available more restrictive assumption about the smoothness of the function $\varphi(r)$. Specifically, if $\varphi(r)$ is four times differentiable and satisfies the following properties (Tolstov, 1960)

1. $\varphi(0)=\varphi'(0)=\varphi''(0)=\varphi'''(0)=0$,
2. $\varphi^{(4)}(r)$ is bounded (it may not exist at some discrete points),

3. $\varphi(1)=\varphi'(1)=\varphi''(1)=0$,

then

$$| c_k |\leq \frac{C}{\mu_k^{7/2}},\ \mu_k\gg 1, \tag{18.17}$$

where C is a certain positive constant. Therefore, the whole analysis of the problem is valid.

The above analysis shows that it is sufficient for the roughness distribution $\varphi(r)$ to be four times differentiable function. For twice differentiable function $\varphi(r)$ such that $\varphi(0)=\varphi'(0)=0$, $\varphi(1)=0$ the estimate for coefficients c_k has the form (Tolstov, 1960) $| c_k |\leq \frac{C_1}{\mu_k^{3/2}}$, $\mu_k\gg 1$, where C_1 is a certain constant. It is sufficient for convergence of the series (18.5) for $\varphi(r)$ but not sufficient for term by term differentiation of this series. Therefore, it is sufficient for the roughness function $\varphi(r)$ to be four times differentiable for the whole analysis to be valid. On the other hand, it seems that three times differentiability of functions $\varphi(r)$ would be sufficient for the validity of the analysis.

Using the relations from (18.3) and (18.12) as well as the fact that the integral $I(\mu,r)$ reaches its maximum at $r=0$ and equal $\int\limits_0^1 \frac{\sin(\mu s)}{s}ds$ we obtain an estimate for a load P in dimensional variables which would be sufficient for providing a full contact in the form

$$P>E'R\lambda\sum\limits_{k=1}^{\infty} | c_k | \max | \psi_k(\mu_k,r) |,$$
$$\max | \psi_k(\mu_k,r) |=1-\frac{\sin\mu_k}{\mu_k}+\mu_k \Big| \int\limits_0^1 \frac{\sin(\mu_k s)ds}{s} \Big|. \tag{18.18}$$

The microphotographs of real surface roughness show (Kudish et al, 2012) that the requirement that the surface roughness distribution $f(r)$ is three/four times differentiable is not an unreasonable one.

18.3 Some Generalizations

The roughness distribution can be also represented by a Shlemilch series in the form (Watson, 1952)

$$\varphi(r)=\frac{c_0}{2}+\sum\limits_{k=1}^{\infty} c_k J_0(\pi kr),$$
$$c_0=2\varphi(0)+2\pi\int\limits_0^1 v\int\limits_0^{\pi/2}\varphi'(\pi v\sin\alpha)d\alpha dv, \tag{18.19}$$
$$c_k=2\pi\int\limits_0^1 v\cos(k\alpha\pi)\int\limits_0^{\pi/2}\varphi'(\pi v\sin\alpha)d\alpha dv,$$

where c_k are called the coefficients of the Shlemilch series. In some cases this representation of roughness distribution may be more beneficial as it usually imposes less limitations on the roughness distribution function $\varphi(r)$. In this case the results are quite similar to the case of the representation of function $\varphi(r)$ by the Fourier-Bessel series (18.5).

The described above analysis can be extended on punches with curved bottom. Let us assume that $z = f(r)$ is the shape of a rigid punch bottom of radius R which is indented by a normal force P in a rough elastic half-space made of a homogeneous material. The punch has sharp edges so that the nominal contact region is a circle of radius R. Then for this punch a problem similar to the one considered above will take the form (Kudish and Covitch, 2010)

$$\lambda\varphi(r) + \frac{8}{\pi E'}\int\limits_0^R \frac{\rho}{\rho+r}K\Big(\frac{2\sqrt{\rho r}}{\rho+r}\Big)p(\rho)d\rho = \delta - f(r), \ \int\limits_0^R \rho p(\rho)d\rho = \frac{P}{2\pi}. \quad (18.20)$$

Under the assumption that the contact is full in dimensionless variables (18.2) the expression for the contact pressure $p(r)$ will have the form (Kudish and Covitch, 2010)

$$p(r) = p_{0f}(r) - \frac{\lambda}{2\pi\sqrt{1-r^2}}\int\limits_0^1 \frac{\varphi'(\rho)d\rho}{\sqrt{1-\rho^2}} + \frac{\lambda}{2\pi}\int\limits_r^1 \frac{1}{\sqrt{s^2-r^2}}\Big\{s\int\limits_0^s \frac{\varphi'(\rho)d\rho}{\sqrt{s^2-\rho^2}}\Big\}'ds,$$

$$\delta_1 = \delta - f(0) - \lambda\varphi(0) = \delta_{0f} + \lambda\int\limits_0^1 \sqrt{1-\rho^2}\varphi'(\rho)d\rho,$$

$$p_{0f}(r) = \frac{1}{2\pi\sqrt{1-r^2}}\Big\{\delta_1 - \int\limits_0^1 \frac{f'(\rho)d\rho}{\sqrt{1-\rho^2}}\Big\} + \frac{1}{2\pi}\int\limits_r^1 \frac{1}{\sqrt{s^2-r^2}}\Big\{s\int\limits_0^s \frac{f'(\rho)d\rho}{\sqrt{s^2-\rho^2}}\Big\}'ds,$$

$$\delta_{0f} = \frac{2\pi}{3} + \int\limits_0^1 \sqrt{1-\rho^2}f'(\rho)d\rho. \quad (18.21)$$

The analysis of the previous section remains the same if $p_{0f}(r) > 0$ for all $r \in [0,1]$. More specifically, the roughness distribution $\varphi(r)$ can be represented by the Fourier-Bessel expansion (18.5). After applying the outlined above analysis the condition for a full contact would look as follows

$$\lambda\sum_{k=1}^{\infty} | c_k || \psi_k(\mu_k, r) | < \delta_{0f}. \quad (18.22)$$

After that an estimate for a load P which would be sufficient for providing a full contact can be obtained in the form similar to (18.18).

In a similar fashion can be considered an axially symmetric problem with free boundary, i.e. when a punch with a characteristic curvature radius R does not have sharp edges and the contact boundary is not fixed. In this case the problem can be formulated as follows (Kudish and Covitch, 2010)

$$\lambda\varphi(r)+\frac{8}{\pi E'}\int\limits_0^a\frac{\rho}{\rho+r}K(\frac{2\sqrt{\rho r}}{\rho+r})p(\rho)\mathrm{d}\rho=\delta-f(r),\ p(a)=0,\ \int\limits_0^a\rho p(\rho)\mathrm{d}\rho=\frac{P}{2\pi},\tag{18.23}$$

where a is the contact radius. Using the dimensionless variables (18.2) in which R is replaced by a and the dimensionless $a'=a/a_H$, $a_H=(\frac{3PR}{4E'})^{1/3}$ the solution can be presented in the form (primes are omitted) (Kudish and Covitch, 2010)

$$\begin{aligned}
p(r)&=p_{OH}(r)+\frac{\lambda}{2\pi a}\int\limits_r^1\frac{1}{\sqrt{s^2-r^2}}\{s\int\limits_0^s\frac{\varphi'(\rho)\mathrm{d}\rho}{\sqrt{s^2-\rho^2}}\}'\mathrm{d}s,\\
\frac{2\pi}{3}\frac{1-a^3H(a)}{a}&=\lambda\int\limits_0^1 s\{s\int\limits_0^s\frac{\varphi'(\rho)\mathrm{d}\rho}{\sqrt{s^2-\rho^2}}\}'\mathrm{d}s,\\
p_{OH}(r)&=\frac{1}{2\pi a}\int\limits_r^1\frac{1}{\sqrt{s^2-r^2}}\{s\int\limits_0^s\frac{f'(a\rho)\mathrm{d}\rho}{\sqrt{s^2-\rho^2}}\}'\mathrm{d}s,\\
H(a)&=\frac{3}{2\pi a^2}\int\limits_0^1 s\{s\int\limits_0^s\frac{f'(a\rho)\mathrm{d}\rho}{\sqrt{s^2-\rho^2}}\}'\mathrm{d}s.
\end{aligned}\tag{18.24}$$

Due to the same structure of all integrals of function $\varphi(r)$ in (18.24) and assuming that $p_{OH}(r)>0$ for all $r\in[0,1]$ the presented analysis can be applied to this problem as well.

References

Abramowitz M, Stegun IA (eds) (1964) Handbook of athematical Functions with Formulas, Graphs and Mathematical Tables. No. 55 in Applied Mathematics Series, National Bureau of Standards

Ciavarella M, Demelio G, Barber JR, , Jang YH (2000) Linear elastic contact of the weierstrass profile. Proc Roy Soc London, Ser A 456:387–405

Gradshtein IS, Ryzhik IM (1971) Tables of Integrals, Sums, Series, and Products. Nauka, Moscow

Kudish II, Covitch MJ (2010) Modeling and Analytical Methods in Tribology. Chapman & Hall/CRC Press

Kudish II, Cohen DK, , Vyletel B (2012) Perfect mechanical sealing in rough elastic contacts: is it possible? Trans ASME Journal of Applied Mechanics 80(1):014,504–1 thru 014,504–6

Kudish II, Cohen DK, , Vyletel B (2014) Surface roughness and contact connectivity. Proc Inst Mech Eng, Part J: Journal of Engineering Tribology 228(7):735–744

Tolstov II (1960) Fourier Series. Fizmatgiz Publisher, Moscow

Watson GN (1952) The Tretise of Bessel Functions. Cambridge University Press, Cambridge

Chapter 19
Geometric Aspects of the Theory of Incompatible Deformations in Growing Solids

Sergei A. Lychev

Abstract Differential-geometrical methods for modeling the incompatible finite deformations in growing solids are developed. Incompatible deformations result in residual stresses and distortion of the geometric shape of a body. These factors determine the critical parameters of modern high-precision technologies and are considered to be essential constituents in corresponding mathematical models. Affine connection on the material manifold represents the intrinsic properties (proper geometry) of the body and is determined by the field of local uniform configurations which performing its "assembly" of identical and uniform infinitesimal "bricks". Uniformity means that the response functional gives for them the same response on all admissible smooth deformations. As a result of assembling, one obtains body, which cannot be embedded in undistorted state into physical manifold. It is an essential feature of residual stressed bodies produced by additive processes. For this reason, it is convenient to use the embedding into a non–Euclidean space (material manifold with non–Euclidean material connection). To this end one can formalize the body and physical space in terms of the theory of smooth manifolds. The deformation is formalized as embedding (or, in special case, as immersion) of former manifold into the latter one.

Key words: Growing solids · Incompatible deformations · Residual stresses · Material manifold · Non-Euclidian geometry · Material connections · Method of the moving frame · Finite deformations · Multilayered elastic cylinder

---

Sergei A. Lychev
Ishlinsky Institute for Problems in Mechanics RAS, Vernadskiy Ave, 101 bldg 1, 119526 Moscow, Russia,
e-mail: lychevsa@mail.ru

H. Altenbach et al. (eds.), *Mechanics for Materials and Technologies*,
Advanced Structured Materials 46, DOI 10.1007/978-3-319-56050-2_19

19.1 Introduction

The geometrical methods in continuum mechanics (Epstein, 2010; Epstein et al, 2007; Maugin, 1993) are based on the fundamentals of modern differential geometry (Postnikov, 1987; Dubrovin et al, 1986; Kobayashi and Nomizu, 1963). The fact that the geometric language allows efficient formalization of various physical phenomena was shown at the end of the 19th century in the pioneering works by Beltrami, Poincare, and Levi-Civita, followed by Einstein, Weyl, Cartan, and others (Lützen, 1995). The fruitfulness of the geometrical approach to the modelling of bodies with defects (dislocations, disclinations, etc.) was demonstrated in Nye (1953); De Wit (1977); Kröner (1958); Kadic and Edelen (1983), as well as to the nonlinear problems for growing solids (Arutyunyan et al, 1987; Yavari, 2010; Lychev and Manzhirov, 2013a; Manzhirov and Lychev, 2012; Lychev and Manzhirov, 2013b; Manzhirov et al, 2013; Lychev, 2011; Sozio and Yavari, 2017).
Note the practical significance of the problem. Currently, we see intense development of technological processes based on the manufacturing of parts by sequential deposition of material on a substrate that may have a very complicated geometric shape. Local polymerization, electrochemical deposition (Choy, 2003), overlaying welding, selective laser sintering, stereolithography, and ion implantation (Nastasi and Mayer, 2006) represent only some of such technologies. They make it possible to manufacture parts with a complicated geometric shape and, theoretically, from any material. The implementation of new technologies is accompanied by specific technological problems (Manzhirov, 2016). Among them is the distortion of the final shape due to material shrinkage during solidification. When thin-walled parts are manufactured, such distortions may exceed the specified precision substantially. One of the methods for decreasing distortions is to compensate them by initial distortions, considered in the design (Huang et al, 2015; Lu et al, 2015). Thus the problem for modeling of the evolution of incompatible deformations caused by technological process is of practical importance.
The typical feature of the bodies, generated during some additive process, is the absence of the natural (stress free) shape. From physical viewpoint it means that the body contains elastic energy that cannot be released under any smooth deformation. This energy can be released only by cutting the body into an innumerable number of disjoined parts, a process that is, in a sense, reversed to additive manufacturing. A similar situation arises in the theory of distributed defect fields (dislocations, disclinations, pores) (Zubov, 1997) and in nonlinear thermomechanics (Ozakin and Yavari, 2010). The effects caused by the absence of natural configurations can be formalized mathematically by different notions, such as incompatibility of deformations (Kröner, 1958; De Wit, 1977), the density of defects (Katanaev, 2005), and the sources of internal stresses (Acharya, 2001). These concepts are interrelated and can be formalized within the representation for deformation as for a geometric transformation in non-Euclidian space.
The first works where the non-Euclidian geometry was used for developing nonlinear continuum mechanics were published in the early 20th century. They proposed relativistic models of deformable continuum, which were stated in terms of Rie-

mannian and non-Riemannian geometry of space and time (Berger, 2003; Eisenhart, 1997, 2005; Petersen, 2006; Cartan, 1962, 1983). Thus, the physical space, containing a deformable body, was equipped with a non-Euclidian structure. The idea that the space which contains the reference unstressed shape of the body should have a non-Euclidean structure, was presented for the first time, apparently, by Eckart (1948) andKondo (1955a, 1963, 1955b, 1964)), although the first studies of solids deformation caused by distortion, not satisfying the conditions of compatibility, go back to Weingarten (1901) and Volterra and Volterra (1960). The idea that the spaces, containing the reference and the actual shape, are not Euclidean, was revolutionary at that time and initiated a self-sustained theoretical line in continuum mechanics, see Bilby et al (1955); Seeger (1961); Nye (1953); Kröner (1996); Noll (1967); Wang (1967); Maugin (1993); Epstein (2010); Epstein et al (2007); Myasnikov and Guzev (2000).

Within the framework of the present paper we consider only simple materials (Noll, 1967). Their response is fully defined by linear transformation, determined by local deformation. Attention should be paid to a formal correspondence between the mathematical description of incompatible deformations of simple materials and the affine connection space geometry. Each elementary volume is transformed to a uniform state by means of linear transformation; these transformations taken together define (by assumption) a continuous field of transformations. The building of the affine connection space by the Cartan method of the moving frame is also determined by the field of linear transformations on the coordinate basis vectors. Thus, the construction of a space, containing arbitrary distorted elementary volumes in a compatible form, reduces to the procedure of building a Cartan space. The specific geometric properties of the thus-built space, i.e., the "inner geometry" of the body, characterize the sources of proper stresses.

19.2 Naive Geometric Motivation

In Noll (1967); Wang (1967); Ozakin and Yavari (2010); Yavari and Goriely (2013); Lychev and Manzhirov (2013a,b); Manzhirov and Lychev (2012); Manzhirov et al (2013) the physical space, containing shapes of the bodies, was assumed to be affine-Euclidean. In the present work, the physical space is considered in general terms as a Riemannian space with a curvature different from zero. Firstly, this makes the theory more complete and consistent, since from the formal point of view the body and the physical space become equivalent geometric objects. Secondly, in modeling two-dimensional bodies, i.e., membranes, images of their configurations can be determined on some fixed surface. Such situation arises, for example, during the deposition of a thin film on the surface of a sphere. In this case, it is convenient to view the deposition surface as a two-dimensional physical space, in general terms, with non-Euclidean connection.

In order to explain such arguments, we shall discuss a naive example of deformation of a membrane, assuming that images of the reference and the actual shapes are known, and that one of them, or both, cannot be embedded into an Euclidean plane. We assume that the observers, describing the "physical reality," are fictional "two-dimensional creatures" populating the membrane, a two-dimensional surface, like inhabitants of Sphereland (Abbott and Burger, 1976). They cannot directly observe deformation of the membrane, but feel changes in the physical properties of their two-dimensional world, caused by deformation. In addition "two-dimensional observers," deformation of the membrane is followed by powerful "three-dimensional observers," who can see the shape of the membrane in its possible embedding in the enveloping three-dimensional space. Lets consider two situations. An image of a uniform configuration of the membrane can not be embedded into an Euclidean plane, but it can be embedded in an enveloping Euclidean plane as part of the sphere (Fig. 19.1). In such embedding, small neighborhoods are in a uniform, or physically indistinguishable, state. Some of them are shown as circumferences in Fig. 19.1. "Two-dimensional observers" cannot say anything about the "spherical" shape of their world and may be convinced that their world is flat. Let the physical space be flat. Any embedding of the membrane in the physical space leads to membrane deformation. In this case, "two-dimensional observers" will find out that their uniform world is not uniform anymore: circumferences, characterizing the neighborhoods in Fig. 19.1, are distorted, and such distortion occurs without any reasons visible for the "two-dimensional observers," because no external force fields have been applied. The distorted circumferences are shown in Fig. 1 by the solid lines. For comparison, undistorted circumferences are shown nearby by the dotted lines. According to Maugin (1993), the membrane acquired material inhomogeneity, remaining materially uniform, i.e., made from one and the same material. It is clear that changes of geometric images can be observed only by the powerful "three-dimensional observers."

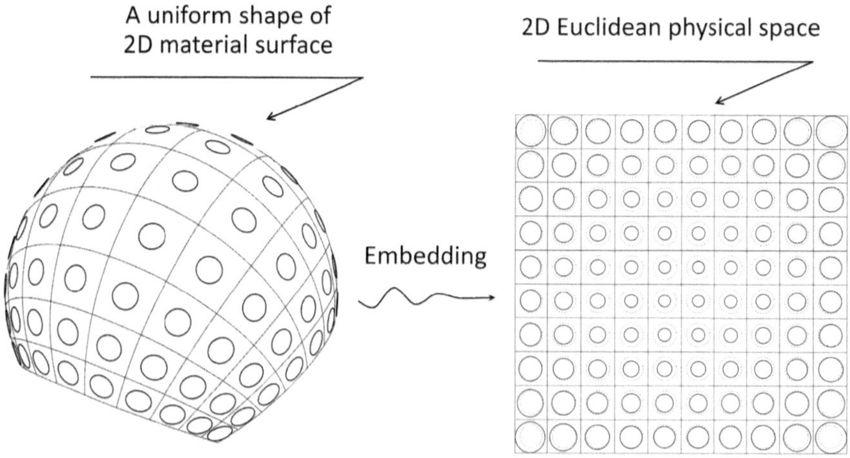

Fig. 19.1: Embedding of a two-dimensional body with a non-Euclidean proper geometry in a flat physical space

Images in Fig. 19.1 are shown from their perspective. "Two-dimensional observers" will notice only changes in the physical properties of their world. For example, the elasticity under small disturbances of various sections of the membrane becomes different, while if the membrane possesses optical properties, it becomes optically inhomogeneous: a sort of a rainbow will appear in the world of "two-dimensional observers."

The circumferences, symbolizing uniform neighborhoods, are transformed into distorted objects from the point of view of "two-dimensional observers" due to reasons that are not related to visible deformation of their two-dimensional world, whereas a "three-dimensional observer" visualizes this deformation explicitly. However, "two-dimensional observers" may assume that, for some reasons, their two-dimensional world changes its geometric properties: its spherical geometry was transformed to a flat geometry. Both points of view can be used to describe the embedding of the body in the physical space; however, appealing to a powerful "multidimensional observer" is often cumbersome and related to difficult-to-interpret geometric images. In the present paper we develop methods of the "internal observer," who can use two notions: deformation of the body and change of the geometry of the space, containing this body. In the discussed example, the space containing a uniform reference of the body shape had a non-Euclidian structure while the physical space was an Euclidean plane. It is easy to imagine an inverted situation: a uniform shape is defined by flat geometry and the physical space is a Riemannian space with nontrivial curvature. Figure 19.2 illustrates this argument: a flat membrane is deposited such that it can slip onto the nondeformable spherical foundation.

Thus, incompatible deformations of the body are characterized by two geometrical concepts: the embedding in the physical space and the structure of the space containing its reference uniform configuration. More details on this issue are reflected in Lychev and Koifman (2016).

19.3 Material Manifold

Intuitive notions of the solid body can be reduced to the following arguments. The body $\mathfrak{B}$ is a set of material points. Each material point neighbors on other material points, continuously filling its neighborhood; this filling being similar to the filling of the interval in $\mathbb{R}$. These arguments lead to formalization of $\mathfrak{B}$ as a topological space (Kuratowski, 1966; Armstrong, 1983). It is assumed that:

1. The cardinality of $\mathfrak{B}$ is equal to the cardinality of continuum ($\mathbb{R}$);
2. $\mathfrak{B}$ is separable (satisfies the Hausdorff separability axiom);
3. $\mathfrak{B}$ is the topological space with a countable base;
4. The local topological dimension of $\mathfrak{B}$ is constant: $\dim\mathfrak{B} = \mathrm{n}$ (for membranes $n = 2$ and for three-dimensional bodies $n = 3$).

Identification of material points requires a method for their analytical description: here it is expedient to use the idea of charting the topological space, developed

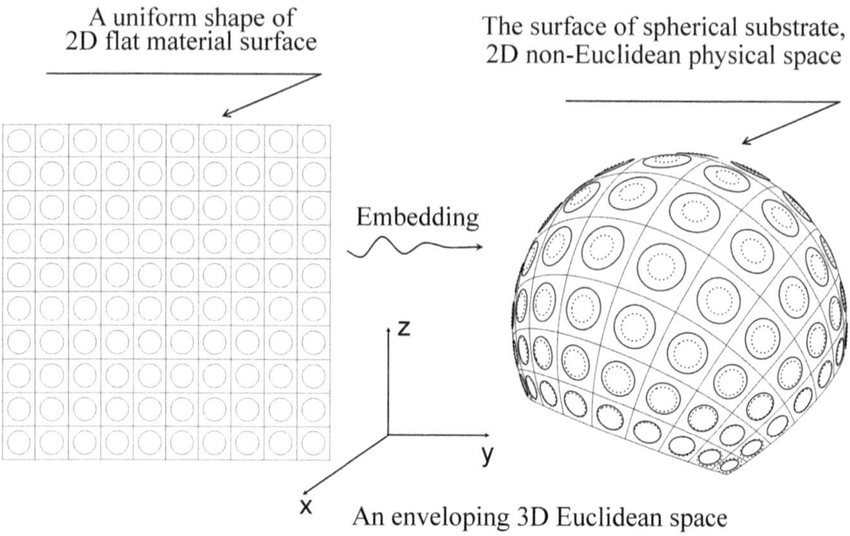

Fig. 19.2: Embedding of a two-dimensional body with an Euclidean proper geometry in a non-Euclidean physical space

in the theory of smooth manifolds (Sternberg, 1964; Hirsch, 1976; Adachi, 2012; Kobayashi and Nomizu, 1963; Nakahara, 2003; Postnikov, 1987; Borisovich et al, 1995; Dubrovin et al, 1986). Charting induces an analytical structure on $\mathfrak{B}$, defining the notion of smoothness on it. As a result of charting, for each material point it is possible to select a neighborhood, homeomorphic to the arithmetic space $\mathbb{R}^n$ (for three-dimensional bodies to $\mathbb{R}^3$, and for membranes to $\mathbb{R}^2$).

Representation of a body as a smooth manifold was used for the first time by Noll (1967); Wang (1967); Gurtin and Murdoch (1975). For instance, Noll introduced an axiomatic definition of a smooth body as a set, equipped with a family of its mappings in the physical space called configurations and satisfying some special system of axioms. Gurtin and Murdoch (1975) gave a self-contained definition of a two-dimensional material surface based on the Noll axioms.

The Noll system of axioms is based on the notion of the configuration $\varkappa$ as a mapping of the set $\mathfrak{B}$, representing a collection of material points constituting the body, to the absolute (Euclidean) physical space $\mathfrak{P}$, i.e., $\varkappa : \mathfrak{B} \to \mathfrak{P}$. Let us assume a class of configurations $\mathscr{C}$. The body $\mathfrak{B}$, is called a continuous body of the class C^p, if the class of configurations $\mathscr{C}$ satisfies the following conditions:

1. Each configuration $\varkappa \in \mathscr{C}$ is a homeomorphism, and the set $\varkappa(\mathfrak{B})$ is an open set in $\mathfrak{P}$, which is called the region, occupied by the body $\mathfrak{B}$ in the configuration $\varkappa$.
2. If $\gamma, \varkappa \in \mathscr{C}$, then the composition

$$\lambda = \gamma \circ \varkappa^{-1} : \varkappa(\mathfrak{B}) \to \gamma(\mathfrak{B})$$

represents a mapping of the class C^p, which is called deformation of the body $\mathfrak{B}$ from the shape $\varkappa(\mathfrak{B})$ to the shape $\gamma(\mathfrak{B})$.

3. If $\varkappa \in \mathscr{C}$ and if $\lambda : \varkappa(\mathfrak{B}) \to \mathfrak{P}$ is a mapping of the class C^p, then $\lambda \circ \varkappa \in \mathscr{C}$. The mapping $\lambda \circ \varkappa$ is called configuration, obtained from the configuration $\varkappa$ by means of deformation λ.

Thus, the smoothness of body is defined by the smoothness of configuration compositions.

According to Noll (1967) the smooth body is the smooth manifold with trivial topological structure, i.e., a trivial manifold, whose atlas contains only one chart (Postnikov, 1987). Complex structures of growing solids lie beyond the theory built on this axiomatics (the need of using nontrivial manifolds, representing the body, is also described in Wang (1967)). In addition, postulating the absolute physical space does not allow taking into account curvilinear substrates as 2D physical spaces.

The Noll axiomatics was used in Lychev and Manzhirov (2013a,b). In the framework of the present paper, we use a more general class of manifolds, representing both the body and the space. We call the topological space $\mathfrak{M}$ a (topological) manifold with the dimension $\dim \mathfrak{M} = n$, when it satisfies the following conditions (Borisovich et al, 1995; Adachi, 2012):

1. $\mathfrak{M}$ satisfies the Hausdorff separability axiom;
2. $\mathfrak{M}$ has a topological countable base;
3. there exists a family $\{U_\alpha\}_{\alpha \in I}$ of open subsets of $\mathfrak{M}$, such that 1) $\mathfrak{M} = \bigcup\limits_{\alpha \in I} U_\alpha$, 2) each of the sets U_α is homeomorphic to $\mathbb{R}^n$, i.e., for any $\alpha \in I$ there exists homeomorphism $\phi_\alpha : \mathbb{R}^n \to U_\alpha$.

Let us explain in more detail the terminology used. The pair (U_α, ϕ_α), where U_α and ϕ_α are defined above, is called a chart in $\mathfrak{M}$; U_α is the chart's domain; the collection of charts $A = \{(U_\alpha, \phi_\alpha)\}_{\alpha \in I}$, where $\{U_\alpha\}_{\alpha \in I}$ is the open cover of $\mathfrak{M}$, is an atlas of the manifold $\mathfrak{M}$. Let $p \in \mathfrak{M}$ be a certain point of the manifold. The chart (U_α, ϕ_α), for which $p \in U_\alpha$, is called a chart containing the point p.

The atlas generates a system of local coordinates on the manifold, namely: any point p belonging to $\mathfrak{M}$ is obtained in, at least, one of the sets U_α; therefore, to it there definitively corresponds the ordered n–tuple $(\xi^1, \ldots, \xi^n) = \phi_\alpha^{-1}(p)$. This ordered set represents local coordinates of the point p. If the intersection $U_\alpha \cap U_\beta$, where U_α, U_β are taken from one atlas, is not empty, then any point p of this intersection has the local coordinates $(\xi^1, \ldots, \xi^n) = \phi_\alpha^{-1}(p)$ and $(\eta^1, \ldots, \eta^n) = \phi_\beta^{-1}(p)$ interrelated by the conversion formulas: the mapping

$$\phi_\beta^{-1} \circ \phi_\alpha|_{\phi_\alpha^{-1}(U_\alpha \cap U_\beta)} : \phi_\alpha^{-1}(U_\alpha \cap U_\beta) \to \phi_\beta^{-1}(U_\alpha \cap U_\beta), \tag{19.1}$$

as a composition of homeomorphisms, is homeomorphism of local coordinates $(\xi^1, \ldots, \xi^n)$ and $(\eta^1, \ldots, \eta^n)$.

The abstract manifold $\mathfrak{M}$ formalizes the continuity property, inherent in the body and the physical space. However, the structure of the topological manifold is not sufficient for correct definition of the smoothness of the mappings, specified on its subsets. Therefore, the manifold is equipped with an additional structure, called differential. It allows us to define invariantly (in relation to atlases in this structure) the notion of a smooth mapping on the manifold. The charts (U_α, ϕ_α), (U_β, ϕ_β) in $\mathfrak{M}$ are C^r–compatible, if either 1) $U_\alpha \cap U_\beta = \varnothing$, or 2) $U_\alpha \cap U_\beta \neq \varnothing$ and homeomorphism (19.1) is C^r–diffeomorphism on $\phi_\alpha^{-1}(U_\alpha \cap U_\beta)$. The atlas in which any two charts are C^r–compatible is called a C^r–atlas.

19.4 Growing Solids

In classical continuum mechanics bodies are treated as fixed sets of material points. Growing bodies are sets of variable composition. The evolution of this set in time can be represented by mapping

$$\mathbb{R} \supset \mathbb{I} \ni \alpha \mapsto \mathfrak{B}_\alpha, \tag{19.2}$$

where α is time–like parameter and $\mathfrak{B}_\alpha$ is a body manifold that represents growing object at instance α. All values of (19.2) constitute a family of bodies

$$\mathfrak{G} = \{\mathfrak{B}_\alpha\}_{\alpha\in\mathbb{I}}. \tag{19.3}$$

Since the family (19.3) represents some evolution process we shall refer to α as to *evolution parameter.*

Define the total body $\overset{*}{\mathfrak{B}}$ and core $\underset{*}{\mathfrak{B}}$ as

$$\overset{*}{\mathfrak{B}}_{\mathbb{I}} = \bigcup_{\alpha\in\mathbb{I}} \mathfrak{B}_\alpha, \quad \underset{*}{\mathfrak{B}}_{\mathbb{I}} = \bigcap_{\alpha\in\mathbb{I}} \mathfrak{B}_\alpha.$$

If the following relation holds $\forall\alpha,\beta \in \mathbb{I} \quad (\alpha < \beta) \Leftrightarrow \left(\mathfrak{B}_\alpha \subseteq \mathfrak{B}_\beta\right)$ then the growing process is said to be monotonic increasing (pure growth). Under the conditions $\forall\alpha,\beta \in \mathbb{I} \quad (\alpha < \beta) \Leftrightarrow \left(\mathfrak{B}_\alpha \supseteq \mathfrak{B}_\beta\right)$ we have monotonic decreasing process (pure removing), and we claim that $\mathfrak{B}$ is the body of permanent composition if $\forall\alpha,\beta \in \mathbb{I} \quad \mathfrak{B}_\alpha = \mathfrak{B}_\beta$ We say that growing process is mixed if no one above relations holds. We also introduce the notion for initial body $\mathfrak{B}_i$ and final body $\mathfrak{B}_f$ which correspond to the instance of growing object at the beginning of growing process and at its end.

For pure growth: $\mathfrak{B}_i = \underset{*}{\mathfrak{B}}, \quad \mathfrak{B}_f = \overset{*}{\mathfrak{B}}$ and for pure removing: $\mathfrak{B}_f = \underset{*}{\mathfrak{B}}, \quad \mathfrak{B}_i = \overset{*}{\mathfrak{B}}$.
In mixed process all defined above quantities are distinguished
According to Lychev and Manzhirov (2013a) we describe the structure of growing body by following statement. The family $\mathfrak{G}$ represents continuous growing body if the set cardinality for $\mathbb{I}$ is continuum and the following relations hold:
Discrete growth. The set $\mathfrak{G}$ is a finite sequence of embedded sets:

$$\mathfrak{G} \,:\, \mathfrak{B}_1 \subset \mathfrak{B}_2 \subset \ldots \subset \mathfrak{B}_N. \tag{19.4}$$

It is obvious that the cardinality of the set $\mathfrak{G}$ is determined by the number N, that is, $|\mathfrak{G}| = N$.
Generalized discrete growth. The set $\mathfrak{G}$ is a transfinite sequence of embedded sets:

$$\mathfrak{G} \,:\, \mathfrak{B}_1 \subset \mathfrak{B}_2 \subset \ldots \subset \mathfrak{B}_\omega \subset \mathfrak{B}_{\omega+1} \subset \ldots, \quad |\mathfrak{G}| = \aleph_0.$$

The family $\mathfrak{G}$ represents continuous growing body if the set cardinality for $\mathbb{I}$ is continuum and the following relations hold:

Continuous growth. The set cardinality for $\mathbb{I}$ is continuum and the following relations hold:

$$\mathbb{I} = \bigcup_{k \in \mathbb{A}} \mathbb{J}_k, \quad \mathrm{Card}\,\mathbb{I} = \aleph_1, \quad \mathrm{Card}\,\mathbb{A} \leq \aleph_0, \quad \forall k,n \in \mathbb{A} \quad (k \neq n) \Rightarrow \mathbb{J}_k \cap \mathbb{J}_n = \varnothing,$$

$$\forall k \in \mathbb{A} \quad \forall \alpha \in \mathbb{J}_k \quad \exists \varepsilon \quad \alpha + \varepsilon \in \mathbb{J}_k, \quad \partial \mathfrak{B}_\alpha = \Psi_\alpha(\Omega_k), \quad \alpha \in \mathbb{J}_k,$$

where Ψ_α is some homeomorphism between two dimensional manifolds, Ω_k is a n_k–connected tori (recall the classification theorem for two dimensional manifolds). The manifolds Ω_k is said to be pre-image of the growing boundary. Thus all considered growing processes are discrete from topological point of view.

Now we introduce conditions for a pure laminar growth:

Laminar growth.

$$\forall \alpha \in \mathbb{I} \quad \exists \varepsilon > 0 \quad \forall \delta \leq \varepsilon \quad \mathfrak{B}_{\alpha+\delta} \setminus \mathfrak{B}_\alpha = \Phi_\alpha \left(\Omega_k \times (\alpha, \alpha+\delta)\right),$$

where Φ_α is a homeomorphism of 3D manifolds.

We say that the manifold $\Phi_\alpha(\Omega_k)$ is the lamina. Under the condition for laminar growth we have the structure of the smooth bundle manifold over $\mathfrak{B}$: the set is a base and laminas $\Phi_\alpha(\Omega_k)$ are fibers, while the manifold Ω_k represents typical fibre.

Let us consider two growing processes, the first one, $\overline{\mathfrak{G}}$, is discrete and the second, $\mathfrak{G}$, is continuous. The discrete process $\overline{\mathfrak{G}}$ is called to be upper bound of $\mathfrak{G}$ if $\forall \alpha_{k-1}, \alpha_k \in \overline{\mathbb{I}} \quad \forall \beta \in \mathbb{I} \quad (\alpha_{k-1} \leq \beta < \alpha_k) \Rightarrow \mathfrak{B}_\beta \subseteq \overline{\mathfrak{B}}_{\alpha_k}$ Consequently, the discrete process $\underline{\mathfrak{G}}$ is called to be lower bound of $\mathfrak{G}$ if $\forall \alpha_{k-1}, \alpha_k \in \underline{\mathbb{I}} \quad \forall \beta \in \mathbb{I} \quad (\alpha_{k-1} \leq \beta < \alpha_k) \Rightarrow \mathfrak{B}_\beta \supseteq \underline{\mathfrak{B}}_{\alpha_k}$ We say that the growing process regular if $\overline{\mathbb{I}} = \underline{\mathbb{I}}, \quad \lim_{\triangle(\mathbb{I}) \to 0} \mu \left(\overline{\mathfrak{B}}_\alpha \setminus \underline{\mathfrak{B}}_\alpha\right) = 0$ where $\triangle\,(\mathbb{I}) = \sup_{\alpha_{k-1}, \alpha_k \in \underline{\mathbb{I}}} |\alpha_k - \alpha_{k-1}|$.

19.5 Mappings Between Manifolds

The density of elastic energy W represents examples of real–valued functions, specified on the manifold. The C^r–structure allows determining on the manifold the classes of smoothness of such mappings. We shall give a detailed explanation (Borisovich et al, 1995).

Let a C^r–structure be specified on $\mathfrak{M}^n$. The function $W : \mathfrak{M}^n \to \mathbb{R}$ is called the function of the class C^r in the neighborhood of the point $p \in \mathfrak{M}^n$, if there is a chart (U_α, ϕ_α) from the atlas of this C^r–structure, containing the point p, such that the mapping $W \circ \phi_\alpha : \mathbb{R}^n \to \mathbb{R}$ is C^r–mapping of $\mathbb{R}^n$ in $\mathbb{R}$. Owing to C^r–compatibleness of the charts, this definition does not depend on the atlas from this structure.

Now consider a continuous mapping $\varkappa : \mathfrak{M}^n \to \mathfrak{N}^m$ of two C^r–manifolds in a point $p \in \mathfrak{M}^n$. Select the chart (U_α, ϕ_α) of the manifold $\mathfrak{M}^n$, containing p, and the chart (V_β, ψ_β) of the manifold $\mathfrak{N}^m$, containing $\varkappa(p)$. Denote $O_{\alpha\beta} = U_\alpha \cap \varkappa^{-1}(V_\beta)$ and consider the mapping

$$\psi_\beta^{-1} \circ \varkappa \circ \phi_\alpha|_{\phi_\alpha^{-1}(O_{\alpha\beta})} : \phi_\alpha^{-1}(O_{\alpha\beta}) \to \psi_\beta^{-1}(\varkappa(O_{\alpha\beta})),$$

which is called a coordinate representation of the mapping $\varkappa$ in the neighborhood of the point p (Borisovich et al, 1995). This is the mapping $\mathbb{R}^n \to \mathbb{R}^m$, and it does not depend on the selection of charts in these C^r–structures on $\mathfrak{M}^n$ and $\mathfrak{N}^m$, since the charts on these manifolds are C^r–compatible.

The coordinate representation makes it possible to determine smoothness of the mapping $\varkappa : \mathfrak{M}^n \to \mathfrak{N}^m$ in the neighborhood of some point. Namely, the continuous mapping $\varkappa : \mathfrak{M}^n \to \mathfrak{N}^m$ is called a C^r–mapping in the neighborhood of the point $p \in \mathfrak{M}^n$, if some coordinate representation of the mapping i in the neighborhood of this point is a C^r–mapping. Hereinafter, $\mathscr{C}^r(\mathfrak{M}^n; \mathfrak{N}^m)$ is used to designate the set of all C^r–mapping from the C^r–manifold $\mathfrak{M}^n$ to the C^r–manifold $\mathfrak{N}^m$. In particular, $\mathscr{C}^r(\mathfrak{M}^n; \mathbb{R})$ is a set of all C^r–mappings from the C^r–manifold $\mathfrak{M}^n$ to the trivial C^r–manifold $\mathbb{R}$. In the framework of these terms, diffeomorphism from the C^r–manifold $\mathfrak{M}^n$ to the C^r–manifold $\mathfrak{N}^m$ shall be any mapping $\varkappa \in \mathscr{C}^r(\mathfrak{M}^n; \mathfrak{N}^n)$, such that 1) there exist $\varkappa^{-1}$ and 2) $\varkappa^{-1} \in \mathscr{C}^r(\mathfrak{N}^n; \mathfrak{M}^n)$.

We assume as a hypothesis that the body and the physical space are equipped with C^r–structures with $r \geqslant 1$. The relationship between the body as a continual set of material points and the set of positions of these points in the physical space is established by means of the mapping $\varkappa : \mathfrak{B} \to \mathfrak{P}$, called configuration. In the present work, this mapping is equipped with additional properties of smoothness and regularity. We explain it in detail.

Let $\mathfrak{B} = \mathfrak{M}^n$ and $\mathfrak{P} = \mathfrak{N}^m$. Among mappings of the class $\mathscr{C}^r(\mathfrak{M}^n; \mathfrak{N}^m)$ we identify the mappings satisfying the requirement of regularity at each point of $\mathfrak{M}^n$: we call the point $p \in \mathfrak{M}^n$ regular point of the mapping $\varkappa \in \mathscr{C}^r(\mathfrak{M}^n; \mathfrak{N}^m)$ (Borisovich et al, 1995), if the coordinate representation $\varkappa$ in the neighborhood of the point p has a Jacobi matrix at the point $\phi_\alpha^{-1}(p)$ (where (U_α, ϕ_α) is a chart, containing p) of the rank $\min\{m, n\}$. In the case of $n \leqslant m$, the mapping $\varkappa \in \mathscr{C}^r(\mathfrak{M}^n; \mathfrak{N}^m)$, regular at each point $p \in \mathfrak{M}^n$, is called C^r–immersion (Adachi, 2012; Borisovich et al, 1995).

The regularity of $\varkappa$ has clear geometrical interpretation. The image of the configuration, as C^r–immersion of $\mathfrak{B}$ on $\mathfrak{P}$, may have self-intersection points. In some cases self-intersections are admissible, while in others to them there corresponds physically impossible self-penetration. To identify the class of configurations, whose images have no self-intersections, we shall narrow the class of all C^r–immersions $\varkappa : \mathfrak{B} \to \mathfrak{P}$, identifying those which are homeomorphic to their own image, i.e., the subspace $\varkappa(\mathfrak{B})$ of the topological space of $\mathfrak{P}$. Such mappings are called C^r–embeddings (Adachi, 2012; Borisovich et al, 1995). By regular configuration $\varkappa$ of the body we understand the C^r–embedding of $\mathfrak{B}$ in $\mathfrak{P}$. Thus, $\varkappa$ is homeomorphism of the body of $\mathfrak{B}$ to its image $\varkappa(\mathfrak{B})$.

Figuratively speaking, in the modeling for deformations of growing body, the main actors are manifolds $\mathfrak{B}_\alpha$, $\mathfrak{P}$, and $\mathbb{R}$. The main actions are embeddings $\mathscr{C} \ni \varkappa_\alpha : \mathfrak{B}_\alpha \to \mathfrak{P}$, and energy distributions $W_\alpha : \mathfrak{B}_\alpha \to \mathbb{R}$, $\mathscr{W}_\alpha : \mathscr{C} \ni \varkappa_\alpha \mapsto W_\alpha$. The latter mapping can be classified as constitutive relation the most general form. For simple elastic material this relation can be significantly simplified: $T\varkappa_\alpha(p) \mapsto W_\alpha(p)$, where $T\varkappa_\alpha$ denotes tangent map, that will be explained below.

19.6 Deformations

Primary geometrical objects that are investigated by continuum mechanics are curves. Curves on $\mathfrak{B}$ represent material fibers and curves on $\mathfrak{P}$ represent trajectories of material particles. In order to define this objects in terms of adopted abstract language one have to introduce the following notions. A curve a passing through the point $p \in \mathfrak{M}$ can be define by mapping ϕ over the interval $\mathbb{J}$, i.e.:

$$a = \{p \in \mathfrak{M} \,|\, \exists \alpha \in \mathbb{J} \quad \phi(\alpha) = p\}, \quad \phi : \mathbb{R} \supset \mathbb{J} \to \mathfrak{M}, \quad \phi(0) = p.$$

For convenience in what follows we suppose that interval $\mathbb{J}$ is symmetric, i.e. $\mathbb{J} = (-\alpha, \alpha)$, where α is a number. All curves passing through the point p are grouped into equivalence classes by equivalence relation $\sim$:

$$(\phi \sim \phi') \Leftrightarrow \left(\frac{d}{dt} \left(\phi_\alpha^{-1} \circ \phi(t)\right)\bigg|_{t=0} = \frac{d}{dt} \left(\phi_\alpha^{-1} \circ \phi'(t)\right)\bigg|_{t=0} \right).$$

We will denote classes by bold Latin symbols, $\mathbf{u}, \mathbf{v}$, etc. The correspondence between equivalence classes and their representations is shown by relation $(\mathbf{u} \cong_p \phi) \Leftrightarrow (\phi \in \mathbf{u})$. Let $\mathbf{u} \mapsto \mathfrak{h}_p(\mathbf{u}) = (u^1, \ldots, u^n) := \left.\frac{d(\phi_\alpha^{-1} \circ \phi(t))}{dt}\right|_{t=0}$ be a bijection. The following relation $\alpha \mathbf{u} + \beta \mathbf{v} = \mathfrak{h}_p^{-1} (\alpha \mathfrak{h}_p(\mathbf{u}) + \beta \mathfrak{h}_p(\mathbf{v}))$ implies that we have a canonical vector space structure. It can be prove that the dimension of this space is equal to $\dim \mathfrak{M}$. We will denote it by $T_p\mathfrak{M}$. Ordered pairs $(p, \mathbf{v})$, where $p \in a$, and $\mathbf{v} \cong_p \phi$ in union determine the section of tangent bundle over a, or in other words, the vector field on this curve.

Now let us introduce the notion of natural frame. Vector $\mathbf{u} \in T_p\mathfrak{M}$ corresponds to directional derivative $\mathscr{L}_u$ for f along $\mathbf{u}$ at point $p \in \mathfrak{M}$ (recall that $\phi(0) = p$)

$$\mathscr{L}_{\mathbf{u}}(f) := \frac{d}{dt} f \circ \phi(t)\bigg|_{t=0}, \quad (p, \phi) \in \mathbf{u}. \tag{19.5}$$

It follows from the definition that the correspondence between equivalence classes and directional derivatives is bijective: $(\mathscr{L}_{\mathbf{u}}(f) = \mathscr{L}_{\mathbf{v}}(f)) \Leftrightarrow (\mathbf{u} = \mathbf{v})$.

Class of special curves, namely coordinate curves

$$c^i = \phi^i(\mathbb{I}) = \{p \in \mathfrak{M} \,|\, p = \phi^i(t) \quad t \in \mathbb{I} \subset \mathbb{R}\},$$

where ϕ^i satisfy the relation $\pi^j \circ \phi_\alpha^{-1} \circ \phi^i(t) = \delta_i^j t$ introduce a coordinate net on $\mathfrak{M}$. Equivalence classes of such curves determine at point $p \in \mathfrak{M}$ a frame $(p, (\mathbf{e}_i)_{i=1}^n)$, $\phi^i \in \mathbf{e}_i$. From definition it follows that $\mathscr{L}_{\mathbf{e}_i}(\pi^j \circ \phi_\alpha^{-1}) = \delta_i^j$. We will refer to directional derivatives along $\mathbf{e}_i$ as to natural basis $\partial_i := \mathscr{L}_{\mathbf{e}_i}$. Note that $\mathscr{L}_{\mathbf{u}}(f) = \alpha^i \mathscr{L}_{\mathbf{e}_i}(f) \equiv \alpha^i \partial_i f$ and $\alpha^i \partial_i$ is an abstract vector that represents both differential operator (19.5) and corresponded class of curves.

The set of all ordered pairs $(p, T_p\mathfrak{M})$ as well as mappings $p \mapsto T_p\mathfrak{M}$ represents tangent bundle over $\mathfrak{M}$. The vector space of linear functionals (covectors) over $T_p\mathfrak{M}$

is denoted by $T_p^*\mathfrak{M}$. The set of all ordered pairs $(p, T_p^*\mathfrak{M})$ as well as mapping $p \mapsto T_p\mathfrak{M}$ represents cotangent bundle. The value of the covector $\mathbf{u} \in T_p^*\mathfrak{M}$ on the vector $\mathbf{v} \in T_p\mathfrak{M}$ is a real number $\langle \mathbf{u}, \mathbf{v} \rangle$. Standard argumentation (canonical isomorphism between vector space and second conjugate space) make the following symmetry relation true

$$\forall \mathbf{u} \in T_p^*\mathfrak{M} \quad \forall \mathbf{v} \in T_p\mathfrak{M} \quad \langle \mathbf{u}, \mathbf{v} \rangle = \langle \mathbf{v}, \mathbf{u} \rangle.$$

Define the Riemannian metrics that is a function $\mathbf{g} : \mathfrak{M} \to T_p\mathfrak{M} \otimes T_p\mathfrak{M}$ which values are definite symmetrical bilinear mappings over $T_p\mathfrak{M}$. The elements of basis in $T_p x^*\mathfrak{M}$ that is conjugated to canonical basis ∂_i are denoted by $\mathbf{d}^j$, i.e. $\langle \partial_i, \mathbf{d}^j \rangle = \delta_i^j$. Thus one can find the vector components by relation $\langle \mathbf{u}, \mathbf{d}^j \rangle = \langle u^i \partial_i, \mathbf{d}^j \rangle = u^i$. Bases-dependent isomorphisms (musical isomorphisms $\sharp, \flat$) between $T_p\mathfrak{M}$ and $T_p^*\mathfrak{M}$ are induced by $\mathbf{g}$:

$$\forall \mathbf{u} \in T_p\mathfrak{M} \quad \exists! \mathbf{v} \in T_p^*\mathfrak{M} \quad \langle \mathbf{u}, \mathbf{v} \rangle = \mathbf{g}(\mathbf{u}, \mathbf{v}^\sharp);$$

$$\forall \mathbf{v} \in T_p\mathfrak{M} \quad \exists! \mathbf{u} \in T_p^*\mathfrak{M} \quad \langle \mathbf{u}, \mathbf{v} \rangle = \mathbf{g}(\mathbf{v}, \mathbf{u}^\flat).$$

Now have completed all preliminary constructions and can focus our attention on the embedding of a body into physical space. Recall that motion is a mapping ϕ whose values are configurations (embeddings), i.e. $\phi : \mathscr{C} \times \mathbb{T} \to \mathfrak{P}$. Note that mapping ϕ with fixed first argument p, $\phi_p(t) = \phi(p,t)\big|_p : \mathbb{R} \to \mathfrak{P}$ is a curve that represents the trajectory of material point p, whereas with fixed second argument t i.e. $\phi_t(p) = \phi(p,t)\big|_t : \mathfrak{B} \to \mathfrak{P}$ is a configuration that represents the shape of deformable body. Configuration gradient $\mathbf{F}$ is a tangent map of configuration ϕ_t, i.e. $\mathbf{F} = T\phi_t : T\mathfrak{B} \to T\mathfrak{P}$. The values of deformation gradient are so-called two point tensors $\mathbf{F}(p) : T_p\mathfrak{B} \to T_{\phi_t(p)}\mathfrak{P}$, that in local charts $\mathfrak{X}^\alpha$ have the following coordinates

$$\mathbf{F}(p) = F^k_{\ \alpha}\, \partial_k \otimes \mathbf{d}^\alpha, \quad F^k_{\ \alpha} = \frac{\partial}{\partial \mathfrak{X}^\alpha} \phi_t^k.$$

One can define the transpose of $\mathbf{F}$ is defined by means of metrics defined in both body and space ($x = \phi_t(p)$):

$$\mathbf{F}^{\mathrm{T}}(x) : T_x\mathfrak{P} \to T_{\phi_t^{-1}x}\mathfrak{B}, \quad (\mathbf{F}\mathbf{V}, \mathbf{v})_G = \left(\mathbf{V}, \mathbf{F}^{\mathrm{T}}\mathbf{v}\right)_g.$$

Right Cauchy-Green deformation tensor $\mathbf{C}(p) : T_p\mathfrak{B} \to T_p\mathfrak{B}$, $\quad \mathbf{C}(p) = \mathbf{F}^{\mathrm{T}}(p)\mathbf{F}(p)$, in components can be written in the form

$$\mathbf{C}(p) = C(p)^\alpha_{\ \beta}\partial_\alpha \otimes \mathbf{d}^\beta, \quad C(p)^\alpha_{\ \beta} = g(\phi(p))_{kp} F(p)^p_{\ \beta} G(p)^{\alpha\beta} F(p)^k_{\ \beta},$$

that corresponds to field representation

$$C^\alpha_{\ \beta} = G^{\alpha\gamma}\left(g_{kp} \circ \phi\right) F^p_{\ \gamma} F^k_{\ \beta} \ : \mathfrak{B} \to \mathbb{R}.$$

The dual of $\mathbf{F}$ in metric independent notion has the form $\mathbf{F}^*(p) : T^*_{\phi(p)}\mathfrak{P} \to T_p^*\mathfrak{B}$, such that $\forall \alpha \in T^*\mathfrak{P}_{\phi(p)} \quad \forall \mathbf{w} \in T_p\mathfrak{B} \quad \langle \mathbf{F}^*(p)\alpha, \mathbf{w} \rangle = \langle \alpha, \mathbf{F}(p)\mathbf{w} \rangle$. In dual bases it can be written as $\mathbf{F}^* = F^\alpha_{\ k}\, \mathbf{d}^k \otimes \partial_\alpha$. Note that $\mathbf{C}^\flat$ is the pull-back of the spatial metrics, i.e. $\mathbf{C}^\flat = \phi^*\mathbf{g} = \mathbf{F}^*\mathbf{g}\mathbf{F}, \quad C_{\alpha\beta} = \left(g_{kp} \circ \phi\right) F^k_{\ \alpha} F^p_{\ \beta}$. Now everything is ready to define left Cauchy-Green deformation tensor. We use the following definition

$$\mathbf{B}(p) \,:\, T_{\phi(p)}\mathfrak{P} \to T_{\phi(p)}\mathfrak{P}, \quad \mathbf{B}(p) = \mathbf{F}(p)\mathbf{F}(p)^{\mathrm{T}}.$$

Corresponding field representation is:

$$B^k_{\ p} = \left(G^{\alpha\beta} \circ \phi^{-1}\right) g_{ps} \left(F^s_{\ \alpha} \circ \phi^{-1}\right) \left(F^k_{\ \beta} \circ \phi^{-1}\right) \,:\, \mathfrak{P} \to \mathbb{R}.$$

Note that $\mathbf{B}^\sharp$ is the push-forward of material metrics $\mathbf{B}^\sharp = \phi_*\mathbf{G}$.
With proper choice of coordinates on the body and the physical space of the density of the elastic energy is a function of two metrics, material and space.

19.7 Material Connection

Recall some facts from the theory of vector bundles. A vector bundle over a C^k-manifold $\mathfrak{M}$ is a family of vector spaces parameterized by $\mathfrak{M}$ (base manifold), i.e. $\mathfrak{M} \ni p \mapsto \mathscr{V}(p)$, such that these vector spaces fit together to form another smooth manifold $\mathscr{E}$ (total manifold) that is related to base manifold via continuous surjection (bundle projection) $\pi \,:\, \mathscr{E} \to \mathfrak{M}$.
A section of vector bundle is a continuous map, $\sigma : \mathfrak{M} \to \mathscr{E}$ such that $\forall p \in \mathfrak{M} \quad \pi \circ \sigma(p) = p$. A section σ is C^k-section, if $\sigma \in C^k(\mathfrak{M}, \mathscr{E})$.
Denote the space of C^k-sections of $\mathscr{E}$ by $\Gamma(\mathscr{E})$. A connection on $\mathscr{E}$ is a linear map

$$\nabla \,:\, \Gamma(\mathscr{E}) \to \Gamma(\mathscr{E} \otimes T\mathfrak{M})$$

such that the Leibniz rule $\nabla(\sigma f) = (\nabla\sigma) f + \sigma \otimes df$ holds for all smooth functions f on $\mathfrak{M}$ and all smooth sections σ of $\mathscr{E}$. If $\mathbf{x}$ is a tangent vector field on $\mathfrak{M}$ (i.e. a section of the tangent bundle $T\mathfrak{M}$) one can define a covariant derivative along $\mathbf{x}$

$$\nabla_{\mathbf{x}} \,:\, \Gamma(\mathscr{E}) \to \Gamma(\mathscr{E}).$$

One can treat ∇ as mappings $\nabla \,:\, \Gamma(\mathscr{E}) \times \Gamma(\mathscr{E}) \to \Gamma(\mathscr{E})$. In fixed frame field $\mathfrak{M} \ni p \mapsto (\mathbf{e}_k)_{k=1}^n$ ∇ determine real–valued functions $\Gamma^i_{\ jk} = \langle \vartheta^i, \nabla_{\mathbf{e}_j}\mathbf{e}_k \rangle$.
Let us obtain the transformation law for $\Gamma^i_{\ jk}$ with respect to changing of frame fields. Let mappings $\Omega^i_{\ j}$ define smooth non degenerated field of $n \times n$ matrices. We believe that they define local transformations infinitesimal neighborhoods of material particles into the stress free state. New frames have the form

$$\mathfrak{M} \ni p \mapsto (\mathbf{e}'_i|_p)_{i=1}^n, \quad \mathbf{e}'_i = \Omega^j_{\ i}\mathbf{e}_j.$$

The corresponding coframe fields $\mathfrak{M} \ni p \mapsto (\vartheta'^i|_p)_{i=1}^n$ are related with previous coframes as $\vartheta'^i = \mho^i_{\ j}\vartheta^j$, $[\mho^i_{\ j}] = [\Omega^i_{\ j}]^{-1}$. Finally let $\mathbf{e}_i = A^j_{\ i}\partial_{x^j}$, where (x^i) are local coordinates on $\mathfrak{M}$. Then

$$\nabla_{\mathbf{e}'_j}\mathbf{e}'_k = \nabla_{(\Omega^s{}_j \mathbf{e}_s)}(\Omega^q{}_k \mathbf{e}_q) = \Omega^s{}_j \nabla_{\mathbf{e}_s}(\Omega^q{}_k \mathbf{e}_q) = \Omega^s{}_j \left\{\Omega^q{}_k \nabla_{\mathbf{e}_s}\mathbf{e}_q + \mathbf{e}_s(\Omega^q{}_k)\mathbf{e}_q\right\},$$

that implies $\Gamma'^i{}_{jk} = \langle \vartheta'^i, \nabla_{\mathbf{e}'_j}\mathbf{e}'_k\rangle = \mho^i{}_m \langle \vartheta^m, \nabla_{\mathbf{e}'_j}\mathbf{e}'_k\rangle$, $\mathbf{e}_s(\Omega^q{}_j) = A^l{}_s \partial_{x^l}\Omega^q{}_j$,

$$\Gamma'^i{}_{jk} = \mho^i{}_m \Omega^s{}_j \left\{\Omega^q{}_k \langle \vartheta^m, \nabla_{\mathbf{e}_s}\mathbf{e}_q\rangle + \langle \vartheta^m, \mathbf{e}_q\rangle A^l{}_s \partial_{x^l}\Omega^q{}_k\right\},$$

With account of equations $\Gamma^m_{sq} = \langle \vartheta^m, \nabla_{\mathbf{e}_s}\mathbf{e}_q\rangle$ $\langle \vartheta^m, \mathbf{e}_q\rangle = \delta^m_q$, we arrive to final expression

$$\Gamma'^i{}_{jk} = \Gamma^m_{sq}\,\mho^i{}_m \Omega^s{}_j \Omega^q{}_k + A^l{}_s \mho^i{}_m \Omega^s{}_j \partial_{x^l}\Omega^m{}_k. \tag{19.6}$$

All frames fields discussed above are of special type, namely coordinate frame fields. Their values $\{\partial_i\}_{i=1}^n$ are vectors, tangential to coordinate net (recall that vector is said to be tangential to a curve if it represents the equivalent class that contain the curve). In general, frame fields gives n linearly independent vectors that are not tangential to any coordinate net. Such frames are called nonholonomic. One can construct field of nonholonomic frames by linear transformations of the values for coordinate frame fields $(\partial_i)_{i=1}^n$ that are represented by matrices $\Omega^j{}_i$ as $\mathfrak{M} \ni x \mapsto (\mathbf{z}_i)_{i=1}^n$, $\mathbf{z}_i = \Omega^j{}_i \partial_j$, where the elements of matrices $\Omega^j{}_i$ are the values of functions from $\mathscr{F}_{\mathfrak{M}}$ which satisfy the following condition $\det \Omega^j{}_i \neq 0$. For coordinate frame $[\partial_i, \partial_j] = 0$ but for the non-coordinate frame field we have $[\mathbf{z}_\alpha, \mathbf{z}_\beta] = -c_{\alpha\beta}{}^\gamma \mathbf{z}_\gamma$ where $c_{\alpha\beta}{}^\gamma$ are components of the object of anhonolomy. Taking into account the property $[f\mathbf{x}, g\mathbf{y}] = fg\,[\mathbf{x}, \mathbf{y}] + f(\mathbf{x}[g])\mathbf{y} - g(\mathbf{y}[f])\mathbf{x}$, obtain the following equation

$$[\mathbf{z}_\alpha, \mathbf{z}_\beta] = \left[\Omega^i{}_\alpha \partial_i, \Omega^j{}_\beta \partial_j\right] = \Omega^i{}_\alpha \left(\partial_i \Omega^j{}_\beta\right)\partial_j - \Omega^j{}_\beta \left(\partial_j \Omega^i{}_\alpha\right)\partial_i = -c_{\alpha\beta}{}^\gamma \Omega^m{}_\gamma \partial_m$$

that gives formula for $c_{\alpha\beta}{}^\gamma$:

$$c_{\alpha\beta}{}^\gamma = -\left(\Omega^i{}_\alpha \left(\partial_i \Omega^m{}_\beta\right) - \Omega^j{}_\beta \left(\partial_j \Omega^m{}_\alpha\right)\right)\mho^\gamma{}_m.$$

Taking into account the relation $\partial_\alpha \mho^j{}_i = -\mho^k{}_i \left(\partial_\alpha \Omega^m{}_k\right)\mho^j{}_m$ we obtain more compact formula for $c_{\alpha\beta}{}^\gamma$, i.e.

$$c_{\alpha\beta}{}^\gamma = \Omega^i{}_\alpha \Omega^j{}_\beta \left(\partial_i \mho^\gamma{}_j - \partial_j \mho^\gamma{}_i\right).$$

To given frame field $\mathfrak{M} \ni p \mapsto (\mathbf{z}_i)_{i=1}^n$ one can derive coframe field $\mathfrak{M} \ni p \mapsto \left(\vartheta^i\right)_{i=1}^n$ from duality relation $\langle \vartheta^i, \mathbf{z}_j\rangle = \delta^i_j$. Obviously, $\vartheta^\gamma = \mho^\gamma{}_m \mathbf{d}^m$.

Suppose that specific connection is chosen. It can be represented in terms of nonholonomic basis by connection 1-form

$$\Gamma \mathbf{z}_\alpha = \mathbf{z}_\gamma \otimes \omega^\gamma{}_\alpha, \quad \Gamma_{\mathbf{z}_\beta}\mathbf{z}_\alpha = \omega^\gamma{}_{\beta\alpha}\mathbf{z}_\gamma, \quad \Gamma_{\mathbf{z}_\beta}\mathbf{z}_\alpha = -\omega^\alpha{}_{\beta\gamma}\vartheta^\gamma.$$

This connection establish absolute parallelism if the following property holds

$$\Gamma \mathbf{z}_\alpha = \omega^\beta{}_\alpha \mathbf{z}_\beta = \mathbf{0}$$

that means $\omega^\beta{}_\alpha = 0$. Relation (19.6) implies that

$$\omega^\gamma{}_{\alpha\beta} = \Omega_\alpha{}^i \Omega_\beta{}^j \left(\mho_m{}^\gamma \Gamma^m{}_{ij} - \partial_i \mho_j{}^\gamma \right).$$

This equation results the connection coefficients in coordinate frame

$$\Gamma^m{}_{ij} = \Omega^m{}_\gamma \partial_i \mho^\gamma{}_j.$$

Incompatible deformation resulting in a growing body define a special connection on a manifold representing the body. This connection can serve as a mathematical description of the incompatibility.

19.8 Example

In order to describe stress-strain of a growing body it is necessary to determine the stress-strain state for a fiber as its structural element. In the case of discrete growth this structural element is a three-dimensional body $\mathfrak{B}_{n+1} \setminus \mathfrak{B}_n$, corresponding to the increment of the sequence (19.4). In the case of continuous growth the material surface plays the role of structural element. Within the present work we assume that each separate body-fiber has a natural configuration immersed in Euclidean space.

We assume that the material of a body-fibers is hyperelastic and incompressible. Then the stress-strain state can be determined analytically by universal solutions of Rivlin–Ericksen type (Lurie, 1990).

Let the image of stress-free (natural) configuration of the body-fiber $\mathfrak{B}_{n+1} \setminus \mathfrak{B}_n$ is embedded into physical (Euclidean) space $\mathscr{E}$. This embedding can be defined by the vector field of placements presented in Cartesian basis $(\mathbf{i}_1, \mathbf{i}_2, \mathbf{i}_3)$ by decomposition $\mathbf{X} = X^m \mathbf{i}_m$. Here (X^1, X^2, X^3) are Cartesian coordinates. Suppose that the deformation of the body-fiber is defined by the map $\mathbf{X} \mapsto \mathbf{x}$. We assume that this map has a symmetry relative to the axial axis of the cylindrical fibers and does not depend on coordinate X^3.

For a more compact formulation of the kinematic relations we use cylindrical coordinates (R, Θ, Z): $X^1 = R\cos\Theta$, $X^2 = R\sin\Theta$, $X^3 = Z$. The cylindrical coordinates define the local basis $(\mathbf{e}_R, \mathbf{e}_\Theta, \mathbf{e}_Z)$ and reciprocal basis $(\mathbf{e}^R, \mathbf{e}^\Theta, \mathbf{e}^Z)$. Elements of them can be presented by the decompositions

$$\mathbf{e}_R = \mathbf{e}^R = \mathbf{i}_1 \cos\Theta + \mathbf{i}_2 \sin\Theta,$$

$$\mathbf{e}_\Theta = -\mathbf{i}_1 R \sin\Theta + \mathbf{i}_2 R\cos\Theta, \mathbf{e}^\Theta = \frac{\mathbf{e}_\Theta}{R^2}, \mathbf{e}^Z = \mathbf{e}_Z = \mathbf{i}_3.$$

The reference positions of material points in simplest form can be written as $\mathbf{X} = R\mathbf{e}_R + Z\mathbf{e}_Z$. Taking into account the central symmetry, the independence with respect to vertical coordinate Z, and the condition of incompressibility $|d\mathbf{x}/d\mathbf{X}| = 1$ we arrive at the following family of mapping (universal deformations)

$$\mathbf{x}(\mathbf{X}) = \mathbf{e}_R\sqrt{(\mathbf{e}^R\cdot\mathbf{X})^2 + a} + \mathbf{e}_Z\otimes\mathbf{e}^Z\cdot\mathbf{X}. \tag{19.7}$$

Here a is a deformation parameter that represents the change of the outer cylindrical surface radius. The deformation gradient $\mathbf{F}$ and left Cauchy–Green tensor $\mathbf{B} = \mathbf{F}\cdot\mathbf{F}^{\mathrm{T}}$ are determined in terms of local basis corresponded to the reference position as follows

$$\mathbf{F} = \frac{R}{\sqrt{R^2+a}}\mathbf{e}_R\otimes\mathbf{e}_R + \frac{\sqrt{R^2+a}}{R^3}\mathbf{e}_\Theta\otimes\mathbf{e}_\Theta + \mathbf{e}_Z\otimes\mathbf{e}_Z.$$

$$\mathbf{B} = \frac{r^2-a}{r^2}\mathbf{e}_r\otimes\mathbf{e}_r + \frac{1}{r^2-a}\mathbf{e}_\theta\otimes\mathbf{e}_\theta + \mathbf{e}_z\otimes\mathbf{e}_z, \tag{19.8}$$

$$\mathbf{e}_R = \mathbf{e}_r, \quad \mathbf{e}_\Theta = \frac{\sqrt{r^2-a}}{r}, \quad \mathbf{e}_\theta, \mathbf{e}_Z = \mathbf{e}_z,$$

If the cylindrical body-fiber is produced from an incompressible material of Mooney–Rivlin type then the strain energy can be presented as a linear function of the first $I_1 = I_1(\mathbf{B})$ and second $I_2 = I_2(\mathbf{B})$ invariants of tensor $\mathbf{B}$, i.e.:

$$W(I_1, I_2) = C_1(I_1 - 3) + C_2(I_2 - 3),$$

$$I_1 = \mathrm{Tr}\mathbf{B} = 3 + \frac{a^2}{r^2(r^2-a)}, \quad I_2 = I_1.$$

Here C_1, C_2 are material constants. Under the conditions of incompressibility we have the following decomposition of Cauchy stress tensor $\mathbf{T} = -p\mathbf{I} + J_1\mathbf{B} + J_{-1}\mathbf{B}^{-1}$, where p is hydrostatic pressure, $J_1 = 2\partial W/\partial I_1 = 2C_1$ and $J_{-1} = -2\partial W/\partial I_2 = -2C_2$ are coefficients of reaction, and $\mathbf{I}$ is a unit tensor. Note that constants C_1, C_2 can be defined by pair of engineering constants μ, β, $C_1 = \mu(1+\beta)/4$, $C_2 = \mu(1-\beta)/4$. Here μ corresponds to the shear modulus and β defines the additional parameter for nonlinear response.
After simple calculations we obtain the following:

$$\mathbf{T} = T^{rr}\mathbf{e}_r\otimes\mathbf{e}_r + T^{\theta\theta}\mathbf{e}_\theta\otimes\mathbf{e}_\theta + T^{zz}\mathbf{e}_z\otimes\mathbf{e}_z, \quad T^{rr} = -p + J_1\frac{r^2-a}{r^2} + J_{-1}\frac{r^2}{r^2-a},$$

$$T^{\theta\theta} = -\frac{p}{r^2} + J_1\frac{1}{r^2-a} + J_{-1}\frac{r^2-a}{r^4}, \quad T^{zz} = -p + J_1 + J_{-1}.$$

Hydrostatic stress component p can be determined by the equilibrium equation $\nabla\cdot\mathbf{T} = \mathbf{0}$. Integrating this equation with respect to r we get

$$T^{rr} = \frac{\mu}{2}\left(\ln\frac{r^2-a}{r^2} - \frac{a}{r^2}\right) + p_0, \tag{19.9}$$

$$T^{\theta\theta} = \frac{T^{rr}}{r^2} + \frac{\mu}{r^2}\left(\frac{r^2}{r^2-a} - \frac{r^2-a}{r^2}\right), \quad T^{zz} = T^{rr} + \mu\, a\frac{r^2-(1+\beta)a/2}{r^2(r^2-a)}, \tag{19.10}$$

where p_0 is the constant of integration.

Consider a finite set of bodies. Let the elements of this set be the circular hollow cylinders of equal height h (in natural configuration). The motion (19.7) transform them to the hollow cylinders of the same height, but of another radii. Such deformation can be realized, e.g. by expanding the hollow cylinder which base lie on the smooth rigid slabs. We assume that the images of the actual configuration of the cylinders are pairwise disjoint and their union is a connected set. The final composite body can be treated as a result of discrete growth because cylindrical parts cannot deform independently after joining.

Let N be the number of cylindrical parts. Assume that the following scenario of growth is realized. On the first step the joining of the 1-st and 2-d body-fibers is performed. A composite body appears which we call the first assembly. Then the third body is joint to the composite body, etc. On the internal $r = r^1_{i,n}$ and the outer boundary $r = r^n_{e,n}$ of this composite bodies the uniformly distributed pressure $p_{i,n}$ and $p_{e,n}$ are defined

$$\mathbf{T}\cdot\mathbf{e}^r\Big|_{r=r^1_{i,n}} = p_{i,n}\mathbf{e}_r, \quad \mathbf{T}\cdot\mathbf{e}^r\Big|_{r=r^n_{e,n}} = p_{e,n}\mathbf{e}_r. \tag{19.11}$$

Index n indicates the number of assembly. The indexing in the notation of intensity of hydrostatic loads $p_{i,n}$, $p_{e,n}$ shows that they may vary during the growing process. Suppose that the contact between body-fibers is ideal, i.e. inner surface of k-th fiber and the outer surface of $k+1$-th fiber in the actual configuration are the same and stresses on them are in equilibrium:

$$\mathbf{T}\cdot\mathbf{e}^r\Big|_{r=r^k_{e,n}} = \mathbf{T}\cdot\mathbf{e}^r\Big|_{r=r^{k+1}_{i,n}}, \quad r^k_{e,n} = r^{k+1}_{i,n}, \; k = 1,2,...n-1. \tag{19.12}$$

The deformation parameters a^k_n and parameters $p^k_{0,n}$, $k = 1,2,\ldots,n$ may be found from the system of $2n$ nonlinear equations (19.11) and (19.12). Taking into account (19.7) and (19.10) we get

$$\begin{gathered}
\frac{\mu}{2}\left[\ln\frac{(R^1_i)^2}{(R^1_i)^2+a^1_n} - \frac{a^1_n}{(R^1_i)^2+a^1_n}\right] + p^1_{0,n} = p_{i,n},\\
\frac{\mu}{2}\left[\ln\frac{(R^n_e)^2}{(R^n_e)^2+a^n_n} - \frac{a^n_n}{(R^n_e)^2+a^n_n}\right] + p^n_{0,n} = p_{e,n},\\
\frac{\mu}{2}\left[\ln\frac{(R^k_e)^2}{(R^k_e)^2+a^k_n} - \frac{a^k_n}{(R^k_e)^2+a^k_n}\right] + p^k_{0,n} =\\
= \frac{\mu}{2}\left[\ln\frac{(R^{k+1}_i)^2}{(R^{k+1}_i)^2+a^{k+1}_n} - \frac{a^{k+1}_n}{(R^{k+1}_i)^2+a^{k+1}_n}\right] + p^{k+1}_{0,n},\\
(R^k_e)^2 + a^k_n = (R^{k+1}_i)^2 + a^{k+1}_n, \; k = 1,2,...,n-1.
\end{gathered} \tag{19.13}$$

The system of equation (19.13) may be transform to simpler form if one introduce the following variables and parameters:
$\alpha_k = 1 + A_k/(R_e^k)^2$, $\beta_k = \gamma_k + A_k/(R_e^k)^2$, $\nu_k = (R_e^1)^2/(R_e^k)^2$, $x_n = a_n^1/(R_e^1)^2$,
$W_n = e^{2\frac{p_{i,n}-p_{e,n}}{\mu}}$ $A_1 = 0$, $A_k = \sum_{p=2}^{k}\left((R_e^{p-1})^2 - (R_i^p)^2\right)$, $k = 2,3,\ldots,n$, $\gamma_k = \left(\frac{R_i^k}{R_e^k}\right)^2$.

Eliminating $p_{0,n}^k$ and potentiating of the left and right hand sides of above equation, we obtain

$$\prod_{k=1}^{n} \gamma_k \frac{\alpha_k + \nu_k x_n}{\beta_k + \nu_k x_n} = W_n \exp\left[\sum_{k=1}^{n}(1-\gamma_k)\frac{\nu_k x_n + \alpha_k - 1}{(\beta_k + \nu_k x_n)(\alpha_k + \nu_k x_n)}\right]. \qquad (19.14)$$

Consider the growing process with known shrinkage of fibers in the moment of accretion (some other similar problems were considered in Manzhirov and Lychev (2015); Lychev et al (2015)). In this case the reference radii of the body-fibers fibers are not known a priori, and the equations (19.14) have to be supplemented by additional equations

$$R_i^n = r_{e,n-1}^{n-1} S_n,$$

where factor S_n determines shrinkage.

To analyze the system of equations firstly allocate in the left and right hand sides of the equation (19.14) the terms corresponding to the n-th body-fiber, i.e.:

$$\gamma_n \frac{\alpha_n + \nu_n x_n}{\beta_n + \nu_n x_n} \prod_{k=1}^{n-1} \gamma_k \frac{\alpha_k + \nu_k x_n}{\beta_k + \nu_k x_n} =$$
$$= W_n \exp\left[\frac{(1-\gamma_n)(\nu_n x_n + \alpha_n - 1)}{(\beta_n + \nu_n x_n)(\alpha_n + \nu_n x_n)} + \sum_{k=1}^{n-1}\frac{(1-\gamma_k)(\nu_k x_n + \alpha_k - 1)}{(\beta_k + \nu_k x_n)(\alpha_k + \nu_k x_n)}\right]. \qquad (19.15)$$

The values of γ_n, α_n, β_n, ν_n, can be defined recursively by relations

$$\gamma_n = \frac{S_n^2(\alpha_{n-1} + x_{n-1}\nu_{n-1})}{\zeta_{n-1}}, \qquad \alpha_n = 1 + \frac{\alpha_{n-1} - S_n^2(\alpha_{n-1} + x_{n-1}\nu_{n-1})}{\zeta_{n-1}},$$
$$\beta_n = \gamma_n + \frac{\alpha_{n-1} - S_n^2(\alpha_{n-1} + x_{n-1}\nu_{n-1})}{\zeta_{n-1}}, \qquad \nu_n = \frac{\nu_{n-1}}{\zeta_{n-1}},$$
$$\zeta_{n-1} = \left(S_n\sqrt{\alpha_{n-1} + x_{n-1}\nu_{n-1}} + \sqrt{\xi_n \nu_{n-1}}\right)^2.$$

Consistent solution of systems of equations completes the solution of problem.

As a result of solving the problem:

- The components of metric tensor associated with each layer are computed. With an increase in the number of layers and reducing their thickness this components approach to continuous functions that defined on Riemannian metrics on $\mathfrak{B}$. Thus one can use approach discussed in section 6.

- The transformations of individual thin cylinders into stress free form are computed. This transformations tends to smooth field of linear transformations, namely Ω, so one can use approach discussed in section 7.

References

Abbott E, Burger D (1976) Flatland and Sphereland (in Russ.). Mir, Moscow

Acharya A (2001) A model of crystal plasticity based on the theory of continuously distributed dislocations. Journal of the Mechanics and Physics of Solids 49(4):761 – 784

Adachi M (2012) Embeddings and Immersions. American Mathematical Soc.

Armstrong MA (1983) Basic Topology. Springer, New York

Arutyunyan NK, Drozdov AD, Naumov VE (1987) Mechanics of Growing Viscoelastoplastic Solids (in Russ.). Nauka, Moscow

Berger M (2003) A Panoramic View of Riemannian Geometry. Springer Science & Business Media, Berlin

Bilby BA, Bullough R, Smith E (1955) Continuous distributions of dislocations: a new application of the methods of non-riemannian geometry. Proceedings of the Royal Society of London A: Mathematical, Physical and Engineering Sciences 231(1185):263–273

Borisovich IG, Blizniakov NM, Izrailevich IA, Fomenko TN (1995) Introduction to Topology (in Russ.). Nauka, Fizmatlit, Moscow

Cartan E (1962) Spaces of Affine, Projective and Conformal Connection. Kazan University Press, Kazan

Cartan E (1983) Geometry of Riemannian Spaces. Math Sci Press, Brookline, MA

Choy K (2003) Chemical vapour deposition of coatings. Progress in Materials Science 48(2):57 – 170

De Wit R (1977) Continual theory of disclinations (in russ.). In: Mekhanika, Novoe Zarubezh. Nauke, Mir, Moscow

Dubrovin BA, Novikov SP, Fomenko AT (1986) Modern Geometry: Methods and Applications (in Russ.). Nauka, Moscow

Eckart C (1948) The thermodynamics of irreversible processes. iv. the theory of elasticity and anelasticity. Phys Rev 73:373–382

Eisenhart LP (1997) Riemannian Geometry. Princeton University Press, Princeton

Eisenhart LP (2005) Non-Riemannian Geometry. Dover, New York

Epstein M (2010) The Geometrical Language of Continuum Mechanics. Cambridge University Press, Cambridge

Epstein M, , Elzanowski M (2007) Material Inhomogeneities and Their Evolution: A Geometric Approach. Springer Science & Business Media, Berlin

Gurtin ME, Murdoch AI (1975) A continuum theory of elastic material surfaces. Archive for Rational Mechanics and Analysis 57(4):291–323

Hirsch MW (1976) Differential Topology. Springer, New York

Huang Q, Zhang J, Sabbaghi A, Dasgupta T (2015) Optimal offline compensation of shape shrinkage for three-dimensional printing processes. IIE Transactions 47(5):431–441

Kadic A, Edelen DGB (1983) A Gauge Theory of Dislocations and Disclinations, Lecture Notes in Physics, vol 174. Springer, Berlin, Heidelberg

Katanaev MO (2005) Geometric theory of defects. Physics-Uspekhi 48(7):675–702

Kobayashi S, Nomizu K (1963) Foundations of Differential Geometry, vol 1. John Wiley & Sons, New York

Kondo K (1955a) Geometry of elastic deformation and incompatibility. In: RAAG Memoirs of the Unifying Study of the Basic Problems in Engineering Science by Means of Geometry, vol Div. C, Gakujutsu Bunken Fukyo-Kai, Tokyo, pp 5–17

Kondo K (1955b) Non-riemannian geometry of imperfect crystals from a macroscopic viewpoin. In: RAAG Memoirs of the Unifying Study of the Basic Problems in Engineering Science by Means of Geometry, vol 1, Division D, Gakujutsu Bunken Fukyo-Kai, Tokyo, p 458–469

Kondo K (1963) Non-Riemannian and Finslerian approaches to the theory of yielding. International Journal of Engineering Science 1(1):71 – 88

Kondo K (1964) On the analytical and physical foundations of the theory of dislocations and yielding by the differential geometry of continua. International Journal of Engineering Science 2(3):219 – 251

Kröner E (1958) Kontinuumstheorie der Versetzungen und Eigenspannungen, Ergebnisse der angewandten Mathematik, vol 5. Springer, Berlin, Heidelberg

Kröner E (1996) Dislocation theory as a physical field theory. Meccanica 31:577–587

Kuratowski K (1966) Topology, vol I. Academic Press, New York

Lu L, Zheng J, Mishra S (2015) A layer-to-layer model and feedback control of ink-jet 3-d printing. IEEE/ASME Transactions on Mechatronics 20(3):1056–1068

Lurie AI (1990) Non-linear Theory of Elasticity. North-Holland

Lützen J (1995) Interactions between mechanics and differential geometry in the 19th century. Archive for History of Exact Sciences 49(1):1–72

Lychev S, Manzhirov A (2013a) The mathematical theory of growing bodies. finite deformations. Journal of Applied Mathematics and Mechanics 77(4):421 – 432

Lychev SA (2011) Universal deformations of growing solids. Mechanics of Solids 46(6):863–876

Lychev SA, Koifman KG (2016) Geometric Aspects of the Theory of Incompatible Deformations. Part I. Uniform Configurations. Nanomechanics Science and Technology: An International Journal 7(3):177–233

Lychev SA, Manzhirov AV (2013b) Reference configurations of growing bodies. Mechanics of Solids 48(5):553–560

Lychev SA, Manzhirov AV, Bychkov PS (2015) Discrete and continuous growth of deformable cylinder. In: Transactions on Engineering Technologies: World Congress on Engineering 2014, Springer Netherlands, pp 239–254

Manzhirov AV (2016) Fundamentals of mechanical design and analysis for {AM} fabricated parts. Procedia Manufacturing 7:59 – 65

Manzhirov AV, Lychev SA (2012) The mathematical theory of growing solids: Finite deformations. Doklady Physics 57(4):160–163

Manzhirov AV, Lychev SA (2015) An approach to modeling of additive manufacturing technologies. In: Transactions on Engineering Technologies: World Congress on Engineering 2014, Springer Netherlands, pp 99–115

Manzhirov AV, Lychev SA, , Gupta NK (2013) Nonlinear models of growing solids. Proc Ind Nat Sci Acad 79(4A):529–545

Maugin GA (1993) Material Inhomogeneities in Elasticity, vol 3. CRC Press, Boca Raton, FL

Myasnikov VP, Guzev MA (2000) Non-euclidean model of materials deformed at different structural levels. Physical Mesomechanics 3(1):5–16

Nakahara M (2003) Geometry, Topology and Physics. CRC Press

Nastasi MA, Mayer JW (2006) Ion Implantation and Synthesis of Materials. Springer, Berlin

Noll W (1967) Materially uniform simple bodies with inhomogeneities. Archive for Rational Mechanics and Analysis 27(1):1–32

Nye J (1953) Some geometrical relations in dislocated crystals. Acta Metallurgica 1(2):153 – 162

Ozakin A, Yavari A (2010) A geometric theory of thermal stresses. Journal of Mathematical Physics 51(3):032,902

Petersen P (2006) Riemannian Geometry, Graduate Texts in Mathematics, vol 171. Springer Science & Business Media, Berlin

Postnikov MM (1987) Lectures in Geometry. Semester III. Smooth Manifolds (in Russ.). Nauka, Moscow

Seeger A (1961) Recent advances in the theory of defects in crystals. physica status solidi (b) 1(7):669–698

Sozio F, Yavari A (2017) Nonlinear mechanics of surface growth for cylindrical and spherical elastic bodies. Journal of the Mechanics and Physics of Solids 98:12 – 48

Sternberg S (1964) Lectures on Differential Geometry, vol 2. Prentice-Hall, Englewood Cliffs, NJ

Volterra V, Volterra E (1960) Sur les distorsions des corps élastiques (théorie et applications). In: Memorial des Sciences Mathematiques, 147, Gauthier-Villars, pp 3–117

Wang CC (1967) On the geometric structures of simple bodies, a mathematical foundation for the theory of continuous distributions of dislocations. Archive for Rational Mechanics and Analysis 27(1):33–94

Weingarten J (1901) Sulle superficie di discontinuita nella teoria della elasticita dei corpi solidi. Rend Reale Accad dei Lincei, classe di sci, fis, mat e nat 10.1(5):57–60

Yavari A (2010) A geometric theory of growth mechanics. Journal of Nonlinear Science 20(6):781–830

Yavari A, Goriely A (2013) Riemann–cartan geometry of nonlinear disclination mechanics. Mathematics and Mechanics of Solids 18(1):91–102

Zubov LM (1997) Nonlinear Theory of Dislocations and Disclinations in Elastic Bodies. Springer Science & Business Media, Berlin

Chapter 20
Free Vibrations of a Transversely Isotropic Plate with Application to a Multilayer Nano-Plate

Nikita F. Morozov, Petr E. Tovstik & Tatyana P. Tovstik

Abstract Free vibrations of a transversely isotropic rectangular plate are investigated within the frameworks of the Timoshenko–Reissner (TR) model including a transversal shear. The dependence of the principal natural frequency on the shear parameter for various variants of boundary conditions is found. The generalized Kirchhoff–Love (GKL) model of fourth differential order for a multilayered plate is proposed. The GKL model is simpler than the TR model, and it gives the same results as the TR model in the most cases. This model is used to describe vibrations of multilayer graphene sheets (MLGS). Explicit formulas for principal frequencies for both multilayer ordinary plates and for MLGS are derived.

Key words: Multilayer plate · Free vibrations · Generalized Kirchhoff-Love model · Boundary conditions effect · Multilayer graphene plate vibration

20.1 Introduction

The classical equation of plate vibrations was derived by Sophie Germain in 1808 with a purpose of explaining Chladni's figures of acoustics. The equation of bending and vibrations of a plate can be obtained on the basis of the Kirchhoff–Love (KL) hypotheses (Kirchhoff, 1876; Love, 1927). More involved and sometimes more exact equations, which take into account the transversal shear and the rotation inertia, can be derived from the Timoshenko–Reissner (TR) hypotheses (Timoshenko, 1921;

Nikita F. Morozov · Petr E. Tovstik
St.Petersburg State University, Universitetskaya nab. 7/9, 199034 St.Petersburg, Russia,
e-mail: morozov@nm1016.spb.edu, peter.tovstik@mail.ru

Tatyana P. Tovstik
Institute of Problems of Mechanical Engineering RAS, Bol'shoi pr. 61, 199178 St. Petersburg, Russia,
e-mail: tovstik_t@mail.ru

H. Altenbach et al. (eds.), *Mechanics for Materials and Technologies*,
Advanced Structured Materials 46, DOI 10.1007/978-3-319-56050-2_20

Reissner, 1945; Mindlin, 1951). An equation of second-order accuracy (SA) with respect to the small thickness parameter $\mu = h/L$ was proposed in Kienzler and Schneider (2014); Tovstik and Tovstik (2014); Morozov et al (2016b). Here, h is the plate thickness and L is the length of waves in the tangential directions.

The SA equations contain the main part, which agrees with the KL equations, and two small additional parts of order μ^2 by comparison with the main part. The first of them, which gives the TR model, consists of summands that account for the transversal shear. The second part describes the effects of the reduce in the plate thickness and has the same asymptotical order μ^2 as the shear part. These summands are not included in the KL and in the TR models. Therefore, for a description of the internal stress state of a thin isotropic plate the TR model has proved to be not more exact (Kienzler and Schneider, 2014; Tovstik, 2007) (in the asymptotic point of view with respect to the small parameter μ) than the KL model, because it does not contain all the summands of the order μ^2.

The asymptotic estimates change for a plate made of a transverse isotropic material with a small transversal shear modulus. We introduce the dimensionless shear parameter $g \sim \mu^2 E/G$, where E and G are the Young and the transversal shear moduli, respectively. The shear summands are of the order of g by comparison with the main bending summands, the orders of the rest summands of the SA model being independent of g. As a result, if $g \geq 1$, then the KL model is unacceptable, and the TR and SA models may be used. The difference between the TR and SA models is of order μ^2, but the TR model is more simple. This is why we study the TR model. Also, problems for multi-layer plates and for functional graded plates are known to be reducible to the TR model (Tovstik and Tovstik, 2014; Morozov et al, 2016b).

3D boundary layers appear near the plate edges; they can be approximately described by the TR model. An analysis of the boundary layers and of their influence on the stress state and on the plate vibrations was extensively studied (see, e.g. Altenbach and Zhilin, 1988; Zhilin, 1992; Ivanova, 1997)). The effect of the rotation inertia involved in the TR model is immaterial for low-frequency vibrations, but it is important for high-frequency vibrations (Ivanova, 1998).

The present paper is concerned with free bending low-frequency vibrations of an elastic homogeneous rectangular plate. A plate may be isotropic or transversely isotropic. The KL and the TR models are used and only the principal natural frequency is discussed for all the main boundary conditions. It is well known that if all the plate edges are simply supported, then the frequency can be written down explicitly. If two opposite edges are simply supported, then the problem admits a separation of variables, the problem is reducible to an one-dimensional problem, and the solution can be simply constructed. This case is studied in the present paper, the dependence of the principal frequency on the parameter g is investigated for various variants of boundary conditions. The KL and the TR models are compared. If the condition that two opposite edges are simply supported is not fulfilled, then the exact analytical solution is absent and only an approximate or numerical solution may be found.

Problems for multi-layer plates and for functional graded plates are known to be reducible to the TR model (Tovstik and Tovstik, 2014; Morozov et al, 2016b). Here, the generalized TR equations for vibrations of a multilayer plate are presented (Morozov et al, 2016b). A continuum model describing vibrations of a crystal graphite nano-plate consisting of graphene layers was proposed in Morozov et al (2016a). The TR model is used, and the explicit formulae for the natural bending frequencies of the nano-plate are delivered.

20.2 Equations of Motion and Their Transformation

Consider a thin homogeneous rectangular plate with sides L_x, L_y and constant thickness h. Free small bending vibrations of a plate within the framework of the TR model are described as:

$$\begin{aligned} \frac{\partial M_1}{\partial x}+\frac{\partial H}{\partial y}-Q_1+\rho J\omega^2\varphi_1&=0,\\ \frac{\partial H}{\partial x}+\frac{\partial M_2}{\partial y}-Q_2+\rho J\omega^2\varphi_2&=0, \qquad J=\frac{h^3}{12},\\ \frac{\partial Q_1}{\partial x}+\frac{\partial Q_2}{\partial y}+\rho h\omega^2 w&=0, \end{aligned} \tag{20.1}$$

where x $(0\le x\le L_x), y$ $(0\le y\le L_y)$ are the Cartesian coordinates in the midplane, $w(x,y)$ is the normal deflection, $\varphi_1(x,y)$ and $\varphi_2(x,y)$ are the average angles of the normal fibre rotation, ω is the natural frequency, ρ is the mass density, M_1, M_2 and H are the bending moments and the torsion moment, respectively, and Q_1 and Q_2 are the transversal shear stress-resultants.

According to the TR model the following elasticity relations are accepted

$$\begin{aligned} &M_1=D(\kappa_1+\nu\kappa_2),\quad M_2=D(\kappa_2+\nu\kappa_1),\quad H=(1-\nu)D\tau,\\ &Q_1=\Gamma\gamma_1,\quad Q_2=\Gamma\gamma_2,\quad \Gamma=kG'h,\quad k=\frac{5}{6},\quad D=\frac{Eh^3}{12(1-\nu^2)}, \end{aligned} \tag{20.2}$$

where

$$\kappa_1=\frac{\partial\varphi_1}{\partial x},\quad \kappa_2=\frac{\partial\varphi_2}{\partial y},\quad 2\tau=\frac{\partial\varphi_1}{\partial y}+\frac{\partial\varphi_2}{\partial x},\quad \gamma_1=\varphi_1+\frac{\partial w}{\partial x},\quad \gamma_2=\varphi_2+\frac{\partial w}{\partial y}. \tag{20.3}$$

Here, E is the Young modulus, ν is the Poisson's ratio, D is the plate bending stiffness, $G'=G_{13}$ is the transversal shear modulus (for a transversely isotropic material the value G' is arbitrary, and for an isotropic material $G'=G=E/(2(1+\nu))$), γ_1 and γ_2 are the shear angles, and k is the correcting multiplier taking into account the distribution of the transversal shear stresses.

In the KL model the shear angles $\gamma_1=\gamma_2=0$, the equations of motion (20.1) remain the same, and the stress-resultants Q_1 and Q_2 are to be found from Eqs. (20.1).

The basic unknowns in the TR model are w, φ_1, φ_2. The differential order of system (20.1) is six, and so three boundary conditions are to be imposed at the plate edges.

We study the following eight variants of boundary conditions at the edges $x = 0$ and $x = L_x$:

$$
\begin{array}{lll}
w = 0 & \text{or} & Q_1 = 0 \\
\varphi_1 = 0 & \text{or} & M_1 = 0 \\
\varphi_2 = 0 & \text{or} & H = 0,
\end{array}
\tag{20.4}
$$

where the first column contains the constrains and the second column contains the corresponding natural boundary conditions. Only one condition is to be taken from each line of Eqs. (20.4). We study only one variant of boundary conditions at the edges $y = 0$ and $y = L_y$:

$$
w = M_2 = \varphi_1 = 0 \quad \text{at} \quad y = 0 \quad \text{and} \quad y = L_y \tag{20.5}
$$

which admits a separation of variables.

The KL system is of fourth order, and so two boundary conditions are to be imposed at the each edge

$$
\begin{array}{l}
w = 0 \ \text{ or } Q_1^* = Q_1 + \dfrac{\partial H}{\partial y} = 0 \\
\varphi_1 = 0 \text{ or } M_1 = 0,
\end{array}
\quad \text{at} \quad x = 0 \quad \text{and} \quad x = L_x \tag{20.6}
$$

The natural frequencies can be found from the variation problem

$$
\omega^2 = \min_{w, \varphi_1, \varphi_2} \frac{\Pi}{T}, \tag{20.7}
$$

where the potential and the kinetic energies are

$$
\begin{aligned}
\Pi &= \frac{1}{2} \int_0^{L_x} \int_0^{L_y} \left(D(\kappa_1^2 + 2\nu\kappa_1\kappa_2 + \kappa_2^2 + 2(1-\nu)\tau^2) + \Gamma(\gamma_1^2 + \gamma_2^2) \right) \mathrm{d}x \mathrm{d}y, \\
T &= \frac{1}{2} \int_0^{L_x} \int_0^{L_y} \left(\rho h w^2 + \rho J(\varphi_1^2 + \varphi_2^2) \right) \mathrm{d}x \mathrm{d}y.
\end{aligned}
\tag{20.8}
$$

The minimum in Eq. (20.7) is to be found under constrains $w = 0$ and/or $\varphi_1 = 0$ and/or $\varphi_2 = 0$ at the plate edges if they are imposed. The functional (20.7) variation leads to Eqs. (20.1) and to the natural boundary conditions in the right column of Eqs. (20.4).

For the KL model we put $\Gamma = J = 0$ in Eq. (20.8).

To simplify the subsequent calculations we introduce the potential functions Ψ and Θ (see Morozov et al, 2016b; Ivanova, 1997):

$$
\varphi_1 = -\frac{\partial \Psi}{\partial x_1} + \frac{\partial \Theta}{\partial x_2}, \qquad \varphi_2 = -\frac{\partial \Psi}{\partial x_2} - \frac{\partial \Theta}{\partial x_1}. \tag{20.9}
$$

As a result, we get the equations for Θ,

$$\frac{1-\nu}{2}D\Delta\Theta - \Gamma\Theta + \rho J\omega^2\Theta = 0, \qquad \Delta(\,) = \frac{\partial^2(\,)}{\partial x^2} + \frac{\partial^2(\,)}{\partial y^2}, \tag{20.10}$$

and the equations for Ψ and for w,

$$-D\Delta^2\Psi + \rho h\omega^2 w - \rho J\omega^2\Delta\Psi = 0, \tag{20.11}$$

$$D\Delta^2 w - \rho h\omega^2 + \left(\frac{D\rho h}{\Gamma} + \rho J\right)\omega^2\Delta w + \frac{J\rho^2 h}{\Gamma}\omega^4 w = 0. \tag{20.12}$$

For an isotropic plate, Eq. (20.10) describes the boundary layer. Equation (20.12) describes the plate bending. Equations (20.10) and (20.12) are connected through the boundary conditions. The summands involving the multiplier J take into account the rotation inertia, and their effect on the principal frequency is very small. Further, we put $J = 0$.

20.3 Principal Natural Frequency in the Dependence of Boundary Conditions

If all plate edges are simply supported, namely, if the conditions

$$w = M_1 = \varphi_2 = 0 \quad \text{at} \quad x = 0,\; x = L_x, \tag{20.13}$$

and (20.5) are imposed, then the function

$$w(x,y) = \sin px \sin qy, \qquad p = \frac{\pi}{L_x}, \quad q = \frac{\pi}{L_y},$$

satisfies all boundary conditions, Eq. (20.12) leading to the principal frequency ω:

$$\omega^2 = \frac{D(p^2+q^2)^2}{\rho h(1+(D/\Gamma)(p^2+q^2))}. \tag{20.14}$$

Let the edges $y = 0$ and $y = L_y$ be simply supported, see Eqs. (20.5). Then we may separate the variables as follows:

$$\begin{aligned} \{w, \varphi_1, M_1, M_2, Q_1\}(x,y) &= \{w, \varphi_1, M_1, M_2, Q_1\}(x)\sin qy, \\ \{\varphi_2, H, Q_2\}(x,y) &= \{\varphi_2, H, Q_2\}(x)\cos qy. \end{aligned} \tag{20.15}$$

We introduce the dimensionless variables

$$\hat{x} = qx,\, \hat{y} = qy,\, \Lambda = \omega_0^2 = \frac{\rho h\omega^2}{Dq^4},\, \mu = qh = \frac{\pi h}{L_y},\, g = \frac{Dq^2}{\Gamma} = \frac{\mu^2 E}{10(1-\nu^2)G'}, \tag{20.16}$$

where Λ is the frequency parameter, ω_0 is the dimensionless frequency, μ is the small thickness parameter, g is the shear parameter. Further the hats will be omitted. In this notation Eq. (20.14) assumes the form

$$\omega_0^2 = \frac{(\hat{p}^2+1)^2}{1+g(\hat{p}^2+1)}, \quad \hat{p} = \frac{p}{q}. \tag{20.17}$$

The parameter g plays an important role in the following investigation. For an isotropic plate, $g = \mu^2/(5(1-\nu)) \ll 1$, and the effect of the transversal shear in Eq. (20.17) is small. In the case with the small transversal shear modulus ($G' \ll E$) its influence may be essential.

Now we write the general solution to Eqs. (20.1) after the separation of variables (20.15) in the dimensionless form. We seek the partial solutions in the form $Y(x) = y_0 e^{\lambda x}$, where Y is any unknown function. For the parameter λ according Eqs. (20.10) and (20.12) we obtain the equation of the sixth order,

$$(\lambda^2 - \lambda_0^2)\cdot((\lambda^2-1)^2 + \Lambda(g(\lambda^2-1)-1)) = 0, \tag{20.18}$$

with $\lambda_0^2 = 1 + 2/(g(1-n))$. The roots of this equation are $\pm\lambda_0,\ \pm\lambda_1,\ \pm\lambda_2$, where $\lambda_{1,2}^2 = 1 - \Lambda g/2 \pm \sqrt{(\Lambda g/2)^2 + \Lambda}$. The roots $\pm\lambda_0,\ \pm\lambda_1$ are real, and the roots $\pm\lambda_2$ may be real or imaginary. For small g the roots $\pm\lambda_0$ are large that generates the boundary layer.

For simplicity, we shall consider the case when at the edges $x = 0$ and $x = L_x$ the identical boundary conditions are given. For convenience we satisfy the boundary conditions at $x = \pm x_0,\ x_0 = L_x/2$. Then the even deflection mode

$$w(x) = C_0 \cosh\lambda_0 x + C_1 \cosh\lambda_1 x + C_2 \cosh\lambda_2 x \tag{20.19}$$

corresponds to the principal frequency, where C_k are arbitrary constants.

Changing the constant C_k we take the following unknown functions involved in the boundary conditions (20.4):

$$\begin{aligned}
w(x) &= C_1(g(\lambda_1^2-1)-1)\cosh\lambda_1 x + C_2(g(\lambda_2^2-1)-1)\cosh\lambda_2 x,\\
\varphi_1(x) &= C_0 \sinh\lambda_0 x + C_1\lambda_1 \sinh\lambda_1 x + C_2\lambda_2 \sinh\lambda_2 x,\\
\varphi_2(x) &= C_0\lambda_0 \cosh\lambda_0 x + C_1 \cosh\lambda_1 x + C_2 \cosh\lambda_2 x,\\
Q_1(x) &= C_0 g^{-1} \sinh\lambda_0 x + C_1(\lambda_1^3 - \lambda_1)\sinh\lambda_1 x + C_2(\lambda_2^3 - \lambda_2)\sinh\lambda_2 x,\\
M_1(x) &= C_0(1-\nu)\lambda_0 \cosh\lambda_0 x + C_1(\lambda_1^2 - \nu)\cosh\lambda_1 x + C_2(\lambda_2^2-\nu)\cosh\lambda_2 x,\\
H(x) &= C_0(g^{-1}+1-\nu)\sinh\lambda_0 x + C_1(1-\nu)\lambda_1 \sinh\lambda_1 x + C_2(1-\nu)\lambda_2 \sinh\lambda_2 x.
\end{aligned} \tag{20.20}$$

We get the frequency equation after equating to zero the determinant of the third order constructed by using Eqs. (20.4). For the imaginary root $\lambda_2 = i\beta$ we take $\cosh\lambda_2 x = \cos\beta x,\ \lambda_2 \sinh\lambda_2 x = -\beta\sin\beta x$.

We also study the generalized KL model (GKL) described by Eq. (20.12) with $J = 0$. In this model, Eq. (20.10) (and simultaneously a boundary layer) is ignored, and the boundary conditions are described by Eqs. (20.6), in which the unknown functions w, f_1, M_1 are given by Eqs. (20.20) with $C_0 = 0$, and

$$Q_1^*(x) = C_1(\lambda_1^3 - \lambda_1(2-\nu))\sinh\lambda_1 x + C_2(\lambda_2^3 - \lambda_2(2-\nu))\sinh\lambda_2 x. \qquad (20.21)$$

If $g = 0$ the GKL model coincides with the ordinary KL model.

20.4 Numerical Results and Their Discussion

We study all 8 possible variants of boundary conditions (20.4) with $x = x_0 = 1$ and with the Poisson ratio $\nu = 0.3$. The shear parameter g changes in the wide limits $0 < g < \infty$. Dependences $\omega_0(g)$ for various variants of boundary conditions are presented in Table 20.1 and in Fig. 20.1.

The numbering of variants of the boundary conditions is chosen so that for moderate g the value $\omega_0(g)$ decreases as the number N increases. In the left part of Fig. 20.1 the functions $\omega_0(g)$ are shown for $0 < g \leq 1$ and in the right part, for $1 \leq g \leq 100$. Some curves in these figures are seen to be very close to each other, the differences between them may be found from Table 20.1.

In the column $g \to 0$ the limiting values $\omega_0(g)$ at $\omega \to 0$ are given. For an isotropic plate the case $g \ll 1$ is investigated in Ivanova (1997); we briefly recall some of the

Table 20.1: Dependences $\omega_0(g)$ for various variants of boudary conditions

N	boundary cond.	$g\to 0$	$g=.001$	$g=.01$	$g=.1$	$g=1$	$g=10$	$g=100$
1	$w=\varphi_1=\varphi_2=0$	6.197	6.156	5.821	4.095	1.735	0.584	0.186
2	$w=\varphi_1=H=0$	6.197	6.156	5.815	4.047	1.699	0.581	0.186
3	$w=M_1=\varphi_2=0$	3.647	3.462	3.407	2.988	1.641	0.581	0.186
4	$w=M_1=H=0$	3.647	3.443	3.352	2.870	1.593	0.578	0.186
5	$Q_1=\varphi_1=\varphi_2=0$	6.197	3.805	2.504	1.578	0.837	0.309	0.100
6	$Q_1=M_1=\varphi_2=0$	3.647	3.049	2.420	1.568	0.819	0.308	0.100
7	$Q_1=\varphi_1=H=0$	1.000	1.000	0.995	0.953	0.707	0.302	0.099
8	$Q_1=M_1=H=0$	0.967	0.967	0.961	0.921	0.691	0.300	0.099
8*	$Q_1^*=M_1=0$	0.967	0.967	0.969	0.932	0.723	0.305	0.100

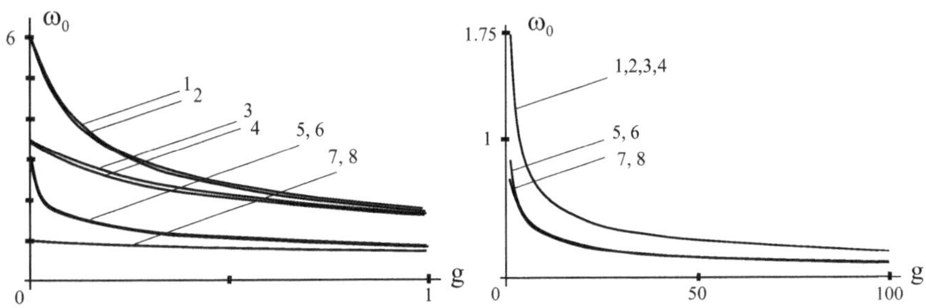

Fig. 20.1: The functions $\omega_0(g)$ for the various variants of boundary conditions at $0 < g \leq 1$ (left), and at $1 \leq g \leq 100$ (right)

results obtained in Ivanova (1997). The limiting values of $\omega_0(+0)$ form four groups: the clamped group (1,2,5), the simply supported group (3,4,6), and the free edge group consisting of two close parts (7) and (8). In accordance with the variation principle (20.7) and with the relation (20.16), $g = Dq^2/\Gamma$, all the functions $\omega_0(g)$ decrease as g grows. From Table 20.1 and from the left part of Fig. 20.1 it follows that for variants 5 and 6 the functions $\omega_0(g)$ decrease very quickly near the point $g = 0$. It is connected with the constraint $\varphi_2 = 0$ which for the very small values g plays the role of the constraint $w = 0$. The constraint $\varphi_2 = 0$ is absent in the framework of the KL model; its significance is rather theoretical than applied.

For moderate values of g ($0.01 \leq g \leq 1$) four other groups with the close values $\omega_0(g)$ are formed: (1,2), (3,4), (5,6), and (7,8).

For large values of g ($g > 1$) only two groups remain: the group (1,2,3,4) with the constraint $w = 0$, and the group (5,6,7,8) with the condition $Q_1 = 0$. For these groups the following asymptotic formulas hold, respectively:

$$\omega_0(g) \approx \frac{1.862}{\sqrt{g}}, \quad \omega_0(g) \approx \frac{1}{\sqrt{g}} \quad \text{at} \quad g \gg 1. \tag{20.22}$$

The first formula (20.22) follows from Eq. (20.17), because Eq. (20.17) corresponds to variant 3. The second formula in (20.22) can be delivered from the variation relation (20.7) by taking $w(x) = 1$.

For a plate with the sides L_x and L_y Eqs. (20.22) assume the form

$$\omega_0(g) \approx \sqrt{\frac{(L_y/L_x)^2 + 1}{g}}, \quad \omega_0(g) \approx \frac{1}{\sqrt{g}} \quad \text{for} \quad g \gg 1. \tag{20.23}$$

Remark 20.1. From Eqs. (20.22) and (20.23) the important conclusion follows: for the large enough shear parameter g it is not necessary to distinguish the details of boundary conditions, and it is enough to look at the presence of a constraint $w = 0$ among them.

In a framework of the GKL model the frequencies at three variants $w = \varphi_1 = 0$, $w = M_1 = 0$ and $Q_1 = \varphi_1 = 0$ of boundary conditions coincide with variants 2, 4, and 7 of the TR model, because in these variants the boundary layer does not appear and $C_0 = 0$ in Eqs. (20.20) (Ivanova, 1997). The variant $M_1 = Q_1^* = 0$ of the GKL model gives frequencies, which are slightly larger than those in variant 8 of the TR model (compare lines 8 and 8* in Table 20.1).

Variants 1 and 3 differ from variants 2 and 4, respectively, by the additional constraint $\varphi_2 = 0$ that leads to the small frequency increase.

It should be noted that all results in Table 20.1 (excluding lines 8 and 8*) are independent of the Poisson ratio ν.

The conclusion of this analysis is as follows: to find the principal bending frequency we recommend to use the GKL model (Eq. (20.12) with $J = 0$ and with two boundary conditions) if the shear parameter g is not very small. Sometimes this approach coincides with the TR model, and in the other cases the error is not large (see Table 20.1).

20.5 The Generalized Kirchhoff–Love (GKL) Model for a Multilayer Plate

Consider a plate of thickness h, consisting of n homogeneous isotropic layers of thicknesses h_k, $k = 1,2,\dots,n$ $(h = \sum h_k)$. Let E_k, ν_k, ρ_k be the Young moduli, the Poisson ratios and the mass densities of layers, respectively.

To reduce the vibration problem to the GKL model

$$D\Delta^2 w - \rho h\omega^2 + \frac{D\rho h}{\Gamma}\omega^2 \Delta w = 0 \tag{20.24}$$

it is necessary to find the equivalent values for a multilayer plate: the bending stiffness D, the shear stiffness Γ, and the mass per unit area ρh. We get (see Morozov et al, 2016b,a)

$$D = \int_0^h (z-a)^2 E_0 \mathrm{d}z, \quad \frac{1}{\Gamma} = \frac{1}{D^2}\int_0^h \frac{1}{G}\left(\int_0^z (z-a)E_0 \mathrm{d}z\right)^2 \mathrm{d}z, \quad \rho h = \int_0^h \rho\, \mathrm{d}z. \tag{20.25}$$

where a is the coordinate of the neutral plane (for a symmetric plate $a = h/2$), $E_0 = E/(1-\nu^2)$, $G = E/(2(1+\nu))$, and E, ν, ρ are the piecewise functions of z. If there are layers with small Young moduli, then the equivalent shear stiffness Γ is also small.

Consider the same rectangular plate as in Sect. 20.2, and let the boundary condition (20.5) be fulfilled. Then the principal bending frequency ω may be written as:

$$\omega = \sqrt{\frac{Dq^4}{\rho h}}\,\omega_0(g,r), \qquad g = \frac{Dq^2}{\Gamma}, \quad q = \frac{\pi}{L_y}, \quad r = \frac{L_y}{L_x}. \tag{20.26}$$

Here, the dimensionless frequency ω_0 depends on the boundary conditions at $x = 0$ and $x = L_x$, and also on the parameters g and r. If the boundary conditions at $x = 0$ and $x = L_x$ are identical, and the aspect ratio $r = \pi/2$, then the value of ω_0 is given in Sect. 20.4.

20.6 Continuum Model of a Multilayer Graphene Sheet (MLGS) Vibrations

Let a rectangular MLGS consists of $n+1$ graphene layers. We model each single-layer graphene sheet (SLGS) as a thin isotropic plate with the extension stiffness K_0 and the bending stiffness D_0 (Morozov et al, 2016a). We model the intermediate layers of thicknesses h_0 between graphene layers as isotropic elastic fictitious layers with the small stiffness. In these layers, the van-der-Waals (vdW) forces act. Hence, we get a multilayer plate (see Fig. 20.2). We model it as an one-layered homogeneous GKL plate (see Sect. 20.5), and calculate the equivalent bending stiffness D and the shear stiffness Γ by Eqs. (20.25). In calculating the integrals in Eqs. (20.25) we neglect the thickness of graphene layers compared with h_0 (then $h = nh_0$), neglect the shear compliance of graphene layers compared with the shear compliance of fictitious layers, and also neglect the extension stiffness of fictitious layers. As a result, we have

$$E_0(z) = K_0 \sum_{k=0}^{n} \delta(z - kh_0), \tag{20.27}$$

where $\delta(z)$ is the Dirac delta-function. The equivalent stiffness D and Γ are found to be

$$\begin{aligned} &D = \alpha_n h_0^2 K_0 + (n+1)D_0, \ \alpha_n = \frac{n(n+1)(n+2)}{12}, \\ &\frac{1}{\Gamma} = \frac{\beta_n h_0^3 K_0^2}{D^2 G_{13}} = \frac{\gamma_n}{h_0 G_{13}}, \quad \beta_n = \frac{\alpha_n(n^2+2n+2)}{10}, \ \gamma_n = \frac{6(n^2+2n+2)}{5n(n+1)(n+2)}. \end{aligned} \tag{20.28}$$

Here, G_{13} is the transversal shear stiffness of the fictitious layers. In the formula for the bending stiffness D the first summand is much larger than the second one and hence may be omitted (the error is estimated in Sect. 20.7). Namely, for an MLGS we may ignore the bending stiffness of separate layers and put $D = \alpha_n h_0^2 K_0$.

Now we may use Eqs. (20.26) to calculate the principal frequency of free vibrations of a MLGS.

Fig. 20.2: An ordinary plate with 5 layers (a), a graphite plate with 3 graphene layers (b), a graphene plate model (c).

20.7 Identification of Graphite and Graphene Parameters and some Numerical Results

A survey of theoretical and experimental investigations dedicated to the definition of graphite and graphene parameters may be found in Goldshtein et al (2009); Berinskii et al (2014). In Table 20.2 the elastic moduli E_{ij}, G_{ij} (also expressed in terms of C_{ij}) of a crystal graphite as a transversely isotropic material are given in Goldshtein et al (2009) (on the basis of Blakslee et al, 1970; Bosak et al, 2007; Cousins, 2003). In the last column the values used in this paper for calculations are given.

The distance between graphene layers is $h_0 = 0.334$ (nm). The longitudinal stiffness of a SLGS is found to be $K_0 = E_0 h_0 = 362$ (N/m). The bending stiffness D_0 of a SLGS lies in the range $D_0 = (0.13 - 0.58) \cdot 10^{-18}$ (J) (Berinskii et al, 2014). We take $D_0 = 0.2 \cdot 10^{-18}$ (J) and estimate its influence on the MLGS bending stiffness D. We re-write Eq. (20.28) in the form $D = \alpha_n h_0^2 K_0 (1 + \xi_n)$. The relative influence ξ_n of D_0 decreases with increasing the number of layers, it is found to be $\xi_1 = 0.02$, $\xi_2 = 0.007$, $\xi_3 = 0.004$.

In Eq. (20.26) $\rho h = (n+1)\rho_0$, where $\rho_0 = 7.608 \cdot 10^{-7}$ (kg/m^2) is the mass density of a SLGS per unit area.

As an example calculation of the principal frequency with the help of Eq. (20.26) we consider a rectangular MLGS with $L_x = 5.25$ nm, $L_y = 3.34$ nm and with the same aspect ratio r as in Sect. 20.4. The number $n+1$ of graphene layers will vary. Suppose that the opposite sides $y = 0$ and $y = L_y$ are simply supported. Then we may use Eq. (20.26) with the shear parameter

$$g = g_n = \frac{Dq^2}{\Gamma} = \frac{\alpha_n \gamma_n \pi^2 h_0 K_0}{G_{13} L_y^2} = 2.38\,(n^2 + 2n + 2). \tag{20.29}$$

Equation (20.26) gives the principal frequency of a MLGS

$$\omega = \omega_n = 1.86 \cdot 10^{12} \sqrt{n(n+2)}\, \omega_0(g_n)\, (1/\mathrm{s}), \quad n = 1, 2, \ldots \tag{20.30}$$

where the dimensionless frequency $\omega_0(g)$ is found in Sect. 20.4.

For a MLGS with 2–5 graphene layers Table 20.3 gives the shear parameter g and the frequencies for three variants of boundary conditions at the edges $x = 0$ and $x = L_x$: clamped, simply supported, and free.

Table 20.2: Elastic moduli of graphite (in GPa)

		Blakslee et al (1970)	Bosak et al (2007)	Cousins (2003)	we use
E_{11}	C_{11}	1060	1109	1050	1085
E_{12}	C_{12}	180	139	168.5	
E_{13}	C_{13}	15	0	7.9	10
E_{33}	C_{33}	36.5	38.7	36.5	37.6
G_{13}	C_{44}	4.0	4.95	5	4.5

The shear parameter g being large, and hence, according to Remark 20.1, the frequencies at clamped and at simply supported boundary conditions are almost coincide. All frequencies slowly decrease as the number of layers grows.

20.8 Numerical Results and Their Discussion

At a discrete approach for a SLGS bending the potential DREIDING (Mayo et al, 1990; Wackerfuß, 2009) may be used. This potential consists of four parts, taking into account deflections of neighboring atoms: the in-plane extension and sheet and two out-of-plane potentials. For a MLGS the additional potential taking into account the vdW forces is to be included in calculations (Wackerfuß, 2009).

At a continuum approach, the ordinary KL plate model may be used (Berinskii et al, 2014; Morozov et al, 2014) for a SLGS bending. The main difficulty here is the bending stiffness D_0 is poorly defined ($D_0 = (0.13 - 0.58) \cdot 10^{-18}$ (J) Berinskii et al, 2014). In the relation

$$D_0 = \frac{K_0 h_s^2}{12}, \qquad K_0 = \frac{E h_s}{1 - \nu^2}, \tag{20.31}$$

(used, for example, in Natasuki, 2015) the longitudinal stiffness K_0 of a SLGS is comparatively well known (see Table 20.2), but the effective thickness h_s of a SLGS is unknown, and rather the h_s is to be estimated from Eq. (20.31): $h_s = (12D_0/K_0)^{1/2}$.

The more complex approach to a SLGS bending consists in the use of the non-local theory of elasticity (Natasuki, 2015; Lu et al, 2007). In this approach the problem of the bending stiffness D_0 definition remains the same as for the ordinary KL model. It occurs that the additional summands connected with the non-local model are similar to the summands connected with a transversal shear (see Eq. (20.24)), but their influence on the principal frequency is small.

In contrary to SLGSs the continuous bending models of MLGSs are used in a limited number of works. Here, a difficulty of the SLGS bending stiffness D_0 definition disappears, because the effect of the D_0 on the stiffness D of a MLGS is small (see the estimates of ξ_n in Sect. 20.7). The problem is to correctly describe the interaction of the neighboring layers (the vdW forces). In Natasuki (2015) these forces are reduced to the normal stresses

Table 20.3: Principal frequencies of a MLGS (in TGz)

$n+1$	g	Clamped	Supported	Free
2	11.88	0.2738	0.2736	0.1423
3	23.77	0.3179	0.3178	0.1679
4	40.41	0.3347	0.3347	0.1779
5	61.80	0.3427	0.3427	0.1829

$$p_{i,i+1} = c(w_{i+1} - w_i), \tag{20.32}$$

where w_i and w_{i+1} are the normal deflections of the neighboring layers, and c is the elastic constant similar to the constant in the Winkler foundation model.

In our paper the interlayer medium is modeled by a fictitious isotropic elastic layer with small stiffness. It is supposed that the transversal shear stiffness G_{13} of this layer is equal to the shear stiffness of a crystal graphite (see Table 20.2).

It is shown (Morozov et al, 2016b) that for an ordinary multilayer plate an influence in the plate deflection of a shear compliance is much larger than the shrinkage of normal fibers. Based on this result we suppose that for a MLGS the effect of shear deformations of fictitious layers which model vdW forces also is much larger than the changing of a distance between the neighboring layers because the elastic moduli satisfy inequality $E_{33} \gg G_{13}$ (see Table 20.2). Namely, we think that the proposed model is better than the model based on Eq. (20.32) which ignore the more essential effect. The case when layers move in the opposite directions is not studied here.

Unfortunately, we were unable to find experimental results that allow us to assess the correctness of the proposed model.

Acknowledgements The work is supported by grants RFBR 16.01.00580-a and 16.51.52025 MHT-a, and by grant St.Petersburg Univ. 6.38.337.2015.

References

Altenbach H, Zhilin PA (1988) The general theory of elastic simple shells. Advances in Mechanics 4:107–148

Berinskii IE, Krivtsov AM, Kudarova AM (2014) Bending stiffness of a graphene sheet. Physical Mesomechanics 17(4):356–364

Blakslee OL, Proctor DG, Seldin EJ, Spence GB, Wang T (1970) Elastic constants of compression-annealed pyrolytic graphite. J Appl Phys 41(8)

Bosak A, Krisch M, Mohr M, Maultzsch J, Thomsen C (2007) Elasticity of single-crystalline graphite: Inelastic x-ray scattering study. Phys Rev B 75:153,408

Cousins CSG (2003) Elasticity of carbon allotropes. iv. rhombohedral graphite: Elasticity, zone-center optic modes, and phase transformation using transferred keating parameters. Phys Rev B 67:024,110

Goldshtein RV, Gorodtsov VA, Lisovenko VS (2009) To description of multilayer nano-tubes in the framework of models of a cylindrical anisitropic elasticity. Physical Mesomechanics 12(5):5–14

Ivanova EA (1997) Comparative analysis of low-frequency free vibrations of rectangular plates. Mechanics of Solids 32(6):125–134

Ivanova EA (1998) Asymptotic and numerical analysis of high-frequency free vibrations of rectangular plates. Mechanics of Solids 33(2):139–149

Kienzler R, Schneider P (2014) Comparison of various linear plate theories in the light of a consistent second order approximation. In: Pietraszkiewicz W, Górski J (eds) Shell Structures: Theory and Applications. Proc. 10th SSTA 2013 Conf., Taylor & Francis, London, vol 3, pp 109–112

Kirchhoff G (1876) Vorlesungen über Matematische Physik. Mechanik. Teubner, Leipzig

Love AEH (1927) A treatise on the mathematical theory elasticity. Dover, New York

Lu P, Zhang PQ, Lee HP, Wang CM, Reddy JN (2007) Non-local elastic plate theories. Proc R Soc A 463(2088):3225–3240

Mayo SL, Olafson BD, Goddard III WA (1990) DREIDING: A generic force field for molecular simulations. Proc Chem 94:8897–8909

Mindlin RD (1951) Influence of rotatory inertia and shear on flexural motion of isotropic elastic plates. Trans ASME, J Appl Mech 18:31–38

Morozov NF, Tovstik PE, Tovstik TP (2014) Continuum model of a graphene deformation. Vestnik St Petersburg Univ, Mathematics 47(1):47–55

Morozov NF, Tovstik PE, Tovstik TP (2016a) Continuum model of a multi-layer nanosheet. Doklady Physics 61(11):567–570

Morozov NF, Tovstik PE, Tovstik TP (2016b) The Timoshenko–Reissner generalized model for a plate highly nonuniform in thickness. Doklady Physics 61(8):394–398

Natasuki T (2015) Theoretical analysis of vibration frequency of graphene sheets used as nanomechanical mass sensor. Electronics 4:732–738

Reissner E (1945) The effect of transverse shear deformation on the bending of elastic plates. Trans ASME, J Appl Mech 12:69–77

Timoshenko SP (1921) On the correction for shear of the differential equation for transverse vibrations of prismatic bars. Philos Mag Ser 6 41(245):774–776

Tovstik PE (2007) On the asymptotic character of approximate models of beams, plates and shells. Vestnik St Petersburg Univ, Mathematics 40(3):188–192

Tovstik PE, Tovstik TP (2014) A thin-plate bending equation of second-order accuracy. Doklady Physics 59(8):389–392

Wackerfuß J (2009) Molecular mechanics in the context of the finite element method. International Journal for Numerical Methods in Engineering 77(7):969–997

Zhilin PA (1992) On the Poison and Kirchhoff theory of plates from the modern point of view. Izv RAN Mekhanika Tverdogo Tela (Mechanics of Solids) 27(3):48–64

Chapter 21
On Thermodynamics of Wave Processes of Heat Transport

Evgenii V. Murashkin & Yuri N. Radayev

Abstract The present paper is devoted to formulations of constitutive equations for the non-linear Green-Naghdi type-III thermoelastic continuum consistent with the principle of thermodynamic (or thermomechanical) orthogonality. Contrary to the original Green-Naghdi model the Lagrange description is employed. The principle of thermodynamic orthogonality originally proposed by Ziegler as a generalization of the Onsager linear irreversible thermodynamics states that the irreversible constituent of thermodynamic flux is orthogonal to the convex dissipative potential level surface in the space of thermodynamic forces for any process of heat transport. The principle of the thermomechanical orthogonality takes its origin from the von Mises maximum principle of the perfect plasticity, where it provides existence of a yield surface, its convexity, and the associated flow rule. Non-linear constitutive laws of heat propagation as of type-III thermoelasticity complying with the principle of thermomechanical orthogonality are discussed. Important for applied thermoelasticity cases covered by type-III theory are studied: GNI/CTE–conventional thermoelasticity based on the Fourier heat conduction law and GNII–dissipationless hyperbolic thermoelasticity. In the latter case the internal entropy production equals zero for any heat transport process having the form of the undamped thermoelastic wave propagating at finite speed.

Evegnii V. Murashkin
Ishlinsky Institute for Problems in Mechanics RAS, Vernadskiy Ave, 101 bldg 1, 119526 Moscow, Bauman Moscow State Technical University, Baumanskaya st. 5 bldg 1, 105005 Moscow, National Research Nuclear University "MEPhI" (Moscow Engineering Physics Institute), Kashirskoye shosse 31, 115409 Moscow, & Moscow Technological University, Vernadsky Ave 78, 119454 Moscow, Russia,
e-mail: murashkin@ipmnet.ru, evmurashkin@gmail.com

Yuri N. Radayev
Ishlinsky Institute for Problems in Mechanics RAS, Vernadskiy Ave, 101 bldg 1, 119526 Moscow, Russia,
e-mail: radayev@ipmnet.ru, y.radayev@gmail.com

H. Altenbach et al. (eds.), *Mechanics for Materials and Technologies*,
Advanced Structured Materials 46, DOI 10.1007/978-3-319-56050-2_21

Key words: Thermoelasticity · GN–thermoelasticity · Thermodiffusion · von Mises maximum principle · Thermodynamic orthogonality · Thermodynamic force · · Thermodynamic flux · Constitutive law.

21.1 Preliminary Remarks

In recent decades, the successful attempts to go beyond the classical linear thermodynamics of irreversible processes have been made in different areas of mechanics and physics Hill (1998); Radayev (2005); Kovalev and Radayev (2010); Bazhin and Murashkin (2012). The main idea of new approaches to the thermodynamics of irreversible processes (known as extended irreversible thermodynamics) is to expand the number of variables representing states of a given thermodynamic system. Additional state variables at the same time address to the generalized thermodynamic fluxes or generalized thermodynamic forces in the dual formulations. The thermodynamic force is usually determined as spatial gradient of standard thermodynamic field variable. Thus the list of standard thermodynamic state variables is complemented by adding their gradients. In this case it is assumed that any given state thermodynamic field variable values and its gradient values are independent and varying independently in a certain range. In this new approach the entropy and state thermodynamic potentials (for example the Helmholtz free energy) are to be functions depending on additional field variables. The same is true for entropy flux vector and entire entropy production.

Note that the *standard* thermodynamic field variables are known as the *slow* variables. Their using as state variables together with the conservation laws in modelling of heat transport leads to a number of paradoxes, such as infinite speed of heat propagation and non-zero internal entropy production in an arbitrary thermodynamic process of heat transport. Additional state variables (*fast* variables) allow to formulate the theory in terms of hyperbolic differential equations. Following this scheme one can get a hyperbolic theory of heat conduction and hyperbolic equations modelling heat transport, which has solutions in the form of heat waves propagating at finite velocity (*second sound waves*) and undamped amplitude. The Green–Naghdi theory of coupled thermoelasticity (Green and Naghdi, 1992, 1993) fully corresponds to the principles of the new thermodynamics of irreversible processes. In this theory the principle thermal *slow* variable is the temperature displacement whereas its referential gradient being the *quick* variable However, in the nonlinear theory of Green–Naghdi still remains quite a wide range of theoretical permissibility of constitutive equations which cannot be reduced to a practically acceptable range without further theory refinements. An important tool suitable for this aim is the thermodynamic orthogonality principle (Ziegler, 1966; Radayev, 2005) historically related to the development of the mathematical theory of plasticity.

The present study is devoted to formulations of constitutive equations for the nonlinear Green-Naghdi type-III thermoelastic continuum consistent with the thermodynamic (or thermomechanical) orthogonality principle. After preliminary remarks

(Sect. 21.1) in Sect. 21.2 thermodynamic orthogonality principle and governing equations of the perfectly plastic solids are discussed. The governing equations are derived by the von Mises maximum principle and associated plastic flow rule. The specifications of the associated plastic flow rule at singular points are discussed by the Koiter rule.

In Sect. 21.3 the internal entropy production for the heat transport processes in thermoelastic continua is found by the consistency principle of thermomechanical formulations.

In Sect. 21.4 type-I and type-II constitutive models as limiting cases of type-III thermoelasticity are analyzed. Type-I theory (conventional thermoelasticity) is based on the classical Fourier law of heat conduction with infinite velocity of propagation of an exponentially decaying heat signal. Type-II theory (hyperbolic thermoelasticity) is characterized by the energy conservation and the finite propagation velocity of thermal waves. The GNIII theory is the most general model unifying the conventional thermodiffusion mechanism of heat transport and wave dissipationless mechanism.

21.2 Thermodynamic Orthogonality and Constitutive Equations of the Perfect Plasticity

The principle similar to that concerning thermodynamic fluxes and forces orthogonality originally appeared in the mathematical theory of perfect plasticity Hill (1998); Radayev (2005). There it is known as the von Mises maximum principle (R. von Mises). The main consequence of this principle is geometric convexity of yield surface $f(\boldsymbol{\sigma}) = 0$ in Haigh-–Westergaard stress space ($\boldsymbol{\sigma}$ is the Cauchy stress tensor) and orthogonality of plastic strain increment $d\boldsymbol{\varepsilon}^P$ to yield surface at its smoothness points for an actual plastic flow process. The principle of geometrical orthogonality in stress space is simultaneously the main constitutive law of the mathematical theory of perfect plasticity known also as associated flow rule

$$d\boldsymbol{\varepsilon}^P = (d\lambda)\frac{\partial f}{\partial \boldsymbol{\sigma}} \qquad (f(\boldsymbol{\sigma}) = 0,\ df(\boldsymbol{\sigma}) = 0), \tag{21.1}$$

where $d\lambda \geqslant 0$ is the undetermined multiplier treated as a Lagrange multiplier appearing while solving extreme problem corresponding to the maximum principle. The indeterminateness of multiplier $d\lambda$ in theory of perfect plasticity is elucidated by the fact that it is not considered as a given function of the thermodynamic state variables and therefore a special constitutive equation need not be formulated. The equations in (21.1) enclosed in parentheses show the criterion of active loading of perfectly plastic solid the implementation of which indicates the presence of irreversible thermodynamic flux $d\boldsymbol{\varepsilon}^P$ only when the actual stresses $\boldsymbol{\sigma}$ equal to yield stresses and moreover when under infinitesimal loading $d\boldsymbol{\sigma}$ stresses $\boldsymbol{\sigma} + d\boldsymbol{\sigma}$ also satisfying to $f(\boldsymbol{\sigma} + d\boldsymbol{\sigma}) = 0$.

The associated flow rule (21.1) states the geometrical orthogonality of thermodynamic flux to yield surface in the stress space at the point corresponding to the actual stresses $\boldsymbol{\sigma}$. In this case, the plastic strain increment $d\boldsymbol{\varepsilon}^P$ is the thermodynamic flux. Shear nature of the plastic flow of metals and rocks is perfectly simulated by solutions of hyperbolic equations of mathematical theory of plasticity with constitutive equations in the form of the associated flow rule and the Coulomb–Tresca yield criterion.

The von Mises maximum principle and associated plastic flow rule are widely discussed in studies deal with the mathematical theory of perfect plasticity (see for example Hill (1998); Radayev (2005)).

The maximum principle seems to be one of the fundamental principles of solid mechanics and in wide sense is universal principle of physics. In conjunction with the Coulomb–Tresca yield criterion it makes possible to formulate a mathematical theory of plasticity by hyperbolic differential equations.

Consider in more detail the maximum principle formulation and its consequences. We need geometrical representation in the stress space. Tensor $d\boldsymbol{\varepsilon}^P$ in this space is represented by a free vector $\overrightarrow{RQ}$; stress tensor $\boldsymbol{\sigma}$ corresponds to vector $\overrightarrow{OS}$.

Actual stresses $\boldsymbol{\sigma}$ and plastic strain increment $d\boldsymbol{\varepsilon}^P$ should satisfy the following equations ($\boldsymbol{\nabla}$ is the Hamiltonian nabla operator, $d\lambda \geqslant 0$)

$$\begin{array}{ll}
\operatorname{div}\boldsymbol{\sigma} = \mathbf{0}, & \\
f(\boldsymbol{\sigma}) \leqslant 0, & \\
d\boldsymbol{\varepsilon}^P = (d\lambda)\dfrac{\partial f}{\partial \boldsymbol{\sigma}} & (f(\boldsymbol{\sigma}) = 0,\ df(\boldsymbol{\sigma}) = 0), \\
d\boldsymbol{\varepsilon}^P = \mathbf{0} & (f(\boldsymbol{\sigma}) = 0,\ df(\boldsymbol{\sigma}) < 0;\ \text{or}\ f(\boldsymbol{\sigma}) < 0), \\
\boldsymbol{\nabla} \times (d\boldsymbol{\varepsilon}^P) \times \boldsymbol{\nabla} = \mathbf{0}, &
\end{array} \tag{21.2}$$

the latter of which is the compatibility equation for the total strain increment.

The irreversible part of the work increment during plastic flow is determined according to

$$dW_p = \operatorname{tr}(\boldsymbol{\sigma} \cdot d\boldsymbol{\varepsilon}^P) \tag{21.3}$$

and satisfies (for actual process) the irreversibility inequality.

$$dW_p \geqslant 0. \tag{21.4}$$

This inequality assures the non-negativity of the internal entropy production during the athermal processes of plastic flow.

Let the actual stress state $\boldsymbol{\sigma}$ lie at the yield surface. The point S representing in the stress space the actual stresses $\boldsymbol{\sigma}$ lies at the yield surface. Along with the actual stresses $\boldsymbol{\sigma}$, consider the stresses of comparison $\boldsymbol{\sigma}^*$, subjecting them to the single constraint: representing point S^* is situated either inside or on the yield surface, i.e.

$$f(\boldsymbol{\sigma}^*) \leqslant 0. \tag{21.5}$$

For actual plastic strain increment $d\boldsymbol{\varepsilon}^P$ we introduce the function

$$dW_p^* = \mathrm{tr}(\boldsymbol{\sigma}^* \cdot d\boldsymbol{\varepsilon}^P), \tag{21.6}$$

which is defined on the set of statically admissible stress tensors and examine it for the maximum. One can obtain dW_p^* as a scalar product by using the geometric representation: $dW_p^* = \overrightarrow{OS^*} \cdot \overrightarrow{RQ}$.
The maximum of dW_p^* is always positive as follows from the equation (21.4) and can not be located at points in the interior of yield surface. In fact, if such point S_0^* were found, then directing a ray from the point of S_0^* parallel to the vector $\overrightarrow{RQ}$ and defining point S_1^* as the intersection of this ray with the yield surface one can verify the inequality $\overrightarrow{OS_1^*} \cdot \overrightarrow{RQ} > \overrightarrow{OS_0^*} \cdot \overrightarrow{RQ}$, which contradicts to the one $\overrightarrow{OS_0^*} \cdot \overrightarrow{RQ} \geqslant \overrightarrow{OS_1^*} \cdot \overrightarrow{RQ}$.
Thus, the search of the maximum should be restricted by the condition that point S^* lies only on the yield surface. Introducing the undetermined Lagrange multiplier $d\lambda$, in order to find extreme values of the function dW_p^* we have the following equation

$$d\boldsymbol{\varepsilon}^P = d\lambda \frac{\partial f(\boldsymbol{\sigma}^*)}{\partial \boldsymbol{\sigma}^*}, \qquad f(\boldsymbol{\sigma}^*) = 0. \tag{21.7}$$

We then prove that if a closed yield surface is smooth and geometrically strictly convex, the solution of equations (21.7) is uniquely determined. Denoting this solution by the $\boldsymbol{\sigma}$ (which is truth because actual stresses $\boldsymbol{\sigma}$ must jointly satisfy the equations (21.7)) one can state that the function dW_p^* reaches the strict global maximum at $\boldsymbol{\sigma}^* = \boldsymbol{\sigma}$ on the set of statically admissible stresses.
This statement can be proved by the following simple rearranging. The point S^* due to (21.7) is such that the normal vector to yield surface at the point S^* should be parallel to the given direction $\overrightarrow{RQ}$. Obviously the point S^* exists if the yield surface is closed and strictly convex. If $\boldsymbol{\sigma}^* = \boldsymbol{\sigma}$ then the both equations (21.7) are valid. Now we need to estimate the difference

$$dW_p^* - dW_p.$$

In geometrical terms we have

$$dW_p^* - dW_p = \overrightarrow{SS^*} \cdot \overrightarrow{RQ} < 0 \qquad (S \neq S^*),$$

since the angle ϑ between the vectors $\overrightarrow{SS^*}$ and $\overrightarrow{RQ}$ is always obtuse in view of the strict convexity of yield surface (see Fig. 21.1). This proves all statements formulated earlier. Note that if the yield surface is convex but not strictly, the solution of the system (21.7) is not unique.
The associated flow rule uniquely defines the direction of the vector representing increments of plastic strain in stress space only at regular points of the yield surface. If the stress strain state is represented by the edge point, or conical singularity on the yield surface, then additional assumptions for the proper derivation of flow rule are needed. Generalization of the maximum principle and the associated flow rule to the case of the yield surface with an singular points have been proposed by Koiter Koiter (1953). This generalization based on the following principle of superposition:

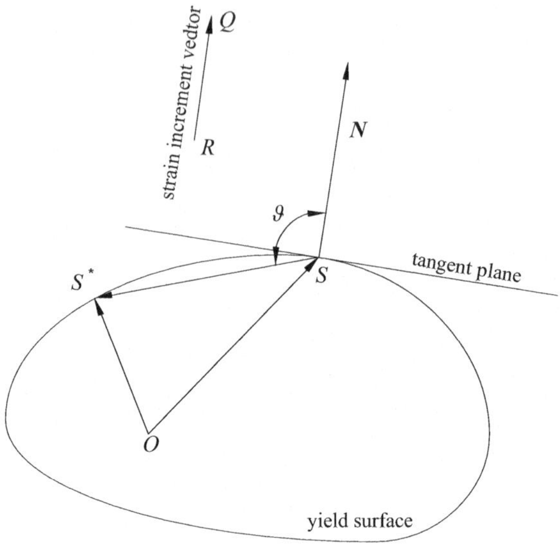

Fig. 21.1 Scalar product negativity $\overrightarrow{SS^*} \cdot \mathbf{N}$ for an arbitrary point $S^* \neq S$ lies on geometrically strict convex yield surface.

singular point of the yield surface are represented as the intersection of a finite number p smooth yield surfaces $f_\gamma(\boldsymbol{\sigma}) = 0$; any of the smooth yield surfaces give an additive contribution (with a corresponding undetermined multiplier) in the total plastic strain increment value.

If the point S is a singular point of the yield surface, then the vector $\overrightarrow{RQ}$ must be located within the cone formed in stress space by normals to the yield surface at the point S.

Active loading, accompanied by change of plastic deformation, is determined by the following conditions

$$\begin{array}{ll} f_\omega(\boldsymbol{\sigma}) = 0, & df_\omega(\boldsymbol{\sigma}) = 0, \\ f_\kappa(\boldsymbol{\sigma}) = 0, & df_\kappa(\boldsymbol{\sigma}) < 0 \text{ or } f_\kappa(\boldsymbol{\sigma}) < 0, \end{array}$$

wherein the indices ω and κ are different and their values together exhaust all values of the index $\gamma = 1, 2, \ldots, p$, and the index ω runs over a non-empty set of values.

Total increment $d\boldsymbol{\varepsilon}^P$ is the sum of increments $d\boldsymbol{\varepsilon}^{P(\omega)}$ indexed by ω

$$d\boldsymbol{\varepsilon}^P = \sum_\omega d\boldsymbol{\varepsilon}^{P(\omega)},$$

wherein each increment $d\boldsymbol{\varepsilon}^{P(\omega)}$ is given according to plastic flow rule with regular yield function f_ω

$$d\boldsymbol{\varepsilon}^{P(\omega)} = \frac{\partial f_\omega(\boldsymbol{\sigma})}{\partial \boldsymbol{\sigma}} d\lambda_\omega.$$

The values of $d\lambda_\omega$ have to be positive.

Finally, the *generalized* associated flow rule takes the following final form

$$
\begin{aligned}
&d\boldsymbol{\varepsilon}^P = \sum_{\gamma=1}^{p} \frac{\partial f_\gamma(\boldsymbol{\sigma})}{\partial \boldsymbol{\sigma}} d\lambda_\gamma, \\
&d\lambda_\gamma > 0 \quad (f_\gamma(\boldsymbol{\sigma}) = 0,\ df_\gamma(\boldsymbol{\sigma}) = 0), \\
&d\lambda_\gamma = 0 \quad (f_\gamma(\boldsymbol{\sigma}) = 0,\ df_\gamma(\boldsymbol{\sigma}) < 0;\ \text{or}\ f_\gamma(\boldsymbol{\sigma}) < 0).
\end{aligned}
\tag{21.8}
$$

21.3 Internal Entropy Production for a Heat Transport Process in Thermoelastic Continua

The von Mises maximum principle can be extended to the processes of heat transport in solids while deriving constitutive equation of thermoelasticity. This scheme can be used to eliminate intrinsic to the classical Fourier theory of heat conduction (CTE) the infinite velocity of thermal wave propagation.

At present mathematical models of the thermoelastic behavior of solids (for example GN-thermoelasticity Green and Naghdi (1992, 1993)) are regularly refined Kovalev and Radaev (2012). GN-thermoelasticity theory itself can be subdivided into the three different types. Type-I theory (GNI/CTE thermoelasticity) is based on the classical Fourier law of heat conduction with infinite velocity of propagation of an exponentially decaying heat signal. Type-II theory (GNII, hyperbolic thermoelasticity) is characterized by the energy conservation and the finite propagation velocity of thermal waves known as *second sound waves* Chester (1963). The known refinements usually aim at derivations of *hyperbolic* partial differential equations of coupled thermoelasticity. The GNIII theory is the most general model combining GNI and GNII theory as limiting cases and unifying the conventional thermodiffusion mechanism of heat trasport and wave dissipationless mechanism.

Hereafter we focus our attention in GNIII theory. Type–III thermoelasticity synthesizes classical thermoelasticity CTE, developed according to Fourier heat conduction law, and non-dissipative theory, assuming zero internal entropy production in a heat transport process in solids and the possibility of undamped thermal wave propagation. Thus for thermodynamically accurate modelling we are going to simultaneously satisfy the following requirements:

1. Finiteness of the velocity of thermal wave propagation, and
2. Existence propagating undamped thermoelastic waves, and
3. Existence of distortionless wave forms similar to the classical d'Alembert type waves.

We use the classical referential description representing the deformation as mapping

$$
\mathbf{x} = \mathbf{x}(\mathbf{X}, t), \tag{21.9}
$$

wherein $\mathbf{x}$ is the actual spatial position, which corresponds to referential state position $\mathbf{X}$.

Hereafter, the referential nabla operator ∇_R is associated with Lagrange variable $\mathbf{X}$. Finite strain tensors are derived from the equation (21.9), using the deformation gradient

$$\mathbf{F} = \nabla_{\mathrm{R}} \otimes \mathbf{x}. \tag{21.10}$$

The system of governing equations of the nonlinear coupled type-III thermoelasticity (GNIII theory) consists of:

- The mass balance equation

$$\left.\frac{\partial \rho_{\mathrm{R}}}{\partial t}\right|_{\mathbf{X}} = 0; \tag{21.11}$$

- The momentum balance equation

$$\rho_{\mathrm{R}} \ddot{\mathbf{x}} = \nabla_{\mathrm{R}} \cdot \mathbf{S}, \tag{21.12}$$

  wherein $\mathbf{S} = J\mathbf{F}^{-\mathrm{T}} \cdot \mathbf{T}$ denotes the first Piola–Kirchhoff stress tensor, $\mathbf{T}$ is the Cauchy stress tensor, $J = \det\mathbf{F}$ is Jacobian of the deformation;

- The internal energy balance equation

$$\dot{e} = -\nabla_{\mathrm{R}} \cdot \mathbf{h}_{\mathrm{R}} + \mathrm{tr}\,(\mathbf{S} \cdot \dot{\mathbf{F}}^{\mathrm{T}}) + \varepsilon, \tag{21.13}$$

  where e is the internal energy density per unit referential volume, $\mathbf{h}_{\mathrm{R}}$ is the referential heat flux vector, ε is the referential volume density of heat sources (radiant heat);

- The entropy balance equation

$$\dot{s} = -\nabla_{\mathrm{R}} \cdot \mathbf{j}_{\mathrm{R}} + \sigma + \xi, \tag{21.14}$$

  where s denotes entropy density per unit referential volume, $\mathbf{j}_{\mathrm{R}}$ is the referential vector of entropy flux, σ is the external entropy production, ξ is the internal entropy production.

The internal entropy production should satisfy the inequality of irreversibility

$$\xi \geqslant 0. \tag{21.15}$$

The entropy balance equation (21.14) is of crucial importance in the irreversible thermodynamics and thermomechanics.
The temperature displacement ϑ is the principal thermal variable in the Green–Naghdi thermoelastic model. In the GNIII theory the complete thermodynamic basis consists of the following thermodynamic state variables:

$$\vartheta, \quad \dot{\vartheta}, \quad \nabla_{\mathrm{R}}\vartheta, \quad \nabla_{\mathrm{R}}\dot{\vartheta}, \quad \mathbf{F}. \tag{21.16}$$

The state variables pointed out are considered to be thermodynamically independent and in this sense form a thermodynamic basis. The choice of the thermodynamic basis is extremely important and decisive step in development any mathematical model related to thermoelastic behaviour.

In this study we shall consider only such processes when the entropy balance equation (21.14) is valid with non-negative internal entropy production (21.15) under differential constraints (21.11)–(21.13). The possible dual formulation is obtained in the following form: the energy balance equation (21.13) is satisfied under constraints (21.11), (21.12), (21.14). Both the initial and dual formulations appear as a manifestation of the *consistency principle* of thermomechanical equations: entropy/energy balance equation should not violate the mass, momentum, and energy/entropy balance equations, i.e. the entropy/energy balance equation must be identically satisfied for any thermodynamic process according to the remaining balance equations.

Taken into account the differential constraints (21.11)–(21.13) by using Lagrange multipliers one can found

$$\theta \mathbf{j}_{\mathrm{R}} = \mathbf{h}_{\mathrm{R}}, \quad \theta \sigma = \varepsilon. \tag{21.17}$$

Wherein θ denotes the absolute temperature of which the inverse is called coldness and acts as the Lagrange multiplier.

The reduced energy balance equation is derived on the basis of equations (21.11)–(21.14) and (21.17)

$$-(\dot{\psi} + s\dot{\theta}) + \mathrm{tr}\,(\mathbf{S} \cdot \dot{\mathbf{F}}^{\mathrm{T}}) - \mathbf{j}_{\mathrm{R}} \cdot \nabla_{\mathrm{R}} \theta = \theta \xi, \tag{21.18}$$

where $\psi = e - s\theta$ is the Helmholtz free energy density per unit referential volume.

Differential equations of thermoelasticity are to be supplemented by constitutive equations. The constitutive equations as of type-III thermoelasticity must read Kovalev and Radaev (2012)

$$\begin{aligned} &\psi = \psi(\vartheta, \dot{\vartheta}, \nabla_{\mathrm{R}} \vartheta, \nabla_{\mathrm{R}} \dot{\vartheta}, \mathbf{F}), \\ &\mathbf{h}_{\mathrm{R}} = \mathbf{h}_{\mathrm{R}}(\vartheta, \dot{\vartheta}, \nabla_{\mathrm{R}} \vartheta, \nabla_{\mathrm{R}} \dot{\vartheta}, \mathbf{F}), \\ &\mathbf{S} = \mathbf{S}(\vartheta, \dot{\vartheta}, \nabla_{\mathrm{R}} \vartheta, \nabla_{\mathrm{R}} \dot{\vartheta}, \mathbf{F}), \\ &s = s(\vartheta, \dot{\vartheta}, \nabla_{\mathrm{R}} \vartheta, \nabla_{\mathrm{R}} \dot{\vartheta}, \mathbf{F}), \\ &\xi = \xi(\vartheta, \dot{\vartheta}, \nabla_{\mathrm{R}} \vartheta, \nabla_{\mathrm{R}} \dot{\vartheta}, \mathbf{F}), \\ &\theta = \theta(\dot{\vartheta}). \end{aligned} \tag{21.19}$$

In this general form the constitutive thermoelastic equations are not practicable in applied problems. Further restrictions on the form of constitutive equations (21.19) can be obtained as follows. The reduced energy balance equation (21.18) with non-negative internal entropy production (21.15) must be satisfied for all thermodynamically admissible processes according to the remaining balance equations. Substituting the constitutive equations (21.19) into the equation (21.18) and noting that the factors $\ddot{\vartheta}$, $\nabla_{\mathrm{R}} \ddot{\vartheta}$, $\dot{\mathbf{F}}$ must vanish, we conclude that the free energy is independent on the state variable $\nabla_{\mathrm{R}} \dot{\vartheta}$, i.e.

$$\psi = \psi(\vartheta, \dot{\vartheta}, \nabla_{\mathrm{R}} \vartheta, \mathbf{F}).$$

In addition, the following equations are proved to be valid

$$s = -\frac{1}{\frac{\partial \theta}{\partial \dot{\vartheta}}} \frac{\partial \psi}{\partial \dot{\vartheta}}, \tag{21.20}$$

$$\mathbf{S} = \frac{\partial \psi}{\partial \mathbf{F}}. \tag{21.21}$$

Therefore the equation for internal entropy production reads

$$\theta\xi = -\left(\mathbf{j}_{\mathrm{R}} + \frac{\partial \psi}{\partial(\nabla_{\mathrm{R}}\vartheta)}\right) \cdot \nabla_{\mathrm{R}}\dot{\vartheta} - \frac{\partial \psi}{\partial \vartheta}\dot{\vartheta}. \tag{21.22}$$

Assuming that the constitutive equations do not explicitly depend on the temperature displacement ϑ

$$\frac{\partial \psi}{\partial \vartheta} = 0$$

we obtain

$$\theta\xi = -\left(\mathbf{j}_{\mathrm{R}} + \frac{\partial \psi}{\partial(\nabla_{\mathrm{R}}\vartheta)}\right) \cdot \nabla_{\mathrm{R}}\dot{\vartheta}. \tag{21.23}$$

The above assumption means that ψ is invariant function under translation of temperature displacement.

21.4 Constitutive Equations for Type-III Thermoelasticity by Virtue of Thermodynamic Orthogonality

Thermodynamic orthogonality principle and its dual formulation have been proposed by Ziegler (1958) as a generalization of the linear Onsager theory (Onsager, 1931a,b); similar topics also discussed in Ziegler (1970, 1981); Germain (1983)).

We introduce the dissipation potential (more precisely, the conjugate dissipation potential) according to equation

$$\theta\xi = \mathscr{D} = \mathscr{D}(\dot{\vartheta}, \nabla_{\mathrm{R}}\vartheta, \mathbf{F}; \nabla_{\mathrm{R}}\dot{\vartheta}) \tag{21.24}$$

and consider $\mathscr{D}$ as a function depending on the thermodynamic force

$$-\nabla_{\mathrm{R}}\dot{\vartheta}. \tag{21.25}$$

In our discussion $\mathbf{j}_{\mathrm{R}}$ plays a role of thermodynamic flux. For the irreversible part of thermodynamic flux one can obtain

$$\mathbf{j}_{\mathrm{R}} + \frac{\partial \psi}{\partial(\nabla_{\mathrm{R}}\vartheta)}, \tag{21.26}$$

wherein the term

$$-\frac{\partial \psi}{\partial(\nabla_R \vartheta)}$$

is the reversible part of the thermodynamic flux. In virtue of the thermomechanical orthogonality principle the referential vector (21.26) is orthogonal to the dissipation potential level surfaces

$$\mathscr{D}(\dot{\vartheta}, \nabla_R \vartheta, \mathbf{F}; \nabla_R \dot{\vartheta}) = \text{const}, \tag{21.27}$$

as stated by the mentioned principle.

In the case of thermoelastic behaviour we arrive at the heat conduction law in the form

$$\mathbf{j}_R + \frac{\partial \psi}{\partial(\nabla_R \vartheta)} = -\lambda^* \frac{\partial \mathscr{D}}{\partial \nabla_R \dot{\vartheta}}, \tag{21.28}$$

or rearranging terms

$$\mathbf{j}_R = -\frac{\partial \psi}{\partial(\nabla_R \vartheta)} - \lambda^* \frac{\partial \mathscr{D}}{\partial(\nabla_R \dot{\vartheta})}. \tag{21.29}$$

The Lagrange multiplier λ^* in the equation (21.28) is expressed in view of the equation (21.27) as

$$\theta \xi = \mathscr{D} = \lambda^* (\nabla_R \dot{\vartheta}) \cdot \frac{\partial \mathscr{D}}{\partial(\nabla_R \dot{\vartheta})},$$

consequently after simple transformations

$$\lambda^* = \frac{1}{(\nabla_R \dot{\vartheta}) \cdot \dfrac{\partial \mathscr{D}}{\partial(\nabla_R \dot{\vartheta})}} \mathscr{D}. \tag{21.30}$$

Thus, if we substitute the equation (21.30) in the equation (21.29) then the heat conduction law for the type-III thermoelasticity complying to the thermodynamic orthogonality principle is formulated as follows:

$$\mathbf{j}_R = -\frac{\partial \psi}{\partial(\nabla_R \vartheta)} - \frac{\mathscr{D}}{(\nabla_R \dot{\vartheta}) \cdot \dfrac{\partial \mathscr{D}}{\partial(\nabla_R \dot{\vartheta})}} \frac{\partial \mathscr{D}}{\partial(\nabla_R \dot{\vartheta})}. \tag{21.31}$$

The irreversibility inequality $\xi \geqslant 0$ is provided by convexity of the level surfaces

$$\mathscr{D}(\dot{\vartheta}, \nabla_R \vartheta, \mathbf{F}; \nabla_R \dot{\vartheta}) = \text{const}.$$

The equation (21.31) shows that the derived type-III thermal conductivity constitutive law comply to the Ziegler orthogonality principle for thermodynamic forces and fluxes.

Note that the case of the zero internal entropy production is thermodynamically admissible. This implies $\mathscr{D} = 0$, and the thermal conductivity law (21.31) takes the form

$$\mathbf{j}_{\mathrm{R}} = -\frac{\partial \psi}{\partial(\nabla_{\mathrm{R}}\vartheta)}, \tag{21.32}$$

intrinsic to the hyperbolic GNII theory capable modelling heat conduction as wave process. The constitutive equation (21.32) for the entropy flux vector is one of the most remarkable in the continuum thermomechanics, as entropy flux and consequently heat flux are determined according to (21.32) by the free energy solely.

The theory of conventional coupled thermoelasticity GNI/CTE is based on the assumption that the free energy ψ is independent on temperature displacement ϑ and referential temperature displacement gradient $\nabla_{\mathrm{R}}\vartheta$. Therefore, the free energy is written as

$$\psi = \psi(\dot{\vartheta}, \mathbf{F}).$$

As a result, the internal entropy production can be found according to

$$\theta\xi = -\mathbf{j}_{\mathrm{R}} \cdot \nabla_{\mathrm{R}}\dot{\vartheta},$$

consequently thermodynamic orthogonality equation generalizing Fourier's law of heat conduction is obtained as

$$\mathbf{j}_{\mathrm{R}} = -\frac{\mathscr{D}}{(\nabla_{\mathrm{R}}\dot{\vartheta}) \cdot \dfrac{\partial \mathscr{D}}{\partial(\nabla_{\mathrm{R}}\dot{\vartheta})}} \frac{\partial \mathscr{D}}{\partial(\nabla_{\mathrm{R}}\dot{\vartheta})}. \tag{21.33}$$

Thus, the entropy flux vector in GNIII theory complying to the thermodynamic orthogonality principle is determined by the equation (21.31) by employing the two potentials: the free energy ψ and the dissipation potential $\mathscr{D}$. It is naturally splitted into two constituents: the reversible part

$$-\frac{\partial \psi}{\partial(\nabla_{\mathrm{R}}\vartheta)}$$

and the irreversible one

$$-\frac{\mathscr{D}}{(\nabla_{\mathrm{R}}\dot{\vartheta}) \cdot \dfrac{\partial \mathscr{D}}{\partial(\nabla_{\mathrm{R}}\dot{\vartheta})}} \frac{\partial \mathscr{D}}{\partial(\nabla_{\mathrm{R}}\dot{\vartheta})}.$$

Equation (21.28) can be generalized at singular points of the level surfaces $\mathscr{D} =$ const by virtue of the Koiter rule (21.8) originally formulated for perfectly plastic solids.

21.5 Conclusions

1. Thermodynamics of wave processes of heat transport based on Green–Naghdi approach has been developed.

2. Constitutive equations for the non-linear Green-Naghdi type-III thermoelastic continuum complying to the thermodynamic (or thermomechanical) orthogonality principle has been obtained. Contrary to the original GNIII theory the derivation of constitutive equations has been realized in terms of Lagrangian description.
3. The thermodynamic orthogonality principle and constitutive equations of the perfect plasticity have been discussed as a background of the constitutive thermoelastic modelling.
4. The internal entropy production for the heat transport processes in thermoelastic continua has been found by the consistency principle of thermomechanical formulations.
5. The constitutive equations for type-III thermoelasticity in virtue of thermodynamic orthogonality have been generalized by the Koiter rule at the singular points of the dissipative potential.
6. Type-I and type-II constitutive models as limiting cases of type-III thermoelasticity have been analyzed. Type-II theory is characterized by the hyperbolic governing equations and consequently admit the finite propagation velocity of thermal waves. The GNIII theory is the most general model unifying the conventional thermodiffusion mechanism of heat trasport and wave dissipationless mechanism.

References

Bazhin AA, Murashkin EV (2012) Creep and stress relaxation in the vicinity of a micropore under the conditions of hydrostatic loading and unloading. Doklady Physics 57(8):294–296

Chester M (1963) Second sound in solids. Phys Rev 131:2013–2015

Germain P (1983) Cours de mécanique des milieux continus, vol 1: Théorie Générale. Masson, Paris

Green AE, Naghdi PM (1992) On undamped heat waves in an elastic soli. Journal of Thermal Stresses 15(2):253–264

Green AE, Naghdi PM (1993) Thermoelasticity without energy dissipation. Journal of Elasticity 31(3):189–208

Hill R (1998) The Mathematical Theory of Plasticity. Oxford University Press, Oxford

Koiter WT (1953) Stress-strain relations, uniqueness and variational theorems for elastic-plastic materials with a singular yield surface. Quarterly of Applied Mathematics 11(3):350–354

Kovalev VA, Radaev YN (2012) Thermomechanical orthogonality in nonlinear type-iii thermoelasticity (gniii). Izv Sarat Univ (NS), Ser Mat Mekh Inform 12(3):72–82

Kovalev VA, Radayev YN (2010) Wave Problems of the Field Theory and Thermomechanics (in Russ.). Samara University Press, Samara

Onsager L (1931a) Reciprocal relations in irreversible processes. i. Phys Rev 37:405–426

Onsager L (1931b) Reciprocal relations in irreversible processes. ii. Phys Rev 38:2265–2279

Radayev YN (2005) Three-Dimensional Problem of the Mathematical Theory of Plasticity (in Russ.). Samara University Press, Samara

Ziegler H (1958) An attempt to generalize onsager's principle, and its significance for rheological problems. Zeitschrift für angewandte Mathematik und Physik ZAMP 9(5):748–763

Ziegler H (1966) Some Extremum Principles In Irreversible Thermodynamics With Application To Continuum Mechanics. North-Holland, Amststerdam

Ziegler H (1970) Proof of an orthogonality principle in irreversible thermodynamics. Zeitschrift für angewandte Mathematik und Physik ZAMP 21(6):853–863

Ziegler H (1981) Discussion of some objections to thermomechanical orthogonality. Ingenieur-Archiv 50(3):149–164

Chapter 22
The Technological Stresses in a Vaulted Structure Built Up on a Falsework

Dmitry A. Parshin

Abstract It is considered a process of gradual building a cylindrical vaulted structure on a rigid base. Building is implemented by means of continuous attaching layers of additional material to the being formed vault external surface. The vault internal surface is supported by a rigid circular falsework during the whole building process and after its completing. The material used for building manifests properties of creep and aging (in particular, concrete can be considered). It is stated a linear quasistatic initial-boundary problem of accreted solids mechanics that describes the process of the vault plane deformation under gravity action for the case of small strains. The analytical solution of this non-classical mathematically two-dimensional problem is built in trigonometric series and quadratures. It is proved a proposition about the structure of technological residual stresses taking place in the completed vault if the falsework is removed from it some time after the end of building. This proposition allows to determine efficiently the stresses mentioned.

Key words: Vaulted structure · Cylindrical vault · Falsework · Gravity · Creep · Aging · Technological stresses · Residual stresses · Analytical solution

22.1 Introduction

Majority of large structures do not appear instantly on the place of their location. They are built there gradually, element by element that is during an accretion process. It is easily seen that the stress-strain state of such structures under gravity action can not be determined only by their final form. Indeed, the weight of every newly added

---

Dmitry A. Parshin
Ishlinsky Institute for Problems in Mechanics RAS, Vernadskiy Ave, 101 bldg 1, 119526 Moscow & Bauman Moscow State Technical University, Baumanskaya st. 5 bldg 1, Moscow, 105005, Russia,
e-mail: parshin@ipmnet.ru

H. Altenbach et al. (eds.), *Mechanics for Materials and Technologies*,
Advanced Structured Materials 46, DOI 10.1007/978-3-319-56050-2_22

material element exerts additional mechanical influence on the present part of the structure being formed and thus excites its additional strain in compliance with the current stiffness of the structure. That means that the stress-strain state of an accreted structure is formed incrementally. As a result the final state of the completely built structure essentially differs in general case from the state this structure would have if it were exposed to gravity action only on completing. As in the last case the structure would deform as a comprehensive whole.

Thus to determine the stress-strain state of a structure being built under gravity action this action is necessary to take into account during the whole building process. The analysis of accreted bodies deformation, that is of bodies gradually formed due to surface inflow of additional material, lets us make the fundamental conclusion that mentioned accounting can not be correctly implemented in the frameworks of classical solid mechanics, even if we consider the traditional equations and boundary conditions in a time-varying region. It is necessary to use special approaches and methods based on accreted solids mechanics concepts Manzhirov (2013, 2014).

Remark that if the structural material used possesses rheological properties than both any actual and the final state of the built structure should considerably depend on the building time mode. During a building process of a structure of such a kind there are two tendencies that are continuously interacting. The first one is the tendency of permanent loading the structure with weight of additional elements and the second one is the stress rearrangement in elements so far added due to strain procrastination under changing structure geometry. The type and the result of this interaction are determined by various particular factors that can be analysed only by solving the corresponding accretion problem Manzhirov (1995, 2015, 2016).

One can inquire about a few solved problems in accreted solids mechanics in Arutyunyan et al (1991); Arutyunyan and Manzhirov (1999); Manzhirov and Chernysh (1992); Manzhirov and Parshin (2006a,b, 2008, 2016a, 2015a,b, 2016b). The present paper is intended to one problem of gradual raising of a heavy semi-circular vaulted structure that manifests properties of creep and aging. Some another problems related to built-up processes of analogous structures were considered in Manzhirov and Chernysh (1992); Manzhirov and Parshin (2015a,b, 2016b).

22.2 Statement of the Problem

Let the process of building a cylindrical vault on a horizontal smooth rigid base start at time $t = t_1$. Building is implemented by means of vault erection due to continuous attachment of uniformly thick elementary layers of additional material to its external surface. During the process the internal surface of the vault is supported by a rigid circular falsework. The very first layer of the material is directly attached to the falsework and adheres to it. The process lasts up to time $t = t_2 > t_1$. During the process and after its completion the vault is attached to the base by a sliding mount that prevents the feet of the vault to be separated from the base but does not prevent their sliding along it.

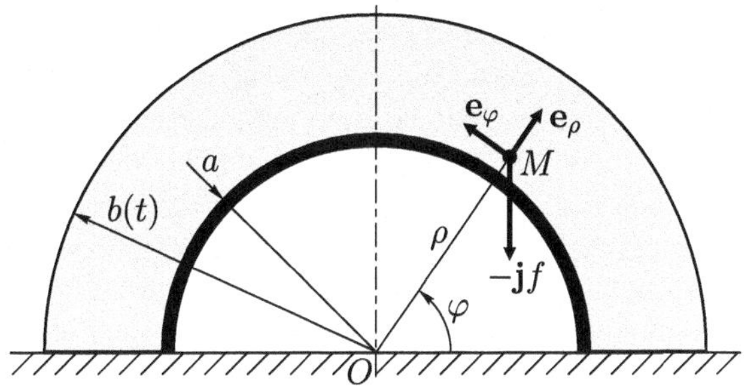

Fig. 22.1 Scheme of a vaulted structure built up by using a rigid falsework

The material used is homogeneous isotropic aging viscoelastic and originally free of stresses. The process of the building vault deformation is assumed to be quasistatic. We perform our consideration for the case of a plane strain state with small strains. It is set a task to study the process of forming the stress-strain state of the vault in question under gravity action.

Let a be the internal vault radius (i.e. the external falsework radius) and $b(t)$ its external radius at time t, $b(t_1) = a$ (Fig. 22.1). Let $b(t) \equiv b_{\text{fin}} = b(t_2)$ when $t > t_2$. We consider the continuous, monotone increasing function $b(t)$ to be known. Suppose this function be piecewise continuously differentiable. We associate with the vault transverse cross-section plane the circular cylindrical coordinate system (ρ, φ, z). Here ρ is the polar radius counted from the arch central axis, φ is the polar angle counted upwards from the base (Fig. 22.1) and z is the longitudinal coordinate. Let the right orthonormal local frame of this system be $\{\mathbf{e}_\rho, \mathbf{e}_\varphi, \mathbf{k}\}$. If we denote the external unit normal to the base by $\mathbf{j}$, the cubic density of the gravity forces acting on the vault will be equal to $\mathbf{f} = -\mathbf{j}f$, where $f = \text{const}$ is the material specific weight (Fig. 22.1).

We describe the material in the framework of the linear theory of viscoelasticity of homogeneously aging isotropic media (Arutyunyan, 1966; Arutyunyan and Manzhirov, 1999), i.e. we start from the state equation

$$\mathsf{T}(\mathbf{r},t) = G(t)\big(\mathscr{I} + \mathscr{N}_{\tau_0(\mathbf{r})}\big)\big[2\,\mathsf{E}(\mathbf{r},t) + (\varkappa - 1)\,\mathbf{1}\operatorname{tr}\mathsf{E}(\mathbf{r},t)\big]. \qquad (22.1)$$

Here T and E are stress and small strain tensors, $\mathbf{1}$ is the unit tensor of rank 2; G is elastic shear modulus, $\varkappa = (1-2\nu)^{-1}$, $\nu = \text{const}$ is Poisson's ratio. The viscoelasticity operator $\mathscr{I} + \mathscr{N}_s$ in (22.1) is determined by the relations:

$$\mathscr{I} + \mathscr{N}_s = (\mathscr{I} - \mathscr{L}_s)^{-1},$$

$$\mathscr{L}_s g(t) = \int_s^t g(\tau) K(t,\tau)\, d\tau, \quad \mathscr{N}_s g(t) = \int_s^t g(\tau) R(t,\tau)\, d\tau,$$

$$K(t,\tau) = G(\tau)\,\frac{\partial \Delta(t,\tau)}{\partial \tau}, \quad \Delta(t,\tau) = G(\tau)^{-1} + \omega(t,\tau),$$

where $\mathscr{I}$ is the identity operator; K and R are creep and relaxation kernels, Δ and ω are specific strain function and creep measure in the case of pure shear. Time t is assumed to be counted from the material fabrication instant; $\tau_0(\mathbf{r})$ is the time at which stresses appear in the point of the body with the position vector $\mathbf{r}$.

Equation (22.1) was suggested by the academician of the Armenian Academy of Sciences N. Kh. Arutyunyan especially for describing creeping processes in concrete and is widely used for this purpose nowadays. Meanwhile, it also well comes to agreement with experimental data for creep of some kinds of polymers, rocks, soil, and ice.

In the problem under consideration stresses appear at points of the body at the time of their joining to it. It means that $\tau_0(\mathbf{r}) \equiv \tau_*(\rho)$ where $\tau_*(\rho)$ is the time of attachment of the layer with radius ρ to the vault. It is clear that the equation $\tau_*(\rho) = t$ describes the external surface $\{\rho = b(t)\}$ of the continuously accreted vault, i.e. its instantaneous accretion surface.

Let us introduce the linear integral operator $\mathscr{H}_s = (\mathscr{I} - \mathscr{L}_s)G(t)^{-1}$ with the real parameter s and its reverse operator $\mathscr{H}_s^{-1} = G(t)(\mathscr{I} + \mathscr{N}_s)$. We denote

$$g^\circ(\mathbf{r},t) = \mathscr{H}_{\tau_0(\mathbf{r})}\, g(\mathbf{r},t) \tag{22.2}$$

for an arbitrary tensor function $g(\mathbf{r},t)$ of point $\mathbf{r}$ of the accreted body and time t. The constitutive relation (22.1) can be written with the denotation (22.2) in the form (Manzhirov, 1995)

$$\mathbf{T}^\circ(\mathbf{r},t) = 2\,\mathbf{E}(\mathbf{r},t) + (\varkappa - 1)\,\mathbf{1}\,\mathrm{tr}\,\mathbf{E}(\mathbf{r},t). \tag{22.3}$$

Note that $g^\circ\big(\mathbf{r},\tau_0(\mathbf{r})\big) = g\big(\mathbf{r},\tau_0(\mathbf{r})\big)/G\big(\tau_0(\mathbf{r})\big)$, so in our case we have

$$g^\circ\big(\mathbf{r},\tau_*(\rho)\big) = g\big(\mathbf{r},\tau_*(\rho)\big)/G\big(\tau_*(\rho)\big). \tag{22.4}$$

In addition, note that by definition of the creep measure (Arutyunyan and Manzhirov, 1999), we have the identity $\omega(\tau,\tau) \equiv 0$, $\tau \geqslant 0$. So

$$\mathscr{H}_\tau 1 = \Delta(t,\tau). \tag{22.5}$$

22.3 Boundary Value Problem for the Built-up Structure

During a process of accretion elements of additional material are added to the growing solid in the course of its deformation motion in space. It is clear that the entire body formed in such a way cannot in general have an unstressed configuration. It is this fact that is decisive in deformation process of any accreted solid and essentially distinguishes the mechanical behavior of such solids from the behavior of solids of constant composition (i.e. classical bodies in continuum mechanics) and of solids whose boundary is variable because of removal of material. Owing to this characteristic property, it is impossible to define the strain measure of an accreted solid by the method adopted in continuum mechanics; therefore, the Cauchy formulas do not hold for the total strain tensor components, and hence the Saint-Venant conditions of their compatibility are not satisfied.

However, note that particles of the material added after their adhesion to the accretion surface continue the motion further in the solid configuration. It means that a sufficiently smooth field of particles velocities is determined in the space region occupied by the entire accreted solid at a given time instant. Therefore, we can expect that the problem on deforming such a solid can be well defined in terms of displacement rates.

Starting from this viewpoint, we write out the analogy of the constitutive relation (22.3) for the rates of the mechanical quantities this relation ties together:

$$\mathsf{S}(\mathbf{r},t) = 2\,\mathsf{D}(\mathbf{r},t) + (\varkappa - 1)\,\mathbf{1}\operatorname{tr}\mathsf{D}(\mathbf{r},t).$$

Here we introduce the tensor $\mathsf{S} = \partial \mathsf{T}^\circ/\partial t$ (see the denotation (22.2)) Manzhirov, 1995) and the strain rate tensor $\mathsf{D} = (\nabla\mathbf{v}^{\mathrm{T}} + \nabla\mathbf{v})/2$ where $\mathbf{v}(\mathbf{r},t) = \mathbf{e}_\rho v_\rho + \mathbf{e}_\varphi v_\varphi + \mathbf{k} v_z$ is the velocity field describing the motion of accreted body particles. In the considered case of plane strain, we have $\mathbf{k}\cdot\mathbf{v} = 0$, $\partial\mathbf{v}/\partial z = \mathbf{0}$.

The equation for the introduced tensor S in the region occupied at the current time by the accreted body can be obtained by applying the linear operator $\mathscr{H}_{\tau_0(\mathbf{r})}$ to the equilibrium equation $\nabla\cdot\mathsf{T} + \mathbf{f} = \mathbf{0}$ and by differentiating the result with respect to time t. Due to the fact that the lower limit in the integral operator mentioned depends on a point of the body this integral operator does not commute with the divergence operator. But one can show (Manzhirov, 1995) that in the accretion processes studied in the present paper we can write $(\nabla\cdot\mathsf{T})^\circ = \nabla\cdot\mathsf{T}^\circ$ for the stress tensor. In this case the following analog of the equilibrium equation is satisfied in the entire accreted solid both during and after the process of its accretion: $\nabla\cdot\mathsf{T}^\circ + \mathbf{f}^\circ = \mathbf{0}$. Differentiate it with respect to t: $\nabla\cdot\mathsf{S} + \partial\mathbf{f}^\circ/\partial t = \mathbf{0}$. By calculating with (22.5) the vector-function $\partial\mathbf{f}^\circ/\partial t = -\mathbf{j}h(\rho,t)$, where

$$h(\rho,t) = f\,\frac{\partial\,\omega\big(t,\tau_*(\rho)\big)}{\partial t}, \tag{22.6}$$

we obtain the equation

$$\nabla\cdot\mathsf{S} = \mathbf{j}h(\rho,t).$$

Note that according to properties of creep measure (Arutyunyan and Manzhirov, 1999) we have $h(\rho,t) \geqslant 0$ when $t > \tau_*(\rho)$.

Consider now a condition that is to be given on the accretion surface of the body. As we suppose the material to be attached is originally free of stresses the complete stress tensor is required to be equal to zero on the accretion surface:

$$\mathsf{T} = \mathbf{0}, \quad \rho = b(t), \quad t \in [t_1,t_2] \quad \Longleftrightarrow \quad \mathsf{T}\big(\mathbf{r},\tau_*(\rho)\big) = \mathbf{0}, \quad \rho \in [a,b_{\mathrm{fin}}]. \tag{22.7}$$

The initial-boundary condition (22.7) can be transformed to the boundary condition for components of the tensor S analogous to a classical force condition on a solid surface (Manzhirov, 1995):

$$\mathbf{e}_\rho \cdot \mathbf{S} = -\mathbf{j}q(t), \quad \rho = b(t), \quad t \in [t_1, t_2], \tag{22.8}$$

where

$$q(t) = f\, b'(t)/G(t) \geqslant 0. \tag{22.9}$$

After we stop building the vault its external surface remains unloaded. Therefore we are to use an ordinary condition of stress absence for $t > t_2$. An evident analog of this condition for the tensor $\mathbf{S}$ is the boundary condition

$$\mathbf{e}_\rho \cdot \mathbf{S} = \mathbf{0}, \quad \rho = b_{\text{fin}}.$$

As we suppose that $b(t) \equiv b_{\text{fin}}$ when $t > t_2$, the written above condition for the tensor $\mathbf{S}$ on the accretion surface remains formally valid even after erecting completing and consequently can be used for any $t \geqslant t_1$. However, it is necessary to understand that the nature of this condition during the vault continuous erecting stage and after its completing is different in essence.

On the internal surface $\{\rho = a\}$ of the vault we have to require meeting the kinematic condition of tough adhesion of the material to the falsework. To set a problem in velocities it should be the condition

$$\mathbf{v} = \mathbf{0}, \quad \rho = a.$$

On the feet $\{\varphi = 0, \pi\}$ of the vault we use the mixed conditions of sliding mount

$$\mathbf{e}_\varphi \cdot \mathbf{T} \cdot \mathbf{e}_\rho = 0, \quad \mathbf{e}_\varphi \cdot \mathbf{v} = 0.$$

Instead of $\mathbf{T}$ we can obviously put here the tensor $\mathbf{S}$.

Thus we get the following boundary value problem which describes the deforming process of the built vault attached to a falsework during and after building:

$$\begin{aligned}
&\nabla \cdot \mathbf{S} = \mathbf{j}h(\rho, t), \quad \rho \in \big(a, b(t)\big), \quad \varphi \in (0, \pi), \quad t > t_1; \\
&\mathbf{S} = 2\mathbf{D} + (\varkappa - 1)\,\mathbf{1}\,\mathrm{tr}\,\mathbf{D}, \quad \mathbf{D} = \frac{\nabla \mathbf{v}^{\mathrm{T}} + \nabla \mathbf{v}}{2}; \quad \mathbf{k} \cdot \mathbf{v} = 0, \quad \frac{\partial \mathbf{v}}{\partial z} = \mathbf{0}; \\
&\mathbf{v}\big|_{\rho=a} = \mathbf{0}; \quad \mathbf{e}_\rho \cdot \mathbf{S}\big|_{\rho=b(t)} = -\mathbf{j}q(t); \\
&\big(\mathbf{e}_\varphi \cdot \mathbf{S} \cdot \mathbf{e}_\rho\big)\Big|_{\varphi=0,\pi} = 0, \quad \big(\mathbf{e}_\varphi \cdot \mathbf{v}\big)\Big|_{\varphi=0,\pi} = 0.
\end{aligned} \tag{22.10}$$

Here the known scalar functions $h(\rho, t)$ and $q(t)$ are defined by formulae (22.6) and (22.9) and depend on the elastic and creeping properties and the specific weight of the material used, as well as on the history of its accretion to the considered built-up structure. It is seen that the problem (22.10) has the form of a classical boundary value problem of linear elasticity theory with a parameter t.

22.4 Analytical Solution of the Problem. Determining the Stresses in the Vault Supported by the Falsework

Let us try the radial and circular components v_ρ and v_φ of the velocity vector field $\mathbf{v}$ solving the problem (22.10) in the form of the trigonometric Fourier expansions. Because of the structure symmetry we have

$$v_\rho\big|_\varphi = v_\rho\big|_{\pi-\varphi}, \quad v_\varphi\big|_\varphi = -v_\varphi\big|_{\pi-\varphi}.$$

Therefore

$$v_\rho = c_0(\rho,t) + \sum_{n=1}^{\infty} c_n(\rho,t)\cos 2n\varphi, \qquad v_\varphi = \sum_{n=1}^{\infty} d_n(\rho,t)\sin 2n\varphi. \tag{22.11}$$

Note that these expansions satisfy the homogenous mixed boundary conditions on the vault feet $\{\varphi = 0, \pi\}$ contained in (22.10).

After we use (22.11) in the Lamé equations that correspond to (22.10) we are to solve systems of linear inhomogeneous ordinary differential equations of the second order with respect to ρ with variable coefficients for unknown functions c_0, c_n, d_n. On solving them we get

$$\frac{c_0}{\rho} = \frac{-\beta B_{40} - B_{10}}{2}, \quad \left\{ \begin{matrix} c_n \\ d_n \end{matrix} \right\} \rho^{-1} = \frac{(n\pm\beta)B_{2n} \mp B_{3n}}{4n-2} \pm \frac{(n\mp\beta)B_{4n} \mp B_{1n}}{4n+2};$$

$$\mathbf{S} = \mathbf{e}_\rho\mathbf{e}_\rho\, S_\rho + \mathbf{e}_\varphi\mathbf{e}_\varphi\, S_\varphi + \mathbf{k}\mathbf{k}\,\nu(S_\rho + S_\varphi) + (\mathbf{e}_\rho\mathbf{e}_\varphi + \mathbf{e}_\varphi\mathbf{e}_\rho)\, S_{\rho\varphi},$$

$$S_{\rho\varphi} = \sum_{n=1}^{\infty} \left[B_{1n} - nB_{2n} + B_{3n} - nB_{4n}\right] \sin 2n\varphi,$$

$$S_{\left\{ \begin{matrix} \rho \\ \varphi \end{matrix} \right\}} = \pm B_{10} - B_{40} + \sum_{n=1}^{\infty} \left[\pm B_{1n} \mp (n\pm 1)B_{2n} \mp B_{3n} \pm (n\mp 1)B_{4n}\right] \cos 2n\varphi.$$

Here $\beta = 1 - 2\nu$ if the left side of an equation contains c, and $\beta = 2(1-\nu)$ if the left side of an equation contains d. We have also introduced the functions

$$\begin{aligned}
B_{1m}(\rho,t) &= K_m^+ \Phi_{m+1}(\rho,t) + \left[b(t)/\rho\right]^{2m+2} \alpha_{1m}(t), \\
B_{2m}(\rho,t) &= \varkappa\, \Phi_m(\rho,t) + \left[b(t)/\rho\right]^{2m} \alpha_{2m}(t), \\
B_{3m}(\rho,t) &= K_m^- \Psi_{m-1}(\rho,t) + (\rho/a)^{2m-2} \alpha_{3m}(t), \\
B_{4m}(\rho,t) &= \varkappa\, \Psi_m(\rho,t) + (\rho/a)^{2m} \alpha_{4m}(t);
\end{aligned}$$

$$\left\{ \begin{matrix} \Phi_m \\ \Psi_m \end{matrix} \right\}(\rho,t) = \frac{\pm k}{2m \mp 1} \int\limits_{\left\{ \begin{smallmatrix} a \\ b(t) \end{smallmatrix} \right\}}^{\rho} (\xi/\rho)^{\pm 2m} h(\xi,t)\, d\xi.$$

Here $K_m^{\pm} = \left[m(2m \pm 3)/(2m \mp 1)\right]\varkappa \pm 1$, $k = 2/\left[\pi(\varkappa + 1)\right]$. The functions $\alpha_{jm}(t)$ are to be found from the system of linear algebraic equations that we obtain on satisfying the boundary conditions in (22.10) on the internal and external vault surfaces. We do not present the expressions for α_{jm} because they are extremely cumbersome. After the problem (22.10) is solved, the tensor $\mathbf{T}^{\circ}$ evolution in every point of the completed vault can be reconstructed from the found rate of this tensor with the initial condition given by (22.7) and (22.4):

$$\mathbf{T}^{\circ}(\mathbf{r},t) = \int\limits_{\tau_*(\rho)}^{t} \mathbf{S}(\mathbf{r},\tau)\,d\tau, \quad t \geqslant \tau_*(\rho).$$

According to the definition of the tensor $\mathbf{T}^{\circ}$ (see (22.2)), by virtue of the known evolution of this tensor it is possible to determine the evolution of the stress tensor $\mathbf{T}$ by solving the integral Volterra equation of the second kind

$$\mathbf{g}(\mathbf{r},t) - \int\limits_{\tau_*(\rho)}^{t} \mathbf{g}(\mathbf{r},\tau)K(t,\tau)\,d\tau = \mathbf{T}^{\circ}(\mathbf{r},t)$$

with respect to time t. In this equation the vector $\mathbf{r}$ is a parameter, and

$$\mathbf{g}(\mathbf{r},t) = \frac{\mathbf{T}(\mathbf{r},t)}{G(t)}$$

is the required tensor-function.

We can write the solution of the mentioned integral equation in the resolvent analytical form:

$$\mathbf{T}(\mathbf{r},t) = \mathscr{H}_{\tau_*(\rho)}^{-1}\mathbf{T}^{\circ}(\mathbf{r},t) = G(t)\left[\mathbf{T}^{\circ}(\mathbf{r},t) + \int\limits_{\tau_*(\rho)}^{t} \mathbf{T}^{\circ}(\mathbf{r},\tau)R(t,\tau)\,d\tau\right], \quad t \geqslant \tau_*(\rho).$$

However, if the expression for the resolvent R of the creep kernel K is too complicated for calculations using this formula, it is more desirable to solve the equation in question numerically, for example, by a quadrature method (Polyanin and Manzhirov, 2008).

22.5 Residual Stresses in the Finished Structure

Now we know the whole evolution of the built vault stress-strain state under gravity action from the instant of building beginning to an arbitrary far instant after the end of building if the falsework is being left attached to the vault on its completing. Let us study the residual stresses taking place in the completed vault after the falsework is removed from its internal surface.

Resting upon the linearity of the problem and applying a known correspondence principle in the theory of aging solids viscoelasticity (Arutyunyan and Manzhirov, 1999) we can prove the following. Let the stress tensors $\mathbf{T}^{\infty}(\mathbf{r})$ and $\mathbf{T}_{\text{res}}(\mathbf{r})$ define the long-term states of a heavy aging viscoelastic accreted vault under consideration after the process of its building (according to a certain time mode) is completed and the vault is left on the falsework or, respectively, the falsework is removed. In the first case

$$\mathbf{t}_a^{\infty}(\varphi) = -\mathbf{e}_\rho \cdot \mathbf{T}^{\infty}\Big|_{\rho=a}$$

is the steady-state falsework reaction, and in the second case we have

$$\mathbf{e}_\rho \cdot \mathbf{T}_{\text{res}}\Big|_{\rho=a} \equiv \mathbf{0}.$$

Suppose we have a weightless elastic vault that is built without residual stresses, has the same dimensions as the completed viscoelastic vault in question and the same elastic characteristics as this vault has after a long aging. The elastic vault mentioned is set on a horizontal rigid base and fixed on it with sliding mount, as the erected viscoelastic vault in question, and loaded on the internal surface by the distributed loading $\mathbf{t}_a^{\infty}$. Let the stress state of such elastic vault be defined by the stress tensor $\mathbf{T}_{\text{el}}(\mathbf{r})$ (it can be found from the solution of the corresponding classical problem of the elasticity theory with zero mass forces). Then

$$\mathbf{T}_{\text{res}} = \mathbf{T}^{\infty} - \mathbf{T}_{\text{el}},$$

i.e. the desired residual stresses can be determined as difference between the stresses corresponding to the above solved non-classical problem of the vault accretion on a falsework and stresses corresponding to a certain classical problem of the elasticity theory with zero mass forces.

Acknowledgements This work is supported by the Russian Foundation for Basic Research (under grants Nos. 17-08-01720-a, 17-51-80010-BRICS_a, and 17-51-45054-IND_a), and by the Department of Energetics, Mechanical Engineering, Mechanics and Control Processes of the Russian Academy of Sciences (Program No. 12 OE).

References

Arutyunyan NK (1966) Some Problems in the Theory of Creep. Pergamon Press, Oxford

Arutyunyan NK, Manzhirov AV (1999) Contact Problems of the Theory of Creep (in Russ.). Izdat. Inst. Mekhaniki NAN RA, Erevan

Arutyunyan NK, Manzhirov AV, Naumov VE (1991) Contact Problems in Mechanics of Growing Solids (in Russ.). Nauka, Moscow

Manzhirov AV (1995) The general non-inertial initial-boundaryvalue problem for a viscoelastic ageing solid with piecewise-continuous accretion. Journal of Applied Mathematics and Mechanics 59(5):805–816

Manzhirov AV (2013) Mechanics of growing solids and phase transitions. Key Engineering Materials 535–536:89–93

Manzhirov AV (2014) Mechanics of growing solids: New track in mechanical engineering. In: ASME 2014 International Mechanical Engineering Congress and Exposition, Montreal, Quebec, Canada, November 14-20, 2014, vol 9: Mechanics of Solids, Structures and Fluids, ASME

Manzhirov AV (2015) Mechanical design of viscoelastic parts fabricated using additive manufacturing technologies. Lecture Notes in Engineering and Computer Science 2218(1):710–714

Manzhirov AV (2016) A method for mechanical design of am fabricated viscoelastic parts. In: Ao Si, Yang GC, Gelman L (eds) Transactions on Engineering Technologies, Springer Singapore, Singapore, pp 223–235

Manzhirov AV, Chernysh VA (1992) Problem of a buried arch structure reinforcement by the accretion method. Izv Akad Nauk Mekh Tverd Tela 27(5):25–37

Manzhirov AV, Parshin DA (2006a) Accretion of a viscoelastic ball in a centrally symmetric force field. Mech Solids 41(1):51–64

Manzhirov AV, Parshin DA (2006b) Modeling the accretion of cylindrical bodies on a rotating mandrel with centrifugal forces taken into account. Mech Solids 41(6):121–134

Manzhirov AV, Parshin DA (2008) Related problems of accreted body mechanics and geomechanics (in Russ.). In: Tkach SM (ed) Nonclassical Problems of Geomechanics. Proc. All-Russia Joint Sci. Session of RAS Sci. Councils in Mechanics of Deformable Solids and in Problems of Rock Sci., Yakutsk, June 16-20, 2008, Izdat. YaNTs SO RAN, Yakutsk, pp 130–137

Manzhirov AV, Parshin DA (2015a) Arch structure erection by an additive manufacturing technology under the action of the gravity force. Mechanics of Solids 50(5):559–570

Manzhirov AV, Parshin DA (2015b) Influence of the erection regime on the stress state of a viscoelastic arched structure erected by an additive technology under the force of gravity. Mechanics of Solids 50(6):657–675

Manzhirov AV, Parshin DA (2016a) Accretion of spherical viscoelastic objects under self-gravity. Lecture Notes in Engineering and Computer Science 2224(1):1131–1135

Manzhirov AV, Parshin DA (2016b) Application of prestressed structural elements in the erection of heavy viscoelastic arched structures with the use of an additive technology. Mechanics of Solids 51(6):692–700

Polyanin AD, Manzhirov AV (2008) Handbook of Integral Equations, 2nd edn. Chapman & Hall/CRC Press, London

Chapter 23
Reversible Plasticity Shape-Memory Effect in Epoxy Nanocomposites: Experiments, Modeling and Predictions

Abishera Ravichandra Rajkumar, Velmurugan Ramachandran, Kondagunta Veeraraghavan Nagendra Gopal & Narinder Kumar Gupta

Abstract Conventionally, shape memory programming involves deforming the material at a temperature higher than the glass transition (T_g) and subsequently cooling the material below T_g while holding the deformation to fix the temporary shape. Alternatively materials with reversible plasticity shape memory (RPSM) property can be programmed at temperatures well below the glass transition temperature which offers several advantages over conventional programing. Here, the RPSM property of multi-walled-carbon-nanotube (MWCNT) reinforced epoxy nanocomposites is investigated. A commercially available epoxy resin is tailored to realize RPSM effect and the properties are enhanced with the addition of nano-fillers in the epoxy matrix. This report systematically investigates the effect of MWCNT addition on the mechanical, thermal and RPSM properties of the epoxy matrix. The RPSM performance under different programming conditions like strain rate, strain level and stress relaxation time is studied. Results reveal that all samples show excellent shape memory properties under various programming conditions. The addition of MWCNT resulted in a significant improvement in mechanical and shape memory properties. Further the RPSM mechanism is explained using a thermo-viscoelastic-viscoplastic model and the model is used to predict the RPSM behavior of the nanocomposite under different programming conditions. As a result, this study shows that by varying the parameters like glass transition temperature, filler content and programming conditions the material can be designed for applications in self-healing systems and smart structures.

Abishera Ravichandra Rajkumar · Ramachandran Velmurugan · Kondagunta Veeraraghavan Nagendra Gopal
Department of Aerospace Engineering, Indian Institute of Technology, Madras, India,
e-mail: ramanv@iitm.ac.in, gopal@ae.iitm.ac.in

Narinder Kumar Gupta
Department of Applied Mechanics, Indian Institute of Technology, Delhi, India,
e-mail: narinder_gupta@yahoo.com

H. Altenbach et al. (eds.), *Mechanics for Materials and Technologies*,
Advanced Structured Materials 46, DOI 10.1007/978-3-319-56050-2_23

Key words: Shape memory polymers · Carbon nanotubes · Cold-Drawing · Thermo-Mechanical properties · Constitutive modeling · Parameter determination

23.1 Introduction

Shape memory polymers (SMP) are a class of smart materials which can be programmed to store large deformations at certain environmental conditions and recover the deformation when a suitable stimulus is applied (Lendlein and Kelch, 2002). SMPs have been designed to respond to stimuli like temperature change, light, electricity, solvent, change in pH, etc (Hu et al, 2012). A conventional thermo-mechanical cycle of a thermally activated SMP involves a programming step and a recovery step. In the programming step, the SMP is deformed at a temperature above the glass transition (T_g) and subsequently cooled below T_g while holding the deformation to fix a temporary shape. The recovery step involves reheating the SMP above T_g to recover the original shape. Contrary to the conventional method, in order to realize Reversible plasticity shape memory (RPSM) effect, a modified programming step is employed wherein the temporary shape can be fixed by plastically deforming the material at a temperature lower than T_g. This modified approach offers several advantages over the conventional approach like simplified programming step, improved recovery stress, faster recovery rates and higher recovery ratios. It has to be noted that RPSM effect is realized only when the material is deformed within a limit where no permanent defects like cracks occur and above the elastic limit (yield point) where there is plastic deformation. Another major drawback is low shape fixity due to the instantaneous recovery of elastic deformation.

Recently, damage-healing of polymer composites has been an active field of research. In this regard, the ability of RPSM materials to recover/heal large plastic deformations can be used for repeated sealing of structural scale damages (Li et al, 2013; Xiao et al, 2010). The previous studies have shown that commercially available thermosetting SMPs, like *Veriflex*, cannot undergo large plastic deformations at room temperature and have low failure strains (Li and Xu, 2011). Also, the thermoplastic SMPs, like polyurethane, are capable of large plastic deformations but have low mechanical properties for structural applications.

Previous studies have shown that, the failure strain is maximum at the onset of glass transition region (Yakacki et al, 2008; Gall et al, 2005; Feldkamp and Rousseau, 2010). So, in order to realize RPSM effect at large plastic deformations the glass transition region must be ideally in the vicinity of room temperature. SMPs have exceptional properties like low cost, easy processing, high recoverable strain, etc. but, their low mechanical properties, and inherent insulating properties limit their applications. Studies have shown that, addition of nano fillers can efficiently improve the properties of bulk polymer (Koo, 2006; Velmurugan and Mohan, 2004, 2009). Among them CNTs with their exceptional thermal, mechanical and electrical properties Thostenson et al (2001) have been found successful in improving the matrix properties while maintaining, sometimes improving the shape memory property of the matrix (Ni et al, 2007; Hashmi et al, 2015; Abishera et al, 2016).

Two main approaches have been followed to model SMPs, namely: Phase transition and thermo-viscoelastic modeling approaches (Nguyen, 2013). The phase transition model was initially proposed by Liu et al (2006). In their model the polymer is idealized as a constrained mixture of two phases: an active phase and a frozen phase whose joint action contributed to the mechanical and SM properties of the material. The volume fraction of the frozen phase which is considered as an internal variable, evolves with the temperature. As the temperature increases the volume fraction of the active phase increases and that of frozen phase decreases. The constitutive theory also incorporates another internal variable called the frozen strain, to characterize the fixed strain in the SM effect. To develop an evolution equation for the frozen strain, it is assumed that during cooling, the elastic strain of the active phase is transformed to stored strain of the frozen phase. This phase transition theory of Liu et al (2006) was further extended to accommodate finite strain (Chen and Lagoudas, 2008), visco-plasticity (Qi et al, 2008), time dependency and hard phase reinforcement (Baghani et al, 2012; Xu and Li, 2010). Although these models capture the SM effect reasonably, parameters like stored strain, frozen phase are considered as nonphysical representation of the glass transition event (Li and Xu, 2011).

To overcome these issues Nguyen et al (2008) proposed a thermo-viscoelastic model incorporating structural and stress relaxation mechanisms. Even though thermo-viscoelastic approach was used in early 1-D models (Tobushi et al, 2001; Morshedian et al, 2005) and Diani et al (2006) 3-D finite deformation model, the constitutive model by Nguyen et al (2008) gives a general approach in defining the material behavior far beyond that of SM effect. Their model was based on the dramatic change in the temperature dependence of the molecular chain mobility induced by the glass transition which drives the thermally activated SM phenomena. This allowed them to develop a physical description of the glass transition event with material parameters which can be determined by standard thermo-mechanical experiments. Although the model neglects heat conduction and pressure on the structural relaxation process, the model was able to predict the thermo-mechanical and SM behavior with reasonable accuracy. Li and Xu (2011) noted that as long as a non-equilibrium configuration can be created and maintained, SMPs can fix a temporary shape at temperatures even below the glass transition temperature ϑ_g, which can be recovered upon heating above ϑ_g. Also they used the thermo-viscoelastic theory to model the cold programming since the approach allows SM behavior to be achieved without any temperature event.

In this study a commercially available structural grade epoxy resin was tailored to realize RPSM effect. MWCNT was added to the epoxy matrix with an objective to improve the mechanical and shape memory properties. The prepared nanocomposites were characterized for their mechanical, thermal, morphological and RPSM properties. A comprehensive study was systematically conducted on the effect of filler content and programming conditions like deformation level, stress relaxation time and strain rate on the RPSM properties and the results are discussed. This paper also attempts to study the influence of MWCNT reinforcement on the RPSM properties of epoxy/MWCNT nanocomposites by using a thermo-viscoelastic-viscoplatic model proposed by Nguyen et al (2008). The influence of MWCNT on the material properties is studied by analyzing the variation in the parameters obtained from modeling neat and filled epoxy. Using the material parameters obtained from thermo-mechanical experiments, numerical predictions were conducted for neat and filled epoxy under various programming conditions and the results are discussed.

23.2 Experimental Methods

23.2.1 Material Selection and Sample Preparation

The SMP was fabricated from Diglycidyl ether of bisphenol A (DGEBA-Araldite LY556), Neopentyl glycol diglycidyl ether (NGDE) and Poly(propylene glycol) bis(2-aminopropyl) ether (Jeffamine230). High purity Multi-Walled-Carbon-Nanotubes were used as nano-fillers. To realize RPSM property at large plastic deformations the onset of glass transition region must be ideally in the vicinity of room temperature. The transition region of the epoxy based SMP can be tuned by introducing varying amount of flexible aliphatic epoxy chains (NGDE) in an amine (Jeffamine230) cured aromatic epoxy system (DGEBA) (Xie and Rousseau, 2009). The following steps are followed to prepare the samples: weighed amounts of DGEBA, NGDE and MWCNT were mixed using a mechanical stirrer at 1500 rpm for 4 hours. After mixing, weighed amount of hardener (Jeffamine 230) was added to the mixture and stirred with the same setting for another 15 minutes. The weight ratio of DGEBA to NGDE to Jeffamine230 was varied until the glass transition region was ideally in the vicinity immediately above the room temperature. The final weight ratio of DGEBA to NGDE to Jeffamine230 with the desired requirements was determined to be 3:2:2. The mixture was poured into a dog-bone shaped (ASTM D630 Type IV) silicone mold and cured at room temperature for 18 hours, de-molded and post cured at 80°C for 2 hours. Samples were designated as E-0-CNT, E-0.5-CNT, E-1-CNT, E-1.5-CNT and E-2-CNT for 0, 0.5, 1, 1.5 and 2 phr (parts per hundred DGEBA-NGDE-Jeffamine230) MWCNT in DGEBA-NGDE-Jeffamine230 system respectively. Figure 23.1 shows the macro-photographs of as-prepared neat (E-0-CNT) and filled (E-1CNT) epoxy.

23.2.2 Material Characterization

The tensile properties of the prepared nanocomposites were studied using a UTM fitted with a thermal chamber of M/s Kalpak Instruments and Controls, India. The thermal properties were studied using Differential Scanning Calorimeter (DSC) of M/s Netzsch DSC200 F3 Maia and a Push-Rod Dilatometer of M/s VB Ceramic

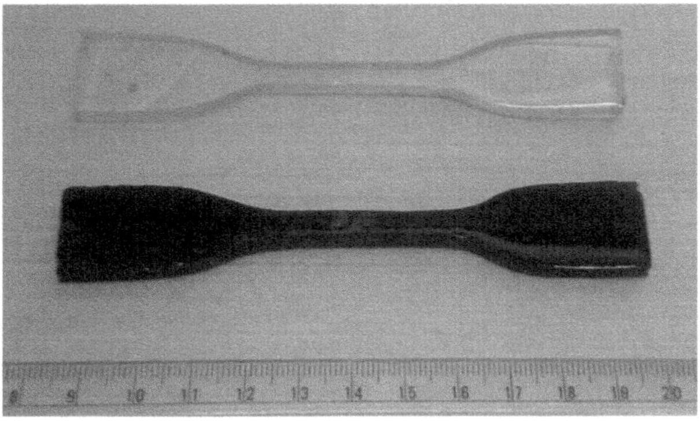

Fig. 23.1: Macro-photographs of neat (E-0-CNT) and filled (E-1CNT) epoxy

Consultants, India. The DSC heating curves of all the samples were obtained for a temperature range from -30 °C to 100 °C in a nitrogen atmosphere. The heating rate was fixed at 10 °C/min. Dilatometry tests were used to evaluate the linear thermal expansion properties. Field Emission Scanning Electron Microscope (FESEM) of M/s FEI Quanta 3D FEG was used to study the morphology of the acquired MWCNTs. The fractured surfaces of the tensile specimens were also studied to evaluate the dispersion of MWCNT in the polymer matrix and the effect of MWCNT addition on tensile fracture.

23.2.3 RPSM Characterization

The steps involved in characterizing the RPSM properties of a sample with φ phr of MWCNT and a gage length of L_o are as follows:

i) the sample was deformed at room temperature T_d to a specific strain ε_d at a constant strain rate $\dot{\varepsilon}$,

ii) the deformation was held at ε_d for time Δt to allow stress relaxation in the sample,

iii) the sample was instantaneously unloaded and the fixed strain, ε_f was measured and finally,

iv) the sample was heated to a temperature, T_r above the glass transition temperature at a heating rate of 3°C/min under stress free conditions to study the strain recovery of the sample.

The steps involved in an RPSM cycle is shown in Fig. 23.2. The effect of φ, ε_d, $\dot{\varepsilon}$ and Δt on the RPSM properties were studied. The shape fixity ratio R_f and the shape recovery R_r were evaluated using the equations $R_f = \varepsilon_f/\varepsilon_d$ and $R_r = \varepsilon_r/\varepsilon_f$ respectively. ε_r is the total strain recovered after free recovery at T_r.

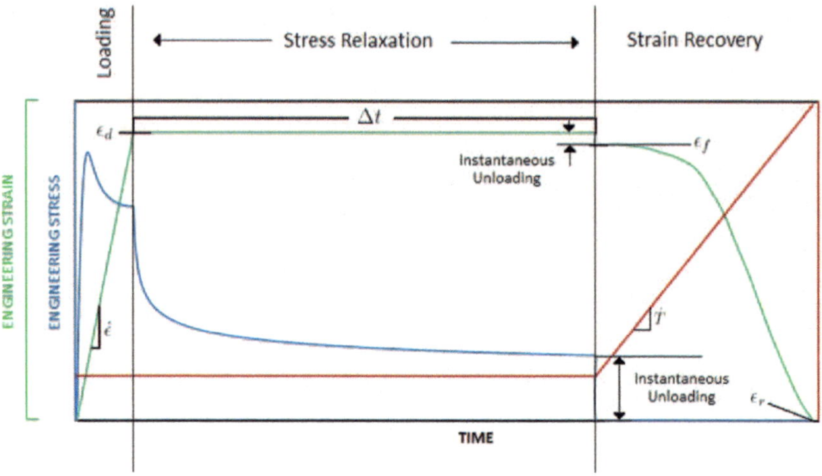

Fig. 23.2: Stress-Strain-Temperature history of a typical RPSM cycle

23.3 Mechanism

The RPSM mechanism can be explained as follows (Abishera et al, 2017). When the material is deformed beyond the initial yield, the molecular segments are reorganized which subsequently change its configurational entropy (Step *i*). In the glassy state (below T_g) the plastic deformation is a rate-dependent process, so the relaxation time allows some reorientation of the deformed molecular structure where the stress reduces asymptotically to a non-zero value (Step *ii*). When the material is elastically unloaded, a relaxed stress-free configuration is obtained. The plastic deformation is fixed due to the high material viscosity and reduced segmental mobility in the glassy state (Step *iii*). When the sample is heated above T_g, the segmental mobility increases and the material recovers back to its original configuration driven by the entropic elasticity of the rubbery state (Step *iv*). This mechanism can be effectively captured using a thermo-viscoelastic-viscoplastic model proposed by Nguyen et al (2008).

23.4 Model Description

23.4.1 Kinematics

Let B_0 and B denote, respectively, the undeformed reference configuration and the deformed and/or heated spatial configuration of the continuum solid under study. For an arbitrary material point X in the reference configuration, let $\chi = \chi(\mathbf{X},t)$

be the motion, at time t, from reference to current configuration. The velocity ($\mathbf{v}$), deformation gradient ($\mathbf{F}$) and the volumetric deformation (J) are defined as follows:

$$\mathbf{v} = \frac{\partial \chi}{\partial t}, \quad \mathbf{F} = \frac{\partial \chi}{\partial \mathbf{X}}, \quad J = \det(\mathbf{F}) \tag{23.1}$$

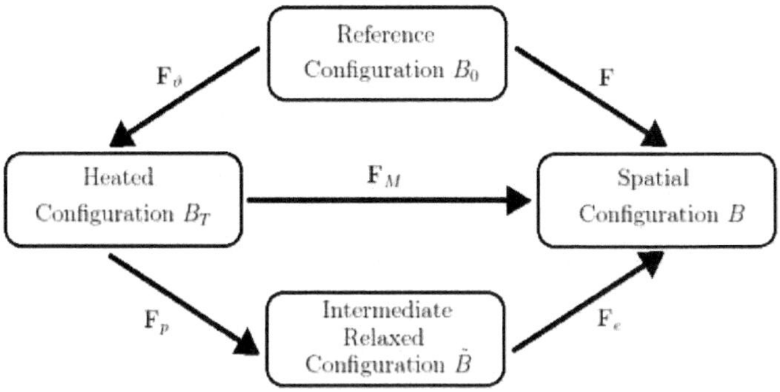

Fig. 23.3: Schematic representation of the decomposition scheme for the deformation gradient $\mathbf{F}$ and $\mathbf{F}_M$

Schematic representation of the decomposition scheme employed for the deformation gradient is shown in Fig. 23.3. The deformation gradient $\mathbf{F}$ can be multiplicative decomposed into thermal and mechanical components (Lion, 1997) as

$$\mathbf{F} = \mathbf{F}_M \mathbf{F}_T \tag{23.2}$$

where, $\mathbf{F}_M$ and $\mathbf{F}_T$ are the mechanical and the thermal deformation gradient respectively. Under the assumption that the thermo-mechanical response is isotropic, the thermal part of the deformation tensor with volumetric deformation $J_T = \det(\mathbf{F_T})$ can be expressed as

$$\mathbf{F}_T = J_T^{1/3} \mathbf{I} \tag{23.3}$$

The mechanical component of the deformation gradient can be further decomposed into elastic and plastic parts Anand and Gurtin (2003) as follows,

$$\mathbf{F}_M = \mathbf{F}_e \mathbf{F}_p \tag{23.4}$$

The elastic part of the deformation gradient can be further decomposed by polar decomposition into a stretch ($\mathbf{V}_e$) and a rotation tensor ($\mathbf{R}_e$) as $\mathbf{F}_e = \mathbf{V}_e \mathbf{R}_e$.
The total volumetric deformation is given by $J = J_M J_T$. With the mechanical component (J_M) given by, under isochoric plastic flow, $J_M = J_e$. The velocity gradient of the plastic deformation tensor is given by,

$$\mathbf{L}_p = \mathbf{D}_p + \mathbf{W}_p \tag{23.5}$$

where $\mathbf{D}_p$ and $\mathbf{W}_p$ are the symmetric and asymmetric parts of $\mathbf{L}_p$ representing the plastic stretch and the plastic spin respectively.

23.4.2 Structural Relaxation and Thermal Deformation

An internal variable called the fictive temperature T_f, by Tool (1946) is used to model the glass transition phenomenon and the Scherer–Hodge nonlinear formulation (Scherer, 1984; Hodge, 1997) of the Adam–Gibbs model (Adam and Gibbs, 1965) is used for the temperature and structure dependence of the structural relaxation time τ_s. T_f is defined as the temperature at which the non-equilibrium structure at T is in equilibrium. The evolution equation for T_f is given by

$$\frac{\mathrm{d}T_\mathrm{f}}{\mathrm{d}t} = -\frac{1}{\tau_\mathrm{s}}(T_\mathrm{f} - T) \tag{23.6}$$

The Scherer–Hodge equation for structural relaxation time was re-written in terms of WLF constants (C_1, C_2) and can be expressed as follows,

$$\tau_\mathrm{s}(T, T_\mathrm{f}) = \tau_\mathrm{s}^o \exp\left[-C_1 . \frac{C_2(T - T_\mathrm{f}) + T(T_\mathrm{f} - T_\mathrm{g})}{T(C_2 + T_\mathrm{f} - T_\mathrm{g})}\right] \tag{23.7}$$

where, τ_s^o is the reference structural relaxation time measured at T_g. It should be noted that at equilibrium $T_\mathrm{f} = T$. Finally, the thermal deformation can be evaluated for a temperature change T_o to T using the following expression,

$$J_T(T, T_\mathrm{f}) = 1 + \alpha_\mathrm{r}(T_\mathrm{f} - T_o) + \alpha_g(T - T_\mathrm{f}) \tag{23.8}$$

where, α_g and α_r are the coefficients of thermal expansion at glassy and rubbery states respectively.

23.4.3 Constitutive Equations for Stress

A rheological analogy for the thermo-viscoelastic-viscoplastic model can be schematically shown as in Fig. 23.4. The stress response ($\mathbf{T}$) can be additively decomposed into two parts namely: $\mathbf{T}_n$ representing the time-independent rubbery behaviour above T_g and $\mathbf{T}_p = \mathbf{T}_e$ representing the time-dependent visco-plastic behavior of the amorphous glassy polymer below T_g. The constitutive relations for the equilibrium stress response can be given by the eight-chain network model of Arruda and Boyce (1993) as follows,

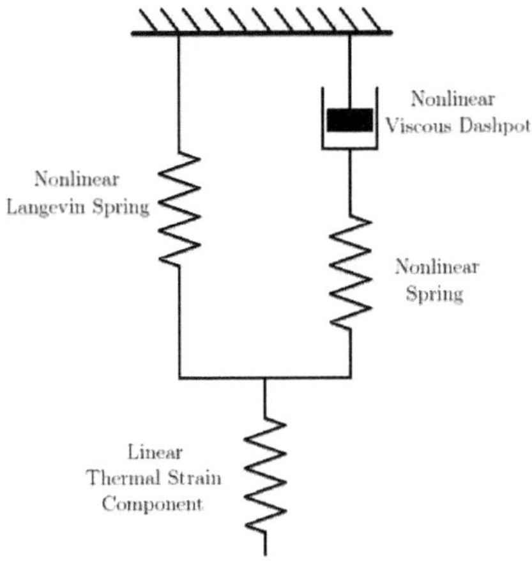

Fig. 23.4 One dimensional rheological analogy of the thermo-viscoelastic-viscoplastic model

$$\mathbf{T}_n = \frac{1}{J_n}\mu_r \frac{\lambda_L}{\lambda_{\text{chain}}}\mathscr{L}^{-1}\left(\frac{\lambda_{\text{chain}}}{\lambda_L}\right)\bar{\mathbf{B}}' + \kappa(J-1)\mathbf{I} \tag{23.9}$$

where μ_r is the initial hardening modulus, λ_L is the locking stretch and κ representing the bulk modulus. With,

$$\lambda_{\text{chain}} = \sqrt{\frac{\text{tr}(\bar{\mathbf{B}})}{3}}, \quad \bar{\mathbf{B}}' = \bar{\mathbf{B}} - \frac{\text{tr}(\bar{\mathbf{B}})}{3}\mathbf{I}, \quad \bar{\mathbf{B}} = \bar{\mathbf{F}}_n\bar{\mathbf{F}}_n^T, \quad \bar{\mathbf{F}}_n = J_n^{-1/3}\mathbf{F}_n,$$

$$J_n = \det(\mathbf{F}_n) \quad \text{and} \quad \mathscr{L}(\beta) = \coth(\beta) - \frac{1}{\beta} \tag{23.10}$$

where, λ_{chain} is the effective stretch on each chain in the eight-chain network model, $\bar{\mathbf{F}}_n$ is the deviatoric portion of $\mathbf{F}_n$ given through the split formulationSimo et al (1985) of the volumetric and deviatoric parts, $\bar{\mathbf{B}}$ is the isochoric left Cauchy-Green tensor and $\mathscr{L}$ is the Langevin function. The constitutive relation for the non-equilibrium stress response is taken as,

$$\mathbf{T}_p = \mathbf{T}_e = \frac{1}{J_e}\mathbf{L}^e(\ln \mathbf{V}_e) \tag{23.11}$$

where, the fourth order isotropic elastic tensor $\mathbf{L}^e$ is given by $\mathbf{L}^e = 2G\mathbf{J} + \lambda\mathbf{I}\otimes\mathbf{I}$. G and λ are the Lamé's constants, $\mathbf{J}$ and $\mathbf{I}$ are the fourth and second order identity tensors.

23.4.4 Flow Rule

Following Nguyen et al (2008), an effective plastic shear stretch rate $\dot{\gamma}_p$ is used to obtain the constitutive relations for the plastic stretch rate $\mathbf{D}_p$. The theory is based on the assumption that the shear viscosity and stress relaxation time is related to the structural relaxation time and in turn can depend on the temperature T and polymer structure through T_f. The stress activated, temperature dependent viscous flow was modeled by extending Eyring's equation to accommodate the dependence on fictive temperature. Under these conditions the following equations hold

$$\mathbf{D}_p = \dot{\gamma}_p \mathbf{n}_p \tag{23.12}$$

$$\dot{\gamma}_p = \frac{s}{\eta_g}\frac{T}{Q}\exp\left[C_1.\frac{C_2(T-T_f)+T(T_f-T_g)}{T(C_2+T_f-T_g)}\right]\sinh\left(\frac{Q}{T}\frac{\bar{\tau}}{s}\right) \tag{23.13}$$

with

$$\mathbf{n}_p = \frac{\mathbf{T}'_p}{\sqrt{\mathbf{T}'_p \cdot \mathbf{T}'_p}}, \qquad \bar{\tau} = \frac{||\mathbf{T}'_p||}{\sqrt{2}} \tag{23.14}$$

where, $\mathbf{n}_p$ is the plastic flow direction, η_g is the shear viscosity at T_g, Q is an activation parameter, $\bar{\tau}$ is the equivalent shear stress and $\mathbf{T}'_p$ is the deviatoric portion of the non-equilibrium stress. s is an internal variable which represents the athermal yield strength which evolves according to the phenomenological model given by (Boyce et al, 1989) as follows,

$$s = h\left(1-\frac{s}{s_s}\right)\dot{\gamma}_p, \qquad s(t=0) = s_o, \qquad s_s < s_o \tag{23.15}$$

where, h is the flow softening modulus and s_s is the saturation value of the shear strength. It should be noted that the viscoplastic flow will not occur when the equivalent shear stress becomes 0. Hence, equation (23.13) can be further modified as,

$$\dot{\gamma}_p = \frac{s}{\eta_g}\frac{T}{Q}\exp\left[C_1.\frac{C_2(T-T_f)+T(T_f-T_g)}{T(C_2+T_f-T_g)}\right]\left\{\sinh\left(\frac{Q}{T}\frac{\bar{\tau}}{s}\right)-1\right\} \tag{23.16}$$

This modification is necessary because, when the plastically deformed configuration is heated above T_g under stress free conditions strain recovery will not occur if $\dot{\gamma} = 0$.

23.5 Results and Discussions

23.5.1 Mechanical Properties

The effect of MWCNT addition on the tensile properties of epoxy nanocomposites is shown in Fig. 23.5. The loading condition is a constant strain rate of 0.0013 s^{-1}

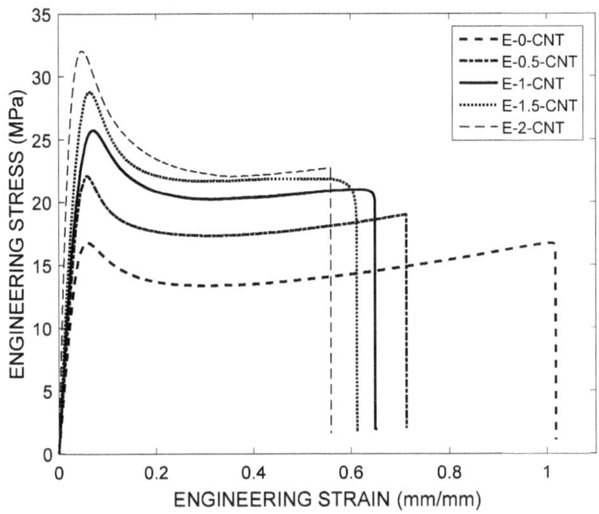

Fig. 23.5: Effect of MWCNT reinforcement on tensile properties

at room temperature (27°C). The stress-strain curve of all the samples show the following regions:

i) an initial linear elastic portion,

ii) the linear portion reaches a maximum point (yield point) after which strain softening is observed which indicates necking.

iii)Finally, there is a (nearly) constant stress region resulting from the propagation of the neck throughout the gage length culminating in failure.

It can be observed that addition of MWCNT results in a increase in tensile modulus and yield strength indicating an effective reinforcement. The improvement in tensile properties could be attributed to effective inter-facial load transfer between the filler and the matrix. For instance, the tensile modulus and strength obtained for neat epoxy was 450.2 MPa and 16.77 MPa respectively and for E-2-CNT it was found to be 867.8 MPa and 32.06 MPa respectively. After the yield point strain softening occurs due to a drop in inter-chain contributions to the stress (Chui and Boyce, 1999). It is observed that the strain softening is steeper in the nanocomposites when compared to pure epoxy (E-0-CNT). This trend increases with increase in MWCNT content suggesting that due to the collapse in inter-chain interactions there is a reduction in effective load transfer between the matrix and MWCNTs leading to steeper strain softening in the stress-strain curve. A significant reduction in the failure strain is observed even with a small addition of MWCNT indicating a reduction in ductility with an increase in filler content. This can be attributed to the mismatch between matrix and filler properties especially in regions with agglomerations, which might have served as stress concentration regions leading to failure. Further increase in

MWCNT (i.e. $>$ 2 phr) drastically reduces the failure strain due to the increased filler agglomerations.

23.5.2 Thermal Properties

Figure 23.6 gives the DSC thermo-graphs of various samples. All samples were tested at a constant heating rate of 10°C/min in a nitrogen atmosphere from -30°C to 100°C. The glass transition temperatures were evaluated from the DSC curves with the aid of Netzsch-Proteus software version 6.1.0. The glass transition temperature of pure epoxy was found to be 42.1°C. It should be noted that the high failure strain of the studied samples at room temperature was a result of the glass transition region being closer to room temperature. The effect of MWCNT reinforcement on the glass transition region can be observed even with the addition of 0.5phr MWCNT. There was a shift in T_g towards a higher temperature with the addition of MWCNT. With further addition of MWCNT resulted in gradual increase of the glass transition temperature with the T_g reaching 44.8°C for an MWCNT content of 2phr. The effect of MWCNT on the glass transition temperature of the epoxy nanocomposite is still under extensive research with researchers reporting an increase, decrease or no change in T_g (Allaoui and El Bounia, 2009). The studies have also shown that use of unmodified MWCNT directly in to the epoxy matrix resulted in an increase or no change in T_g . The glass transition region is characterized by an increase in

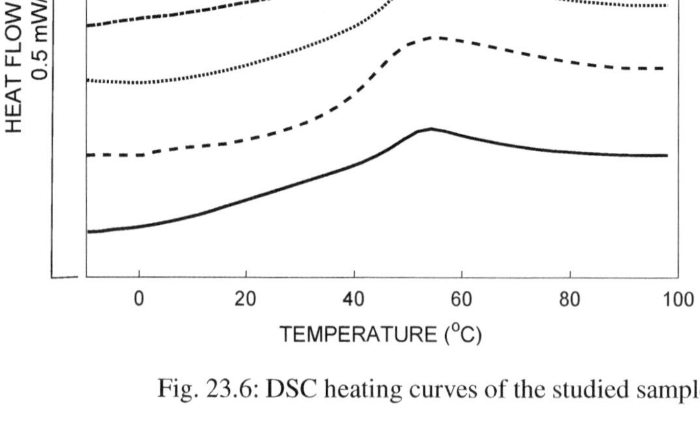

Fig. 23.6: DSC heating curves of the studied samples

Table 23.1: Linear Coefficient of Thermal Expansion (LCTE) and glass transition temperature (T_g) of different samples

Sample	LCTE below T_g (10^-5/°C)	LCTE above T_g (10^-5/°C)	T_g(°C) (from DSC curves)
E-0-CNT	6.04	18.92	42.1
E-0.5-CNT	5.84	18.16	43.6
E-1-CNT	5.97	18.67	44.1
E-1.5-CNT	5.72	18.79	44.6
E-2-CNT	5.56	17.67	44.8

mobility of the polymer chains. Introducing MWCNTs into the epoxy matrix might have reduced the mobility in the polymer chains resulting in an increase of T_g. Table 23.1 gives the linear coefficient of thermal expansion (LCTE) of the studied samples above and below T_g obtained from dilatometry and T_g values obtained from DSC curves. The tests were conducted at two temperature ranges separately namely 27°C to T_g and T_g to 70°C and the slope of strain-temperature diagrams of the respective tests gave the LCTE below and above T_g. As shown in Table 23.1, all samples show a several fold increase in the LCTE above T_g as compared to the LCTE below T_g, a characteristic of amorphous polymers. At low filler content, MWCNT had little or no effect on LCTE for the values both above and below the T_g. A considerable effect is observed in the sample reinforced with 2phr MWCNT (E-2-CNT) where the LCTE decreases for both above and below T_g. This decrease may be attributed to low LCTE of the carbon nanotubes as reported in several studies Schelling and Keblinski (2003); Shirasu et al (2015).

23.5.3 Morphological Properties

FESEM micro-graphs of pristine MWCNTs and fractured surface of unreinforced epoxy is shown in Fig. 23.7. The MWCNTs were found to be highly pure ($>95\%$ according to the supplier) with no visible impurities. The outer diameter was 20-40nm with an average length of 7.5μm yielding an average aspect ratio of 250. To evaluate the dispersion state and degree of agglomerations the fractured surface of the samples were analyzed. The studied samples were fractured in room temperature at a strain rate of 0.0013/s. The fractography of neat epoxy shows a relatively smooth fracture surface indicating a ductile failure mode. Figure 23.8 shows the fracture surface of the nanocomposites at two different magnifications. At low magnifications a considerable difference in fracture surface can be observed. The depth in the river-bed pattern increases with the increase in MWCNT content indicating a decrease in ductility of the nanocomposites. Even at low filler content, agglomerations can be observed as bright white spots in the low magnification micro-graphs. The size and concentration of agglomerations increase with the increase in MWCNT content with a corresponding decrease in failure strain and ductility Montazeri and Montazeri (2011). Though the agglomerations were unavoidable due to the prolonged

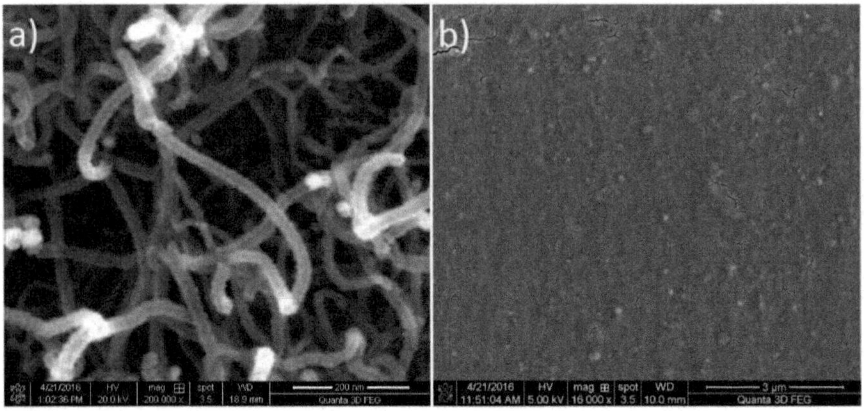

Fig. 23.7: FESEM micrographs of a) pristine MWCNT and b) fracture-surface of pure Epoxy, fractured at a strain rate of 0.0013/s in room temperature.

curing time at room temperature, the average size of the agglomerates was found to be less than $1\mu m$ and were uniformly distributed. The inset images show enlarged micro-graph of an agglomerate in respective nanocomposites.

The high magnification micro-graphs in Fig. 23.8 show individually pulled out MWCNTs protruding from the fracture surface. Wong et al. Wong et al (2003) studied the physical interactions of nanotube and polymer interface for epoxy and polystyrene matrices. They showed that these polymers adhered well to CNT surface at a nanometer scale due to electrostatic and Van-der-Waals forces resulting in a higher inter-facial shear strength which was about an order of magnitude higher than the composite. As seen in Fig. 23.8 all nanotubes including agglomerated MWCNTs were completely covered by epoxy indicating a good adhesion between MWCNT and the matrix resulting in better mechanical properties. The average effective fiber diameter of epoxy coated MWCNT was found to be 85nm.

23.5.4 RPSM Properties

The reversible plasticity shape memory effect of the nanocomposites was characterized within a strain limit of 30%. Though neat epoxy and nanocomposites with low filler content can exhibit RPSM characteristics at much higher strain levels, 30% strain was chosen in order to uniformly compare all samples and to avoid inflicting any permanent damages to nanocomposites with higher MWCNT content. Figure 9 shows the effect of MWCNT reinforcement on the RPSM properties of epoxy nanocomposites under a constant strain-rate (0.0013/s), stress relaxation time (1800s) and deformation strain (30%). It can be observed that under given conditions all samples show excellent shape memory properties. Though the samples

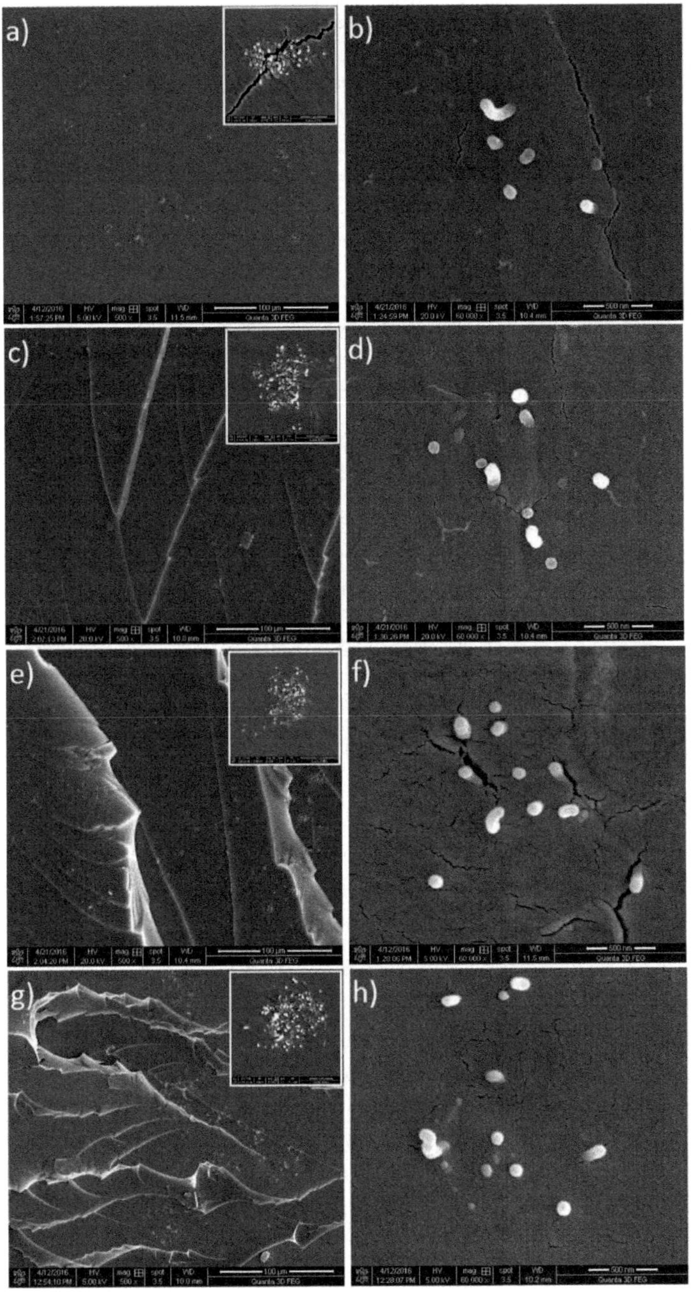

Fig. 23.8: FESEM micro-graphs of fracture-surface of E-0.5-CNT(a,b), E-1-CNT(c,d), E-1.5-CNT(e,f) and E-2-CNT(g,h) under different magnifications. All samples were fractured at a strain rate of 0.0013/s in room temperature. Inset: Enlarged image of an agglomerate in respective nanocomposites.

have deformed to very high strains the strain recovery of all samples were 100% when heated above T_g. In the thermo-mechanical tests conducted by Li and XuLi and Xu (2011) they reported that, at large deformations some permanent irreversible damage occurred with a programming temperature of $T_g - 42°$. In our experiments the deformation temperature is $T_g - 15°$ where little or no irreversible damage might have occurred during the cold drawing step. There was an improvement in the strain fixity with an increase in MWCNT content. The plastic deformation is fixed through the reorganization of segments which is held by the decreased segmental mobility in glassy state Zhang et al (2016). The improvement in fixity can be attributed to a further decrease in segmental mobility in the glassy state with the addition of MWCNT in the polymer matrix. The stress levels of the nanocomposites were higher throughout the thermo-mechanical cycle indicating an efficient reinforcement of MWCNT in the epoxy matrix. It has been reportedGall et al (2005); Ping et al (2005) that the strain recovery started at temperatures below T_g when the programming temperature is below T_g. A similar trend was observed in our experiments with neat epoxy. This might be attributable to the increase in segmental mobility with an increase in temperature which would have enabled the material to recover. In E-0-CNT, the recovery occurred over a range of temperature starting well before the glass transition and achieving a complete recovery well above T_g. This was not observed in nanocomposites where the recovery began only after the temperature reached T_g and the stored strain was recovered completely just above T_g as indicated by the strain recovery curves in Fig. 23.9. This indicates that the MWCNTs were able to successfully hold the fixed strain by restraining the segmental mobility until the temperature reaches T_g. Once the nanocomposite reaches rubbery state the material starts to recover under entropic elasticity of the matrix, assisted by the elastic energy stored in the MWCNT resulting in a faster recovery.

Figure 23.10 shows a representative three-dimensional curve depicting the steps involved in a RPSM cycle and the macro-photographs of samples after various steps. Each cycle in the three dimensional plots distinctly show the following steps; i) loading step: loading at a constant strain rate up to a predetermined strain $(\sigma_0, \varepsilon_0, t_0)$, ii) relaxation step: the strain is kept constant until the stress reaches an asymptotic value $(\sigma_1, \varepsilon_1 = \varepsilon_0, t_1)$, iii) unloading step: the load is removed instantly $(\sigma_2 = 0, \varepsilon_2, t_2 = t_1)$ and iv) strain recovery step: the sample is heated to a temperature higher than T_g at a constant heating rate where the sample recovers completely $(\sigma_3 = 0, \varepsilon_3 = 0, t_3)$. The model parameters used in this study are given in Table 23.2. The method to obtain the parameters is given in the appendix. The constitutive model was coded and numerically solved using MATLAB®. The MATLAB program was used to numerically simulate RPSM cycles under various programing conditions and the simulation results were compared with the experimental studies.

Figure 23.11 presents the comparisons between experiments and simulations for samples loaded at different strain rates. The simulation results were qualitatively in good agreement with the experiments. The significance of strain rate can be observed in the relaxation curves. The time taken to reach ε_0 increases and the time taken to relax from σ_0 to σ_1 decreases i.e the time taken for the stress to reach an asymptotic value decreases. In E-0-CNT, there is no significant difference in the

Table 23.2: Material parameters used in the preliminary constitutive model

Model Parameter	E-0-CNT	E-1-CNT (Effective Properties)
Glass transition temperature, ϑ_g	42.1°C	44.1°C
Programming temperature, ϑ_o	27°C	27°C
Volumetric CTE of below ϑ_g, α_g	$1.8*10^{-4}/°C$	$1.8*10^{-4}/°C$
Volumetric CTE of above ϑ_g, α_r	$5.6*10^{-4}/°C$	$5.6*10^{-4}/°C$
Glassy state Shear Modulus, G	160.8 MPa	191.5 MPa
Glassy state Lame Constant, λ	643.2 MPa	766 MPa
Rubbery state Shear Modulus, μ_r	2.184 MPa	4.260 MPa
Bulk Modulus, κ	750.4 MPa	893.7 MPa
Locking stretch. λ_L	0.67	0.71
Shear viscosity at ϑ_g, η_g	10123 MPa-s	24000 MPa-s
Initial Shear strength, s_o	16.77 MPa	25.77 MPa
Saturation value of Shear strength, s_s	13.35 MPa	18 MPa
Flow activation Parameter, Q	1899 K	2216.2 K
Flow softening modulus, h	250 MPa	300 MPa
First WLF constant, C_1	33.9	35
Second WLF constant, C_2	76.1°C	150 °C
Structural relaxation time at ϑ_g, τ_s^o	96.3 s	90 s

asymptotic value of the stress under different strain rates. Whereas, in E-1-CNT when the strain rate decreases the asymptotic value of the stress increases significantly i.e. when loaded at low strain rates MWCNT reinforced epoxy nanocompos-

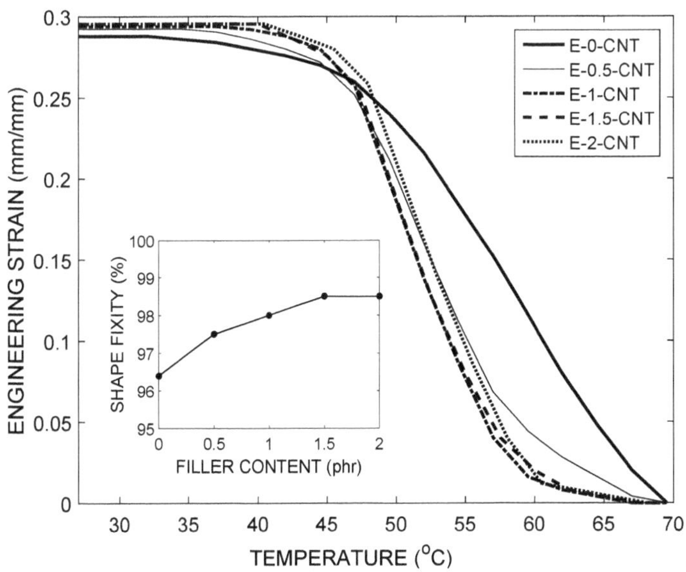

Fig. 23.9: Unconstrained strain recovery curves of the studied nanocomposites. Inset: Effect of MWCNT loading on strain fixity

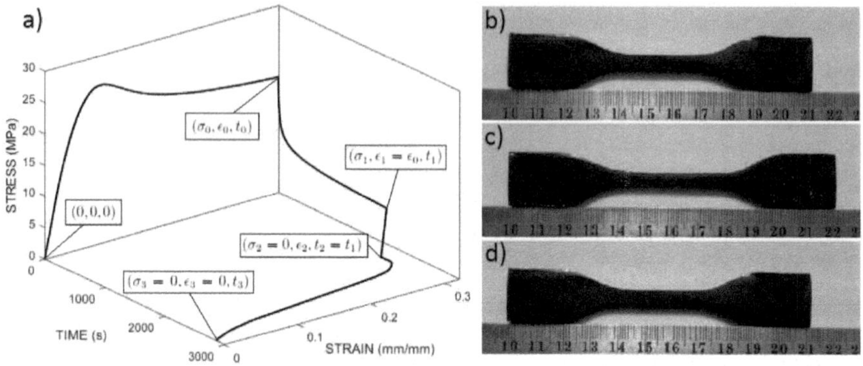

Fig. 23.10: a) Representative 3D plot showing the steps involved in a RPSM cycle. b) As prepared sample (E-1-CNT)–Permanent Shape c) After cold drawing and stress relaxation–Temporary Shape d) After stress-free recovery–Recovered Shape

ite samples exhibited lesser stress relaxation than the samples which were loaded at high strain rates. A possible explanation can be given as follows, at low strain rates an effective load transfer between the MWCNT and matrix might have occurred with minimal inter-facial failure enabling the MWCNTs to store the stress as elastic energy, leading to a lower stress relaxation.

The comparison between experimental and theoretical results for samples deformed to different deformation levels is shown in Fig. 23.12. The samples were tested for RPSM properties at three different strain levels namely; 10, 20 and 30%. Below 10%, for example at 5% strain level the deformation is predominantly in the elastic region, hence little or no strain fixity is observed regardless of the relaxation time for both neat and reinforced epoxy. It can be observed that the σ_0 value decreases as the deformation length increases, due to strain softening of the tensile specimen during cold drawing. This decrease in σ_0 ultimately leads to a decrease in the asymptotic value (σ_1). Once the deformation crosses onto inelastic region all samples show a decent amount of shape fixity. The pre-strain value has a considerable effect on the shape fixity at zero relaxation time, with an increase in pre-strain leading to an increase in shape fixity. No significant effect is observed in shape fixity when the samples are allowed to relax until the stress reaches an asymptotic value. So it can be concluded that good shape fixity can be obtained for samples that have been deformed above the yield strain, if the samples are allowed to relax until the stress reaches an asymptotic value. To understand the effect of stress relaxation on RPSM properties the samples were tested at different relaxation times and the results are presented in Fig. 23.13 along with simulation results. The model was able to capture the thermo-mechanical behavior quite efficiently. The samples were tested for four relaxation times namely; 0, 5, 10 and 30 minutes. It can be observed that with an increase in stress relaxation time a significant improvement in shape fixity can be obtained. For example the shape fixity in neat epoxy after relaxation times of 0, 5, 10 and 30 minutes is 85.2%, 88.1%, 94% and 96% respectively. It can be observed

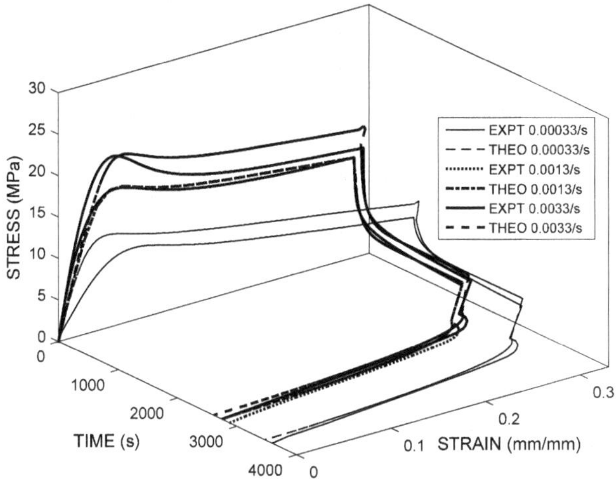

(a) E-0-CNT

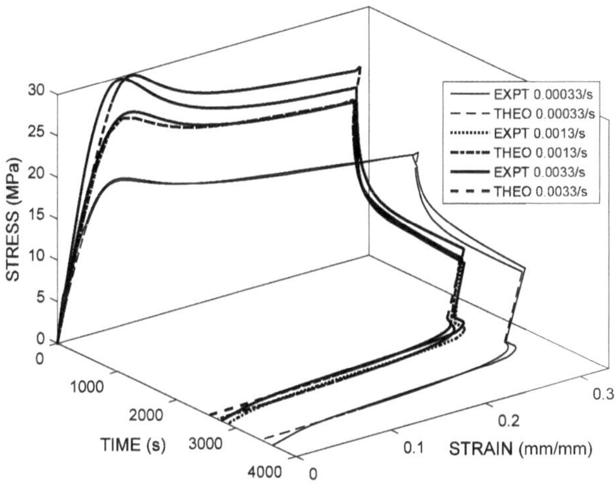

(b) E-1-CNT

Fig. 23.11: Comparison between experimental and predicted three dimensional curves of RPSM thermo-mechanical cycles under different strain rates

that there is very little improvement in fixity for 20 and 30 minutes relaxation time since the stress has relaxed to an asymptotic value in 20 minutes. It should be noted the shape fixity reaches a saturation value above which no amount of relaxation time will improve the shape fixity any further. Also, unlike conventional programming, 100% shape fixity cannot be obtained by cold drawing because a certain amount of strain is always instantaneously recovered during elastic unloading. For all relaxation times the shape fixity in nanocomposites were higher than neat epoxy. Also,

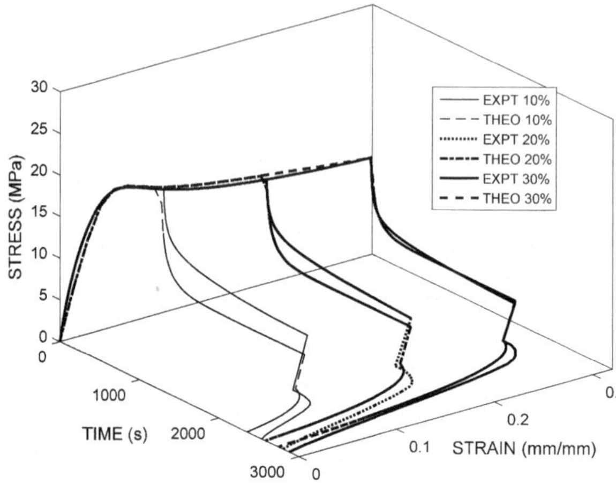

(a) E-0-CNT

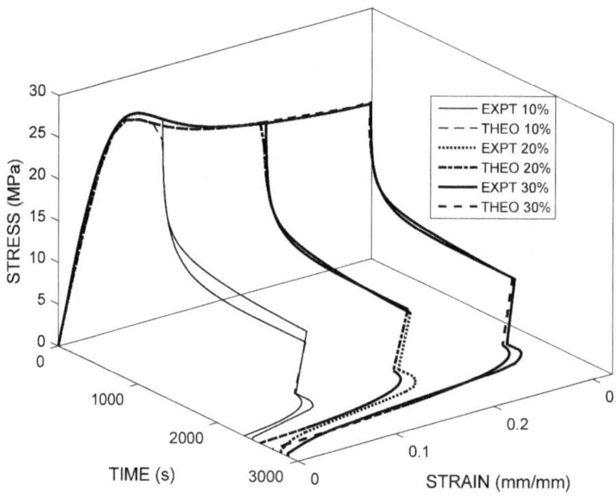

(b) E-1-CNT

Fig. 23.12: Comparison between experimental and predicted three dimensional curves of RPSM thermo-mechanical cycles at different deformation levels

since stress relaxation is a characteristic of the matrix, it has qualitatively similar effect on the nanocomposites.

From the above results and discussions it can be observed that by varying various parameters like MWCNT content, transition region and programming conditions the material can be effectively designed as per requirements. For instance, in a self healing system, given the loading conditions like strain rate, the MWCNT content and the transition region can be optimized to ensure that the maximum strain is below the

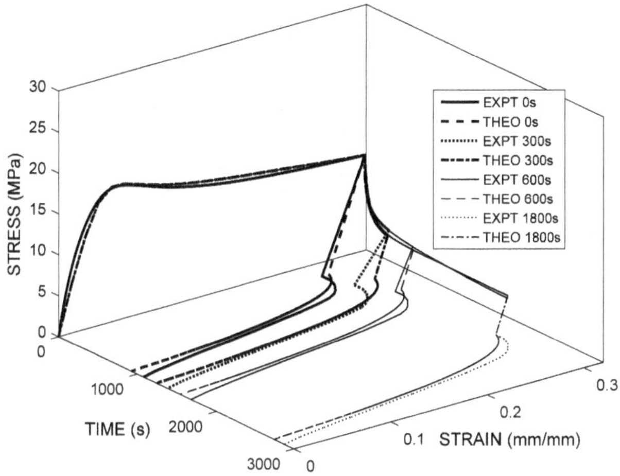

(a) E-0-CNT

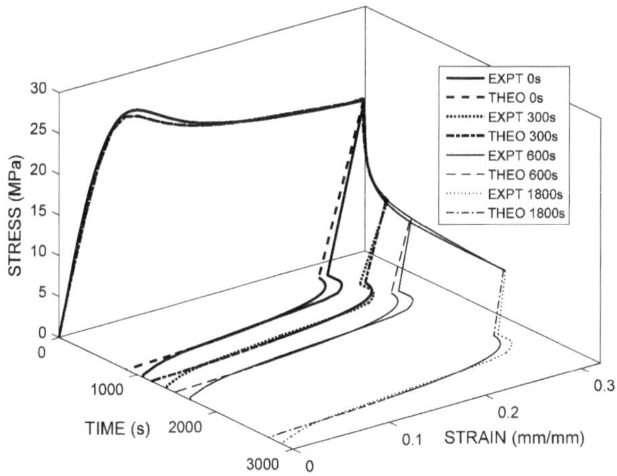

(b) E-1-CNT

Fig. 23.13: Comparison between experimental and predicted three dimensional curves of RPSM thermo-mechanical cycles after different relaxation times

failure strain and that the plastic deformation is completely recovered. Likewise in an actuator or a morphing system, given the recovery conditions like morphing configuration (recovery strain) and work to be done during actuation (recovery stress), the parameters can be effectively tuned to obtain desired results.

23.6 Conclusion

In conclusion, an epoxy based SMP was tailored in order to achieve reversible plasticity shape memory properties. The properties of SMP was further improved by addition of multi-walled carbon nanotubes. The nanocomposites have been characterized for mechanical, thermal and morphological properties and the results were discussed. All samples showed excellent RPSM properties. The effect of MWCNT content and programming conditions like strain rate, deformation level and stress relaxation on the RPSM effect were studied systematically. The results proved that the addition of MWCNTs in the epoxy matrix significantly increases the mechanical and RPSM properties. The results also show that the material was able to display RPSM effect under various programming conditions with 100% strain recovery in each case. A thermo-viscoelastic-viscoplastic model was used to simulate the reversible plasticity shape memory effect of epoxy and MWCNT reinforce epoxy nanocomposites. The model parameters were determined from standard thermo-mechanical tests enabling a close correlation between theory and experiments. The effective material parameters thus obtained were able to simulate the RPSM behaviour of both neat and filled epoxy satisfactorily. The model offered a simple and effective method in understanding the underlaying thermo-mechanical mechanisms and an efficient tool for designing smart structures. This work also shows that by optimizing parameters like MWCNT content and glass transition region, the material can be effectively designed for applications in smart structures with shape memory and/or self-healing properties.

18.A Appendix

18.A.1 Parameter Determination and Effect of MWCNT on the Material Parameters

The material parameters were systematically obtained from basic thermo-mechanical experiments. In order to maintain computational simplicity a homogenization concept is employed. The nanocomposites are treated as a single phase continuum while obtaining the effective properties from experimental methods. The effective parameters thus obtained from the experimental methods for neat and reinforced epoxy were compared with existing theoretical models and experimental studies to understand the effect of MWCNT on the effective material properties.

18.A.2 Determination of μ_r, κ, G and λ

The glassy and rubbery moduli were obtained from the initial modulus measured by uniaxial tension tests at a constant strain rate of $0.0013/s$ in room temperature and at $80°C$ respectively. The effective Young's modulus (E^g_{eff}) obtained at room temperature along with a typical Poisson's ratio of 0.4 for glassy amorphous polymers yielded κ, G and λ. The effective Young's modulus (E^r_{eff}) obtained at $80°C$ along with a typical Poisson's ratio of 0.5 for rubbery polymers yielded μ_r. In the past, several micro-mechanics based methods have been employed to compute the effective properties of particulate composites (Thostenson and Chou, 2003; Halpin, 1969; Fisher et al, 2003; Yeh et al, 2006; Srinivasulu et al, 2015; Velmurugan et al, 2014). In this study, Halpin (1969) equation was employed to understand the effect of MWCNT reinforcement on the effective Young's modulus of the nanocomposites. The Halpin–Tsai equation is as follows,

$$E_{\text{eff}} = \frac{(1+c\eta v_f)}{(1-\eta v_f)} E_m, \qquad \eta = \frac{(\alpha E_f/E_m)-1}{(\alpha E_f/E_m)+c} \tag{23.17}$$

where, E_m and E_f are the Young's modulus of the matrix and MWCNT respectively. v_f is the volume fraction of MWCNTs and $c = p(2L/d)$ is the constant shape factor related to the reinforcement length L and diameter d of MWCNT. The parameter p is a semi-empirical parameter relating to the curved nature of the nanotubes and the degree of agglomerations (Park and Bandaru, 2010). At higher v_f the increase in modulus becomes non-linear where p should be taken as a non-linear function of v_f. α is Cox's orientation parameter (Yeh et al, 2006) which accounts for the randomness of discontinuous fibers. For samples with thickness much higher than the length of MWCNT, as is the case in this study the value $\alpha = 1/6$ is used which represents a random orientation in three dimensions. The density of MWCNTs was calculated from the density of graphite (2.25g/cm^3 (Montazeri et al, 2010)) and considering the ratio of outer to inner radius is 3 (as per the supplier). The density of MWCNTs was obtained as 2g/cm^3. The Young's modulus for MWCNTs was taken as 950 GPa (Yu et al, 2000). The value for p was obtained by fitting the experimental modulus values to the Halpin–Tsai equation seperately at room temperature and at 80°C. It was found to be 0.21 and 0.35 respectively. Although a temperature dependent function can be used to model the variation of moduli below and above ϑ_g (Srivastava et al, 2010), for computational simplicity the moduli in this study are considered to be temperature independent. Fig. 23.14 shows the ability of Halpin–Tsai equation to predict the modulus with different filler content.

By similar reasoning, the Halphin–Tsai equations can also be used to determine the tensile strengths of the nanocomposites using following equations,

$$S_{\text{eff}} = \frac{(1+c\eta v_f)}{(1-\eta v_f)} S_m, \qquad \eta = \frac{(\alpha S_f/S_m)-1}{(\alpha S_f/S_m)+c} \tag{23.18}$$

where S_m and S_f are the tensile strength of the matrix and MWCNT respectively. The tensile strength of MWCNTs was taken as 63 GPa (Yu et al, 2000).The value for p was obtained by fitting the experimental strength values to the Halpin–Tsai equation and was found to be 0.21. Figure 23.15 shows the ability of Halpin–Tsai equation to predict the strength at different filler content.

18.A.3 Determination of C_1, C_2, τ_s^o, α_g and α_r

The WLF parameters C_1, C_2 and the reference structural relaxation time τ_s^o appearing in the evolution equation for ϑ_f can be determined from the glass transition region of the DSC curves.The DSC heating curves of all the samples were obtained for a temperature range from -30°C to 100°C in a nitrogen atmosphere. The heating rate was fixed at 10°C/min. Assuming that the enthalpy of equilibrium material at 0K is zero and the heat capacity of the rubbery and glassy state are independent of temperature, the enthalpy, H of an amorphous material can be given by Simon and McKenna (2000),

$$H = \Delta C_p(\vartheta_f - \vartheta) + C_{pr}\vartheta = \Delta C_p \vartheta_f + C_{pg}\vartheta \tag{23.19}$$

where C_{pg} and C_{pr} are the heat capacities of glassy and rubbery state respectively and $\Delta C_p = C_{pr} - C_{pg}$. The C_p at the onset and end of the transition region are taken as C_{pg} and C_{pr} respectively. The equation for heat flow (P) can be obtained by taking the first derivative of the enthalpy equation as follows,

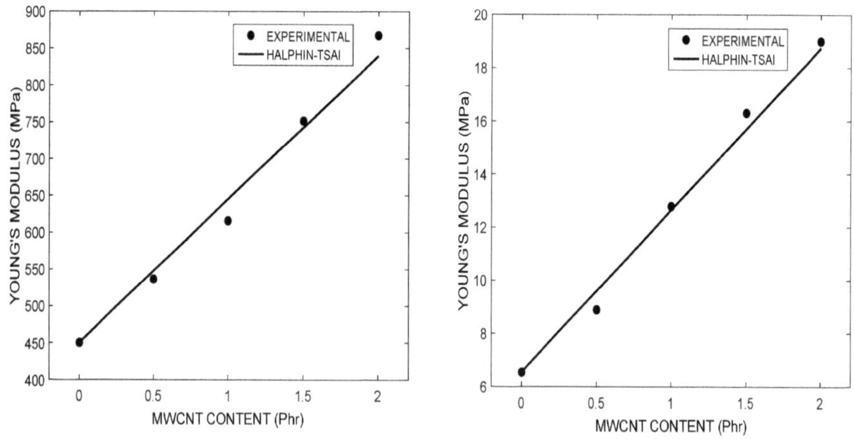

(a) (b)

Fig. 23.14: Experimental and theoretical values of a) Modulus at room temperature and b) Modulus at 80°C

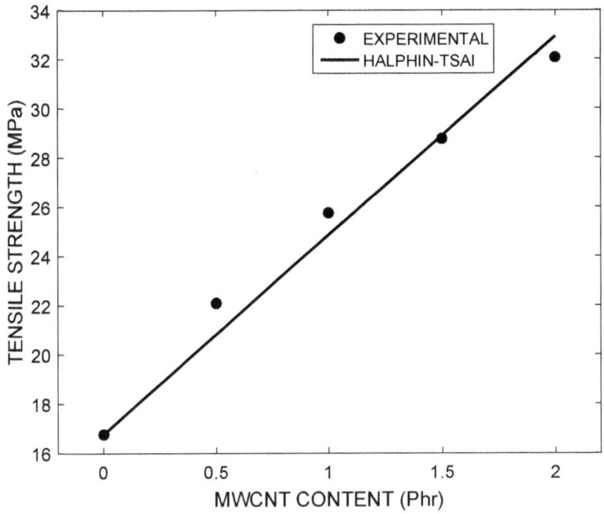

Fig. 23.15: Experimental and theoretical values of tensile strength at room temperature.

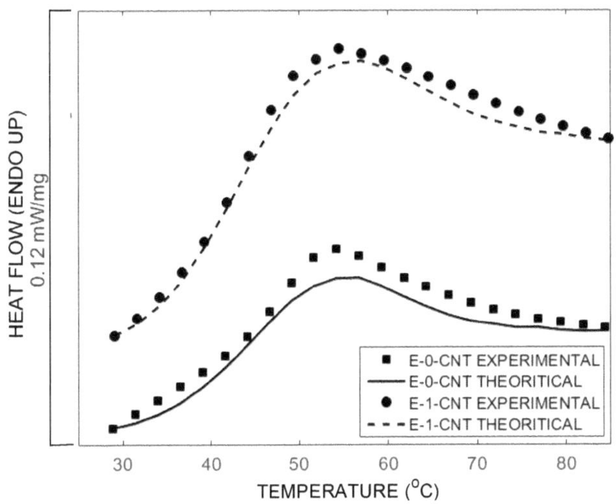

Fig. 23.16: Experimental and theoretical DSC curves

$$P = \frac{dH}{dt} = \Delta C_{\mathrm{p}} \frac{d\vartheta_{\mathrm{f}}}{dt} + C_{\mathrm{pg}} \frac{d\vartheta}{dt} \tag{23.20}$$

Curve fitting the above equation to the DSC curves gives the values of C_1, C_2 and τ_s^o as shown in Fig. 23.16. The values obtained by this method are given in Table. 23.2. The values of C_1 and C_2 obtained from this method vary largely from that of pseudo universal WLF constants ($C_1 \approx 17$, $C_2 \approx 50°C$). Montazeri et al (2012) obtained similar variations in the WLF parameters and concluded that the variation is considerably larger in network structures and composites than for linear polymers. Using this method, there was no considerable change in C_1 value with the addition of MWCNTs but an significant increase in C_2 was obtained. Only a slight variation in the characteristic structural relaxation time was obtained with the addition of MWCNTs. The volumetric coefficient of thermal expansion (CTE) was approximated as three times the linear CTE obtained from dilatometry tests. There was no significant change in CTE at low volume fractions of MWCNTs. Also, from the recovery curves it can be observed that the strain response is entropy dominated and the influence of thermal expansion is less. Considering these factors same volumetric CTEs were taken for filled and unfilled epoxy both above and below ϑ_g.

18.A.4 Determination of η_g, s_s, Q and h

Following Nguyen et al (2008) the initial guess of Q was obtained as follows; the flow stress, s and the effective stretch rate, $\dot{\gamma}$ were evaluated from uniaxial tension tests as $s = \sigma/\sqrt{3}$ and $\dot{\gamma} = \dot{\varepsilon}\sqrt{3}$ where σ and $\dot{\varepsilon}$ are the tensile yield stress and strain rate. The initial guess of the activation parameter was obtained from the following equation.

$$Q \approx s_0 \vartheta \frac{\ln \dot{\gamma}_1 - \ln \dot{\gamma}_1}{s_1 - s_2} \tag{23.21}$$

where $\dot{\gamma}_1$ and $\dot{\gamma}_2$ are obtained from two different strain rates and s_1 and s_2 were obtained from respective tensile yield stresses. Three values of Q were obtained from the three strain rates tested and the average was taken as the initial guess for the activation parameter. It can be seen from Table 23.2 that the value of Q is higher for MWCNT filled epoxy than neat epoxy, thereby resulting in a higher flow stress in the glassy state of the nanocomposites. Other parameters namely η_g, s_s, λ_L and h were fitted to the uniaxial tensile and relaxation testes for different strain rates and temperatures. The final values are given in Table 23.2. It can be seen that s_s/s_o for the nanocomposite is lesser than that of neat epoxy indicating a steeper strain softening in the nanocomposites. This behavior is further established with the increase in flow softening modulus, h. An increase in η_g obtained from the curve fitting procedure indicates an increase in shear viscosity in the nanocomposites, indicating a decrease in ductility with the addition of MWCNTs.

References

Abishera R, Velmurugan R, Gopal KN (2016) Reversible plasticity shape memory effect in carbon nanotubes reinforced epoxy nanocomposites. Composites Science and Technology 137:148–158

Abishera R, Velmurugan R, Gopal KVN (2017) Reversible plasticity shape memory effect in epoxy/cnt nanocomposites – a theoretical study. Composites Science and Technology p Accepted

Adam G, Gibbs JH (1965) On the temperature dependence of cooperative relaxation properties in glass-forming liquids. The journal of chemical physics 43(1):139–146

Allaoui A, El Bounia NE (2009) How carbon nanotubes affect the cure kinetics and glass transition temperature of their epoxy composites?–a review. Express Polymer Letters 3(9):588–594

Anand L, Gurtin ME (2003) A theory of amorphous solids undergoing large deformations, with application to polymeric glasses. Int J Solids Struct 40(6):1465–1487

Arruda EM, Boyce MC (1993) A three-dimensional constitutive model for the large stretch behavior of rubber elastic materials. Journal of the Mechanics and Physics of Solids 41(2):389–412

Baghani M, Naghdabadi R, Arghavani J, Sohrabpour S (2012) A thermodynamically-consistent 3d constitutive model for shape memory polymers. International Journal of Plasticity 35:13–30

Boyce MC, Weber G, Parks DM (1989) On the kinematics of finite strain plasticity. Journal of the Mechanics and Physics of Solids 37(5):647–665

Chen YC, Lagoudas DC (2008) A constitutive theory for shape memory polymers. part i: large deformations. Journal of the Mechanics and Physics of Solids 56(5):1752–1765

Chui C, Boyce MC (1999) Monte carlo modeling of amorphous polymer deformation: Evolution of stress with strain. Macromolecules 32(11):3795–3808

Diani J, Liu Y, Gall K (2006) Finite strain 3d thermoviscoelastic constitutive model for shape memory polymers. Polymer Engineering & Science 46(4):486–492

Feldkamp DM, Rousseau IA (2010) Effect of the deformation temperature on the shape-memory behavior of epoxy networks. Macromolecular Materials and Engineering 295(8):726–734

Fisher F, Bradshaw R, Brinson L (2003) Fiber waviness in nanotube-reinforced polymer composites-I: Modulus predictions using effective nanotube properties. Composites Science and Technology 63(11):1689–1703

Gall K, Yakacki CM, Liu Y, Shandas R, Willett N, Anseth KS (2005) Thermomechanics of the shape memory effect in polymers for biomedical applications. Journal of Biomedical Materials Research Part A 73(3):339–348

Halpin JC (1969) Effects of environmental factors on composite materials. Tech. rep., DTIC Document

Hashmi S, Prasad HC, Abishera R, Bhargaw HN, Naik A (2015) Improved recovery stress in multi-walled-carbon-nanotubes reinforced polyurethane. Materials & Design 67:492–500

Hodge IM (1997) Adam-gibbs formulation of enthalpy relaxation near the glass transition. Journal of Research-National Institute of Standards and Technology 102:195–206

Hu J, Zhu Y, Huang H, Lu J (2012) Recent advances in shape–memory polymers: Structure, mechanism, functionality, modeling and applications. Progress in Polymer Science 37(12):1720–1763

Koo JH (2006) Polymer nanocomposites. McGraw-Hill Professional Pub.

Lendlein A, Kelch S (2002) Shape-memory polymers. Angewandte Chemie International Edition 41(12):2034–2057

Li G, Xu W (2011) Thermomechanical behavior of thermoset shape memory polymer programmed by cold-compression: testing and constitutive modeling. Journal of the Mechanics and Physics of Solids 59(6):1231–1250

Li G, Ajisafe O, Meng H (2013) Effect of strain hardening of shape memory polymer fibers on healing efficiency of thermosetting polymer composites. Polymer 54(2):920–928

Lion A (1997) On the large deformation behaviour of reinforced rubber at different temperatures. Journal of the Mechanics and Physics of Solids 45(11):1805–1834

Liu Y, Gall K, Dunn ML, Greenberg AR, Diani J (2006) Thermomechanics of shape memory polymers: uniaxial experiments and constitutive modeling. International Journal of Plasticity 22(2):279–313

Montazeri A, Montazeri N (2011) Viscoelastic and mechanical properties of multi walled carbon nanotube/epoxy composites with different nanotube content. Materials & Design 32(4):2301–2307

Montazeri A, Javadpour J, Khavandi A, Tcharkhtchi A, Mohajeri A (2010) Mechanical properties of multi-walled carbon nanotube/epoxy composites. Materials & Design 31(9):4202–4208

Montazeri A, Pourshamsian K, Riazian M (2012) Viscoelastic properties and determination of free volume fraction of multi-walled carbon nanotube/epoxy composite using dynamic mechanical thermal analysis. Materials & Design 36:408–414

Morshedian J, Khonakdar HA, Rasouli S (2005) Modeling of shape memory induction and recovery in heat-shrinkable polymers. Macromolecular theory and simulations 14(7):428–434

Nguyen TD (2013) Modeling shape-memory behavior of polymers. Polymer Reviews 53(1):130–152

Nguyen TD, Qi HJ, Castro F, Long KN (2008) A thermoviscoelastic model for amorphous shape memory polymers: incorporating structural and stress relaxation. Journal of the Mechanics and Physics of Solids 56(9):2792–2814

Ni QQ, Zhang CS, Fu Y, Dai G, Kimura T (2007) Shape memory effect and mechanical properties of carbon nanotube/shape memory polymer nanocomposites. Composite Structures 81(2):176–184

Park S, Bandaru P (2010) Improved mechanical properties of carbon nanotube/polymer composites through the use of carboxyl-epoxide functional group linkages. Polymer 51(22):5071–5077

Ping P, Wang W, Chen X, Jing X (2005) Poly (ε-caprolactone) polyurethane and its shape-memory property. Biomacromolecules 6(2):587–592

Qi HJ, Nguyen TD, Castro F, Yakacki CM, Shandas R (2008) Finite deformation thermo-mechanical behavior of thermally induced shape memory polymers. Journal of the Mechanics and Physics of Solids 56(5):1730–1751

Schelling P, Keblinski P (2003) Thermal expansion of carbon structures. Physical Review B 68(3):035,425

Scherer GW (1984) Use of the adam-gibbs equation in the analysis of structural relaxation. Journal of the American Ceramic Society 67(7):504–511

Shirasu K, Yamamoto G, Tamaki I, Ogasawara T, Shimamura Y, Inoue Y, Hashida T (2015) Negative axial thermal expansion coefficient of carbon nanotubes: Experimental determination based on measurements of coefficient of thermal expansion for aligned carbon nanotube reinforced epoxy composites. Carbon 95:904–909

Simo J, Taylor R, Pister K (1985) Variational and projection methods for the volume constraint in finite deformation elasto-plasticity. Computer Methods in Applied Mechanics and Engineering 51(1):177–208

Simon SL, McKenna GB (2000) Quantitative analysis of errors in tmdsc in the glass transition region. Thermochimica Acta 348(1):77–89

Srinivasulu G, Velmurugan R, Jayasankar S (2015) A hybrid method for computing the effective properties of composites containing arbitrarily shaped inclusions. Computers & Structures 150:63–70

Srivastava V, Chester SA, Anand L (2010) Thermally actuated shape-memory polymers: Experiments, theory, and numerical simulations. Journal of the Mechanics and Physics of Solids 58(8):1100–1124

Thostenson ET, Chou TW (2003) On the elastic properties of carbon nanotube-based composites: modelling and characterization. Journal of Physics D: Applied Physics 36(5):573

Thostenson ET, Ren Z, Chou TW (2001) Advances in the science and technology of carbon nanotubes and their composites: a review. Composites Science and Technology 61(13):1899–1912

Tobushi H, Okumura K, Hayashi S, Ito N (2001) Thermomechanical constitutive model of shape memory polymer. Mechanics of materials 33(10):545–554

Tool AQ (1946) Relation between inelastic deformability and thermal expansion of glass in its annealing range. Journal of the American Ceramic society 29(9):240–253

Velmurugan R, Mohan T (2004) Room temperature processing of epoxy-clay nanocomposites. Journal of materials science 39(24):7333–7339

Velmurugan R, Mohan T (2009) Epoxy-clay nanocomposites and hybrids: Synthesis and characterization. Journal of Reinforced Plastics and Composites 28(1):17–37

Velmurugan R, Srinivasulu G, Jayasankar S (2014) Influence of fiber waviness on the effective properties of discontinuous fiber reinforced composites. Computational Materials Science 91:339–349

Wong M, Paramsothy M, Xu X, Ren Y, Li S, Liao K (2003) Physical interactions at carbon nanotube-polymer interface. Polymer 44(25):7757–7764

Xiao X, Xie T, Cheng YT (2010) Self-healable graphene polymer composites. Journal of Materials Chemistry 20(17):3508–3514

Xie T, Rousseau IA (2009) Facile tailoring of thermal transition temperatures of epoxy shape memory polymers. Polymer 50(8):1852–1856

Xu W, Li G (2010) Constitutive modeling of shape memory polymer based self-healing syntactic foam. Int J Solids Struct 47(9):1306–1316

Yakacki CM, Willis S, Luders C, Gall K (2008) Deformation limits in shape-memory polymers. Advanced Engineering Materials 10(1-2):112–119

Yeh MK, Tai NH, Liu JH (2006) Mechanical behavior of phenolic-based composites reinforced with multi-walled carbon nanotubes. Carbon 44(1):1–9

Yu MF, Lourie O, Dyer MJ, Moloni K, Kelly TF, Ruoff RS (2000) Strength and breaking mechanism of multiwalled carbon nanotubes under tensile load. Science 287(5453):637–640

Zhang X, Tang Z, Guo B (2016) Reversible plasticity shape memory polymers: Key factors and applications. Journal of Polymer Science Part B: Polymer Physics 54(14):1295–1299

Chapter 24
The Dynamics of an Accreting Vibrating Rod

Michael Y. Shatalov, Stephan V. Joubert, Igor A. Fedotov, Masodi Lesiba Glen Lekalakala & Judith Nomantande Maureen Bidie

Abstract A linear model of longitudinal vibration is formulated for a viscoelastic rod subjected to external harmonic excitation within the framework of the classical theory of vibrating rods. It is assumed that the rod has a time-dependent variable length and cross-section. A mixed problem of dynamics is formulated, which contains non-conventional fixed-free boundary conditions with the coordinate on the right-hand side of the rod being dependent on time. A special transformation of variables eliminates the dependence of the right-hand side coordinate of the boundary conditions on time. The transformation substantially simplifies the boundary conditions, converting them to the classical fixed-free boundary conditions. The simplification of the boundary conditions is, in turn, exacerbated by the equation of rod motion because it becomes a linear partial differential equation with variable coefficients containing some additional terms. The proposed solution of this equation is built in terms of a trigonometric series with time-dependent coefficients, where the spatial components satisfy the boundary conditions. In this case the original partial differential equation is converted into an infinite system of coupled ordinary differential equations with corresponding initial conditions. Truncation of the system produces an initial problem which is solved numerically. The corresponding truncated trigonometric series rapidly converges to the solution. The solutions are built for different combinations of the parameters of the varying rod. It is shown that for lightly damped rods, the amplitudes of different modes are mainly defined by free solutions of the initial problem. The notion of generated equations of the system is introduced. Free solutions can be obtained from the generating equations of the coupled system of ordinary differential equations. Moreover, exact solutions

Stephan V. Joubert · Michael Y. Shatalov · Igor A. Fedotov · Judith Nomantande Maureen Bidie
Department of Mathematics and Statistics, Tshwane University of Technology, South Africa,
e-mail: JoubertSV@tut.ac.za, ShatalovM@tut.ac.za, FedotovI@tut.ac.za, bidiejnm@tut.ac.za

Masodi Lesiba Glen Lekalakala
Department of Mathematics, Tshwane University of Technology, South Africa,
e-mail: glen@vut.ac.za

H. Altenbach et al. (eds.), *Mechanics for Materials and Technologies*,
Advanced Structured Materials 46, DOI 10.1007/978-3-319-56050-2_24

of the generating equations are built in terms of the elementary Kummer and confluent Heun functions. These exact solutions give one proper insight into the dynamic processes governing vibrations of the varying lightly damped rods. In the case of heavily damped coefficients, free vibration of the rod is rapidly suppressed and the amplitude behaviour of the modes on a finite time interval is defined by the excitation force. For example, in the case of a linearly growing rod of constant volume, the amplitude of the equivalent excitation force also grows proportionally to time. Owing to this effect, the amplitudes of the particular modes, in turn, are linearly increased with time.

Key words: Longitudinal vibration of a rod · Rod with variable length · Rod with variable cross-section · Resonance · Heun function · Kummer function

24.1 Introduction

The development of mathematical theories describing the processes of modern additive manufacturing technologies is a subsection of the topical problems of theoretical and applied mechanics (Manzhirov et al, 2011; Gupta et al, 2013). The dynamics of accreting bodies is an important part of these theories. This was mainly developed in the pioneering work of Prof. A.V. Manzhirov and his collaborators (Manzhirov and Lychev, 2015, 2014; Manzhirov and Parshin, 2015b,a; Manzhirov, 2014; Manzhirov and Lychev, 2013; Levitin et al, 2012, 2014; Lychev and Fedotov, 2015; Kuznetsov et al, 2012; Manzhirov et al, 2012; Lychev et al, 2011; Manzhirov and Lychev, 2009; Shatalov et al, 2012). The fundamentally new mathematical models describing the strain-stress evolutions in the process of additive manufacturing were developed in Manzhirov and Lychev (2015) and Manzhirov and Lychev (2014). The influence of the force of gravity in the problem of an arch structure erected by the additive technologies was analysed in Manzhirov and Parshin (2015b) and Manzhirov and Parshin (2015a). New mathematical models of continuously growing bodies in the case of small and finite deformations were presented in Manzhirov (2014) and Manzhirov and Lychev (2013). Non-stationary dynamical problems of growing bodies are now of special interest. A transient dynamical problem for an accreted thermoelastic parallelepiped was considered in Levitin et al (2012) and Levitin et al (2014). Transient oscillations of a spherical growing shell were analysed in Lychev and Fedotov (2015). Non-stationary heat conduction problems in a growing ball were treated in Kuznetsov et al (2012) and Manzhirov et al (2012). The unsteady lateral vibration of a circular plate with growing thickness was investigated in Lychev et al (2011). A unified variational approach for the mathematical modelling of growth processes in nature and engineering was outlined in Manzhirov and Lychev (2009).

The present paper is devoted to the longitudinal vibration of a slender rod with varying geometrical parameters such as length and cross-section. Previously analogous problems were considered in the works of some of the co-authors of the present paper (Shatalov et al, 2012; Lekalakala et al, 2015b,a, 2016, 2015c). A Kelvin viscoelastic rod of initially unit length, which is actuated by a harmonic force, is considered in the framework of the classical theory of rods. It is assumed that the left end of the rod is fixed and the free right end moves with small linear velocity so that the length of the rod increases linearly. Moreover, it is assumed that the cross-section area of the rod simultaneously increases or decreases in accordance with the power law. The corresponding mixed (boundary-initial) problem is formulated mathematically so that the coordinate of the right-hand side boundary condition depends on time. A transformation of coordinates is proposed, which converts the given unconventional boundary-value problem into the standard one. This transformation simplifies the boundary conditions but simultaneously substantially complicates the governing partial differential equation coefficients which become time dependent. Assuming the damping effects are small, we neglect the effects of second-order smallness as well as the second-order effects of small longitudinal growth rate. The simplified equation obtained is approximately solved by its transformation into a truncated system of ordinary differential equations with corresponding initial conditions. The assumed truncated series form of the solution is formulated so that the boundary conditions are automatically satisfied. A notion of generating equation is introduced and exact solutions of these equations are obtained in some special cases. For example, it is shown that the solution of the undamped rod simultaneously varying in longitudinal ($l(t) = l_0(1+\varepsilon t)$) and the lateral ($A(t) = A_0(1+\varepsilon\eta t)$) directions, is expressed in terms of elementary functions. If viscoelastic damping effects are not neglected, the exact solution is expressed in terms of the confluent Heun functions (if $\eta \neq 1$) or Kummer functions (if $\eta = 1$). The physical interpretations of the results obtained are discussed. Numerical simulations of the vibrating rod with variable geometric dimensions are considered in the most important cases: from free vibrations of the viscoelastic rod to forced vibrations of an equivoluminally growing rod with light and heavy damping.

24.2 Equations of Motion and Their Transformations

In this section we describe the mathematical model of a longitudinally vibrating cylindrical medium's geometrical parameters, which are changed in both the longitudinal and/or lateral directions. This model can be used to describe either an elastic solid rod or an ideal compressible fluid (acoustic medium) enveloped in a rigid cylindrical shell with variable geometry and a soft membrane on the right-hand side. In all of these cases we will refer to the deformed and vibrating medium as a "rod". The equation of motion of a longitudinally vibrating rod actuated by force $f_0 \sin(\omega t)$ in the Ox-direction is

$$\frac{\partial}{\partial t}\left(A(t)\frac{\partial u}{\partial t}\right) - 2dA(t)\frac{\partial^3 u}{\partial t\,\partial x^2} - c^2 A(t)\frac{\partial^2 u}{\partial x^2} = f_0 \sin(\omega t) \tag{24.1}$$

Where $u = u(t,x)$ is the longitudinal (particle) displacement of the rod and $A(t) = A_0(1+\varepsilon\eta t)^{\alpha}$ is its time-dependent cross-sectional area . It is assumed that the change of the length of the rod is proportional to time: $l(t) = l_0(1+\varepsilon t)$. Let us assume that $l_0 = 1$, $\eta = const$, $\alpha$is positive, negative or zero constant and ε is positive or negative "small" constant ($0 < |\varepsilon| << 1$). The physical meaning of ε is the rate of growth (if $\varepsilon > 0$) or shrinkage (if $\varepsilon < 0$) of the rod's length. The dissipative factor d which is also assumed to be small is proportional to the viscoelastic damping in the Kelvin-Voight model. The term $2dA(t)\frac{\partial^3 u}{\partial t\,\partial x^2}$ describes wave dissipation in the viscous heat transfer medium. In this case $2d = \frac{\rho}{A_0}\left[\frac{4\eta}{3}+\zeta+\kappa\left(\frac{1}{c_v}-\frac{1}{c_p}\right)\right]$, where η, ζ are correspondingly volumetric and shear viscosities, $\kappa$is thermal conductivity, and c_v, c_p are heat capacities of the medium at constant volume and pressure, respectively. The phase velocity of the longitudinal wave propagation is $c = \sqrt{\frac{E}{\rho}}$, where E is the modulus of elasticity and ρ is the mass density of the rod. The amplitude of periodic excitation of the rod, f_0, is proportional to the excitation force, and ω is the (angular) frequency of periodic excitation. It is also possible to consider a compressible fluid stem oscillating in a longitudinal direction in the framework of equation (24.1) with $c = \sqrt{\frac{K}{\rho}}$, where K is the bulk modulus of the fluid and ρ is its mass density.

Boundary conditions for the rod are as follows:

$$\begin{aligned} x = 0: &\quad u(t, x = 0) = 0 \\ x = 1+\varepsilon t: &\quad \frac{\partial u}{\partial x}(t, x = 1+\varepsilon t) = 0 \end{aligned} \tag{24.2}$$

The initial conditions are

$$t = 0: \; u(t = 0, x) = 0, \; \frac{\partial u}{\partial t}(t = 0, x) = 0 \tag{24.3}$$

Let us make a change of variables $(t,x) \to (\tau, y)$ in the mixed problem (24.1) – (24.3) as follows:

$$t = \tau, \; x = y(1+\varepsilon\tau) \tag{24.4}$$

After making the change of variables (24.4) in equation (24.1) and after its division by$A(\tau)$, the equation of motion is:

$$\begin{aligned} &\frac{\partial^2 v}{\partial \tau^2} - \frac{2d}{(1+\varepsilon\tau)^2}\frac{\partial^3 v}{\partial \tau\,\partial y^2} - \frac{2\varepsilon y}{1+\varepsilon\tau}\frac{\partial^2 v}{\partial \tau\,\partial y} + \frac{2\varepsilon d y}{(1+\varepsilon\tau)^3}\frac{\partial^3 v}{\partial y^3} + \frac{2\varepsilon\eta\alpha}{1+\varepsilon\eta\tau}\frac{\partial v}{\partial \tau} \\ &\qquad + 2\varepsilon^2\left[\frac{1}{(1+\varepsilon\eta\tau)^2} - \frac{\eta\alpha}{(1+\varepsilon\tau)(1+\varepsilon\eta\tau)}\right] y\frac{\partial v}{\partial y} \\ &\quad - \left[\frac{c^2-\varepsilon^2 y^2}{(1+\varepsilon\tau)^2} - \frac{4\varepsilon d}{(1+\varepsilon\tau)^3}\right]\frac{\partial^2 v}{\partial y^2} + \frac{2\varepsilon^2\eta^2\alpha(\alpha-1)}{(1+\varepsilon\eta\tau)^2}v = \frac{f_0}{A_0}\frac{\sin(\omega\tau)}{(1+\varepsilon\eta\tau)^{\alpha}} \end{aligned} \tag{24.5}$$

where $v = v(\tau, y) = u(t = \tau, x = y(1+\varepsilon\tau))$. The new boundary conditions are

$$
\begin{aligned}
y=0: &\quad v(\tau,y=0)=0 \\
y=1: &\quad \tfrac{\partial v}{\partial y}(\tau,y=1)=0
\end{aligned}
\tag{24.6}
$$

and new initial conditions are

$$\tau=0:\ v(\tau=0,y)=0,\ \tfrac{\partial v}{\partial \tau}(\tau=0,y)=0 \tag{24.7}$$

Equation (24.5) contains several terms of higher than $O(\varepsilon)$ and $O(d)$ order of smallness. Neglecting terms $O(\varepsilon^2)$ and $O(\varepsilon d)$ in equation (24.5), the simplified equation is written as follows:

$$
\begin{aligned}
&\frac{\partial^2 v}{\partial \tau^2}-\frac{2\varepsilon y}{1+\varepsilon\tau}\frac{\partial^2 v}{\partial\tau\partial y}-\frac{2d}{(1+\varepsilon\tau)^2}\frac{\partial^3 v}{\partial\tau\partial y^2}+\frac{2\varepsilon\eta\alpha}{1+\varepsilon\eta\tau}\frac{\partial v}{\partial\tau}-\frac{c^2}{(1+\varepsilon\tau)^2}\frac{\partial^2 v}{\partial y^2} \\
&\approx\frac{f_0}{A_0}\frac{\sin(\omega\tau)}{(1+\varepsilon\eta\tau)^{\alpha}}
\end{aligned}
\tag{24.8}
$$

The boundary and initial conditions for (24.8) are the same for equations (24.6) and (24.7). Let us now consider methods of solution for the formulated mixed problem given by equations (24.6) to (24.8).

24.3 Theoretical Treatment: Solution of Mixed Problem (24.6) to (24.8)

(24.8) is as follows:

$$v=v(\tau,y)=\sum_{m=1}^{\infty}C_m(\tau)\sin\left[\frac{(2m-1)\pi}{2}y\right], \tag{24.9}$$

This series obviously satisfies boundary conditions (24.6). Substituting (24.9) in (24.8), multiplying the result by $2\sin\left[\frac{(2n-1)\pi}{2}y\right]$, $(n=1,2,\ldots)$ and integrating with respect to y from 0 to 1, the following infinite system of ordinary differential equations is obtained:

$$
\begin{aligned}
&\frac{d^2C_m}{d\tau^2}+\left[\frac{d\pi^2(2m-1)^2}{2(1+\varepsilon\tau)^2}+\frac{2\varepsilon\eta\alpha}{1+\varepsilon\eta\tau}-\frac{\varepsilon}{1+\varepsilon\tau}\right]\frac{dC_m}{d\tau}+\frac{c^2\pi^2(2m-1)^2}{4(1+\varepsilon\tau)^2}C_m \\
&+\frac{\varepsilon}{1+\varepsilon\tau}\sum_{\substack{n=1\\(n\neq m)}}^{\infty}\frac{(-1)^{m+n+1}\cdot(2n-1)^2}{(n-m)(m+n-1)}\frac{dC_n}{d\tau}\approx\frac{4f_0}{(2m-1)\pi A_0(1+\varepsilon\eta\tau)^{\alpha}}\sin(\omega\tau)
\end{aligned}
\tag{24.10}
$$

where $m=1,2,\ldots$. Note that at $n=n_0=const:\ \lim\limits_{m\to\infty}\left[\frac{(2n-1)^2}{(m-n)(m+n-1)}\right]=0$, and at $m=m_0=const:\ \lim\limits_{n\to\infty}\left[\frac{(2n-1)^2}{(n-m)(m+n-1)}\right]=4$. At low frequency excitation of the rod (the classical theory of longitudinal vibration of rods describes exclusively the low frequency vibrations of rods), the system of equations (24.10) converges rapidly to a solution, and it is thus possible to consider a truncated system of ordinary differential equations ($m = 1, 2, \dots, N$). The explicit form of system (24.10) at $N = 4$ is as follows:

$$\begin{aligned} &\frac{d^2C_1}{d\tau^2} + \left[d\frac{\pi^2}{2(1+\varepsilon\tau)^2} + \frac{2\varepsilon\eta\alpha}{1+\varepsilon\eta\tau} - \frac{\varepsilon}{1+\varepsilon\tau} \right] \frac{dC_1}{d\tau} + \frac{c^2\pi^2}{4(1+\varepsilon\tau)^2} C_1 \\ &+ \frac{\varepsilon}{1+\varepsilon\tau} \left(\frac{9}{2}\frac{dC_2}{d\tau} - \frac{25}{6}\frac{dC_3}{d\tau} + \frac{49}{12}\frac{dC_4}{d\tau} \right) \approx \frac{4f_0}{\pi A_0 (1+\varepsilon\eta\tau)^\alpha} \sin(\omega\tau) \end{aligned}$$

$$\begin{aligned} &\frac{d^2C_2}{d\tau^2} + \left[d\frac{9\pi^2}{2(1+\varepsilon\tau)^2} + \frac{2\varepsilon\eta\alpha}{1+\varepsilon\eta\tau} - \frac{\varepsilon}{1+\varepsilon\tau} \right] \frac{dC_2}{d\tau} + \frac{9c^2\pi^2}{4(1+\varepsilon\tau)^2} C_2 \\ &+ \frac{\varepsilon}{1+\varepsilon\tau} \left(-\frac{1}{2}\frac{dC_1}{d\tau} + \frac{25}{4}\frac{dC_3}{d\tau} - \frac{49}{10}\frac{dC_4}{d\tau} \right) \approx \frac{4f_0}{3\pi A_0 (1+\varepsilon\eta\tau)^\alpha} \sin(\omega\tau) \end{aligned}$$

$$\begin{aligned} &\frac{d^2C_3}{d\tau^2} + \left[d\frac{25\pi^2}{2(1+\varepsilon\tau)^2} + \frac{2\varepsilon\eta\alpha}{1+\varepsilon\eta\tau} - \frac{\varepsilon}{1+\varepsilon\tau} \right] \frac{dC_3}{d\tau} + \frac{25c^2\pi^2}{4(1+\varepsilon\tau)^2} C_3 \\ &+ \frac{\varepsilon}{1+\varepsilon\tau} \left(\frac{1}{6}\frac{dC_1}{d\tau} - \frac{9}{4}\frac{dC_2}{d\tau} + \frac{49}{6}\frac{dC_4}{d\tau} \right) \approx \frac{4f_0}{5\pi A_0 (1+\varepsilon\eta\tau)^\alpha} \sin(\omega\tau) \end{aligned}$$

$$\begin{aligned} &\frac{d^2C_4}{d\tau^2} + \left[d\frac{49\pi^2}{2(1+\varepsilon\tau)^2} + \frac{2\varepsilon\eta\alpha}{1+\varepsilon\eta\tau} - \frac{\varepsilon}{1+\varepsilon\tau} \right] \frac{dC_4}{d\tau} + \frac{49c^2\pi^2}{4(1+\varepsilon\tau)^2} C_4 \\ &+ \frac{\varepsilon}{1+\varepsilon\tau} \left(-\frac{1}{12}\frac{dC_1}{d\tau} + \frac{9}{10}\frac{dC_2}{d\tau} - \frac{25}{6}\frac{dC_3}{d\tau} \right) \approx \frac{4f_0}{7\pi A_0 (1+\varepsilon\eta\tau)^\alpha} \sin(\omega\tau) \end{aligned} \tag{24.11}$$

The corresponding homogeneous generating components of all equations from (24.10) to (24.11) are:

$$\frac{d^2C_m}{d\tau^2} + \left[\frac{d\pi^2 (2m-1)^2}{2(1+\varepsilon\tau)^2} + \frac{2\varepsilon\eta\alpha}{1+\varepsilon\eta\tau} - \frac{\varepsilon}{1+\varepsilon\tau} \right] \frac{dC_m}{d\tau} + \frac{c^2\pi^2 (2m-1)^2}{4(1+\varepsilon\tau)^2} C_m = 0 \tag{24.12}$$

To understand the nature of vibrations in the rod it is worthwhile to consider a particular case when damping is negligibly small so that it is possible to assume $d \approx 0$. Let factor $\eta = 1$ so that equation (24.12) is as follows:

$$\frac{d^2C_m}{d\tau^2} - \frac{\chi\varepsilon}{1+\varepsilon\tau}\frac{dC_m}{d\tau} + \frac{c^2\pi^2 (2m-1)^2}{4(1+\varepsilon\tau)^2} C_m = 0 \tag{24.13}$$

where $\chi = 1 - 2\alpha$. In this case the exact solution of (24.13) represents almost periodic oscillations with period changing according to a logarithmic law and the amplitude changing at a rate proportional to $(1+\varepsilon t)^{\frac{\chi+1}{2}}$:

$$C_m(\tau) = \{S_1 \cos[w_m \ln(1+\varepsilon t)] + S_2 \sin[w_m \ln(1+\varepsilon t)]\} (1+\varepsilon t)^{\frac{\chi+1}{2}} \tag{24.14}$$

where $w_m = \sqrt{\frac{c^2\pi^2(2m-1)^2}{4\varepsilon^2} - \left(\frac{\chi+1}{2}\right)^2}$. In particular, if $\chi = -1$ (this corresponds to the case of simultaneous growth of the length of the rod and its cross-sectional area with the same rate ε), then the amplitude of vibration remains constant. If $\chi = 0$ (in this case the cross-section of the rod grows proportionally to the square root of its longitudinal growth, $\alpha = 0.5$), then the amplitude of vibration grows proportionally to $\sqrt{1+\varepsilon t}$. At $\chi = 1$ (the longitudinal growth of the rod is not accompanied by change of its cross-section, $\alpha = 0$), the amplitude of its vibration grows linearly with the longitudinal growth of the rod $(1+\varepsilon t)$. The energy considerations help us to understand the nature of the above-mentioned behaviour of the rod. For example, if in the last case we calculate how the total energy $E = \frac{A_0}{2}\int_0^1 \left[\left(\frac{\partial u}{\partial t}\right)^2 + c^2\left(\frac{\partial u}{\partial x}\right)^2\right] dy$ of the first mode $u = S_1 \cos\left[w_m \ln(1+\varepsilon t)\right] \sin\left(\frac{\pi y}{2}\right)$ changes with time (in a non-autonomous system the law of energy conservation does not hold in general), we obtain

$$E(t) = \frac{A_0}{16}\pi^2 c^2 S_1^2 \left[1 + \frac{\varepsilon}{c}\sin\left(\frac{\pi c}{\varepsilon}\ln(1+\varepsilon\tau)\right) + O\left(\varepsilon^2\right)\right] \approx \frac{A_0}{16}\pi^2 c^2 S_1^2 \left[1 + \frac{\varepsilon}{c}\sin(\pi c\tau)\right]. \quad (24.15)$$

Hence the mean value of the total energy is not changed and the energy circulates between the rod and energy source. This supports the constant rate of linear growth of the rod. Analogous considerations are valid in all other particular cases.

If the damping effects are not neglected $(d \neq 0)$ (see equation (12)), then the qualitative behaviour of vibrations is changed. The general solution of equation (24.12) is as follow:

$$C_m(\tau) = \left[\begin{array}{l} S_1 \cdot HC\left(\bar{a}, 2\bar{b}, \bar{c}, \bar{d}, \bar{e}, \frac{1-\eta^{-1}}{1+\varepsilon\tau}\right)(1+\varepsilon\tau)^{-\bar{b}} + \\ S_2 \cdot HC\left(\bar{a}, -2\bar{b}, \bar{c}, \bar{d}, \bar{e}, \frac{1-\eta^{-1}}{1+\varepsilon\tau}\right)(1+\varepsilon\tau)^{\bar{b}} \end{array}\right] \cdot e^{\frac{d_m\pi^2}{2\varepsilon(1+\varepsilon\tau)}} (1+\varepsilon\tau)^{1-\alpha} \quad (24.16)$$

where $HC(\ldots)$ are confluent Heun functions of six arguments (see Appendix A). The corresponding parameters in (24.16) are $\bar{a} = \frac{d\pi^2\eta}{2\varepsilon(\eta-1)}$, $\bar{b} = i\sqrt{\left(\frac{w_m}{\varepsilon}\right)^2 - (\alpha-1)^2}$, $\bar{c} = 2\alpha - 1$, $\bar{d} = \frac{3d\pi^2\eta}{4\varepsilon(\eta-1)}$ and $\bar{e} = 2\left(\alpha^2 - \frac{3}{2}\alpha + \frac{1}{4}\right) - \frac{d\eta\pi^2}{2\varepsilon(\eta-1)}\left(\alpha - \frac{3}{2}\right)$, $(i^2 = -1)$. It follows from the general solution (24.16) that at initial time interval $[0, \tau]$, the behaviour of the solution is defined by the product of the confluent Heun functions with the exponent $e^{\frac{d_m\pi^2}{2\varepsilon(1+\varepsilon\tau)}}$ (see the next paragraph devoted to numerical simulations of the processes), and at $\tau \to \infty$ this product tends to 1 and both fundamental solutions change at a rate proportional to $(1+\varepsilon\tau)^{1-\alpha}$. Qualitatively, this behaviour of the rod can be explained from equation (24.12), where the factors responsible for amplitude change of vibrations $\left(\frac{2\varepsilon\eta\alpha}{1+\varepsilon\eta\tau} - \frac{\varepsilon}{1+\varepsilon\tau}\right)\frac{dC_m}{d\tau}$ are inversely proportional to time and other terms. Indeed, the viscous damping term, $\frac{d\pi^2(2m-1)^2}{2(1+\varepsilon\tau)^2}\frac{dC_m}{d\tau}$ is inversely proportional to the square of time and hence decreases faster than the previous terms.

It is obvious that solution (24.16) is applicable to $\eta \neq 1$. If $\eta = 1$, the solution of equation (24.12) can be expressed in terms of the hypergeometric functions. But it is simpler to express this solution in terms of the Kummer functions as follows:

$$C_m(\tau) = \left[S_1 \cdot M\left(-\tilde{a}, \tilde{b}, \frac{d}{\varepsilon(1+\varepsilon\tau)}\right) + S_2 \cdot U\left(-\tilde{a}, \tilde{b}, \frac{d}{\varepsilon(1+\varepsilon\tau)}\right)\right] \cdot (1+\varepsilon\tau)^{\tilde{a}} \tag{24.17}$$

where $M\left(-\tilde{a}, \tilde{b}, \frac{d}{\varepsilon(1+\varepsilon\tau)}\right)$, $U\left(-\tilde{a}, \tilde{b}, \frac{d}{\varepsilon(1+\varepsilon\tau)}\right)$ are fundamental solutions of Kummer's equation $\tilde{a} = (1-\alpha) - i\sqrt{\left(\frac{w_m}{\varepsilon}\right)^2 - (1-\alpha)^2}$, $\tilde{b} = 1 + 2i\sqrt{\left(\frac{w_m}{\varepsilon}\right)^2 - (1-\alpha)^2}$. The general solution of the non-homogeneous equation corresponding to (24.12) with the right-hand side equal to $f_m(t)$ is calculated as follows:

$$C_m(t) = C_{m,0}(t) + \int_0^t G_m(t,\tau) f_m(\tau) d\tau \tag{24.18}$$

where $C_{m,0}(t) = S_1 \varphi_{m,1}(t) + S_2 \varphi_{m,2}(t)$ is the general solution (24.16) of the homogeneous equation (24.12) and

$$\varphi_{m,1,2}(t) = HC\left(\bar{a}, \pm\bar{b}, \bar{c}, \bar{d}, \bar{e}, \frac{1-\eta^{-1}}{1+\varepsilon t}\right)(1+\varepsilon t)^{\mp i\sqrt{\left(\frac{w_m}{\varepsilon}\right)^2 - (\alpha-1)^2}} e^{\frac{d_m \pi^2}{2\varepsilon(1+\varepsilon t)}} (1+\varepsilon t)^{1-\alpha} \tag{24.19}$$

are fundamental solutions of (24.12), if $\eta \neq 1$. If $\eta = 1$, $\varphi_{m,1}(t) = M\left(-\tilde{a}, \tilde{b}, \frac{d}{\varepsilon(1+\varepsilon\tau)}\right)$, $\varphi_{m,2}(t) = U\left(-\tilde{a}, \tilde{b}, \frac{d}{\varepsilon(1+\varepsilon\tau)}\right)$ and

$$G_m(t,\tau) = \frac{\varphi_{m,1}(\tau)\varphi_{m,2}(t) - \varphi_{m,1}(t)\varphi_{m,2}(\tau)}{W_m(\tau)} \tag{24.20}$$

is Green's function of equation (24.12) with

$$W_m(\tau) = W_{m,0} \frac{(1+\varepsilon\eta\tau)^{2\alpha}}{1+\varepsilon\tau} e^{\frac{d\pi^2(2m-1)^2}{2(1+\varepsilon\tau)}\tau} \tag{24.21}$$

the Wronskyan of functions $\varphi_{m,1}(t), \varphi_{m,2}(t)$ and $W_{m,0} = const$.

24.4 Numerical Simulations and Discussions

In this section we perform several numerical simulations of initial problems for system (24.10) (the first four equations of this system are represented by (24.11)). In all subsequent simulations we used a truncated system of ten equations but graphically illustrated the behaviour of only the first four modes on the time interval, $\tau \in [0, 250]$. Numerical integration was performed using the Fehlberg fourth-fifth-order Runge-Kutta method with degree four interpolant. For the following examples we selected some characteristic values of parameters $c,\ d,\ \varepsilon,\ A_0,\ \alpha,\ \eta,\ f_0$ and ω.

In the *first example*, free vibrations of the linearly growing lightly damped rod are considered with parameters $c = 1,\ d = 0.05,\ \varepsilon = 0.05,\ A = 1,\ \alpha = 0,\ f_0 = 0$ (values of η and ω are not important in these simulations). Initial conditions are $C_1(\tau = 0) = 1,\ C_2(\tau = 0) = \ldots = C_{10}(\tau = 0) = 0,\ \left.\frac{dC_1}{d\tau}\right|_{\tau=0} = \ldots = \left.\frac{dC_{10}}{d\tau}\right|_{\tau=0} = 0.$ The solution to this problem is shown in Fig. 24.1.

Fig. 24.1 illustrates the qualitative behaviour of a vibrating rod. An exponential drop of amplitudes of the modes is observed at the initial time interval due to the forces of viscoelastic damping. This short time tendency is changed by the following linear growth of amplitudes of the modes (the so-called "quasi resonance effect of a growing rod"). The results of simulation are in full agreement with the qualitative behaviour of all individual modes, which have been described in the theoretical part of the paper (see solutions (24.16) – (24.17) and subsequent considerations). These graphs also illustrate generation of higher modes at the excitation of the first mode and the fast convergence of the solution, because the amplitudes of higher modes $C[m] = C_m(\tau)$ rapidly decrease as the number m grows.

As a *second example*, we consider forced oscillations of a linearly growing rod. The amplitude of the excitation force is $f_0 = 1$ and excitation frequency is $\omega_0 = \frac{4}{3}$. Other parameters are the same as in the *first example*. Initial parameters are zero: $C_1(\tau = 0) = \ldots = C_{10}(\tau = 0) = 0,\ \left.\frac{dC_1}{d\tau}\right|_{\tau=0} = \ldots = \left.\frac{dC_{10}}{d\tau}\right|_{\tau=0} = 0.$ The results of simulation are shown in Fig. 24.2.

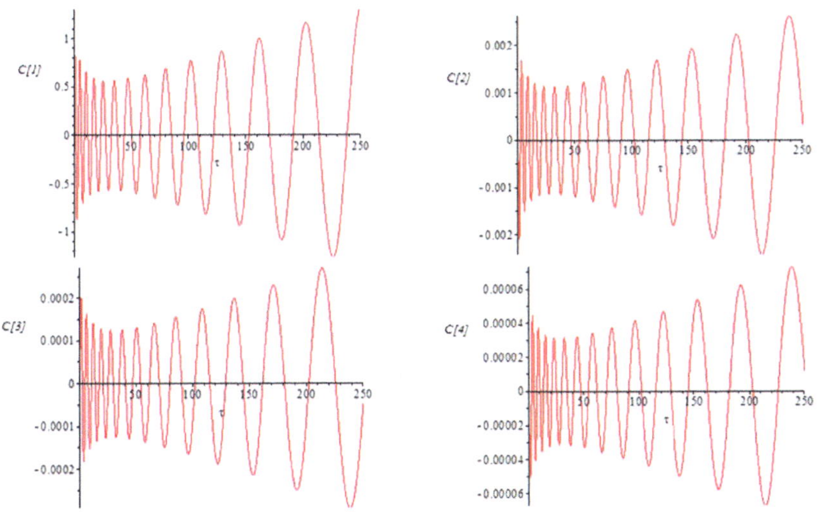

Fig. 24.1: Four modes of free vibration of a viscoelastic growing rod with light damping ($c = 1,\ d = 0.05,\ \varepsilon = 0.05,\ A_0 = 1,\ \alpha = 0,\ f_0 = 0$)

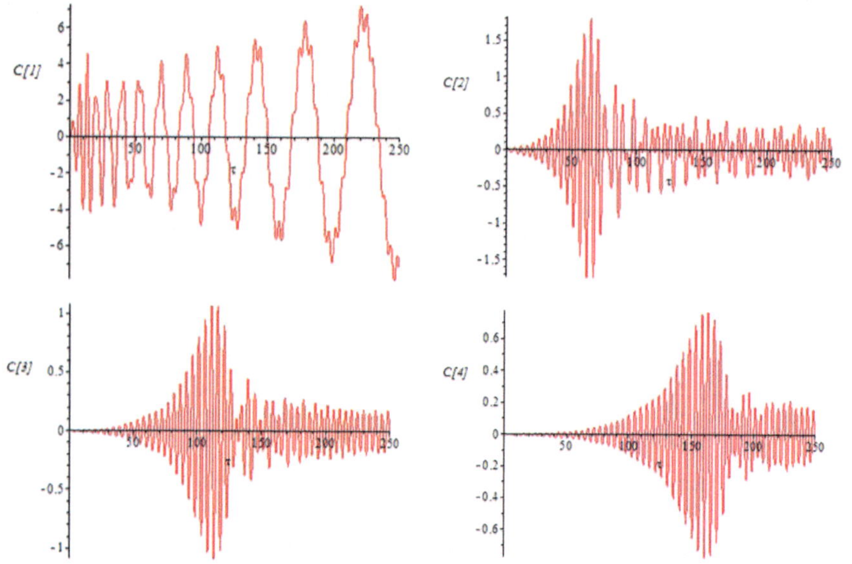

Fig. 24.2: Four modes of forced vibration of viscoelastic growing rod with light damping ($c = 1$, $d = 0.05$, $\varepsilon = 0.05$, $A_0 = 1$, $\alpha = 0$, $f_0 = 1$, $\omega = \frac{4}{3}$)

In this example it was assumed that excitation frequency $\omega = \frac{4}{3}$ is less than the lowest eigenvalue ($\omega_1 = \frac{\pi c}{2 \cdot l(\tau=0)} = \frac{\pi}{2} > \frac{4}{3} = \omega$) of the rod at its initial state. The eigenvalues of a rod which does not grow longitudinally are equal to $\omega_n = \frac{\pi c(2n-1)}{2 \cdot l}$, where $n = 1, 2, \cdots$. In the present example it is assumed that the rod grows linearly according to the law: $l = l(\tau) = 1 + \varepsilon\tau$. Hence, at $c = 1$ and at a particular time instant $\tau = \tilde{\tau}$, the eigenvalues of the "frozen" rod are distributed according to $\tilde{\omega}_n = \frac{\pi(2n-1)}{2 \cdot (1+\varepsilon\tilde{\tau})}$. There exists an m where the eigenvalue is equal to the excitation frequency $\left(\omega = \frac{4}{3}\right)$ at $\tau = \tau_m = \frac{1}{\varepsilon}\left[\frac{\pi(2m-1)}{2 \cdot \omega} - 1\right] = \frac{1}{\varepsilon}\left[\frac{3\pi(2m-1)}{8} - 1\right]$. Hence, the time interval between two neighbouring resonance peaks of m-th and $(m+1)$-th modes of the rod is equal to $\Delta\tau = \tau_{m+1} - \tau_m = \frac{3\pi}{4\varepsilon} \simeq 47.1$ time units. This resonance behaviour of a growing rod excited by an harmonic force is illustrated in Fig. 2. In this figure one can see that the first four modes pass through a point in time (called a point of resonance) where resonance occurs. Another interesting aspect of the considered light damping vibration is that the first mode has its amplitude linearly increasing after passing through the resonance point while the higher modes appear not to have increasing amplitude after this point. This is explained by the fact that the effective viscose damping factor of m-th mode is proportional to $(2m-1)^2$, and hence, its natural vibrations are damped substantially higher than those of the lower modes (see equations (24.10) – (24.12)). It is possible to show that at longer time intervals, the effect

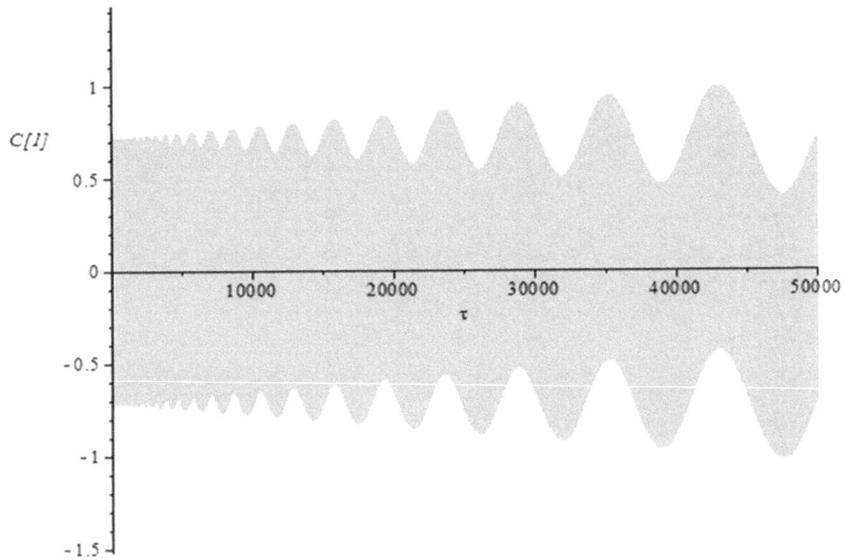

Fig. 24.3: First mode of forced vibration of viscoelastic growing rod with heavy damping ($c = 1$, $d = 0.25$, $\varepsilon = 0.05$, $A_0 = 1$, $\alpha = 0$, $f_0 = 1$, $\omega = \frac{4}{3}, \tau \in [0, 50000]$)

of modulation of the external vibration with increasing amplitude is observed due to coupling of the higher modes with the first mode in the equations of motion.

Let us consider a *third example* where the damping factor is higher. In this case we consider the same parameters as in the *second example* but the damping factor is equal to$d = 0.25$. In this case simulations demonstrate the sum of excited and linearly growing natural vibrations at the first harmonic of a bar with increasing length. This process is depicted in (Fig. 24.3) for the time interval, $\tau \in [0, 50000]$.

In the *fourth example* we consider forced oscillations of a lightly damped rod with a simultaneously increasing length and proportionally decreasing area of cross-section so that the volume of the rod remains constant ("equivoluminal growth" of the rod). In this case the parameters of the rod are assumed to be equal to $c = 1$, $d = 0.05$, $\varepsilon = 0.05$, $A_0 = 1$, $\alpha = -1$, $\eta = 1$, $f_0 = 1$, $\omega = \frac{4}{3}$. The solution to the problem at time interval $\tau \in [0, 750]$ is shown in Fig. 24.4.

It follows from this simulation that the first mode starts to increase amplitude of its free oscillations according to a quadratic law as it is prescribed by solution (24.17). One can observe that this mode is represented as a superposition of quadratically growing natural oscillations and growing forced oscillations at excitation frequency ω. Equations (24.10) and (24.11) suggest that the amplitude of the excited vibrations grows linearly due to the linear growth of excitation force at $\alpha = -1$. Higher harmonics also demonstrate vibrations with increasing amplitude, but the rate of amplitude growth is less than that of the first harmonic due to the increase (proportionally to $(2m-1)^2$) of the equivalent damping factor.

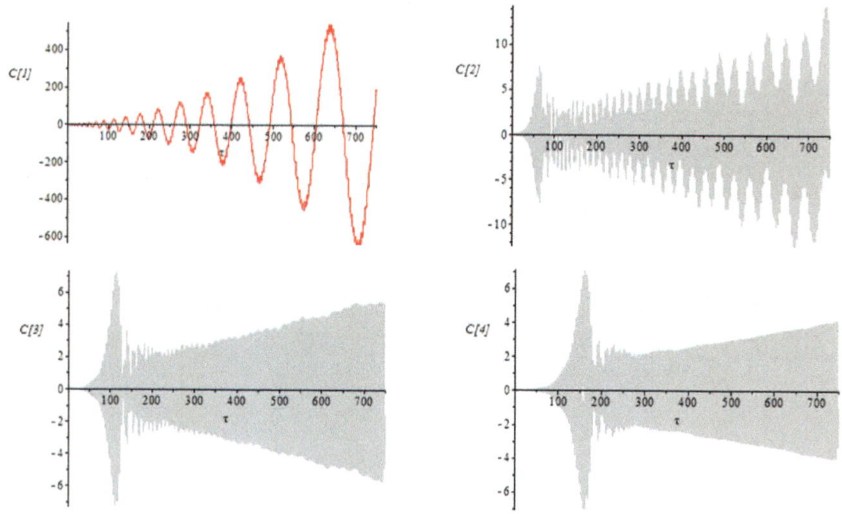

Fig. 24.4: Four modes of forced vibration of "equivoluminally growing" rod with light damping ($c = 1$, $d = 0.05$, $\varepsilon = 0.05$, $A_0 = 1$, $\alpha = -1$, $\eta = 1$, $f_0 = 1$, $\omega = \frac{4}{3}, \tau \in [0, 750]$)

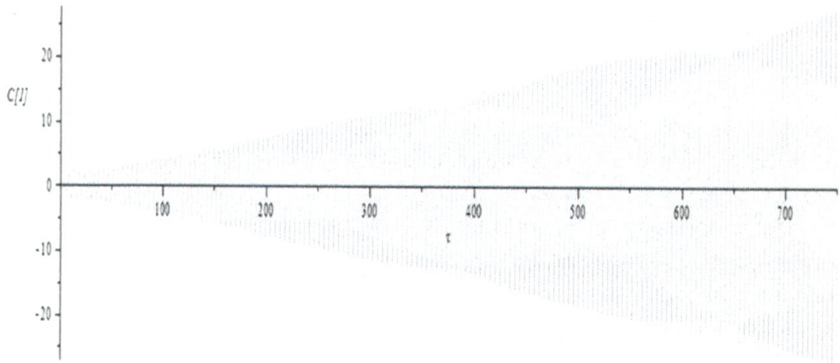

Fig. 24.5: First mode of forced vibration of viscoelastic "equivoluminally growing" rod with heavy damping ($c = 1$, $d = 0.25$, $\varepsilon = 0.05$, $A_0 = 1$, $\alpha = -1$, $\eta = 1$, $f_0 = 1$, $\omega = \frac{4}{3}, \tau \in [0, 750]$)

In the final, *fifth example*, the "equivoluminal growth" of a substantially damped rod subject to forced oscillations is considered. In this case it is assumed that $d = 0.25$. Other parameters are the same as in the previous example. The solution in this case is depicted in Fig. 24.5 for the first mode.

The qualitative difference between the behaviour of the first mode in the cases of light and heavy damping is the rate of development of natural oscillations. In Fig. 24.4 ("light damping" case) an increase in amplitude is observed mainly due to the quadratic increase of free oscillation amplitude. In Fig. 24.5 ("heavy damping" case) amplitude growth is linear, which is mainly explained by the increase of the amplitude of the excitation force.

24.5 Conclusion

A linear model of longitudinal vibration is formulated for a viscoelastic rod with variable length and variable cross-section in the classical framework. It is assumed that the rod is subjected to external harmonic excitation. A mixed problem of dynamics is formulated which contains fixed-free non-conventional boundary conditions where the coordinate of the right end of the rod is time dependent. A transformation of variables is proposed, which eliminates the dependence of the right-hand side coordinate of the boundary conditions on time. This transformation substantially simplifies the boundary conditions converting them to the classical ones. The trade-off of the boundary condition simplification leads to a substantial complication of the equation of rod motion, which becomes a linear partial differential equation with variable coefficients. A solution of this equation is sought in terms of trigonometric series with time dependent coefficients, where the spatial components satisfy the boundary conditions. By means of this representation of the solution, the original partial differential equation is converted into an infinite system of ordinary differential equations with corresponding initial conditions which are solved numerically. Truncation of the system generates a truncated trigonometric series which rapidly converges to the solution. Solutions are built for various combinations of the parameters of the varying rod. It is shown that for lightly damped rods, amplitudes of different modes are mainly defined by free solutions of the initial problem. These free solutions can be obtained from the generating equations of the coupled system of ordinary differential equations. Exact solutions of the generating equations are built in terms of elementary, Kummer and confluent Heun functions. These exact solutions give proper insight into the dynamical processes governing vibrations of varying lightly damped rods. In the case of heavy damping coefficients, free vibration of the rod is suppressed rapidly, and the amplitude behaviour of the various modes on a finite time interval is defined by the excitation force. For instance, in the case of a linearly growing rod of constant volume, the amplitude of the equivalent excitation force also grows proportionally to time. Owing to this effect, in turn, the amplitudes of certain modes increase linearly with time.

Acknowledgements This material is based upon the work supported financially by the Tshwane University of Technology (TUT) and the National Research Foundation (NRF) of South Africa (NRF grant reference number 81643). Any opinions, findings and conclusions or recommendations expressed in this material are those of the authors and therefore the TUT and the NRF do not accept any liability in regard thereto.

References

Gupta N, Manzhirov A, Velmurugan R (eds) (2013) Topical Problems in Solid and Fluid Mechanics. Elite Publishing House Pvt Ltd, New Dehli

Kuznetsov S, Manzhirov A, Fedotov I (2012) Heat conduction problem for a growing ball. Mechanics of Solids 46(6):929–936

Lekalakala M, Shatalov M, Fedotov I (2015a) The behaviour of the elastic accreting rod subjected to free longitudinal vibrations: Rayleigh-Love model. In: Topical Problems of Continuous Mechanics Proceedings of IV International Conference, Tsakhkadzor, Armenia

Lekalakala M, Shatalov M, Fedotov I (2015b) The behavioural pattern of the longitudinally vibrating rod that is accreting in cross-sectional area. In: IUTAM Symposium on Growing Solids, Institute for Problems in Mechanics, Moscow

Lekalakala M, Shatalov M, Fedotov I, Joubert S (2016) On the Rayleigh-Love rod accreting in both length and cross-sectional area: Forced and damped vibrations. The Australian Journal of Mathematical Analysis and Applications (AJMAA) 13(article 18):1–9

Lekalakala MLG, Shatalov M, Fedotov I (2015c) Oon the classical rod that is accreting in cross-sectional area and length subjected to longitudinal vibrations. IOSR Journal of Engineering 1(12):20–24

Levitin A, Lychev S, Manzhirov A, Shatalov M (2012) Nonstationary vibrations of a discretely accreted thermoelastic parallelepiped. Mechanics of Solids 47(6):677–689

Levitin A, Lychev S, Saifutdinov I (2014) Transient dynamical problem for a accreted thermoelastic parallelepiped. In: Proceedings of the World Congress on Engineering 2014, WCE 2014, July 2 - 4, 2014, London, U.K., vol II

Lychev S, Fedotov I (2015) Transient oscillations of a growing spherical shell. In: Proceedings of ICSV22, Florence , Italy, 12-16 July, 2015

Lychev S, Lycheva T, Manzhirov A (2011) Unsteady vibration of a growing circular plate. Mechanics of Solids 46(2):325–333

Manzhirov A (2014) Mechanics of growing solids: New track in mechanical engineering. In: ASME 2014 International Mechanical Engineering Congress and Exposition, Montreal, Quebec, Canada, November 14–20, 2014, vol 9: Mechanics of Solids, Structures and Fluids

Manzhirov A, Lychev S (2009) Mathematical modeling of growth processes in nature and engineering: A variational approach. Journal of Physics: Conference Series 181:012,018

Manzhirov A, Lychev S (2013) The mathematical theory of growing solids: Finite deformations. Doklady Physics 57:160–163

Manzhirov A, Lychev S (2014) Mathematical modeling of additive manufacturing technologies. In: Proceedings of the World Congress on Engineering 2014, WCE 2014, July 2 - 4, 2014, London, U.K., vol II

Manzhirov A, Lychev S (2015) An approach to modeling of additive manufacturing technologies. In: Yang GC, Ao SI, Gelman L (eds) Transactions on Engineering Technologies, Springer Netherlands, pp 99–115

Manzhirov A, Parshin D (2015a) Arch structure erection by an additive manufacturing technology under the action of the gravity force. Mechanics of Solids 50(5):559–570

Manzhirov A, Parshin D (2015b) Influence of the erection regime on the stress state of a viscoelastic arched structure erected by an additive technology under the force of gravity. Mechanics of Solids 50(6):657–675

Manzhirov A, Gupta N, Indeitsev D (eds) (2011) Topical Problems in Theoretical and Applied Mechanics. Elite Publishing House Pvt Ltd, New Dehli

Manzhirov A, Lychev S, Fedotov I, Joubert S (2012) Nonstationary heat transfer problem for a growing ball. In: Topical Problems of Continuous Mechanics, Armenia, pp 14–18

Shatalov M, Manzhirov A, Fedotov I (2012) On the resonance behavior of longitudinally vibrating accreting rod. In: 8th South African Conference on Computational and Applied Mechanics (SACAM 2012), Johannesburg, South Africa, 3–5 September 2012

Chapter 25
A New, Direct Approach Toward Modeling Rate-Dependent Fatigue Failure of Metals

Si-Yu Wang, Lin Zhan, Zheng-Nan Yin & Heng Xiao

Abstract The main objective of this study is to propose a new finite J_2−flow elastoplastic model toward directly modeling metal fatigue behavior over the whole deformation range up to failure. A departure from usual elastoplastic models is that the concept of yielding is rendered irrelevant with a gradual, smooth transition from the elastic to the plastic state in a more realistic sense and, moreover, asymptotic loss of the strength is incorporated. It is demonstrated that the failure behavior may be derived as a direct consequence of the proposed model, without involving any additional failure criteria and any additional variables. Moreover, a direct and explicit approach is suggested to identify each rate-dependent parameter based on suitable test data. Results are presented for rate-dependent failure effects for both monotone and cyclic uniaxial tensile loadings from low to high strain rates. Model predictions compare well with test data.

Key words: Metal fatigue · Finite strain · Elastoplasticity · New model · Direct simulation

25.1 Introduction

It is of great interest to develop an effective methodology to assess reliability and safety of materials and structures associated with fatigue, fracture and failure. Numerous results in these respects have been obtained from various standpoints in the past decades. Usually, the fracture-mechanics-based approach and the damage-mechanics-based approach are used, in conjunction with additional failure criteria.

Si-Yu Wang, Lin Zhan, Zheng-Nan Yin & Heng Xiao
Shanghai Institute of Applied Mathematics and Mechanics, Shanghai University, Yanchang Road 149, 200072 Shanghai, China
e-mail: 343396676@qq.com, 302974155@qq.com, znyin@shu.edu.cn, xiaoheng@shu.edu.cn

H. Altenbach et al. (eds.), *Mechanics for Materials and Technologies*,
Advanced Structured Materials 46, DOI 10.1007/978-3-319-56050-2_25

Below, certain representative samples of most recent results are mentioned in three respects, separately.

Firstly, certain recent results in the study of fracture behavior may be found in, e.g., Bronkhorst et al (2006) for a study of the localization behavior of tantalum and stainless steel; in Shyam et al (2007) for a model of small fatigue crack growth in metallic materials; in Kim et al (2011); Stoughton and Yoon (2011) for a study of the shear fracture and for a failure criterion for sheet metals; in Jansen et al (2013) for an anisotropic stress-based criterion to predict the fracture mechanism etc.; Khan and Liu (2012a,b) for latest advances concerning strain rate and temperature effects, and many others.

Secondly, for certain recent results in continuum damage mechanics and its applications, reference may be made to, e.g., Brünig (2003a,b) for a ductile damage model with irreversible thermodynamics and for numerical treatment; Bonora et al (2005) for ductile damage evolution under multi-axial stress states; Brünig and Ricci (2005) for a nonlocal model for anisotropically damaged metals; Brünig et al (2008); Brünig and Gerke (2011) for a ductile damage criterion under multi-axial stress states and for damage evolution simulation under dynamic loading; Chung et al (2011); Malcher et al (2012) for a modified damage model and for a study of ductile fracture under various cases of triaxial stress; Brünig et al (2013) for micromechanical studies in triaxial stress cases; Shojaei et al (2013) for a study of brittle to ductile damage based on viscoplastic models; and many others.

Thirdly, results in the study of fatigue failure seem immense. For certain representative samples, refer to Susmel (2008, 2014); Rozumek and Macha (2009); Argon (2011); Macha and Nieslony (2012) for energy-based and critical-plane-based approaches and others and to Zhang (2016) for thermomechanical fatigue of single crystal superalloys, as well as the references therein.

In the existing studies, results are usually obtained based either on augmented constitutive structures with additional damage-like variables or on ad hoc criteria for fracture and failure etc.. In a most recent study (Xiao, 2014) a new, direct approach has been proposed to model fatigue, fracture and failure behaviors of metals and alloys etc. The central idea is to establish new elastoplastic models into which the fatigue failure behavior is incorporated as inherent constitutive features. To this goal, it has been indicated (Xiao et al, 2014; Xiao, 2014) that the usual concept of yielding should be abandoned toward a gradual, smooth transition from the elastic to the plastic state in a more realistic sense. In addition, the fatigue failure behavior may be automatically derived as a direct consequence of certain simple asymptotic properties of the strength quantities incorporated.

In this study, we are going to extend the above new elastoplasticity model to a broader case with rate-dependent effects and, then, simulate rate-dependent failure effects of a metal sample under monotone and cyclic uniaxial tensile loadings from low to high strain rates. Results will be presented for a simple case of the proposed model. It will be shown that good agreement with test data may be achieved even with this simple case.

The main context of this contribution will be arranged as follows. Section 25.2 will establish the new rate-dependent elastoplastic model and introduce a simple case of this model; Sect. 25.3 will study the failure behavior of a uniaxial sample under both monotone and cyclic tensile loadings; Sect. 25.4 will present numerical results, and, finally, Sect. 25.5 will give some remarks on the main results.

25.2 New Rate-Dependent Elastoplasticity Model

Consider a finite deforming elastoplastic body. Let $\boldsymbol{D}$, $\boldsymbol{F}$ and $\boldsymbol{L}$ be the stretching tensor, the vorticity tensor and the velocity gradient, respectively. Moreover, $\boldsymbol{\tau}$ and $\boldsymbol{\sigma}$ are the Kirchhoff stress and the Cauchy stress (true stress), namely,

$$\boldsymbol{D} = \frac{1}{2}(\boldsymbol{L}+\boldsymbol{L}^{\mathrm{T}}), \qquad \boldsymbol{W} = \frac{1}{2}(\boldsymbol{L}-\boldsymbol{L}^{\mathrm{T}}), \qquad \boldsymbol{L} = \boldsymbol{F}\cdot\boldsymbol{F}^{-1}, \tag{25.1}$$

$$\boldsymbol{\tau} = J\boldsymbol{\sigma}, J = \det\boldsymbol{F}. \tag{25.2}$$

In the above, $\boldsymbol{F}$ and J are the deformation gradient and the deformation Jacobian.

The self-consistent Eulerian rate formulation of finite elastoplasticity is used ((cf.,e.g. Xiao et al, 2006). The starting point is the additive decomposition of the stretching $\boldsymbol{D}$, as shown below:

$$\boldsymbol{D} = \boldsymbol{D}^e + \boldsymbol{D}^p. \tag{25.3}$$

In the above, the $\boldsymbol{D}^e$ and $\boldsymbol{D}^p$ are the elastic and the plastic part of $\boldsymbol{D}$, respectively. An objective Eulerian rate equation of Hookean type is used to relate the elastic part $\boldsymbol{D}^e$ and an objective stress rate $\mathring{\boldsymbol{\tau}}$:

$$\boldsymbol{D}^e = \frac{1}{2G}\mathring{\boldsymbol{\tau}} - \frac{\nu}{E}(\mathrm{tr}\mathring{\boldsymbol{\tau}})\boldsymbol{I} \tag{25.4}$$

Here, G, ν and E are the shear modulus, Poisson ratio and Young's modulus as in classical Hooke's law, with $E = 2G(1+\nu)$.

Prior to the yielding (hence $\boldsymbol{D}^e = \boldsymbol{D}$), the elastic rate Eq. (25.4) should be exactly integrable to really deliver a hyper-elastic stress-strain relation based on an elastic strain-energy function. As demonstrated in Xiao et al (1999, 2000a,b, 2007); Bruhns et al (1999, 2003), the just-mentioned requirement is fulfilled if and only if the stress rate $\mathring{\boldsymbol{\tau}}$ in Eq. (25.4) is the logarithmic rate of the stress $\mathring{\boldsymbol{\tau}}$ as defined below (Xiao et al, 1996, 1997, cf., e.g.,):

$$\mathring{\boldsymbol{\tau}} = \dot{\boldsymbol{\tau}} + \boldsymbol{\tau}\cdot\boldsymbol{\Omega} - \boldsymbol{\Omega}\cdot\boldsymbol{\tau}, \tag{25.5}$$

with the logarithmic spin:Logarithmic strain, logarithmic spin and logarithmic rate

$$\boldsymbol{\Omega} = \boldsymbol{W} + \sum_{r\neq s=1}^{m} \left(\frac{1+(\frac{b_r}{b_s})}{1-(\frac{b_r}{b_s})} + \frac{2}{ln(\frac{b_r}{b_s})}\right)\boldsymbol{B}_r\cdot\boldsymbol{D}\cdot\boldsymbol{B}_s, \tag{25.6}$$

where the b_i and the $\boldsymbol{B}_i$ are the m distinct eigenvalues and the corresponding eigenprojections of the Cauchy-Green tensor $\boldsymbol{B} = \boldsymbol{F} \cdot \boldsymbol{F}^T$, respectively. Details may be found in Xiao et al (1997).

Moreover, the rate equation prescribing the plastic stretching $\boldsymbol{D}^p$ is given by a normality flow rule of the following form Bruhns et al (1999, 2003, 2005); Xiao et al (2000b, 2007):

$$\boldsymbol{D}^p = \rho\zeta\frac{\partial f}{\partial \boldsymbol{\tau}}. \tag{25.7}$$

In the above, the function f and the loading factor ζ will be given slightly later. According to the classical theory of elastoplasticity, the plastic indicator ρ in Eq. (25.7) is associated with the concept of yielding and represents the very feature of plastic behavior in an idealized sense. In fact, it is assumed that no plastic deformation would be induced, namely, $\rho = 0$ prior to yielding, whereas plastic deformation would be induced, namely, $\rho = 1$ at yielding. Specifically, the loading-unloading conditions associated with the yield condition should be introduced to specify whether the plastic indicator ρ takes value 0 or takes value 1.

The above representation of plastic behavior in the idealized or simplified sense results in certain issues (Xiao et al, 2014; Xiao, 2014). Here, a relevant point is that any fatigue failure effects under cyclic loadings would be excluded. However, it is noted that, for realistic elastoplastic behavior, the plastic deformation would be induced at any stress level with a non-vanishing plastic indicator $0 \leq \rho \leq 1$. Therefore, a more realistic characterization may be presented by specifying $0 \leq \rho \leq 1$ in the following manner: The value of ρ is growing up to 1 whenever the stress point $\boldsymbol{\tau}$ is going close to the yield surface $f = 0$, whereas the value of ρ becomes vanishingly small whenever the stress point is going far away from the yield surface $f = 0$.

With the above idea in mind, a new flow rule may be proposed as follows (Xiao et al, 2014; Xiao, 2014):

$$D^p = \frac{g}{r}\exp\left[-m\left(1-\frac{g}{r}\right)\right]\frac{\zeta+|\zeta|}{2}\frac{\partial f}{\partial \boldsymbol{\tau}}. \tag{25.8}$$

In the above, $m \geq 0$ is a non-negative dimensionless parameter and f is the von Mises function of the form :

$$\begin{cases} f = g - r, \\ g = \frac{1}{2}J_2 = \frac{1}{2}tr\tilde{\boldsymbol{\tau}}^2, \\ r = \frac{1}{3}q^2, \end{cases} \tag{25.9}$$

where the $\tilde{\boldsymbol{\tau}} = \boldsymbol{\tau} - \frac{1}{3}(\mathrm{tr}\boldsymbol{\tau})\boldsymbol{I}$ is the deviatoric part of $\boldsymbol{\tau}$ and the q, called the stress limit, relies on both the plastic work κ and the stretching magnitude d, namely,

$$q = q(\kappa, d), \tag{25.10}$$

where

$$\dot{\kappa} = \boldsymbol{\tau} : \boldsymbol{D}^p , \tag{25.11}$$

$$d = \sqrt{\frac{2}{3}\mathrm{tr}\boldsymbol{D}^2} . \tag{25.12}$$

Moreover, the ζ in Eq. (25.8) is given by

$$\zeta = \frac{\breve{f}}{\breve{h}} \tag{25.13}$$

with

$$\breve{f} = 2G\tilde{\boldsymbol{\tau}} : \boldsymbol{D} - \frac{2}{3}q\frac{\partial q}{\partial d}\dot{d}, \tag{25.14}$$

$$\breve{h} = \frac{2}{3}J_2(3G + qq') . \tag{25.15}$$

Here and henceforward, the notation q' is used to denote the derivative of q with respect to κ, viz.

$$q' = \frac{dq}{d\kappa} . \tag{25.16}$$

Equations (25.3), (25.5) and (25.8)-(25.13) together produce

$$\frac{\mathring{\boldsymbol{\tau}}}{2G} = \frac{\nu}{1-2\nu}(\mathrm{tr}\boldsymbol{D}\boldsymbol{I} + \boldsymbol{D} - \frac{1.125q^{-2}}{3G+qq'}\exp\left[-m\left(1-\frac{3J_2}{2q^2}\right)\right](\breve{f} + |\breve{f}|)\tilde{\boldsymbol{\tau}}, \tag{25.17}$$

where the plastic work κ is given by

$$\dot{\kappa} = \frac{1.125J_2q^{-2}}{3G+qq'}\exp\left[-m\left(1-\frac{3J_2}{2q^2}\right)\right](\breve{f} + |\breve{f}|). \tag{25.18}$$

Equations (25.17)-(25.18) with Eqs. (25.10)-(25.12) and (25.14) provide a new rate-dependent elastoplasticity model, into which the yielding behavior and the loading-unloading behavior are not imposed as extrinsic coercive conditions but incorporated into intrinsic physical features. Now, plastic flow can be induced at any non-zero stress level. The plastic strain will become appreciable when the stress meets the yield condition $f = 0$, whereas it may be negligibly small in the case when $f < 0$. Detailed explanations of the consequences implied by such a new model may be found in Xiao et al (2014); Xiao (2014). Here, the proposed model is a rate-dependent extension of the rate-independent models in the previous study just mentioned. It is worth pointing out that the rate-independent case is smoothly incorporated as the particular case of the new model when d tends to vanish.

Here, of particular interest is a study of the fatigue failure behavior for metals. According to the result presented in a broad sense in a previous study (Xiao, 2014) that, as an inherent constitutive feature of the new model, the fatigue failure may be represented by the following asymptotic property:

$$\lim_{\kappa \leftarrow +\infty} q(\kappa, d) = 0. \tag{25.19}$$

As such, the fatigue failure will be automatically predicted from the new model with a stress limit of the above asymptotic property. Note that this property just represents the following basic fact: the strength, characterized here by the stress limit q, should be gradually reduced and eventually become vanishing with full development of plastic flow.

In the subsequent development, a simple form of the stress limit $q = q(\kappa, d)$ with the asymptotic property Eq. (25.19) will be taken into consideration and given below:

$$q = \frac{1}{2} q_0 [1 + \tanh \beta (\kappa_c - \kappa)]. \tag{25.20}$$

In the above, q_0, β and κ_c are three positive material parameters and, in general, rate-dependent, namely,

$$q_0 = q_0(d), \quad \beta = \beta(d), \quad \kappa_c = \kappa_c(d). \tag{25.21}$$

The stress limit q given in Eqs. (25.20)-(25.21) is a rate-dependent extension of the counterpart given in a previous study for the rate-independent case (Wang and Xiao, 2015). It exhibits asymptotic loss of the stress-bearing capacity, since it tends to vanish as the plastic work κ goes to infinity, namely, the condition given by Eq. (25.19) is met.

The main features of the stress limit given and the parameters therein are explained as follows: For a fairly large β, the stress limit q given by Eq. (25.16) actually yields a constant value, i.e., q_0, before the plastic work κ reaches a critical value slightly smaller than κ_c, whereas it goes rapidly to vanish after κ exceeds a critical value slightly greater than κ_c.

In the next section, results for both high and low cycle fatigue behaviors of metals at different strain rates will be derived based on the proposed model.

25.3 Failure Under Monotone and Cyclic Loadings

25.3.1 Governing Equations in the Uniaxial Case

Consider a cylindrical bar undergoing uniaxial deformations. Let τ be the axial Kirchhoff stress and let a and l be the stretch ratios in the axial direction and the lateral direction, respectively. Then, from Eqs. (14) and (17)-(18) the rate equations governing the axial Hencky strain $h = \ln a$ and the plastic work κ may be derived as follows:

$$\dot{h} = \frac{q_0}{E} \left(1 + \frac{E\phi}{3G(1-\phi) + qq'} \right) \dot{\tau}, \tag{25.22}$$

$$\dot{\kappa} = \frac{q_0^2 \phi \bar{\tau}}{3G(1-\phi) + qq'} \dot{\bar{\tau}}, \tag{25.23}$$

where

$$\phi = \frac{\bar{\tau}^2 q_0^2}{q^2} [\bar{\tau}\dot{h}] \exp\left[-m\left(1 - \frac{\bar{\tau}^2 q_0^2}{q^2}\right)\right], \tag{25.24}$$

$$q' = -\frac{1}{2}\beta q_0 \frac{1}{\cosh^2 \beta(\kappa - \kappa_c)}. \tag{25.25}$$

Moreover, by means of the fact that the volumetric strain is elastic, the lateral stretch ratio l is determined by

$$\ln l = -\frac{h}{2} + \frac{1-2\nu}{2} \frac{q_0}{E} \bar{\tau}. \tag{25.26}$$

Here and henceforth we denote

$$\bar{\tau} = \frac{\tau}{q_0},$$

$$[x] = \frac{1}{2}(1 + \mathrm{sgn}\,(x)) = \begin{cases} 1 \text{ for } x > 0, \\ 0 \text{ for } x \leq 0. \end{cases}$$

In deriving Eqs. (25.22)-(25.24), the axial strain rate $\dot{h}$ is taken to be constant and, therefore, $\dot{d} = \ddot{h} = 0$.
For any given loading process, namely, $\tau = \tau(t)$, the response of the axial Hencky strain $h = h(t)$ may be derived from Eqs. (22)-(26) and, vice versa. In what follows, the monotone strain case up to failure and cyclic loading case up to failure will be taken into account, separately.

25.3.2 Parameter Identification with Monotone Strain Data

For the purpose of identifying the rate-dependent parameter incorporated, stress-strain data under monotone tensile strain up to failure are needed at different strain rates. At each given strain rate, the axial Hencky strain h constantly grows from zero to infinity till eventual failure. In this process, the axial stress increases from zero to a maximum value and, after that, decreases from this maximum to zero. With the values of the parameters m, q_0, β and κ_c for any given strain rate $\dot{h}$, the stress response as a function of the axial strain h may be determined from Eqs. (25.22)-(25.23) with $[\tau\dot{h}] = 1$, namely, from the following equations

$$\frac{d\bar{\tau}}{dh} = \frac{E}{q_0} \frac{3G(1-\phi) + qq'}{3G + (2\nu - 1)G\phi + qq'}, \tag{25.27}$$

$$\frac{d\kappa}{dh} = \frac{q_0^2 \phi \bar{\tau}}{3G(1-\phi) + qq'} \frac{d\bar{\tau}}{dh}, \tag{25.28}$$

with $[\tau\dot{h}] = 1$ in Eq. (25.24) and the initial value

$$\tau|_{h=0} = 0, \quad \kappa|_{h=0} = 0, \quad h \in [0, +\infty).$$

Then, the values of the foregoing parameters may be identified in such a sense that the stress response fits the stress-strain data given at the strain rate at issue as closely as possible. Given data sets at several strain rates, several sets of parameter values may accordingly be available. Each rate-dependent parameter is then obtained by standard interpolating procedures, as will be shown in the next section.

25.3.3 Predictions for Fatigue Failure Under Cyclic Loadings

With the model parameters identified, fatigue failure under cyclic loadings may be predicted from Eqs. (25.22)-(25.23). Here, the following type of stress cycle is treated, namely, the stress τ grows from zero to a certain value and then goes back to zero.

Since the plastic working $\dot{\kappa}$ is positive for $\tau\dot{h} > 0$ and the stress limit $q = q(\kappa, d)$ is of the asymptotic property Eq. (25.19), a critical state for failure will be attained with accumulation of the plastic work as the cycle number is increasing. This state is prescribed by the vanishing of the denominator in Eq. (25.23), namely, by the following condition:

$$3G(1-\phi) + qq' = 0, \tag{25.29}$$

with $[\tau\dot{h}] = 1$ in Eq. (25.24). Immediately following this critical state, the strain will grow indefinitely large as the stress goes down to vanish. This implies that indefinitely growing strain will be induced by vanishingly small stress and, therefore, eventual failure will emerge following the critical state in the foregoing.

On account of the above facts, two deformation stages need be treated separately. Prior to the critical state, at each cycle the strain h as a function of the stress τ is determined from Eqs. (25.22)-(25.23) with $[\tau\dot{h}] = 1$ as the stress grows from zero to a given value. Note that the value at the start of the $s-$th cycle is just the value at the end of the $(s-1)-$th cycle.

As the critical state is attained, i.e., the condition Eq. (27) is met, the stress cycle can no longer be performed as before but, instead, the stress will go down to vanish. In this case, the stress τ should be determined as a function of the strain h from Eqs. (25.22)-(25.23) with $[\tau\dot{h}] = -1$, namely, from the following equations:

$$\frac{d\bar{\tau}}{dh} = \frac{E}{q_0}\frac{3G(1+\phi) + qq'}{3G + (1-2\nu)G\phi + qq'}, \tag{25.30}$$

$$\frac{d\kappa}{dh} = \frac{-q_0^2\phi\bar{\tau}}{3G(1+\phi) + qq'}\frac{d\bar{\tau}}{dh}, \tag{25.31}$$

with $[\tau\dot{h}] = 1$ in Eq. (25.24).

The above two stages demarcated by a critical state will be shown in the numerical results in the next section.

25.4 Numerical Results

Recently, Giroux et al. (2010) have obtained tensile stress-strain data of P92 steel under monotone strain up to failure at the four strain rates $2.5\times10^{-6}/s$, $2.5\times10^{-5}/s$, $2.5\times10^{-4}/s$ and $2.5\times10^{-3}/s$. Data in this paper were given for the axial engineering strain and engineering stress, denoted ε and P. These data are converted to those for the axial Hencky strain h and Kirchhoff stress τ via the following relations:

$$\begin{cases} h = \ln(1+\varepsilon), \\ \tau e^{-(1-2\nu)\tau/E} = (1+\varepsilon)P. \end{cases} \tag{25.32}$$

Since the Poison ratio ν is about $\frac{1}{3}$ and the magnitude of order of τ/E is about 0.01 for the greatest possible values of the stress τ, the exponential term in the second relation above may be set to be 1 with sufficient accuracy. As a result, the foregoing two relations may be reduced to

$$\begin{cases} h = \ln(1+\varepsilon), \\ \tau = (1+\varepsilon)P. \end{cases} \tag{25.33}$$

The material parameters, including E, m, q_0, β and κ_c etc., may be determined by simulating a uniaxial stress-strain curve with monotonically increasing strain up to failure and then numerical results for fatigue failure may be obtained by integrating the system of the two equations (25.22) and (25.22), as explained in Subsect. 25.3.2 and Subsect. 25.3.3. These will be done separately.

25.4.1 Failure Under Monotone Strain

The material parameters are found by fitting the four experimental curves in the foregoing reference. Results are presented in Table 25.1 for the parameter values and in Fig. 25.1 through Fig. 25.4 for comparisons with the test data for the four strain rates, separately. Table 25.1 shows that the maximum value of the stress limit and the critical plastic work, i.e., q_0 and κ_c, display rate effects. As indicated in Subsect. 25.3.2, these rate-dependent parameters may be given directly by interpolating functions, as shown below:

Table 25.1: Parameter values at four strain rates

Rates (s^{-1})	m	ν	E (GPa)	β(MPa^{-1})	κ_c(MPa)	q_0(MPa)
2.5×10^{-6}	0.1	0.3	120	0.05	58	340
2.5×10^{-5}	0.1	0.3	120	0.05	87	395
2.5×10^{-4}	0.1	0.3	120	0.05	92	440
2.5×10^{-3}	0.1	0.3	120	0.05	82	480

Fig. 25.1 Uniaxial stress-strain curve at strain rate $2.5 \times 10^{-3}\mathrm{s}^{-1}$, with parameters $\beta = 0.05$ / MPa, $\nu = 0.3$, $E = 120000$ MPa, $m = 0.1$, $\kappa_c = 82$ MPa, $q_0 = 480$ MPa.

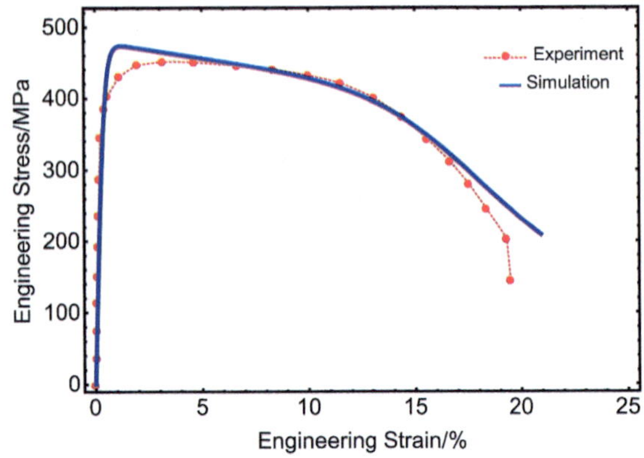

Fig. 25.2 Uniaxial stress-strain curve at strain rate $2.5 \times 10^{-4}\mathrm{s}^{-1}$, with parameters $\beta = 0.05$/MPa, $\nu = 0.3$, E = 120000 MPa, m = 0.1, $\kappa_c = 92$ MPa, $q_0 = 440$ MPa

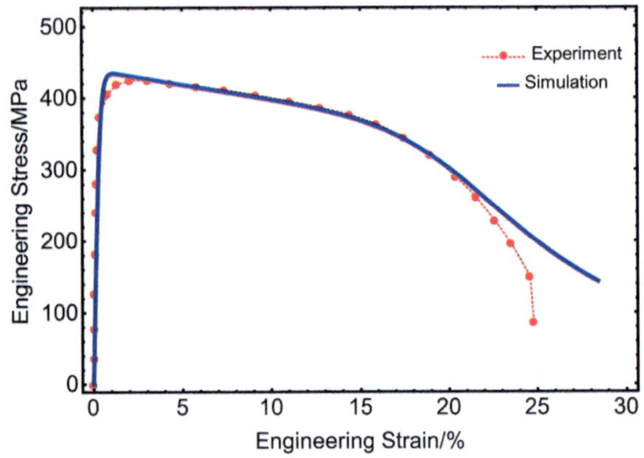

$$\kappa_c = 54.426 + 144.382\bar{d} - 56.853\bar{d}^2 + 2.045\bar{d}^3,$$

$$q_0 = 333.266 + 271.881\bar{d} - 100.679\bar{d}^2 + 3.601\bar{d}^3,$$

where $\bar{d} = 1000d$. With these polynomial interpolating functions, the two parameters exactly provide those values given in Table 25.1 for the four strain rates.

25.4.2 Predictions for Fatigue Failure Under Cyclic Loadings

With the parameters identified, the model predictions for the fatigue failure behavior of the metal at issue may be derived under various types of cyclic loadings. Here, the stress cycle indicated in Subsect. 25.3.3 is taken into consideration. The fatigue life, i.e., the cycle number N to failure, may be determined by Eq. (25.29) for the four strain rates. Also, the stress-strain curves over the whole strain range up to failure may be obtained following the procedures in Subsect. 25.3.3. Results for the four strain rates are shown in Fig. 25.5 through Fig. 25.8 for the stress-strain

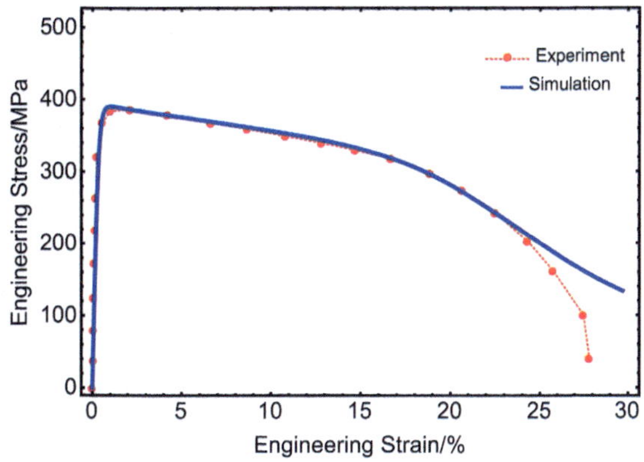

Fig. 25.3 Uniaxial stress-strain curve at strain rate $2.5 \times 10^{-5} s^{-1}$, with parameters $\beta = 0.05$/MPa, $\nu = 0.3$, E = 120000 MPa, m = 0.1, $\kappa_c = 87$ MPa, $q_0 = 395$ MPa

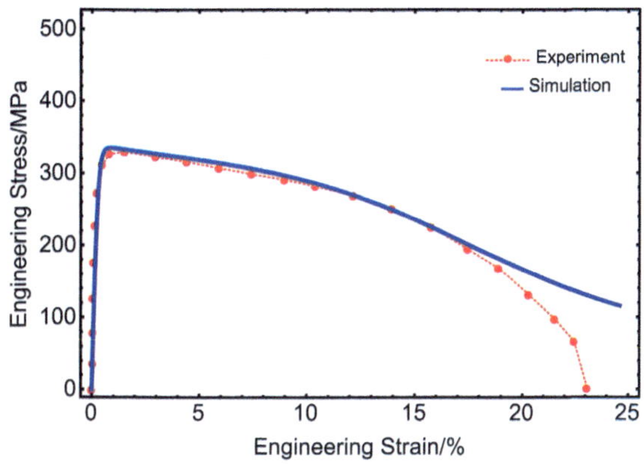

Fig. 25.4 Uniaxial stress-strain curve at strain rate $2.5 \times 10^{-6} s^{-1}$, with parameters $\beta = 0.05$/MPa, $\nu = 0.3$, E = 120000 MPa, m = 0.1, $\kappa_c = 58$ MPa, $q_0 = 340$ MPa

curves for the stress amplitude of 150MPa and in Fig. 25.9 for the curves of the fatigue life versus the stress amplitude. The falling parts in Figs. 25.5-25.5 clearly show the emergence of both the critical state and the eventual failure, namely, the

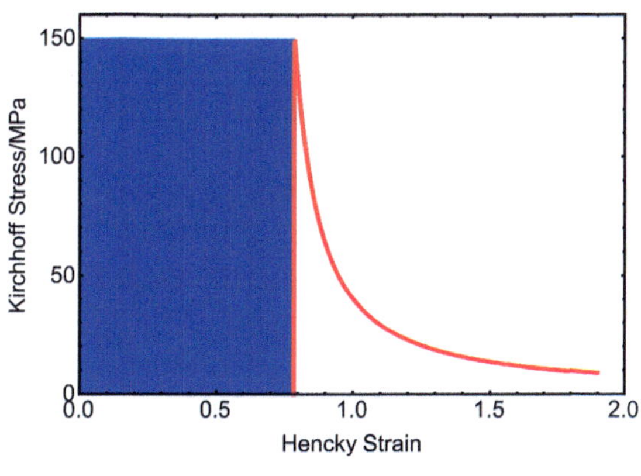

Fig. 25.5 Fatigue failure under the cyclic loadings at strain rate $2.5 \times 10^{-3} s^{-1}$, $N = 18095$

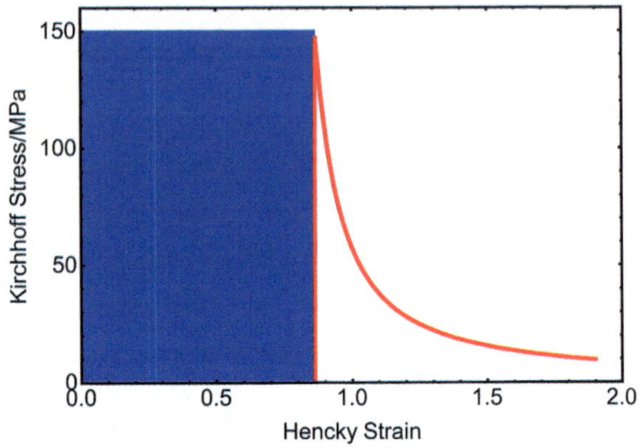

Fig. 25.6 Fatigue failure under the cyclic loadings at strain rate $2.5 \times 10^{-4} s^{-1}$, $N = 17098$

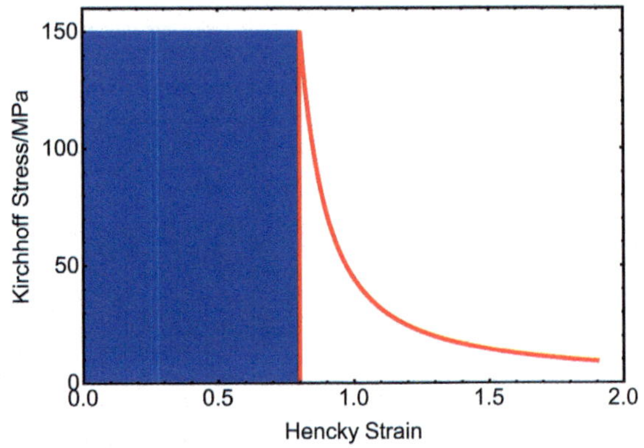

Fig. 25.7 Fatigue failure under the cyclic loadings at strain rate $2.5 \times 10^{-5} s^{-1}$, $N = 12573$

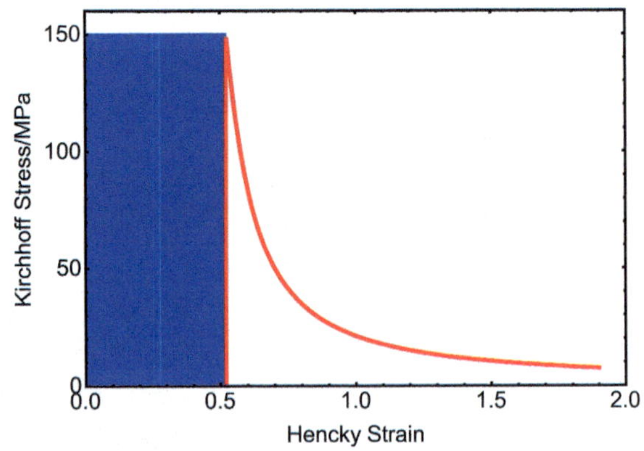

Fig. 25.8 Fatigue failure under the cyclic loadings at strain rate $2.5 \times 10^{-6} s^{-1}$, $N = 5353$

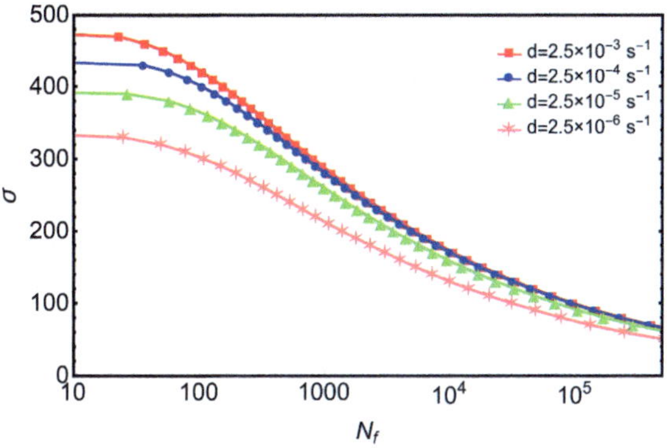

Fig. 25.9 Fatigue characteristic curves of the metal at different strain rates (σ and N_f are the stress amplitude and the cycle number to failure)

strain grows rapidly to infinity, and, in the meantime, the stress goes sharply down to vanish. This is predicted automatically from the model, without involving any ad hoc failure criteria. The model predictions for the fatigue life under the stress cycle considered agree with the salient feature of the fatigue failure for metals. As shown in Fig. 25.9, the fatigue life, i.e., the cycle number to failure, increases with increasing strain rate.

25.5 Concluding Remarks

The fatigue failure behavior of metals under different strain rates is simulated based on the new elastoplastic model proposed, in a direct sense without involving any additional variables and any ad hoc failure criteria. It is noted that the parameters characterizing the plastic behavior, in particular, the parameters m, q_0 and κ_c, play significant roles in simulating the fatigue failure behavior. The numerical simulation results are in agreement with fatigue failure behavior of realistic materials. The proposed model is for general multi-axial deformations. Here, simulation results are given for the uniaxial case. Moreover, comparison of the model prediction with test data has been done merely for the monotone strain case up to failure, whereas that is not the case for the fatigue life due to lack of data for the test sample considered. Further study will be needed to compare model predictions with test data for the fatigue life under various types of cyclic loadings and, furthermore, to treat deformation modes with rotational effects, such as simple shear etc. Results will be reported elsewhere.

Acknowledgements This study was completed the joint support of the fund from the Natural Science Foundation of China (No. 11372172; No. 11472164; No. 11542020) and of the start-up fund from the Education Committee of China through Shanghai University (No.: S.15-B002-09-032) and the fund from the Innovation Program of Shanghai Municipal Education Commission (No.: 14YZ028).

References

Argon A (2011) Topics in Fracture and Fatigue. Springer, Berlin

Bonora N, Gentile D, Pirondi A, Newaz G (2005) Ductile damage evolution under triaxial state of stress: theory and experiments. International Journal of Plasticity 21(5):981 – 1007

Bronkhorst C, Cerreta E, Xue Q, Maudin P, Masson T, Gray G (2006) An experimental and numerical study of the localization behavior of tantalum and stainless steel. Int J Plast 22:1304–1335

Bruhns O, Xiao H, Meyers A (1999) Self-consistent eulerian rate type elasto-plasticity models based upon the logarithmic stress rate. International Journal of Plasticity 15:479 – 520

Bruhns O, Xiao H, Meyers A (2003) Some basic issues in traditional eulerian formulations of finite elastoplasticity. International Journal of Plasticity 19:2007 – 2026

Bruhns O, Xiao H, Meyers A (2005) A weakened form of ilyushin's postulate and the structure of self-consistent eulerian finite elastoplasticity. International Journal of Plasticity 21:199 – 219

Brünig M (2003a) An anisotropic ductile damage model based on irreversible thermodynamics. International Journal of Plasticity 19(10):1679 – 1713

Brünig M (2003b) Numerical analysis of anisotropic ductile continuum damage. Computer Methods in Applied Mechanics and Engineering 192(26 - 27):2949 – 2976

Brünig M, Gerke S (2011) Simulation of damage evolution in ductile metals undergoing dynamic loading conditions. International Journal of Plasticity 27(10):1598 – 1617

Brünig M, Ricci S (2005) Nonlocal continuum theory of anisotropically damaged metals. International Journal of Plasticity 21(7):1346 – 1382

Brünig M, Chyra O, Albrecht D, Driemeier L, Alves M (2008) A ductile damage criterion at various stress triaxialities. International Journal of Plasticity 24(10):1731 – 1755, special Issue in Honor of Jean-Louis Chaboche

Brünig M, Gerke S, Hagenbrock V (2013) Micro-mechanical studies on the effect of the stress triaxiality and the lode parameter on ductile damage. International Journal of Plasticity 50:49 – 65

Chung K, Ma N, Park T, Kim D, Yoo D, Kim C (2011) A modified damage model for advanced high strength steel sheets. International Journal of Plasticity 27(10):1485 – 1511

Jansen Y, Logé RE, Milesi M, Massoni E (2013) An anisotropic stress based criterion to predict the formability and the fracture mechanism of textured zinc sheets. Journal of Materials Processing Technology 213(6):851 – 855

Khan AS, Liu H (2012a) A new approach for ductile fracture prediction on al 2024-t351 alloy. International Journal of Plasticity 35:1 – 12

Khan AS, Liu H (2012b) Strain rate and temperature dependent fracture criteria for isotropic and anisotropic metals. International Journal of Plasticity 37:1 – 15

Kim JH, Sung JH, Piao K, Wagoner R (2011) The shear fracture of dual-phase steel. International Journal of Plasticity 27(10):1658 – 1676

Macha E, Nieslony A (2012) Critical plane fatigue life models of materials and structures under multiaxial stationary random loading: The state-of-the-art in opole research centre {CESTI} and directions of future activities. International Journal of Fatigue 39:95 – 102

Malcher L, Pires FA, de Sá JC (2012) An assessment of isotropic constitutive models for ductile fracture under high and low stress triaxiality. International Journal of Plasticity 30 - 31:81 – 115

Rozumek D, Macha E (2009) A survey of failure criteria and parameters in mixed-mode fatigue crack growth. Materials Science 45(2):190 – 210

Shojaei A, Voyiadjis GZ, Tan P (2013) Viscoplastic constitutive theory for brittle to ductile damage in polycrystalline materials under dynamic loading. International Journal of Plasticity 48:125 – 151

Shyam A, Allison JE, Szczepanski CJ, Pollock TM, Jones JW (2007) Small fatigue crack growth in metallic materials: A model and its application to engineering alloys. Acta Materialia 55(19):6606 – 6616

Stoughton TB, Yoon JW (2011) A new approach for failure criterion for sheet metals. International Journal of Plasticity 27(3):440 – 459

Susmel L (2008) The theory of critical distances: a review of its applications in fatigue. Engineering Fracture Mechanics 75(7):1706 – 1724

Susmel L (2014) Four stress analysis strategies to use the modified wöhler curve method to perform the fatigue assessment of weldments subjected to constant and variable amplitude multiaxial fatigue loading. International Journal of Fatigue 67:38 – 54

Wang ZL, Xiao H (2015) A study of metal fatigue failure as inherent features of elastoplastic constitutive equations. In: Holm Altenbach DO Tetsuya Matsuda (ed) From Creep Damage Mechanics to Homogenization Methods - A Liber Amicorum to celebrate the brithday of Nobutada Ohno, Advanced Structural Materials, vol 64, Springer, Heidelberg, pp 529–540

Xiao H (2014) Thermo-coupled elastoplasticity models with asymptotic loss of the material strength. International Journal of Plasticity 63:211 – 228

Xiao H, Bruhns OT, Meyers A (1996) A new aspect in kinematics of large deformations. In: Gupta N (ed) Plasticity and Impact Mechanics, New Age Publ. Ltd., New Dehli, pp 100–1009

Xiao H, Bruhns O, Meyers A (1997) Logarithmic strain, logarithmic spin and logarithmic rate. Acta Mech 124:89–105

Xiao H, Bruhns O, Meyers A (1999) Existence and uniqueness of the integrable-exactly hypoelastic equation $\dot{\boldsymbol{\tau}}^* = \lambda(\mathrm{tr}\boldsymbol{d})\boldsymbol{i}$ and its significance to finite inelasticity. Acta Mech 138:31–50

Xiao H, Bruhns O, Meyers A (2000a) The choice of objective rates in finite elastoplasticity: general results on the uniqueness of the logarithmic rate. Proc Roy Soc London A 456:1865–1882

Xiao H, Bruhns O, Meyers A (2000b) A consistent finite elastoplasticity theory combining additive and multiplicative decomposition of the stretching and the deformation gradient. Int J Plasticity 16:143–177

Xiao H, Bruhns O, Meyers A (2006) Elastoplasticity beyond small deformations. Acta Mech 182:31–111

Xiao H, Bruhns O, Meyers A (2007) The exact integrability criterion in finite elastoplasticity and its constitutive implications. Acta Mech 188:227–244

Xiao H, Bruhns O, Meyers A (2014) Free rate-independent elastoplastic equations. ZAMM 94(6):461–476

Zhang WJ (2016) Thermal mechanical fatigue of single crystal superalloys: Achievements and challenges. Materials Science and Engineering: A 650:389 – 395

Printed by Printforce, the Netherlands